Reading Essentials
Course 2

The McGraw-Hill Companies

Send all inquiries to:
McGraw-Hill Education
8787 Orion Place
Columbus, OH 43240-4027

ISBN: 978-0-07-889389-6
MHID: 0-07-889389-5

Printed in the United States of America.

1 2 3 4 5 6 7 8 9 10 REL 15 14 13 12 11 10

Table of Contents

To the Student

In today's world, a knowledge of science is important for thinking critically, solving problems, and making decisions. But understanding science can sometimes be a challenge.

Reading Essentials takes the stress out of reading, learning, and understanding science. This book covers important concepts in science, offers ideas for how to learn the information, and helps you review what you have learned.

In each lesson you will find:

Before You Read

- **What do you think?** asks you to agree or disagree with statements about topics that are discussed in the lesson.

Read to Learn

This section describes important science concepts with words and graphics. In the margins you can find a variety of study tips and ideas for organizing and learning information.

- **Study Coach and Mark the Text** offer tips for finding the main ideas in the text.
- **Foldables® Study Organizers** help you divide the information into smaller, easier-to-remember concepts.
- **Reading Check** questions ask about concepts in the lesson.
- **Think It Over** elements help you consider the material in-depth, giving you an opportunity to use your critical thinking skills.
- **Visual Check** questions relate specifically to the art and graphics used in the text. You will find questions that get you actively involved in illustrating the information you have just learned.
- **Math Skills** reinforces the connection between math and science.
- **Academic Vocabulary** defines important words that help you build a strong vocabulary.
- **Word Origin** explains the English background of a word.
- **Key Concept Check** features ask the Key Concept questions from the beginning of the lesson.
- **Interpreting Tables** includes questions or activities that help you interact with the information presented.

After You Read

This final section reviews key terms and asks questions about what you have learned.

- The **Mini Glossary** assists you with science vocabulary.
- Review questions focus on the key concepts of the lesson.
- **What do you think now?** gives you an opportunity to revisit the *What do you think?* statements to see if you changed your mind after reading the lesson.

See for yourself—***Reading Essentials*** makes science enjoyable and easy to understand.

Scientific Explanations

Understanding Science

Read to Learn

What is science?

The last time that you watched squirrels play in a park or in your yard, did you realize that you were practicing science? Every time you observe the natural world, you are practicing science. **Science** *is the investigation and exploration of natural events and of the new information that results from those investigations.*

When you observe the natural world, you might form questions about what you see. While you are exploring those questions, you probably use reasoning, creativity, and skepticism to help you find answers to your questions. These behaviors are the same ones that scientists use in their work and that other people use in their daily lives to solve problems.

Scientists use a reliable set of skills and methods in different ways to find answers to questions. After reading this chapter, you will have a better understanding of how science works, the limitations of science, and scientific ways of thinking. In addition, you will recognize that when you practice science at home or in the classroom, you probably use scientific methods to answer questions just as scientists do.

Branches of Science

No one person can study all the natural world. Therefore, people tend to focus their efforts on one of the three fields or branches of science—life science, Earth science, or physical science. Then people or scientists can seek answers to specific problems within one field of science.

Life Science Biology, or life science, is the study of all living things. For example, a forest ecologist is a life scientist who studies interactions in forest ecosystems. Biologists ask questions such as the following: How do plants produce their own food? Why do some animals give birth to live young and others lay eggs? How are reptiles and birds related?

Key Concepts 
- What is scientific inquiry?
- What are the results of scientific investigations?
- How can a scientist prevent bias in a scientific investigation?

Study Coach

Building Vocabulary Write each vocabulary term in this lesson on an index card. Shuffle the cards. After you have studied the lesson, take turns picking cards with a partner. Each of you should define the term using your own words.

Reading Check

1. Describe the behaviors scientists use in their work.

Earth Science The study of Earth, including Earth's landforms, rocks, soil, and forces that shape Earth's surface, is Earth science. Earth scientists ask questions such as the following: How do rocks form? What causes earthquakes? What substances are in soil?

Physical Science The study of chemistry and physics is physical science. Physical scientists study the interactions of matter and energy. They ask questions such as these: How do substances react and form new substances? Why does a liquid change to a solid? How are force and motion related?

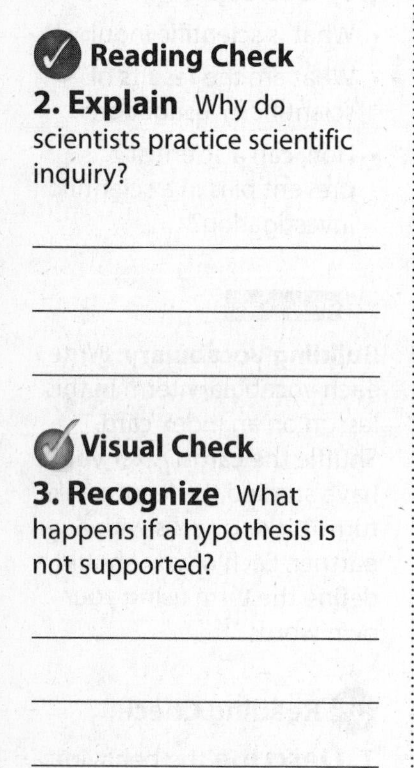

Scientific Inquiry

As scientists study the natural world, they ask questions about what they observe. To find the answers to these questions, they use certain skills, or methods. The figure below and at the bottom of the next page shows a sequence of the skills that a scientist might use in an investigation. Sometimes, not all of these skills are performed in an investigation or are performed in this order. Scientists practice scientific inquiry—a process that uses a variety of skills and tools to answer questions or to test ideas about the natural world.

Ask Questions

Like a scientist, you use scientific inquiry in your life, too. Suppose you decide to plant a vegetable garden. As you plant, you water some seeds more than others. You weed part of the garden and mix fertilizer into some of the soil.

Steps in Scientific Investigation

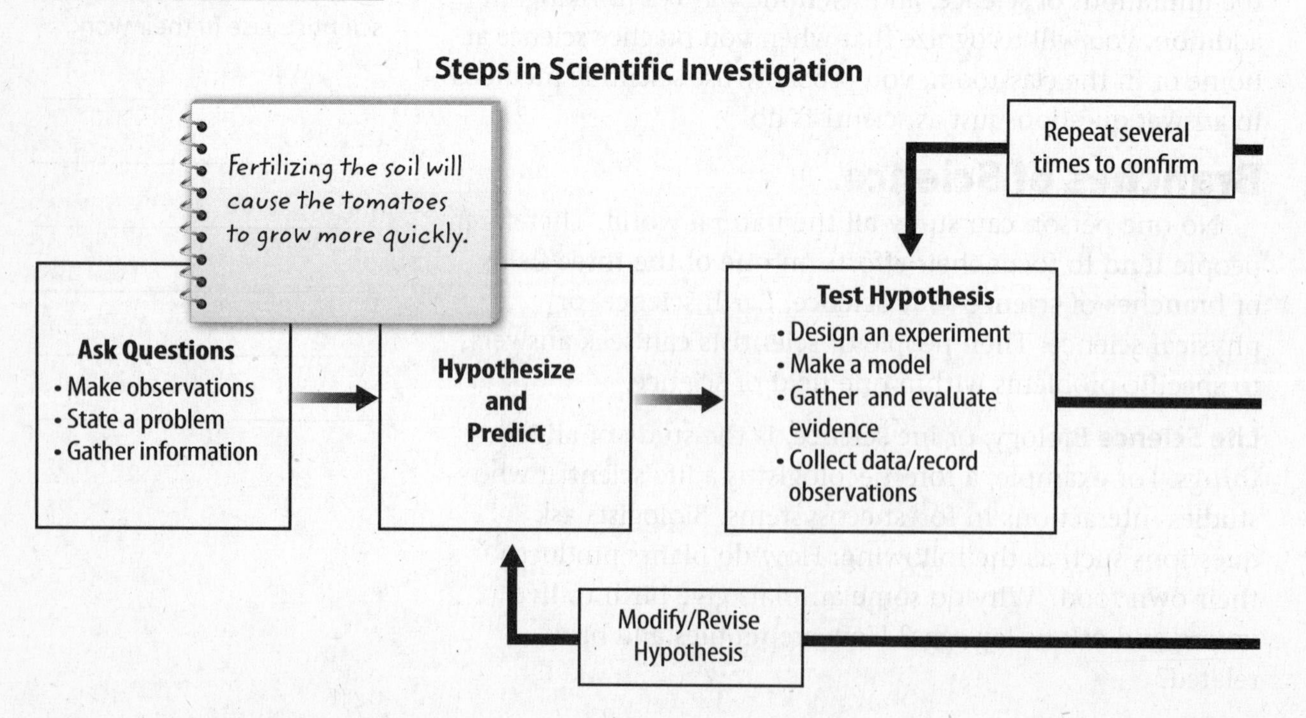

Observation After a few weeks, you observe that some plants are growing better than others. An ***observation*** *is using one or more of your senses to gather information and take note of what occurs.* Observations often are the beginning of the process of inquiry and can lead to questions such as "Why are some plants growing better than others?"

Inferring As you are making observations and asking questions, you recall that plants need water and light to grow. You infer that perhaps some vegetables are receiving more water or sunlight than others and, therefore, are growing better. *An* ***inference*** *is a logical explanation of an observation that is drawn from prior knowledge or experience.*

Hypothesize

After making observations and inferences, you are ready to develop a hypothesis and investigate why some vegetable plants are growing better than others. *A possible explanation about an observation that can be tested by scientific investigations is a* ***hypothesis.*** Your hypothesis might be: Some plants are growing larger and more quickly than others because they are receiving more water and sunlight. Or, your hypothesis might be: The plants that are growing quickly have received fertilizer because fertilizer helps plants grow.

Predict

After you state a hypothesis, you might make a prediction to help you test your hypothesis. *A* ***prediction*** *is a statement of what will happen next in a sequence of events.* For instance, based on your hypotheses, you might predict that if some plants receive more water, sunlight, or fertilizer, then they will grow larger.

Reading Check

4. Consider How does the process of inquiry usually begin?

____

____

____

Reading Check

5. Analyze What is the purpose of a hypothesis?

____

____

____

____

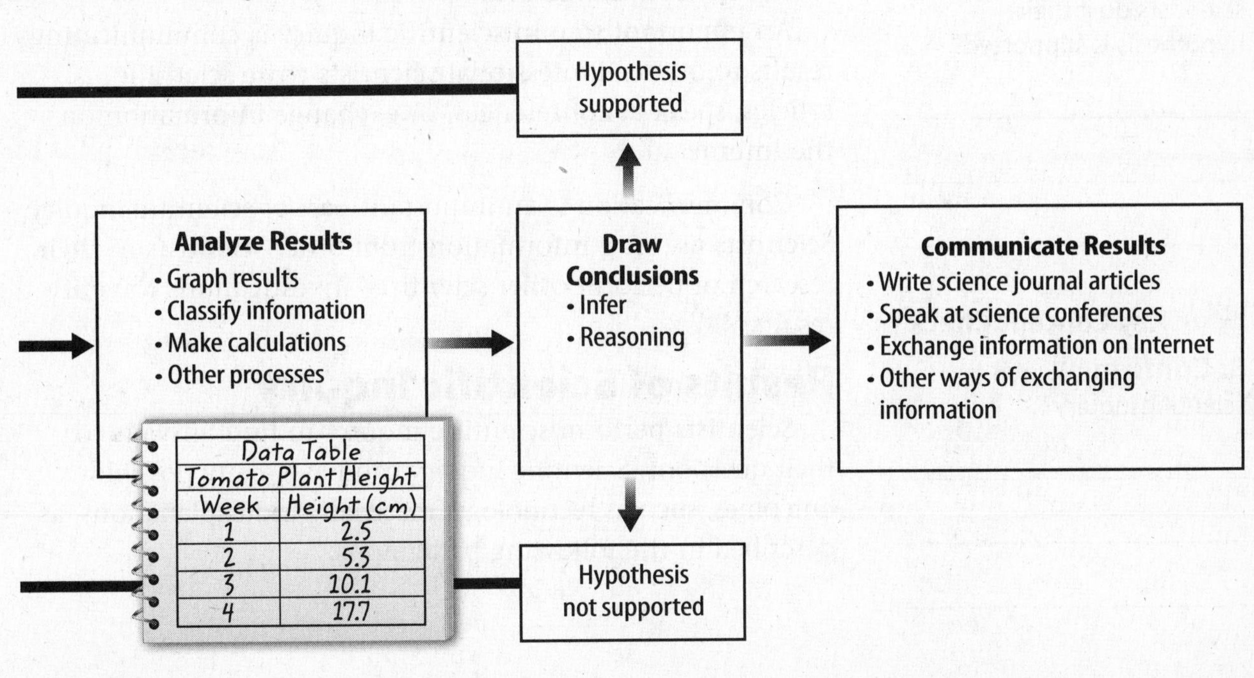

Data Table
Tomato Plant Height

Week	Height (cm)
1	2.5
2	5.3
3	10.1
4	17.7

Test Your Hypothesis

When you test a hypothesis, you often are testing your predictions. For example, you might design an experiment to test you hypothesis on the fertilizer. You set up an experiment in which you plant seeds and add fertilizer to only some of them. Your prediction is that the plants that get the fertilizer will grow more quickly. If your prediction is confirmed, your hypothesis is supported. If your prediction is not confirmed, you hypothesis might need revision.

Analyze Results

As you are testing your hypothesis, you are probably collecting data about the plants' growth rates and how much fertilizer each plant receives. At first, it might be difficult to recognize patterns and relationships in data. Your next step might be to organize and analyze your data. You can create graphs, classify information, or make models and calculations. After the data are organized, you can more easily study the data and draw conclusions.

Reading Check
6. Relate What are three ways to organize data?

Draw Conclusions

Now you must decide whether your data support your hypothesis and then draw conclusions. A conclusion is a summary of the information gained from testing a hypothesis. You might make more inferences when drawing conclusions. If your hypothesis is supported, you can repeat your experiment several times to confirm your results. If your hypothesis is not supported, you can modify it and repeat the scientific inquiry process.

Reading Check
7. Explain What should scientists do if their hypothesis is supported?

Communicate Results

An important step in scientific inquiry is communicating results to others. Professional scientists write scientific articles, speak at conferences, or exchange information on the Internet.

Communication is an important part of scientific inquiry. Scientists use new information from other scientists in their research or perform other scientists' investigations to verify results.

Key Concept Check
8. Confirm What is scientific inquiry?

Results of Scientific Inquiry

Scientists perform scientific inquiry to find answers to their questions. Scientific inquiry can have many possible outcomes, such as technology, materials, and explanations, as described in the following paragraphs.

Technology New technology is one possible outcome of scientific inquiry. **Technology** *is the practical use of scientific knowledge, especially for industrial or commercial use.* Televisions, MP3 players, and computers are examples of technology.

New Materials The creation of new materials is another possible outcome of an investigation. For example, scientists have developed a bone bioceramic. A bioceramic is a natural calcium-phosphate mineral complex that is part of bones and teeth. This synthetic bone mimics natural bone's structure. Its porous structure allows a type of cell to grow and develop into new bone tissue.

The bioceramic can be shaped into implants that are treated with certain cells from the patient's bone marrow. It then can be implanted into the patient's body to replace missing bone.

Possible Explanations Many times, scientific investigations answer the questions who, what, when, where, and how. For example, who left fingerprints at a crime scene? When should fertilizer be applied to plants? What organisms live in rain forests?

Scientific Theory and Scientific Laws

Scientists often repeat scientific investigations many times to verify, or confirm, that the results for a hypothesis or a group of hypotheses are correct. This can lead to a scientific theory.

Scientific Theory The everyday meaning of the word *theory* is "an untested idea or an opinion." However, in science, *theory* has a different meaning. *A* **scientific theory** *is an explanation of observations or events based on knowledge gained from many observations and investigations.*

For example, about 300 years ago, scientists began looking at samples of trees, water, and blood through the first microscopes. They noticed that all these organisms were made of tinier units, or cells. As more scientists observed cells in other organisms, their observations became known as the cell theory.

The cell theory explains that all living things are made of cells. A scientific theory is assumed to be the best explanation of observations unless it is disproved. The cell theory will continue to explain the makeup of all organisms until an organism is discovered that is not made of cells.

FOLDABLES

Make a two-column chart book to organize your notes on scientific investigations.

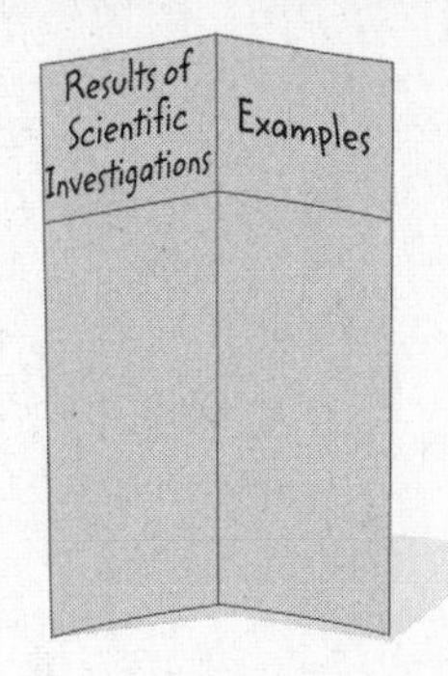

Key Concept Check

9. Define What are the results of scientific investigations?

____

____

____

____

Reading Check

10. Point Out What can change a scientific theory?

____

____

____

Scientific Laws Scientific laws are different from societal laws, which are an agreement on a set of behaviors. *A* **scientific law** *describes a pattern or an event in nature that is always true.* A scientific theory might explain how and why an event occurs. But a scientific law states only that an event in nature will occur under specific conditions.

Example of Scientific Law The law of conservation of mass states that the mass of materials will be the same before and after a chemical reaction. This scientific law does not explain why this occurs—only that it will occur. The table below compares a scientific theory and a scientific law.

Comparing Scientific Theory and Scientific Law

Scientific Theory	Scientific Law
A scientific theory is based on repeated observations and scientific investigations.	Scientific laws are observations of similar events that have been observed repeatedly.
If new information does not support a scientific theory, the theory will be modified or rejected.	If new observations do not follow the law, the law is rejected.
A scientific theory attempts to explain why something happens.	A scientific law states that something will happen.
A scientific theory usually is more complex than a scientific law and might contain many well-supported hypotheses.	A scientific law usually contains one well-supported hypothesis that states that an event will occur.

Interpreting Tables

11. Consider How do scientific theories and laws compare?

____

____

____

____

Skepticism in Media

When you see scientific issues in the media, such as newspapers, radio, television, and magazines, it is important to be skeptical. When you are skeptical, you question information that you read or hear, or events you observe. Is the information truthful? Is it accurate? It also is important that you question statements made by people outside their area of expertise and claims that are based on vague statements.

Reading Check

12. Summarize Why is it important to question information in the media?

____

____

____

Evaluating Scientific Evidence

Critical thinking is an important skill in scientific inquiry. **Critical thinking** *is comparing what you already know with the information you are given in order to decide whether you agree with it.* Identifying and preventing bias also is important when conducting scientific inquiry. To prevent bias in an investigation, sampling, repetition, and blind studies can be helpful, as shown in the figure on the next page.

Key Concept Check

13. State How can a scientist prevent bias in a scientific investigation?

____

____

____

____

Identifying and Preventing Bias

Sampling
A method of data collection that involves studying small amounts of something in order to learn about the larger whole is sampling. A sample should be a random representation of the whole.

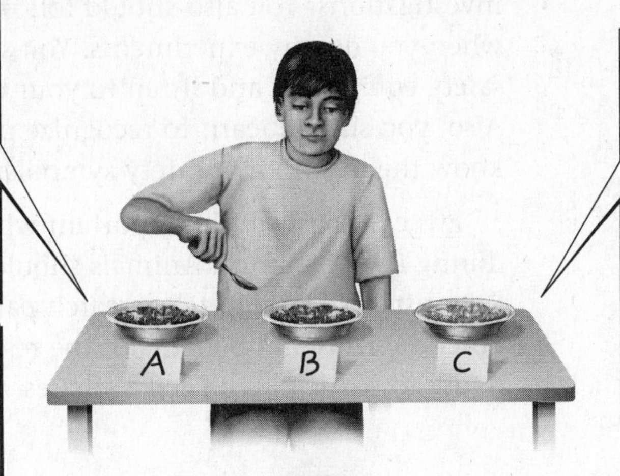

Blind Study
A procedure that can reduce bias is a blind study. The investigator, subject, or both do not know which item they are testing. Personal biases cannot affect an investigation if participants do not know what they are testing.

Bias
It is important to remain unbiased during scientific investigations. Bias is intentional or unintentional prejudice toward a specific outcome. Sources of bias in an investigation can include equipment choices, hypothesis formation, and prior knowledge.

Suppose you were part of a taste test for a new cereal. If you knew the price of each cereal, you might think that the most expensive one tastes the best. This is a bias.

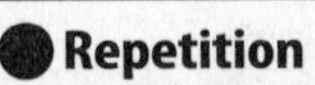

Repetition
If you get different results when you repeat an investigation, then the original investigation probably was flawed. Repetition of an experiment helps reduce bias.

Most Popular Brand $2.00

Mediocre Brand $1.50

Least Popular Brand $0.50

Science cannot answer all questions.

You might think that any question can be answered through a scientific investigation. But some questions cannot be answered using science.

For example, science cannot answer a question such as, Which paint color is the prettiest? Questions about personal opinions, values, beliefs, and feelings cannot be answered scientifically. However, some people use scientific evidence to try to strengthen their claims about these topics.

Visual Check

14. Identify What are some sources of bias?

Safety in Science

Scientists follow safety procedures when they conduct investigations. You also should follow safety procedures when you do any experiments. You should wear appropriate safety equipment and listen to your teacher's instructions. Also, you should learn to recognize potential hazards and know the meaning of safety symbols.

ACADEMIC VOCABULARY
ethics
(noun) rules of conduct or moral principles

Ethics are especially important when using living things during investigations. Animals should be treated properly. Scientists also should tell research participants about the potential risks and benefits of the research. Anyone can refuse to participate in scientific research.

After You Read

Mini Glossary

critical thinking: comparing what you already know with the information you are given in order to decide whether you agree with it

hypothesis: a possible explanation about an observation that can be tested by scientific investigations

inference: a logical explanation of an observation that is drawn from prior knowledge or experience

observation: using one or more of your senses to gather information and take note of what occurs

prediction: a statement about what will happen next in a sequence of events

science: the investigation and exploration of natural events and of the new information that results from those investigations

scientific law: a pattern or an event in nature that is always true

scientific theory: an explanation of observations or events based on knowledge gained from many observations and investigations

technology: the practical use of scientific knowledge, especially for industrial or commercial use

1. Review the terms and their definitions in the Mini Glossary. Write a sentence that describes the importance of critical thinking.

2. Use the graphic organizer below to list the three branches of science.

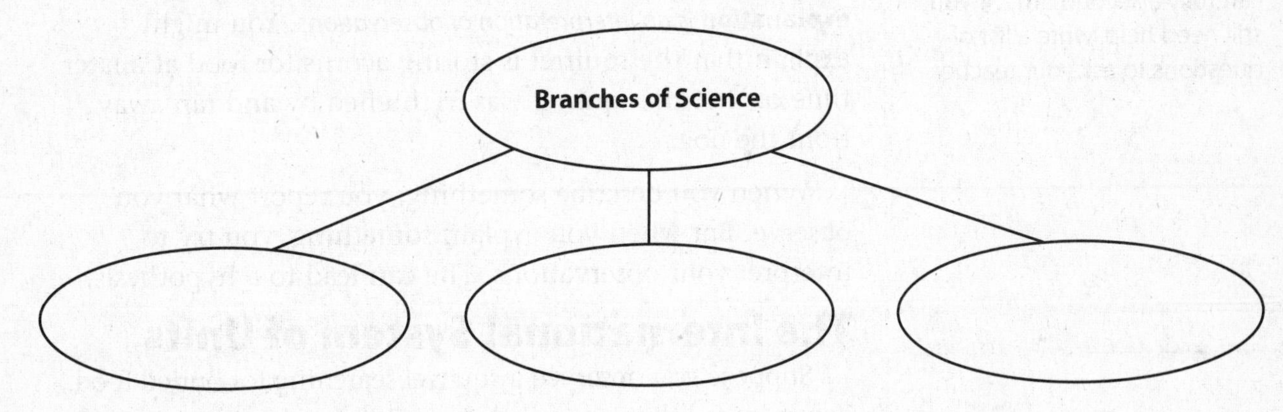

3. Discuss what types of questions cannot be answered scientifically and why.

Log on to ConnectED.mcgraw-hill.com and access your textbook to find this lesson's resources.

Nature of Science
LESSON 2

Scientific Explanations

Measurement and Scientific Tools

Key Concepts 

- What is the difference between accuracy and precision?
- Why should you use significant digits?
- What are some tools used by life scientists?

Mark the Text

Sticky Notes As you read, use sticky notes to mark information that you do not understand. Read the text carefully a second time. If you still need help, write a list of questions to ask your teacher.

Reading Check

1. Define What is the International System of Units?

____

____

____

Read to Learn

Description and Explanation

How would you write a description of a squirrel's activity? *A* ***description*** *is a spoken or written summary of observations.*

Your description might include information such as the squirrel buried five acorns near a large tree or that the squirrel climbed the tree when a dog barked. A qualitative description uses your senses (sight, sound, smell, touch, taste) to describe an observation. *A large tree* is a qualitative description.

A quantitative observation uses numbers to describe the observation. *Five acorns* is a quantitative description. You can use measuring tools, such as a ruler, a balance, or a thermometer to make quantitative descriptions.

How would you explain the squirrel's activity? *An* ***explanation*** *is an interpretation of observations.* You might explain that the squirrel is storing acorns for food at a later time or that the squirrel was frightened by and ran away from the dog.

When you describe something, you report what you observe. But when you explain something, you try to interpret your observations. The can lead to a hypothesis.

The International System of Units

Suppose you observed a squirrel searching for buried food. You recorded that it traveled about 200 feet from its nest.

Someone who measures distances in meters might not understand how far the squirrel traveled. The scientific community solved this problem in 1960. It adopted *an internationally accepted system for measurement called the* ***International System of Units (SI).***

SI Base Units and Prefixes

Like scientists and many others around the world, you probably use the SI system in your classroom. All SI units are derived from the seven base units listed in the table on the left below. For example, the base unit for length, or the unit most commonly used to measure length, is the meter. You have probably made measurements in kilometers or millimeters before. Where do these units come from?

A prefix can be added to a base unit's name to indicate either a fraction or a multiple of that base unit. The prefixes are based on powers of ten, such as 0.01 and 100, as shown below on the right. One centimeter (cm) is one-hundredth of a meter and a kilometer (km) is 1,000 meters. ✓

✓ Reading Check

2. Name What is added to the name of a base unit to indicate a fraction or a multiple of the base unit?

____

SI Base Units

Quantity Measured	Unit (Symbol)
Length	meter (m)
Mass	kilogram (kg)
Time	second (s)
Electric current	ampere (A)
Temperature	Kelvin (K)
Substance amount	mole (mol)
Light intensity	candela (cd)

Prefixes

Prefix	Meaning
Mega– (M)	1,000,000 (10^6)
Kilo– (k)	1,000 (10^3)
Hecto– (h)	100 (10^2)
Deka– (da)	10 (10^1)
Deci– (d)	0.1 (10^{-1})
Centi– (c)	0.01 (10^{-2})
Milli– (m)	0.001 (10^{-3})
Micro– (μ)	0.000 001 (10^{-6})

Conversion It is easy to convert from one SI unit to another. You multiply or divide by a power of ten. You also can use proportion calculations to make conversions. For example, a biologist measures an Emperor goose in the field. Her triple-beam balance shows that the goose has a mass of 2.8 kg. She could perform the calculation below to find the goose's mass in grams, *x*. ✓

$$\frac{x}{2.8\text{ kg}} = \frac{1{,}000\text{ g}}{1\text{ kg}}$$

$$(1\text{ kg})x = (1{,}000\text{ g})(2.8\text{ kg})$$

$$x = \frac{(1{,}000\text{ g})(2.8\text{ }\cancel{\text{kg}})}{1\text{ }\cancel{\text{kg}}}$$

$$x = 2{,}800\text{ g}$$

Notice that the answer has the correct units.

Interpreting Tables

3. Identify What unit measures mass?

____

✓ Reading Check

4. State How do you convert one SI unit to another?

____

____

____

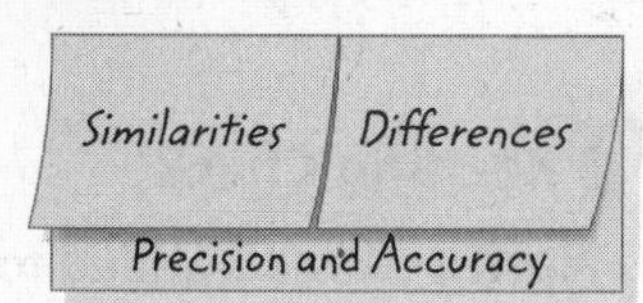

Make a horizontal two-tab book with a top tab to compare precision and accuracy.

Precision and Accuracy

Suppose your friend Simon tells you that he will call you in one minute, but he calls you a minute and a half later. Sarah tells you that she will call you in one minute, and she calls exactly 60 seconds later. What is the difference? Sarah is accurate, and Simon is not. **Accuracy** *is a description of how close a measurement is to an accepted or true value.* However, if Simon always calls about 30 seconds later than he says he will, then Simon is precise. **Precision** *is a description of how similar or close measurements are to each other,* as shown in the figure below.

Accuracy and Precision

Accurate

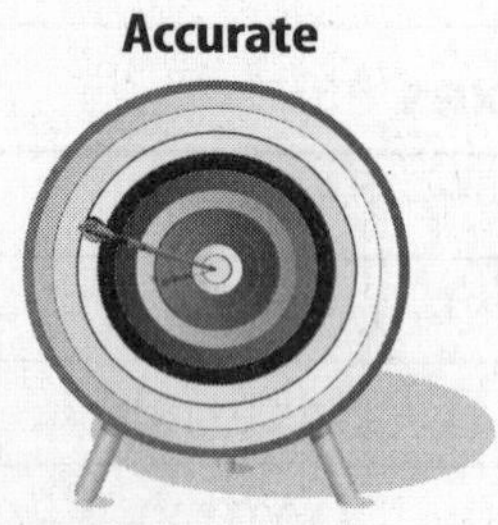

An arrow in the center indicates high accuracy.

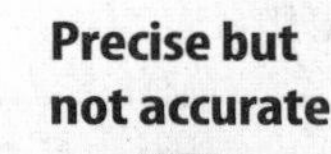

Precise but not accurate

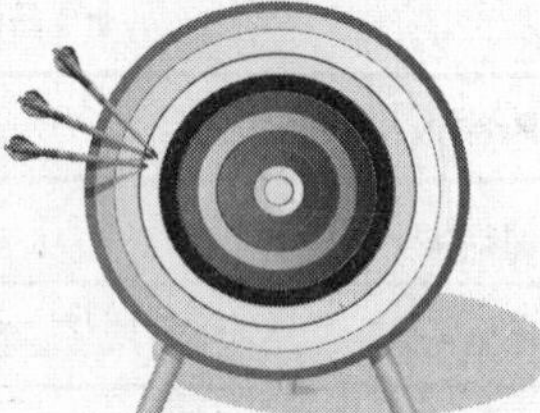

Arrows far from the center indicate low accuracy. Arrows close together indicate high precision.

Accurate and precise

Arrows in the center indicate high accuracy. Arrows close together indicate high precision.

Not accurate or precise

Arrows far from the center indicate low accuracy. Arrows far apart indicate low precision.

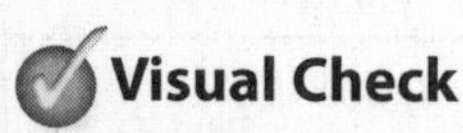

Visual Check

5. Interpret What do arrows close together in the target indicate?

____

____

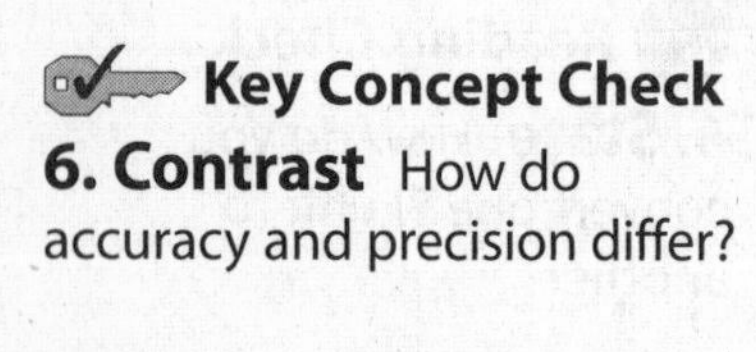

Key Concept Check

6. Contrast How do accuracy and precision differ?

____

____

____

The table at the top of the next page illustrates the difference between precise and accurate measurements.

Students were asked to find the melting point of sucrose, or table sugar. Each student took three temperature readings and calculated the mean, or average, of his or her data.

Refer to the table at the top of the next page. As the recorded data in the table shows, student A had the most accurate data. That student's melting point mean, 184.7°C, is closest to the scientifically accepted melting point, 185°C. Although not accurate, student C's measurements are the most precise because they are similar in value.

Student Melting Point Data			
	Student A	**Student B**	**Student C**
Trial 1	183.5°C	190.0°C	181.2°C
Trial 2	185.9°C	183.3°C	182.0°C
Trial 3	184.6°C	187.1°C	181.7°C
Average	184.7°C	186.8°C	181.6°C
Sucrose Melting Point (accepted value) 185°C			

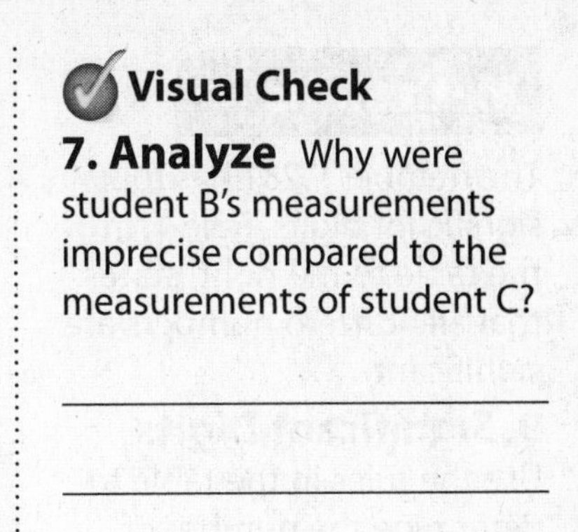

Visual Check

7. Analyze Why were student B's measurements imprecise compared to the measurements of student C?

Measurement and Accuracy

The tools used to take measurements can limit the accuracy of the measurements. Suppose you are measuring the temperature at which sugar melts, and the thermometer's measurements are divided into whole numbers. If your sugar sample melts between 183°C and 184°C, you can estimate the temperature between these two numbers. But, if the thermometer's measurements are divided into tenths, and your sample melts between 183.2°C and 183.3°C, your estimate between these numbers would be more accurate.

Significant Digits

In the second example above, you know that the temperature is between 183.2°C and 183.3°C. You could estimate that the temperature is 183.25°C. When you take any measurement, you know some digits for certain and you estimate some digits. **Significant digits** *are the number of digits in a measurement that are known with a certain degree of reliability*. The significant digits in a measurement include all digits you know for certain, plus one estimated digit. Therefore your measurement of 83.25°C would contain five significant digits. The figure below shows how to round to three significant digits. Since the ruler is divided into tenths, you know the rod is between 5.2 cm and 5.3 cm. You can estimate that the rod is 5.25 cm.

Measurement Tools: Known and Estimated Digits

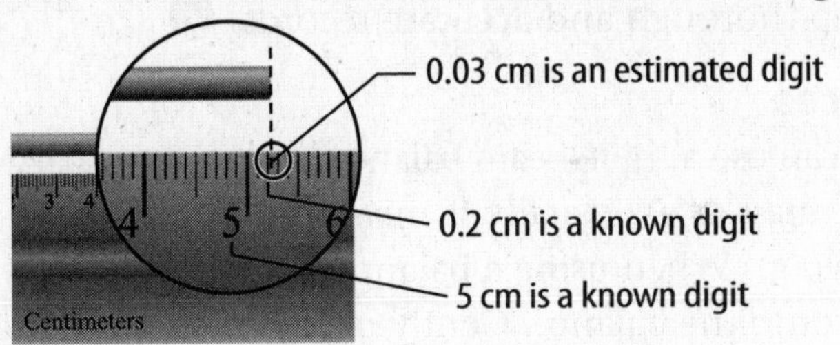

Visual Check

8. Interpret Is 4.5 in the figure a known or an estimated digit?

Math Skills

The number 5,281 has four significant digits. Rule 1, in the table to the right, states that all nonzero numbers are significant.

9. Significant Digits

Use the rules in the table to determine the number of significant digits in each of the following numbers: 2.02; 0.0057; 1,500; and 0.500.

Key Concept Check

10. Explain Why should you use significant digits?

Reading Check

11. Describe Why would you use a science journal?

Significant Digits		
Rules **1.** All nonzero numbers are significant. **2.** Zeros between nonzero digits are significant. **3.** Final zeros used after the decimal are significant. **4.** Zeros used solely for spacing the decimal are not significant. The zeros indicate only the position of the decimal. * The bold numbers in the examples are the significant digits.		
Example Number	**Significant Digits**	**Applied Rules**
1.234	4	1
1.2	2	1
0.0**23**	2	1, 4
0.**200**	3	1, 3
1,002	4	1, 2
3.07	3	1, 2
0.00**1**	1	1, 4
0.0**12**	2	1, 4
50,600	3	1, 2, 4

Using significant digits lets others know how certain your measurements are. The rules for using significant digits are shown in the table above.

Scientific Tools

Scientific inquiry often requires the use of tools. Scientists, including life scientists, might use the tools listed below and on the next page. You might use one or more of them during a scientific inquiry, too.

Science Journal

In a science journal, you can record descriptions, explanations, plans, and steps used in a scientific inquiry. A science journal can be a spiral-bound notebook or a loose-leaf binder. It is important to keep your science journal organized so you can find information when you need it. Make sure you keep thorough and accurate records.

Balances

You can use a triple-beam balance or an electric balance to measure mass. Mass usually is measured in kilograms (kg) or grams (g). When using a balance, do not let objects drop heavily onto the balance. Gently remove an object after you record its mass.

Thermometer

A thermometer measures the temperature of substances. The Kelvin (K) is the SI unit for temperature. However, in the science classroom, you measure temperature in degrees Celsius (°C). Use care when you place a thermometer into a hot substance so you do not burn yourself. Handle glass thermometers gently so they do not break. If a thermometer does break, tell your teacher immediately. Do not touch the broken glass or the thermometer's liquid. Never use a thermometer to stir anything.

 **Reading Check**

12. State In a science classroom, what unit of measure do you use for temperature?

____

Glassware

Laboratory glassware is used to hold, pour, heat, and measure liquids. Most labs have many types of glassware. For example, flasks, beakers, petri dishes, test tubes, and specimen jars are used as containers. To measure the volume of a liquid, you use a graduated cylinder. The unit of measure for liquid volume is the liter (L) or milliliter (mL).

 **Reading Check**

13. Identify Which unit of measure do you use for liquid volume?

____

____

Compound Microscope

Microscopes enable you to observe small objects that you cannot observe with just your eyes. Usually, two types of microscopes are in science classrooms—dissecting microscopes and compound light microscopes. Microscopes have either a single eyepiece or two eyepieces to observe a magnified image of a small object or an organism.

Microscopes can be damaged easily. It is important to follow your teacher's instructions when carrying and using a microscope.

Computers—Hardware and Software

Computers process information. In science, you can use computers to compile, retrieve, and analyze data for reports. You also can use them to create reports and other documents, to send information to others, and to research information.

The physical components of computers, such as monitors and keyboards, are called hardware. The programs that you run on computers are called software. These programs include word processing, spreadsheet, and presentation programs. When scientists write reports, they use word processing programs. They use spreadsheet programs for organizing and analyzing data. Presentation programs can be used to explain information to others.

 **Reading Check**

14. Name What are the physical components of a computer called? What are computer programs called?

____

____

Tools Used by Life Scientists

Life scientists often use the tools described below.

Magnifying Lens

A magnifying lens is a hand-held lens that magnifies, or enlarges, the image of an object. It is not as powerful as a microscope and is useful when great magnification is not needed. Magnifying lenses also can be used outside the lab where microscopes might not be available.

Reading Check
15. Describe how you might use a magnifying lens.

Slide

To observe an item using a compound light microscope, you first must place it on a thin, rectangular piece of glass called a slide. You must handle slides gently to avoid breaking them.

Dissecting Tools

Scientists use dissecting tools, such as scalpels and scissors, to examine tissues, organs, or prepared organisms. Dissecting tools are sharp, so always use extreme caution when handling them.

Pipette

A pipette is similar to a eyedropper. It is a small glass or plastic tube used to draw up and transfer liquids.

Key Concept Check
16. Name What are some tools used by life scientists?

After You Read

Mini Glossary

accuracy: a description of how close a measurement is to an accepted or true value

description: a spoken or written summary of observations

explanation: an interpretation of observations

International System of Units (SI): the internationally accepted system for measurement

precision: a description of how similar or close measurements are to each other

significant digits: the number of digits in a measurement that are known with a certain degree of reliability

1. Review the terms and their definitions in the Mini Glossary. Write a sentence that explains the importance of accuracy in measurement.

____

____

2. Complete the following table to identify and describe SI base units.

Unit (symbol)	Quantity Measured
ampere (A)	
	light intensity
meter (m)	
	temperature
mole (mol)	

3. Explain the difference between a description and an explanation.

____

____

Log on to ConnectED.mcgraw-hill.com and access your textbook to find this lesson's resources.

Scientific Explanations

Case Study

Key Concepts

- How do independent and dependent variables differ?
- How is scientific inquiry used in a real-life scientific investigation?

Study Coach

Identify the Main Ideas As you read, organize your notes in two columns. In the left column, write the main idea of each paragraph. In the right column, write details that support each main idea. Review your notes to help you remember the details of the lesson.

Read to Learn

Biodiesel from Microalgae

For several centuries, fossil fuels have been the main sources of energy for industry and transportation. However, scientists have shown that burning fossil fuels negatively affects the environment. Also, some people are concerned about eventually using up the world's reserves of fossil fuels.

During the past few decades, scientists have explored using protists to produce biodiesel. Biodiesel is a fuel made primarily from living organisms. Protists are a group of tiny organisms that usually live in water or moist environments. Some of these protists are plantlike because they make their own food using a process called photosynthesis. Microalgae are plantlike protists.

Designing a Controlled Experiment

Scientists use scientific inquiry to investigate the use of protists to make biodiesel. They designed controlled experiments to test their hypotheses. The tables in this lesson are examples of how scientists practiced inquiry and the skills you read about in Lesson 1.

A controlled experiment is a scientific investigation that tests how one variable affects another variable. *A* ***variable*** *is any factor in an experiment that can have more than one value.* Controlled experiments have two types of variables. *The* ***dependent variable*** *is the factor measured or observed during an experiment. The* ***independent variable*** *is the factor that you want to test. It is changed by the investigator to observe how it affects a dependent variable.* ***Constants*** *are the factors in an experiment that remain the same.*

A controlled experiment has an experimental group and a control group. The experimental group is used to study how a change in the independent variable changes the dependent variable. The control group contains the same factors as the experimental group, but the independent variable is not changed. Without a control, it is difficult to know whether your experimental observations result from the variable you are testing or from another factor.

Key Concept Check

1. Differentiate How do independent and dependent variables differ?

____

____

____

____

Biodiesel

The idea of running diesel engines on fuel made from plant or plantlike sources is not entirely new. Rudolph Diesel invented the diesel engine. He used peanut oil to demonstrate how the engine worked. However, once petroleum was introduced as a diesel fuel source, it was preferred over peanut oil because of it was cheaper. ✓

Oil-rich food crops such as soybeans can be used to produce biodiesel. However, some people are concerned that crops grown for fuel use will replace crops grown for food. If farmers grow more crops for fuel, then the amount of food available worldwide will be reduced. Because of food shortages in many parts of the world, replacing food crops with fuel crops is not a good solution. ✓

Aquatic Species Program

In the late 1970s, the U.S. Department of Energy began funding its Aquatic Species Program (ASP) to investigate ways to remove air pollutants. Coal-fueled power plants produce carbon dioxide (CO_2), a pollutant, as a by-product. In the beginning, the study examined all aquatic organisms that use CO_2 during photosynthesis—their food-making process. These included large plants, commonly known as seaweeds, plants that grow partially underwater, and microalgae.

Scientists hoped that these organisms might remove excess CO_2 from the atmosphere. During the studies, however, the project leaders noticed that some microalgae produced large amounts of oil. The program's focus soon shifted to using microalgae to produce oils that could be processed into biodiesel. The scientists' observations and prediction are summarized in the table below. When referring to the examples in tables in this lesson, recall that a hypothesis is a tentative explanation that can be tested by scientific investigation. A prediction is a statement of what someone expects to happen next in a sequence of events.

Observation and Prediction

Scientific investigations often begin when someone observes an event in nature and wonders why or how it occurs.

Observation	Prediction
While testing microalgae to discover if they would absorb carbon pollutants, ASP project leaders saw that some species of microalgae had high oil content.	If the right conditions are met, then plants and plantlike organisms can be used as a source of fuel.

✓ **Reading Check**

2. Name What did Rudolph Diesel use as fuel?

✓ **Reading Check**

3. Explain Why is there a concern that crops grown for fuel use will replace crops grown for food?

Interpreting Tables

4. Relate How do scientific investigations often begin?

✔ Reading Check
5. Define What are microalgae?

Interpreting Tables
6. State What was the goal of the screening test?

Interpreting Tables
7. Identify What is the dependent variable in this example?

Which Microalgae?

Microalgae are microscopic organisms that live in marine (salty) or freshwater environments. Like many plants and plantlike organisms, they use photosynthesis and make sugar. The process requires light energy. Microalgae make more sugar than they can use. They convert excess sugar to oil. Scientists focused on these microalgae because their oil then could be processed into biodiesel. ✔

Scientists started by collecting and identify promising microalgae species. The search focused on microalgae in shallow, inland, saltwater ponds. Scientists predicted that these microalgae were more resistant to changes in temperature and salt content in the water.

Design an Experiment and Collect Data
One way to test a hypothesis is to design an experiment and collect data.
The ASP scientists developed a rapid screening test to discover which microalgae species produced the most oil.

By 1985, a test was in place for identifying microalgae with high oil content. Two years later, 3,000 microalgae species had been collected. Scientists checked these samples for tolerance to acidity, salt levels, and temperature. From the samples, 300 species were selected. Of these 300 species, green algae and diatoms showed the most promise. However, it was obvious that no one strain was going to be perfect for all climates and water types.

Hypotheses and Predictions
During a long investigation, scientists form many hypotheses and conduct many tests.
Hypothesis: Microalgae species in shallow, saltwater ponds are most resistant to variations in temperature and salt content.
Prediction: Microalgae species most resistant to variations in temperature and salt content will be the most useful species in producing biodiesel.
Hypothesis: Microalgae grown with inadequate amounts of nitrogen alter their growth processes and produce more oil.
Independent Variable: amount of nitrogen available
Dependent Variable: amount of oil produced
Constants: the growing conditions of algae (temperature, water quality, exposure to the Sun, etc.)

Oil Production in Microalgae

Scientists also began researching how microalgae produce oil. Some studies suggested that starving microalgae of nutrients, such as nitrogen, could increase the amount of oil they produced. However, starving the microalgae also caused them to be smaller, resulting in no overall increase in oil production. ✓

Outdoor Testing v. Bioreactors

By the 1980s, the ASP scientists were growing microalgae in outdoor ponds in New Mexico. However, outdoor conditions were very different from those in the laboratory. The cooler outdoor temperatures resulted in smaller microalgae. Native algae species also invaded the ponds, forcing out the high-oil-producing laboratory microalgae species.

The scientists continued to focus on growing microalgae in open ponds. Many scientists still believe that these open ponds are better for producing large quantities of biodiesel from microalgae. But some researchers are now growing microalgae in closed glass containers called bioreactors. Inside these bioreactors, organisms live and grow under controlled conditions. This method avoids many of the problems associated with open ponds. However, bioreactors are more expensive than open ponds.

A biofuel company in the western United States has been experimenting with a low-cost bioreactor. A scientist at the company explained that they examined the ASP program and hypothesized that they could use long plastic bags instead of closed glass containers. However, microalgae grown in plastic bags are more expensive to harvest. ✓

Why So Many Hypotheses?

According to Dr. Richard Sayre, a biofuel researcher, all the ASP research was based on forming hypotheses. He says, "It was hypothesis-driven. You just don't go in and say 'Well, I have a feeling this is the right way to do it.' You propose a hypothesis. Then you test it." Dr. Sayre added, "Biologists have been trained over and over again to develop research strategies based on hypotheses. It's sort of ingrained into our culture. You don't get research support by saying, 'I'm going to put together a system, and it's going to be wonderful.' You have to come up with a question. You propose some strategies for answering the question. What are your objectives? What outcomes do you expect for each objective?" ✓

✓ **Reading Check**

8. Explain Why didn't starving microalgae of nutrients provide an overall benefit?

____

____

____

____

✓ **Reading Check**

9. Relate What is a disadvantage of bioreactors?

____

____

____

✓ **Reading Check**

10. Describe Why is it important for a scientific researcher to develop a good hypothesis?

____

____

____

____

Increasing Oil Yield

Scientists from a biofuel company in Washington State thought of another way to increase oil production. Researchers knew microalgae use light energy, water, and carbon dioxide and they make sugar. The microalgae eventually convert sugar into oil. The scientists wondered if they could increase microalgae oil production by distributing light to all microalgae, including those below the surface.

Interpreting Tables
11. State How did scientists think they could get microalgae to produce more oil?

Hypothesis and Prediction
Scientists hypothesize that they can increase microalgae oil production by distributing light to greater depths.
Hypothesis: If the top layer of microalgae blocks light from reaching microalgae beneath them, then they produce less oil because light is not distributed evenly to all the microalgae.
Prediction: If light is distributed more evenly, then more microalgae will grow, and more biodiesel will be produced.

Bringing Light to Microalgae

Normally microalgae grow near the surface of a pond. Any microalgae about 5 cm below the pond's surface will grow less. Why is this? First, water blocks light from reaching deep into the pond. Second, microalgae at the top of a pond block light from reaching microalgae below them. Only the top part of a pond is productive.

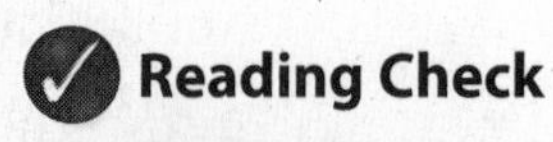

Reading Check
12. State Where do microalgae normally grow within a pond?

Experimental Group

Researchers decided to assemble a team of engineers to design a light distribution system. Light rods distribute artificial light to microalgae in a bioreactor. The bioreactor controls the environmental conditions that affect how the microalgae grow. These conditions include temperature, nutrient levels, carbon dioxide level, airflow, and light.

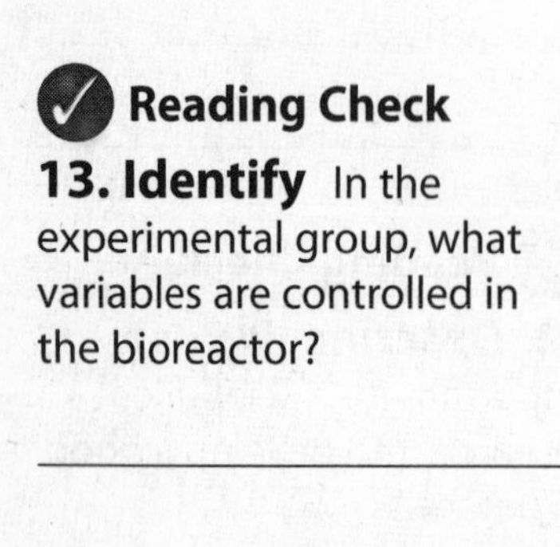

Reading Check
13. Identify In the experimental group, what variables are controlled in the bioreactor?

Data from their experiments showed scientists how their microalgae in well-lit environments grow compared to how microalgae grow in dimmer environments. Using solar data for various parts of the country, the scientists concluded that the light rod would significantly increase microalgae growth and oil production in outdoor ponds. These scientists next plan to use the light-rod growing method in outdoor ponds.

Field Testing

Scientists plan to take light to the microalgae instead of moving microalgae to light. Dr. Jay Burns is chief microalgae scientist at a biofuel company. He said, "What we are proposing to do is to take the light from the surface of a pond and distribute it throughout the depth of the pond. Instead of only the top 5 cm being productive, the whole pond becomes productive."

Note that research scientists and scientists in the field rely on scientific methods and scientific inquiry to solve real-life problems. When a scientific investigation lasts for several years and involves many scientists, such as this study, many hypotheses can be tested. Some hypotheses are supported, and other hypotheses are not.

Regardless of which hypotheses are supported, information is gathered and lessons are learned. Hypotheses are refined and tested many times. This process of scientific inquiry results in a better understanding of the problem and the possible solutions.

Analysis and Conclusion
Scientists tested their hypothesis, collected data, analyzed the data, and drew conclusions.
Analyze Results: The experimental results showed that microalgae would produce more oil using a light-rod system than by using just sunlight.
Draw a Conclusion: The researchers concluded that the light-rod system greatly increased microalgae oil production.

Another Way to Bring Light to Microalgae

Light rods are not the only way to bring light to microalgae. Paddlewheels can be used to keep changing the location of the microalgae. Paddlewheels continuously rotate microalgae to the surface so the organisms are exposed to more light.

Reading Check

14. Summarize What is the benefit of the light-distribution system?

Interpreting Tables

15. Express What did the researchers conclude would increase algae yield?

Key Concept Check

16. Explain Describe three ways in which scientific inquiry was used in this case study.

Why Grow Microalgae?

Although the focus of this case study is microalgae growth for biodiesel production, growing microalgae has other benefits.

Some of the benefits of growing algae are shown in the figure below. Power plants that burn fossil fuels release carbon dioxide into the atmosphere. Evidence indicates that this contributes to global warming. During photosynthesis, microalgae use carbon dioxide and water, release oxygen, and produce sugar, which they convert to oil. Not only do microalgae produce a valuable fuel, they also remove pollutants from and add oxygen to the atmosphere.

Visual Check

17. Analyze What is used as a feedstock for microalgae?

____

Cultivating Microalgae

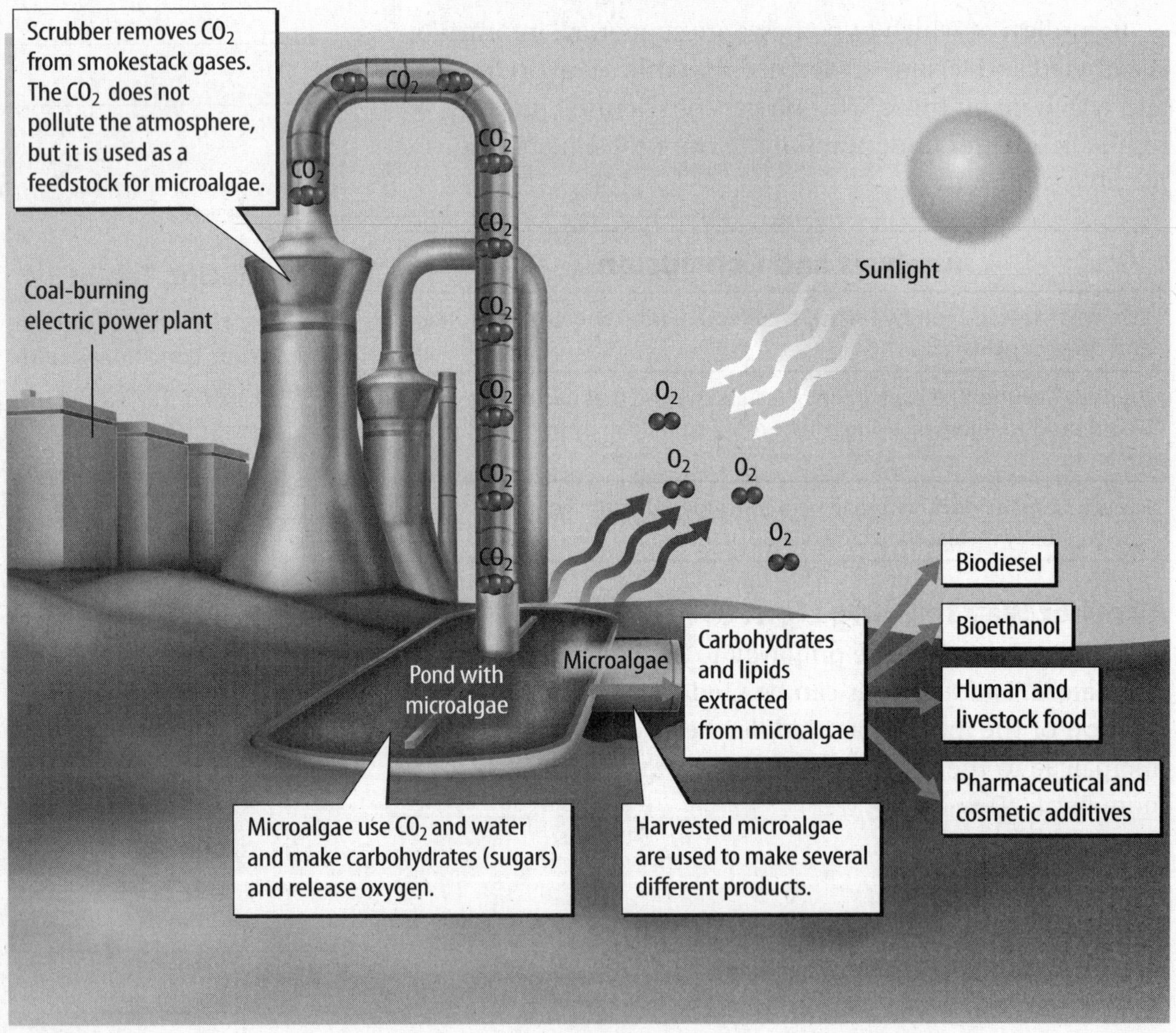

Are microalgae the future?

Scientists face many challenges in their quest to produce biodiesel from microalgae. For now, the costs of growing microalgae and extracting their oils are too high to compete with petroleum-based diesel. However, the combined efforts of government-funded programs and commercial biofuel companies might one day make microalgae-based biodiesel an affordable reality in the United States.

New Plants One company in Israel has a successful test plant in operation. Plans are underway to build a large-scale industrial facility to convert carbon dioxide gases released from an Israeli coal-powered electrical plant into useful microalgae products. If this technology performs as expected, microalgae cultivation might occur near coal-fueled power plants in other parts of the world, too.

Drawing Conclusions Currently, scientists have no final conclusions about using microalgae as a fuel source. As long as petroleum remains relatively inexpensive and available, it probably will remain the preferred source of diesel fuel. However, if petroleum prices increase or availability decreases, new sources of fuel will be needed. Biodiesel made from microalgae might be one of the alternative fuel sources used.

Reading Check

18. Assess What is preventing algae-based biodiesel from competing with petroleum-based diesel?

Reading Check

19. Predict If the technology is successful, what might happen to algae cultivation?

Reading Check

20. State What might cause a demand for biodiesel made from microalgae?

After You Read

Mini Glossary

constant: a factor in an experiment that remains the same

dependent variable: the factor measured or observed during an experiment

independent variable: a factor that you want to test and that is changed by the investigator to observe how it affects a dependent variable

variable: any factor in an experiment that can have more than one value

1. Review the terms and their definitions in the Mini Glossary. Write a sentence that describes the use of variables in controlled experiments.

____

____

2. Complete the following flowchart that shows the ASP's initial study of pollution control.

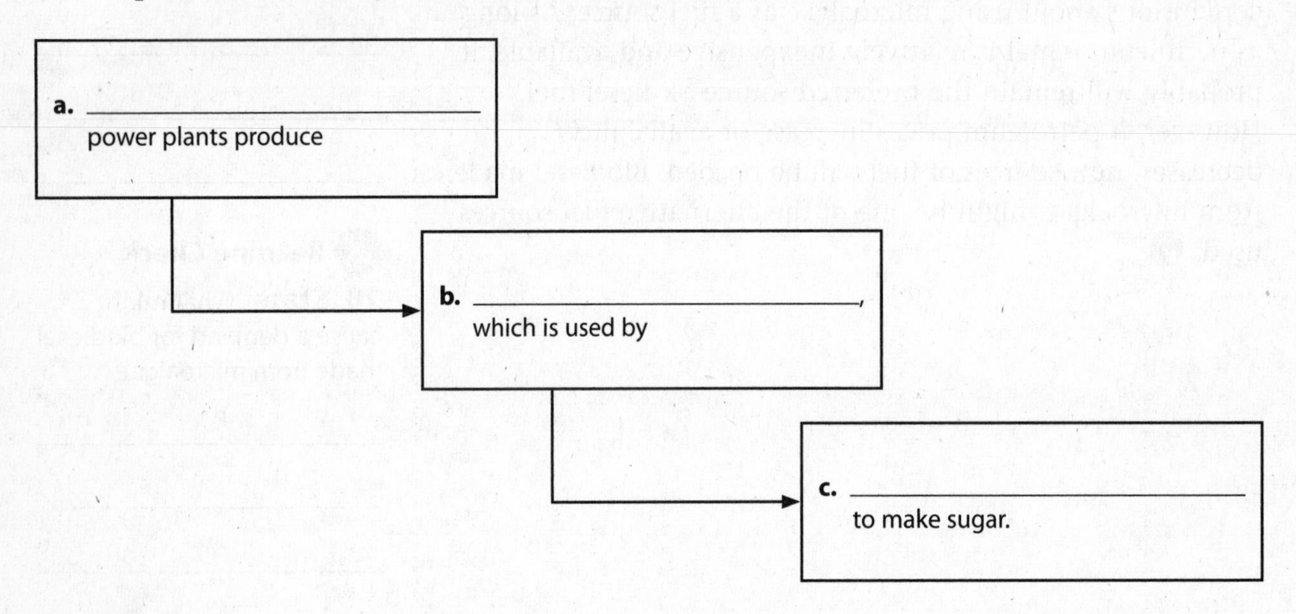

3. Explain how identifying the main idea of each paragraph helped you understand this lesson.

____

____

Log on to ConnectED.mcgraw-hill.com and access your textbook to find this lesson's resources.

Classifying and Exploring Life

Characteristics of Life

Before You Read

What do you think? Read the two statements below and decide whether you agree or disagree with them. Place an A in the Before column if you agree with the statement or a D if you disagree. After you've read this lesson, reread the statements to see if you have changed your mind.

Before	Statement	After
	1. All living things move.	
	2. The Sun provides energy for almost all organisms on Earth.	

Read to Learn

Characteristics of Life

Your classroom is full of nonliving things and living things. Desks, books, and lights are nonliving things. Your classmates, teacher, and plants are living things. What makes people and plants different from desks and lights?

People and plants, like all living things, have all the characteristics of life. All living things are organized. They grow and develop. All living things reproduce. They respond to their environment. All living things maintain certain internal conditions and use energy.

Nonliving things do not have all these characteristics. Books might be organized into chapters. Lights might use energy. But only those things that have all the characteristics of life are living. *Things that have all the characteristics of life are called* **organisms.**

Organization

Your school has organization. The classrooms are for learning and the gym is for sports. Living things are also organized. Their organization involves cells. *A* **cell** *is the smallest unit of life.* An organism might be made of just one cell or of many cells. All organisms have structures with specific functions, or jobs.

Key Concept
- What characteristics do all living things share?

Study Coach

Make Flash Cards Write each boldface word on one side of a flash card. Write the definition on the other side. Use the cards to quiz yourself.

Reading Check

1. Identify How do living things differ from nonliving things?

FOLDABLES®

Make a half-book and use it to organize your notes on the characteristics of living things.

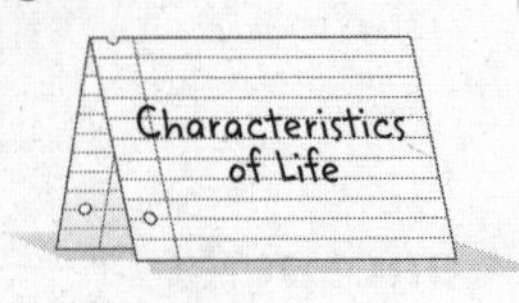

2. Apply You are a multicellular organism. Name one function that groups of your cells carry out.

____

Unicellular Organisms *Living things that are made of only one cell are called* **unicellular** *organisms.* A unicellular organism has structures with specialized functions. Some structures control cell activities. Some take in nutrients. Other structures enable the organism to move.

Multicellular Organisms *Living things that are made of two or more cells are called* **multicellular** *organisms.* Some multicellular organisms only have a few cells, but others have trillions of cells. The cells of a multicellular organism usually do not all do the same things. Instead, groups of cells have specialized functions. These functions might include digestion or movement.

Growth and Development

Think about the tadpole in the figure below. The tadpole does not look like the frog it will become. The tadpole will lose its tail and grow legs. Like all organisms, the tadpole will grow and develop.

Stage 1: Amphibian eggs are laid and fertilized in water.

Stage 2: Fertilized frog eggs are hatched into tadpoles.

Stage 3: Tadpoles begin to grow into adults.

Stage 4: The adult frog can live on land.

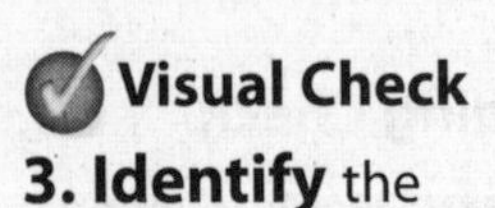

3. Identify the characteristics of life that you see in the figure.

____

____

____

____

How do organisms grow? When an organism grows, it increases in size. A unicellular organism grows as its one cell gets bigger. A multicellular organism grows when more cells are produced.

How do organisms develop? The changes in an organism during its lifetime are called development. A multicellular organism develops as cells become specialized into different cell types, such as skin and muscle cells. Some organisms have amazing developmental changes over their lifetimes. An example is a tadpole developing into a frog. ✓

Reproduction

Reproduction is the process by which one organism makes one or more new organisms. Organisms must reproduce, or they will die out. Not all organisms reproduce, such as family pets. But if a type of organism is going to survive, some organisms of that type must reproduce.

Organisms reproduce in many ways. Some unicellular organisms divide and become two new organisms. Each new organism is just like the original cell. Some organisms must have a mate to reproduce. Other organisms can reproduce without a mate. Organisms produce different numbers of offspring. Humans usually produce only one or two offspring at a time. Other organisms, such as frogs, can produce hundreds of offspring at one time.

Responses to Stimuli

Organisms live in environments that change all the time. These changes are called stimuli (STIHM yuh li). One change is called a stimulus. All organisms respond to stimuli.

Internal Stimuli

Internal stimuli are changes inside an organism. They include hunger, thirst, and pain. If you feel hungry and look for food, you are responding to an internal stimulus—hunger. The feeling of thirst that causes you to look for water is another internal stimulus. ✓

External Stimuli

External stimuli are changes outside an organism. They are usually changes in the environment that the organism lives in. Light and temperature are examples of external stimuli.

✓ **Reading Check**

4. Explain What is development?

____

____

✓ **Reading Check**

5. Name two internal stimuli.

____

____

Think it Over

6. Contrast How do external stimuli differ from internal stimuli?

____

____

____

Light Many organisms respond to changes in light. Many plants will grow toward light. You respond to light, too. If you spend time in sunlight, your skin's response might be to darken, turn red, or freckle.

Temperature How does your body respond to changes in temperature? Like many animals, your body responds by increasing or decreasing the amount of blood flow to your skin. If the temperature gets warmer, your blood vessels respond by widening. Then more blood can flow to your skin. You feel cooler.

7. Apply What word explains why you go to the bathroom more when you drink a lot of water?

Homeostasis

All organisms are able to maintain some internal conditions. **Homeostasis** (hoh mee oh STAY sus) *is an organism's ability to maintain steady internal conditions when outside conditions change*. Have you ever noticed that if you drink more water than usual, you have to go to the bathroom more often than you usually do? Your body is keeping your internal conditions steady.

The Importance of Homeostasis

Cells need certain conditions to function the way they should. Homeostasis makes sure cells can function. If cells cannot function the way they should, an organism might get sick or die.

8. Explain Why is maintaining homeostasis important to organisms?

Methods of Regulation

Humans cannot survive if their body temperature changes more than a few degrees from 37°C. When your outside environment becomes too hot or too cold, your body responds. It sweats, shivers, or changes the flow of blood to maintain the body temperature of 37°C.

Both unicellular and multicellular organisms have ways to maintain homeostasis. Some unicellular organisms have a structure called a contractile vacuole (kun TRAK tul • VA kyuh wohl). It collects and pumps extra water out of the cell.

There is a limit to the amount of change that can occur inside an organism. For example, you could live for only a few hours in very cold water. Your body could not maintain steady internal conditions, or homeostasis, in this environment. Your cells could not function.

9. Generalize What conditions might be too harsh for an organism to maintain homeostasis?

Energy

All organisms use energy. Digesting food, thinking, reading, and sleeping use energy. Cells use energy to transport substances, make new cells, and perform chemical reactions. All of the characteristics of life use energy.

Energy's Origin Where does this energy come from? The energy that most organisms use originally came to Earth from the Sun, as shown below. The energy goes from one organism to another. Energy in the cactus comes from the Sun. The squirrel gets energy from the cactus that it eats. The coyote gets energy from eating the squirrel.

Key Concept Check

10. State What characteristics do all living things share?

Visual Check

11. Interpret Diagrams From which food sources does the badger get energy?

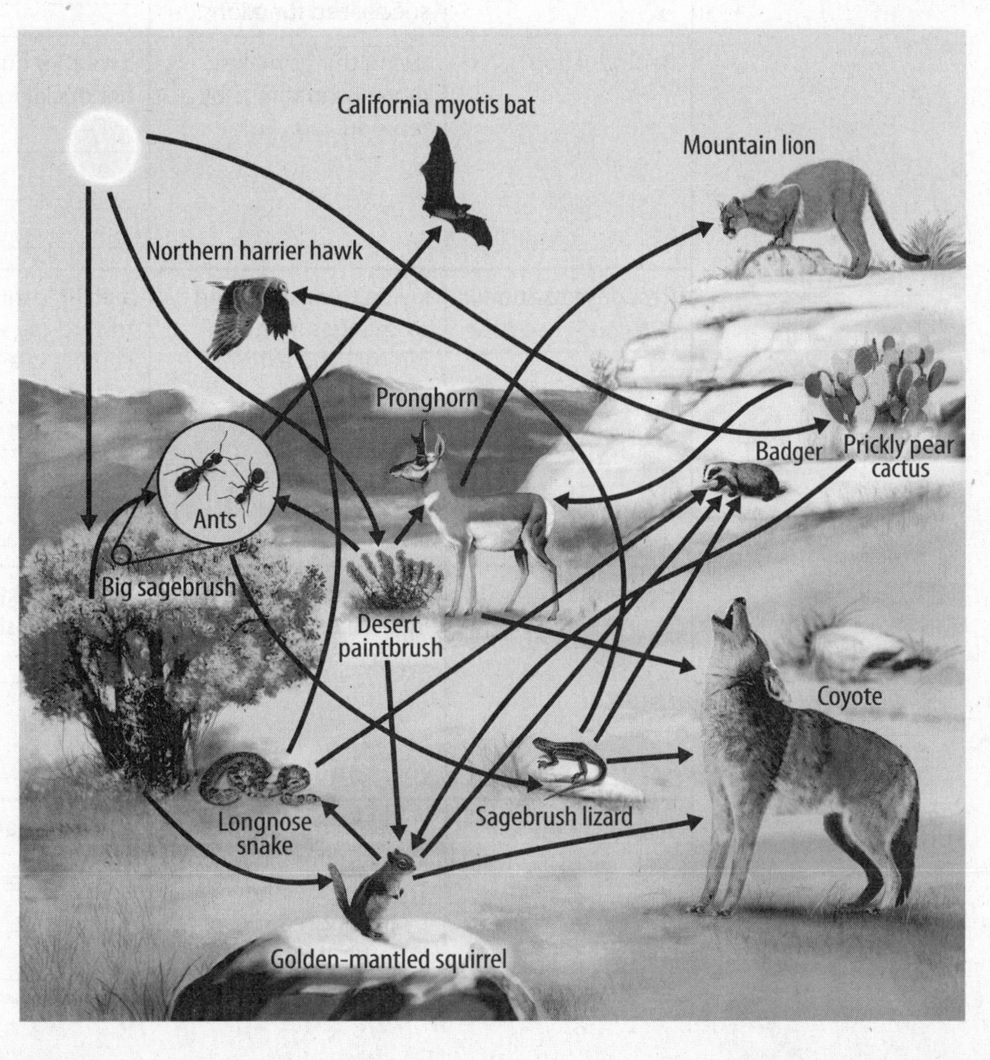

Energy for Life There are six characteristics of life. From the chart below you will learn how each characteristic depends on energy from the Sun.

Visual Check

12. Apply In the chart, add another example for each characteristic of life.

Characteristics of Life

Characteristic	Definition	Example
Organization	Living things have structures with their own functions or jobs. In living things with more than one cell, groups of cells work together. These living things have a higher level of organization than living things with only one cell do.	a leopard running __________
Growth and Development	Living things might grow by increasing cell size. They also might grow by making more cells. Living things develop when cells get specialized functions.	a tadpole changing into a frog __________ __________ __________
Reproduction	Living things make more living things by reproducing.	a mother duck and her ducklings __________ __________
Response to Stimuli	Living things respond to changes in their internal and external environments.	a plant leaning toward the sunlight coming in a nearby window __________ __________ __________
Homeostasis	Living things keep their internal environment stable.	a girl drinking water after exercising __________ __________ __________
Use of Energy	Living things use energy for everything they do. They get their energy by making, eating, or absorbing food.	a squirrel eating a nut __________ __________ __________

After You Read

Mini Glossary

cell: the smallest unit of life

homeostasis (hoh mee oh STAY sus): an organism's ability to maintain steady internal conditions when outside conditions change

multicellular: describes a living thing that is made of two or more cells

organism: a thing that has all the characteristics of life

unicellular: describes a living thing that is made of only one cell

1. Review the terms and their definitions in the Mini Glossary. Write one or two sentences to explain the difference between multicellular and unicellular organisms.

____

____

2. Fill in the table below to identify the six characteristics all living things share. Then add an example for each characteristic from your life or the life of someone you know.

Characteristic of Life	Personal Example

3. How did making flash cards help you learn the important terms in the lesson?

____

____

What do you think NOW?

Reread the statements at the beginning of the lesson. Fill in the After column with an A if you agree with the statement or a D if you disagree. Did you change your mind?

Log on to ConnectED.mcgraw-hill.com and access your textbook to find this lesson's resources.

Classifying and Exploring Life

Classifying Organisms

Key Concepts
- What methods are used to classify living things into groups?
- Why does every species have a scientific name?

Before You Read

What do you think? Read the two statements below and decide whether you agree or disagree with them. Place an A in the Before column if you agree with the statement or a D if you disagree. After you've read this lesson, reread the statements to see if you have changed your mind.

Before	Statement	After
	3. A dichotomous key can be used to identify an unknown organism.	
	4. Physical similarities are the only traits used to classify organisms.	

Mark the Text

Identify Main Ideas As you read this lesson, underline the main idea in each paragraph.

Read to Learn

Classifying Living Things

There have been many different ideas about how to organize, or classify, organisms. First you will learn about some early ideas for classifying organisms. Then you will learn about the system used today.

Greek philosopher Aristotle lived more than 2,000 years ago. He was one of the first people to classify organisms. He placed all organisms into two groups—plants and animals. Animals were classified based on whether or not the animal had "red blood," the animal's environment, and the shape and size of the animal. Plants were classified based on structure and size and on whether the plant was a tree, a shrub, or an herb.

Determining Kingdoms

In the 1700s, Carolus Linnaeus, a Swedish physician and botanist, classified organisms based on similar structures. Linneaus placed all organisms into two main groups, called kingdoms. For the next 200 years, people learned more about organisms and discovered new organisms. In 1969, Robert H. Whittaker, an American biologist, came up with a five-kingdom system for classifying organisms. Those kingdoms are Monera, Protista, Plantae, Fungi, and Animalia.

Science Use v. Common Use
kingdom
Science Use a classification category that ranks above phylum and below domain
Common Use a territory ruled by a king or queen

Determining Domains

The classification method used today is called systematics. It uses everything that is known about organisms to classify them. It looks at an organism's cell type, its habitat, the way it gets food and energy, the structure and function of its features, and the common ancestry of organisms. Systematics also uses molecular analysis—the study of molecules, such as DNA, within organisms.

Scientists using systematics found two distinct groups in Kingdom Monera. They added another classification level called domains. There are three domains—Bacteria, Archaea (ar KEE uh), and Eukarya (yew KER ee uh). They are shown below. All organisms are now classified into one of the three domains and then into one of the six kingdoms.

Key Concept Check
1. Summarize What evidence is used to classify living things into groups?

Domains and Kingdoms

Domain	Bacteria	Archaea	Eukarya			
Kingdom	**Bacteria**	**Archaea**	**Protista**	**Fungi**	**Plantae**	**Animalia**
Characteristics	Bacteria are simple, unicellular organisms.	Archaea are simple, unicellular organisms. They often live in very hot or salty environments.	Protists are unicellular and more complex than bacteria or archaea.	Fungi are unicellular or multicellular and absorb food.	Plants are multicellular and make their own food.	Animals are multicellular and take in their food.

Scientific Names

Suppose you did not have a name. What would people call you? All organisms, just like people, have names. We still use the naming system that Linnaeus created. It is called binomial nomenclature (bi NOH mee ul • NOH mun klay chur).

Visual Check
2. Name Why is a dog in Kingdom Animalia instead of Kingdom Fungi?

Binomial Nomenclature

Linneaus's naming system, **binomial nomenclature,** *gives each organism a two-word scientific name*. For example, the scientific name for the brown bear is *Ursus arctos*. This two-word scientific name is the name of an organism's species (SPEE sheez). *A* **species** *is a group of organisms that have similar traits and are able to produce fertile offspring*. In binomial nomenclature, the first word is the organism's genus (JEE nus) name, such as *Ursus*. *A* **genus** *is a group of similar species*. The second word might describe the way an organism looks or the way it acts.

How are organisms grouped? How do genus and species fit into kingdoms and domains? Similar species are grouped into one genus. (The term for more than one genus is *genera*.) Similar genera are grouped into families. Similar families are grouped into orders. Similar orders are grouped into classes. Similar classes are grouped into phyla. Similar phyla are grouped into kingdoms. And similar kingdoms are grouped into domains. The binomial nomenclature for the brown bear is shown below.

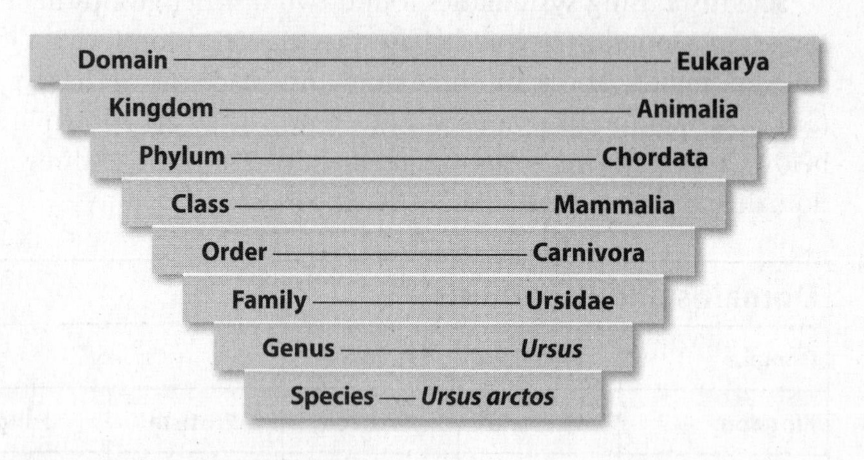

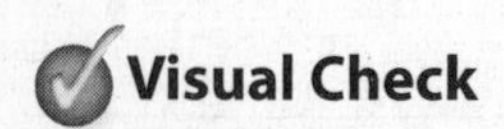

Visual Check

3. Name What domain does the brown bear belong to?

Uses of Scientific Names

Some people would call a large brown bear a brown bear. Others would call it a grizzly bear. But it has only one scientific name: *Ursus arctos*.

A common name might also refer to more than one type of organism. Imagine two different evergreen trees. Even though they are two different species, they have the same common name—pine trees. Scientific names are important for many reasons. Each species has its own scientific name. Scientific names are the same worldwide. This makes communication about organisms easier because everyone uses the same name for the same species.

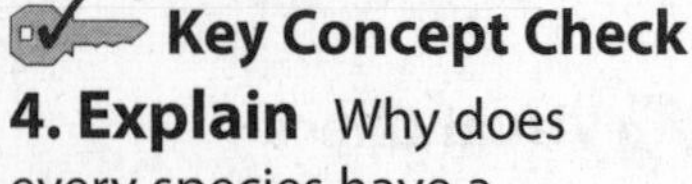

Key Concept Check

4. Explain Why does every species have a scientific name?

Classification Tools

Imagine that you are fishing. You catch a fish that you have never seen before. How can you find out what type of fish you have caught? You can use several tools to identify organisms.

FOLDABLES®

Make a two-tab book to compare two of the tools scientists use to identify organisms—dichotomous keys and cladograms.

Dichotomous Keys

A **dichotomous key** *is a series of descriptions arranged in pairs that lead the user to the identification of an unknown organism.* Each chosen description leads to another description. You keep making choices until you reach the name of the organism. The dichotomous key below identifies some species of mice.

Key to Some Mice of North America	
1. Tail hair	**a.** no hair on tail; scales show plainly; house mouse, *Mus musculus* **b.** hair on tail, go to 2
2. Ear size	**a.** ears small and nearly hidden in fur, go to 3 **b.** ears large and not hidden in fur, go to 4
3. Tail length	**a.** less than 25 mm; woodland vole, *Microtus pinetorum* **b.** more than 25 mm; prairie vole, *Microtus ochrogaster*
4. Tail coloration	**a.** sharply bicolor, white beneath and dark above; deer mouse, *Peromyscus maniculatus* **b.** darker above than below but not sharply bicolor; white-footed mouse, *Peromyscus leucopus*

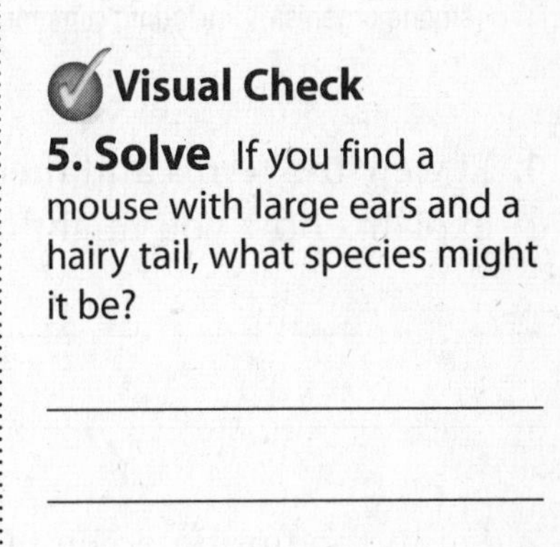

Visual Check

5. Solve If you find a mouse with large ears and a hairy tail, what species might it be?

Cladograms

Have any of your relatives made a family tree? Family trees are branching charts that show how family members are related. Biologists use a similar diagram to show how species are related. It is called a cladogram. *A* **cladogram** *is a branched diagram that shows the relationships among organisms, including common ancestors.*

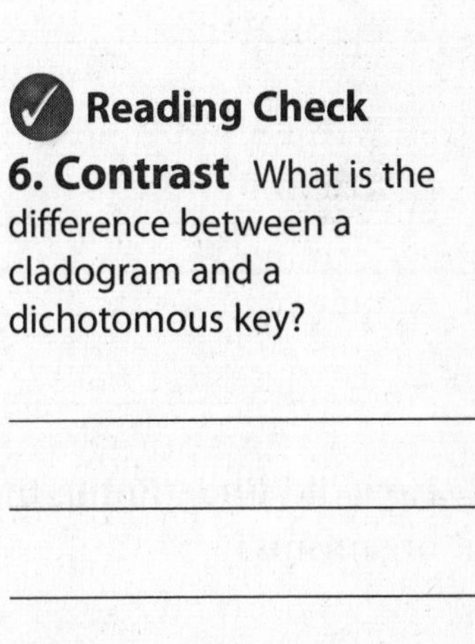

Reading Check

6. Contrast What is the difference between a cladogram and a dichotomous key?

The cladogram below has a series of branches. Each branch follows a new characteristic. Each characteristic can be seen in the species to its right. See what this cladogram tells you about the relationships among the living things that are shown. The salamander, lizard, hamster, and chimpanzee have lungs. The salmon does not have lungs. Therefore, the other animals are more closely related to each other than they are to the salmon.

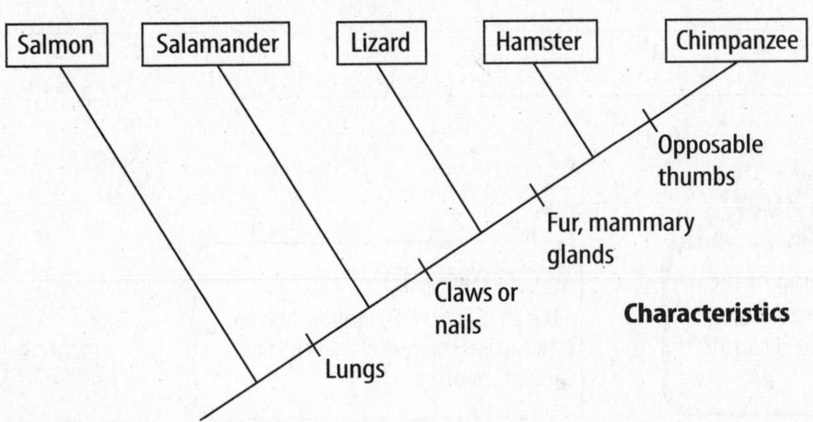

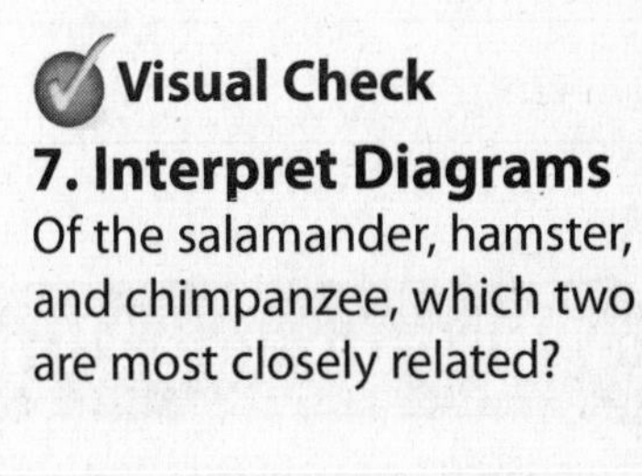

Visual Check

7. Interpret Diagrams Of the salamander, hamster, and chimpanzee, which two are most closely related?

After You Read

Mini Glossary

binomial nomenclature (bi NOH mee ul • NOH mun klay chur): a naming system that gives each organism a two-word scientific name

cladogram: a branched diagram that shows the relationships among organisms, including common ancestors

dichotomous key: a series of descriptions arranged in pairs that lead the user to the identification of an unknown organism

genus (JEE nus): a group of similar species

species (SPEE sheez): a group of organisms that have similar traits and are able to produce fertile offspring

1. Review the terms and their definitions in the Mini Glossary. Write a sentence that explains how you might use a dichotomous key while birdwatching in the woods.

____

____

2. Fill in the upside-down pyramid below to show how living things are classified.

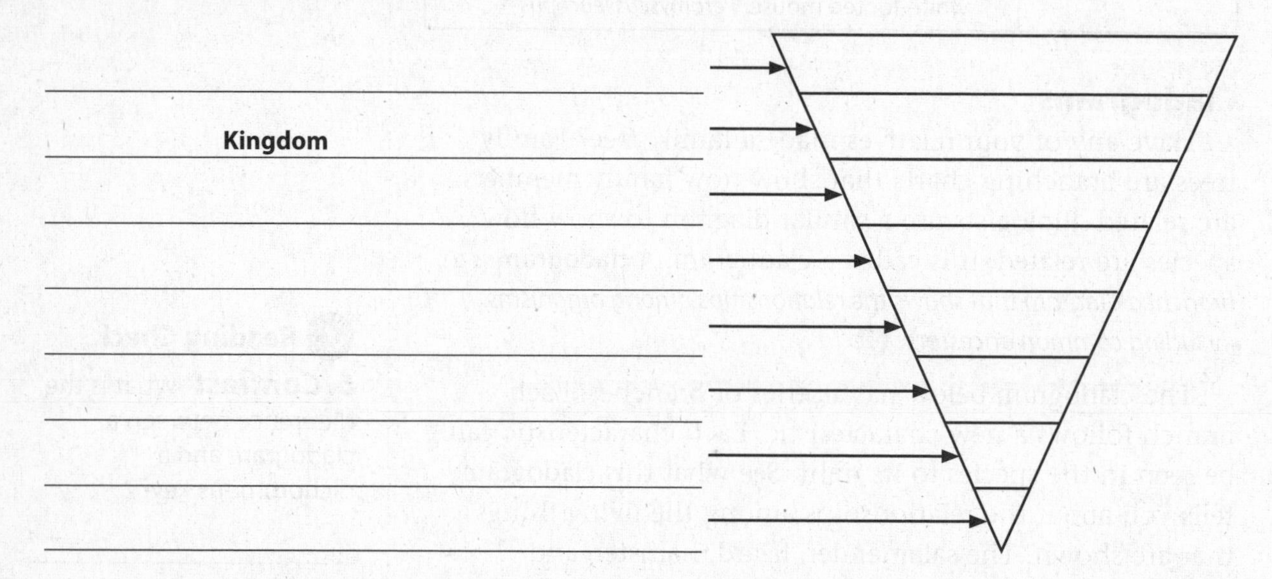

3. How did underlining the main idea in each paragraph help you learn about classifying organisms?

____

____

What do you think NOW?

Reread the statements at the beginning of the lesson. Fill in the After column with an A if you agree with the statement or a D if you disagree. Did you change your mind?

Log on to ConnectED.mcgraw-hill.com and access your textbook to find this lesson's resources.

Classifying and Exploring Life

Exploring Life

Before You Read

What do you think? Read the two statements below and decide whether you agree or disagree with them. Place an A in the Before column if you agree with the statement or a D if you disagree. After you've read this lesson, reread the statements to see if you have changed your mind.

Before	Statement	After
	5. Most cells are too small to be seen with the unaided eye.	
	6. Only scientists use microscopes.	

Key Concepts
- How did microscopes change our ideas about living things?
- What are the types of microscopes, and how do they compare?

Read to Learn

The Development of Microscopes

Have you ever used a magnifying glass, or lens, to see an object in more detail? If so, you have used a tool similar to the first microscope. Using microscopes, people can see details that cannot be seen with the unaided eye. People have used microscopes to discover many things about organisms.

In the late 1600s, Anton van Leeuwenhoek (LAY vun hook) made one of the first microscopes. His microscope had one lens. It made things look about 270 times their original size. In the early 1700s, Robert Hooke observed and named cells while using a microscope. Before microscopes were invented, people did not know that living things are made of cells.

Study Coach

Record Questions As you read this lesson, write down any questions you have about microscopes and how they are used. Discuss these questions and their answers with another student or your teacher.

Types of Microscopes

One characteristic of all microscopes is that they magnify objects. Magnification makes an object look larger than it really is. Another characteristic of microscopes is resolution. Resolution is how clearly a magnified object can be seen.

The two main types of microscopes are light microscopes and electron microscopes. They are different in their magnification and resolution.

Key Concept Check

1. Describe How did microscopes change our ideas about living things?

Math Skills

The magnification of the ocular lens of a compound microscope is 10×. The magnification of the objective lens is also 10×. To determine a microscope's magnification, multiply the power of the ocular lens by the power of the objective lens. A microscope with a 10× ocular lens and a 10× objective lens magnifies an object 10 × 10, or 100 times.

2. Use Multiplication
What is the total magnification of a compound microscope with a 10× ocular lens and a 4× objective lens?

____

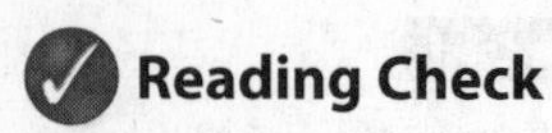

Reading Check

3. Name some ways an object can be examined under a light microscope.

____

____

____

FOLDABLES®

Use a two-column folded chart to organize your notes about how different types of microscopes magnify images.

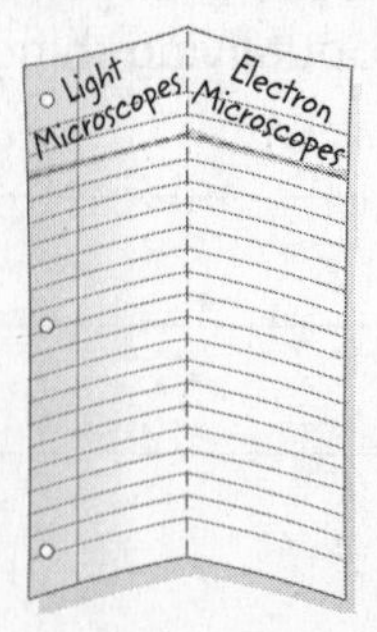

Light Microscopes

Your school might have a light microscope. **Light microscopes** *use light and lenses to enlarge an image of an object.* A simple light microscope has only one lens. *A light microscope that uses more than one lens to magnify an object is called a* **compound microscope.**

In a compound microscope, the ocular lens magnifies the image. That image is further magnified by the objective lens. The total magnification of an object is equal to the magnification of the ocular lens times the magnification of the objective lens.

Some light microscopes can enlarge images as much as 1,500 times their original size. A light microscope has a resolution of about 0.2 micrometers (µm), or two-millionths of a meter. A resolution of 2 µm means that you can clearly see points on an object that are at least 0.2 µm apart.

You can use light microscopes to view living or nonliving things. Sometimes the object is placed under the microscope. For other light microscopes, the object must be mounted on a slide. Some objects must be stained with a dye to see their details. ✓

Electron Microscopes

You might recall that electrons are tiny particles inside atoms. *An* **electron microscope** *uses a magnetic field to focus a beam of electrons through an object or onto an object's surface.* An electron microscope can magnify an image 100,000 times or more. The resolution of an electron microscope can be as small as 0.2 nanometers (nm), or two-billionths of a meter. This resolution is up to 1,000 times greater than a light microscope's.

The two main types of electron microscopes are transmission electron microscopes (TEMs) and scanning electron microscopes (SEMs). Because objects must be mounted in plastic and then sliced very thinly, only dead organisms can be viewed with a TEM.

Transmission Electron Microscopes People use TEMs to study very small things, such as structures inside a cell. The electrons that pass through the object produce an image that can be viewed on a computer.

Scanning Electron Microscopes People use SEMs to study the surface of an object. Electrons bounce off the object, and a computer produces an image of the object in three dimensions.

Using Microscopes

Today's microscopes are useful tools in many fields. They are used in health care, police work, science research, and industry.

Health Care

People who work in health-care fields often use microscopes. Doctors and laboratory technicians find them useful. Surgeons use microscopes in cataract surgery and brain surgery. Microscopes enable doctors to view the surgical area in detail. The area can be shown on a TV screen so other people can watch and learn. Laboratory technicians use microscopes to study body fluids, such as blood and urine.

Other Uses

Health care is not the only field that uses microscopes. These tools are also useful in police work. Have you ever wondered how police figure out how and where a crime happened? Scientists use microscopes to study evidence from crime scenes. For example, a corpse might contain insects that could help identify when and where a crime happened. A microscope can help identify the type and age of the insects.

People who study fossils also use microscopes. They might examine a fossil and other materials from where the fossil was found.

There are many other ways that microscopes are used. People in the steel industry use microscopes to look for impurities in steel. Jewelers use microscopes to identify stones. They can also see markings and impurities in stones that they could not see with their unaided eyes.

Key Concept Check

4. Compare In what ways are TEMs different from SEMs?

Reading Check

5. Recall How might a health-care professional use a microscope?

ACADEMIC VOCABULARY

identify
(verb) to determine the characteristics of a person or thing

Reading Check

6. Summarize List some uses of microscopes.

After You Read

Mini Glossary

compound microscope: a light microscope that uses more than one lens to magnify an object

electron microscope: a microscope that uses a magnetic field to focus a beam of electrons through an object or onto an object's surface

light microscope: a tool that uses light and lenses to enlarge an image of an object

1. Review the terms and their definitions in the Mini Glossary. Write a sentence that describes how a compound microscope and a light microscope are related.

2. Complete the graphic organizer below by explaining how each kind of microscope differs from the previous one.

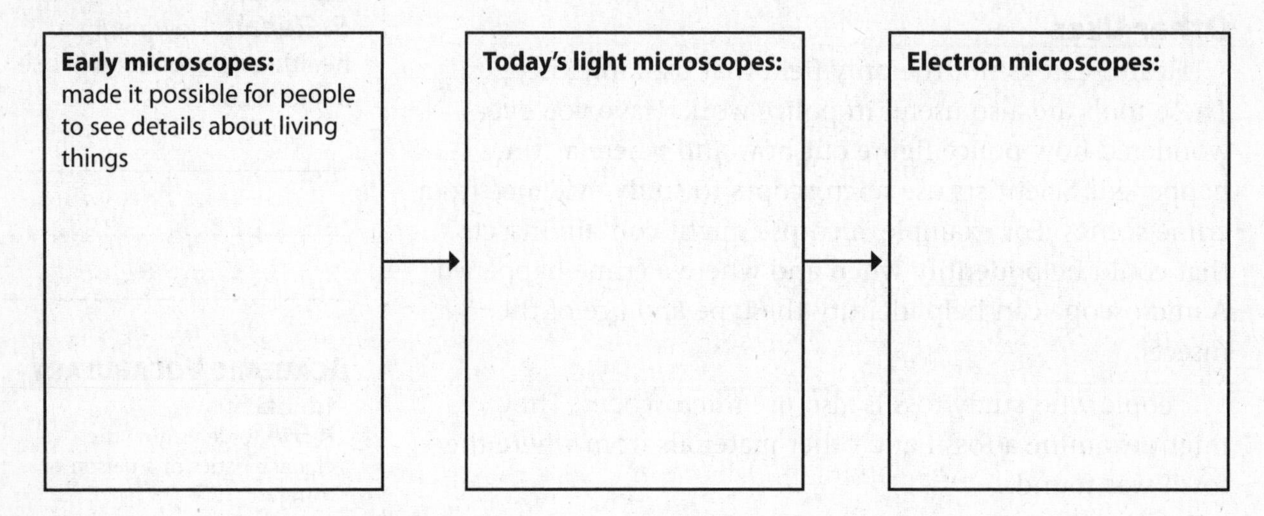

3. How did writing down questions about microscopes and then discussing their answers help you understand what you read?

What do you think NOW?

Reread the statements at the beginning of the lesson. Fill in the After column with an A if you agree with the statement or a D if you disagree. Did you change your mind?

Log on to ConnectED.mcgraw-hill.com and access your textbook to find this lesson's resources.

Cell Structure and Function

Cells and Life

Before You Read

What do you think? Read the two statements below and decide whether you agree or disagree with them. Place an A in the Before column if you agree with the statement or a D if you disagree. After you've read this lesson, reread the statements to see if you have changed your mind.

Before	Statement	After
	1. Nonliving things have cells.	
	2. Cells are made mostly of water.	

Key Concepts 

- How did scientists' understanding of cells develop?
- What basic substances make up a cell?

Read to Learn

Understanding Cells

The cells that make up all living things are very small. Early scientists did not have the tools to see cells until the invention of the microscope. More than 300 years ago, Robert Hooke built a microscope. He used it to look at cork. He saw small openings in the cork similar to the honeycomb shown in the figure below. The openings reminded him of the small rooms, called cells, where monks lived. Hooke named these small structures cells.

Mark the Text

Identify Main Ideas As you read, highlight the main ideas under each heading. After you finish reading, review the main ideas of the lesson.

Visual Check

1. Identify The small openings of the honeycomb look most like which of the following? (Circle the correct answer.)

a. cells

b. plants

c. tiny animals

Think it Over

2. Define What are the three principles of the cell theory?

____

____

____

____

____

____

Key Concept Check

3. Explain how scientists' understanding of cells developed.

____

____

____

Visual Check

4. Identify Match the scientist with his part of the cell theory. Draw a line from the scientist to his observation.

Reading Check

5. Define What are macromolecules?

____

____

____

The Cell Theory

Scientists made better microscopes. They looked for cells in places such as pond water and blood. The newer microscopes made it possible for scientists to see different structures inside cells. A scientist named Matthias Schleiden (SHLI dun) looked at plant cells. Another scientist, Theodore Schwann, studied animal cells. Later, Rudolf Virchow (VUR koh) said all cells come from cells that already exist. The observations made by these scientists, shown in the table below, became known as the cell theory. *The* ***cell theory*** *states that all living things are made of one or more cells, cells are the smallest unit of life, and all new cells come from cells that already exist.*

Cell Theory Matchup

Scientist	Observation
1. Theodore Schwann	**a.** By studying plants, he determined that all living things are made of one or more cells.
2. Rudolf Virchow	**b.** By studying animals, he determined that all living things are made of one or more cells.
3. Matthias Schleiden	**c.** All new cells come from cells that already exist.

Basic Cell Substances

The cell theory raised more questions for scientists. Scientists began to look into what cells are made of. *Cells are made of smaller parts called* ***macromolecules*** *that form when many small molecules join together.*

The Main Ingredient—Water

The main ingredient in every cell is water. Water makes up more than 75 percent of a cell. It is necessary for life. Water also surrounds cells. The water surrounding your cells helps to insulate your body. This helps your body maintain a stable internal environment, or homeostasis.

Water also is useful because it can dissolve other substances, such as salt (sodium chloride). For substances to move into and out of a cell, they must be dissolved in a liquid. In the figure below, the water molecules have a positive end and a negative end.

- The more negative end of a water molecule (–) can attract the positive part of another substance.
- The more positive end of a water molecule (+) can attract the negative part of another substance. With sodium chloride, the sodium (Na) ions and chloride (Cl) ions are more attracted by the water molecules. This attraction is similar to the attraction of magnets.

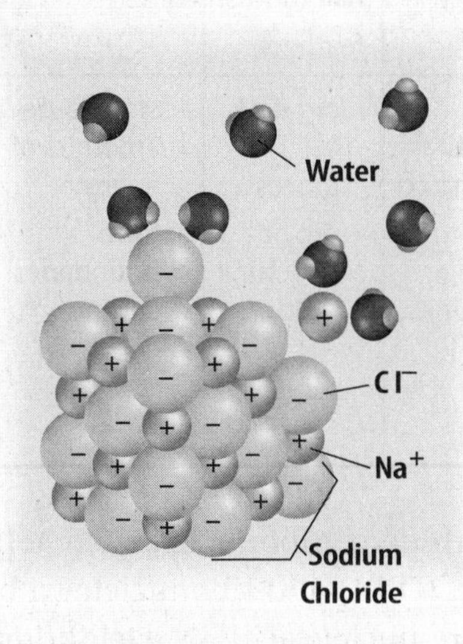

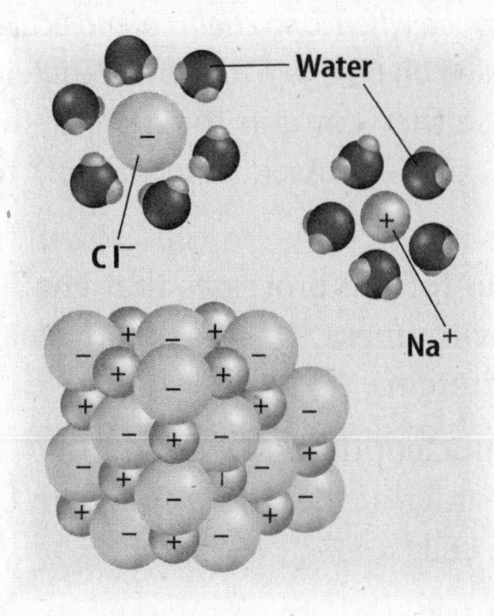

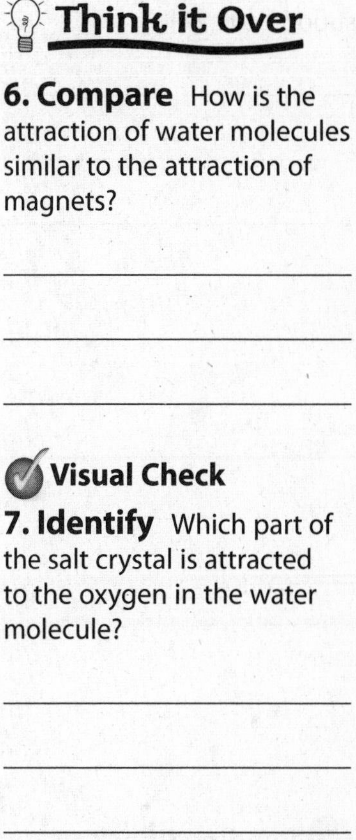

Think it Over

6. Compare How is the attraction of water molecules similar to the attraction of magnets?

____

____

____

Visual Check

7. Identify Which part of the salt crystal is attracted to the oxygen in the water molecule?

____

____

____

![FOLDABLES]

FOLDABLES®

Make a four-page book to organize information about macromolecules in a cell.

Nucleic acids | Proteins | Lipids | Carbohydrates

Macromolecules

All cells contain other substances besides water that help cells do what they do. There are four types of macromolecules in cells. They are nucleic acids, proteins, lipids, and carbohydrates. Each type of macromolecule has its own job, or function, in a cell. These functions range from growth and communication to movement and storage. The table below describes each macromolecule's function.

Macromolecules in Cells

	Nucleic Acids	Proteins	Lipids	Carbohydrates
Elements	carbon oxygen hydrogen nitrogen phosphorus	carbon oxygen hydrogen nitrogen sulfur	carbon oxygen hydrogen phosphorus	carbon hydrogen oxygen
Examples	DNA RNA	enzymes hair (horns, feathers)	fats oils	sugars starch cellulose
Function	• carry hereditary information • used to make proteins	• regulate cell processes • provide structural support	• store large amounts of energy • form boundaries around cells	• supply energy for cell processes • short-term energy storage • provide structural support

Visual Check

8. Identify two macromolecules whose function is to provide support to a cell.

Nucleic Acids Both deoxyribonucleic (dee AHK sih ri boh noo klee ihk) acid (DNA) and ribonucleic (ri boh noo KLEE ihk) acid (RNA) are nucleic acids. ***Nucleic acids** are macromolecules formed when long chains of molecules called nucleotides* (NEW klee uh tidz) *join together*. Nucleic acids are important because they contain the genetic material of a cell. This information is passed from parents to offspring.

DNA includes instructions for cell growth, for cell reproduction, and for cell processes that enable a cell to respond to its environment. DNA is used to make RNA. RNA is used to make proteins.

Reading Check

9. Name three instructions that DNA provides to a cell.

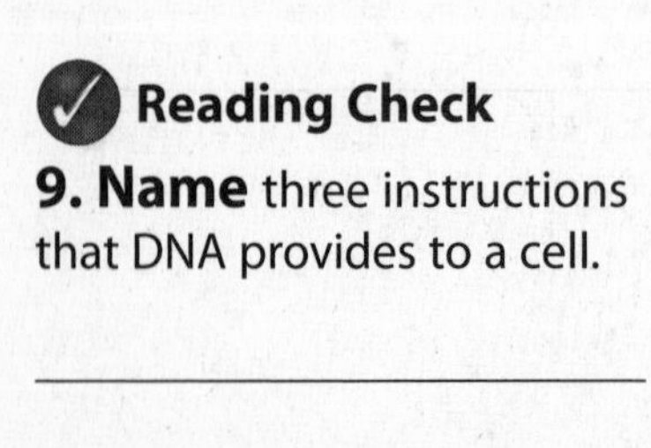

The order of nucleotides in DNA and RNA is important. A change in the order of the nucleotides can change the information in a cell.

Proteins The macromolecules necessary for almost everything cells do are proteins. *A* **protein** *is a macromolecule made of long chains of amino acid molecules.* RNA contains instructions for joining amino acids together.

Cells have hundreds of proteins. Each protein has its own function. Some proteins help cells communicate with other cells. Other proteins move substances around inside cells. Some proteins help to break down nutrients in food. Other proteins, such as keratin (KER uh tun), which is found in hair, horns, and feathers, make up supporting structures.

Lipids *A* **lipid** *is a large macromolecule that does not dissolve in water.* Because lipids do not dissolve in water, they protect cells. Lipids also are a large part of the cell membrane. Lipids store energy for cells and help with cell communication. Cholesterol (kuh LES tuh rawl), phospholipids (fahs foh LIH pids), and vitamin A are lipids.

Carbohydrates *One sugar molecule, two sugar molecules, or a long chain of sugar molecules make up* **carbohydrates** (kar boh HI drayts). Carbohydrates store energy, provide structural support for cells, and help cells communicate. Sugars and starches are carbohydrates that store energy. Fruits contain sugars. Bread and pasta are mostly starch. The energy stored in sugars and starches can be released quickly through chemical reactions in cells. Cellulose is a carbohydrate in the cell walls of plants that provides support.

Reading Check

10. Identify three functions of proteins in cells.

______

______

______

Reading Check

11. Explain Why are lipids important to cells?

______

______

Key Concept Check

12. Name the basic substances that make up a cell.

______

______

______

After You Read

Mini Glossary

carbohydrate: (kar boh HI drayt): one sugar molecule, two sugar molecules, or a long chain of sugar molecules

cell theory: states that all living things are made of one or more cells, the cell is the smallest unit of life, and all new cells come from preexisting cells

lipid: a large macromolecule that does not dissolve in water

macromolecule: a substance that forms by joining many small molecules together

nucleic acid: a macromolecule that forms when long chains of molecules called nucleotides join together

protein: a long chain of amino acid molecules

1. Review the terms and their definitions in the Mini Glossary. Write a sentence that describes a lipid.

2. Fill in the chart below by identifying the different types of macromolecules and giving examples of each.

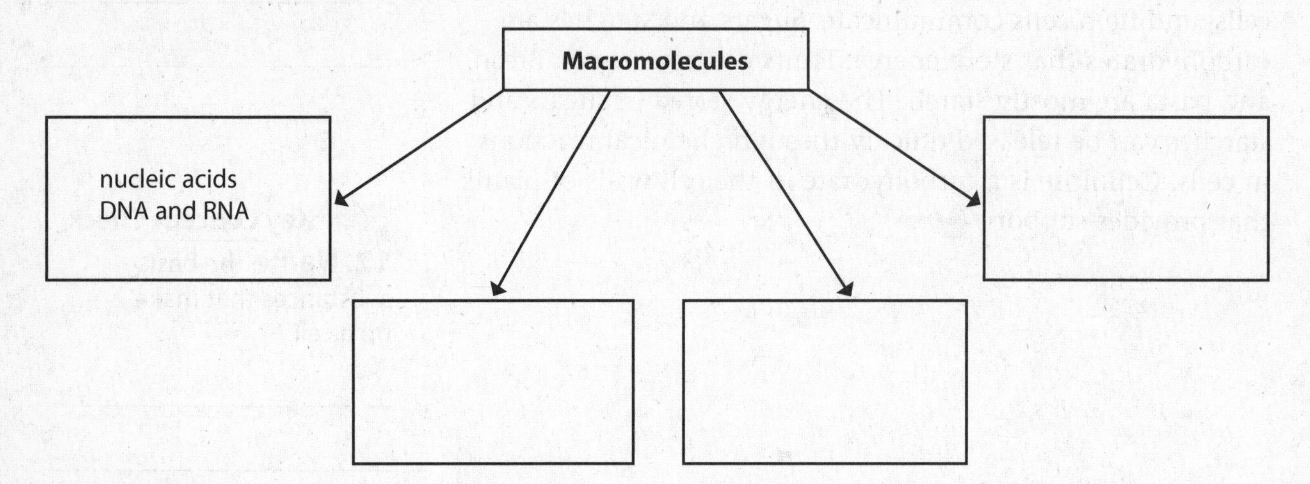

3. How did highlighting the main ideas in each section of this lesson improve your understanding of the cell?

What do you think NOW?

Reread the statements at the beginning of the lesson. Fill in the After column with an A if you agree with the statement or a D if you disagree. Did you change your mind?

Log on to ConnectED.mcgraw-hill.com and access your textbook to find this lesson's resources.

Cell Structure and Function

The Cell

Before You Read

What do you think? Read the two statements below and decide whether you agree or disagree with them. Place an A in the Before column if you agree with the statement or a D if you disagree. After you've read this lesson, reread the statements to see if you have changed your mind.

Before	Statement	After
	3. Different organisms have cells with different structures.	
	4. All cells store genetic information in their nuclei.	

Key Concepts 

- How are prokaryotic cells and eukaryotic cells similar, and how are they different?
- What do the structures in a cell do?

Read to Learn

Cell Shape and Movement

Cells come in many shapes and sizes. The size and shape of a cell is part of the function of the cell. Some cells, such as human red-blood cells, can be seen only by using a microscope. The cells can pass easily through small blood vessels because of their small size. Their disk shapes are important for carrying oxygen. Nerve cells have parts that jut out. These projections on nerve cells can send signals over long distances. Some plant cells are hollow. These hollow cells make up tubelike structures that can carry water and dissolved substances to parts of the plant.

The size and shape of a cell make it possible for the cell to carry out its functions. The parts that make up a cell have their own functions as well. A cell's parts are like the players on a football team who perform different tasks on the playing field. A cell is made up of different parts that perform different functions to keep the cell alive.

Study Coach

Use Prior Knowledge
Before you read this lesson, look at the figures and headings to learn what the lesson is about. Write what you know about the cell on a piece of paper. As you read the lesson, fill in what you learned about the cell.

ACADEMIC VOCABULARY
function
(noun) the purpose for which something is used

Visual Check

1. Describe the location of the cell wall.

____

____

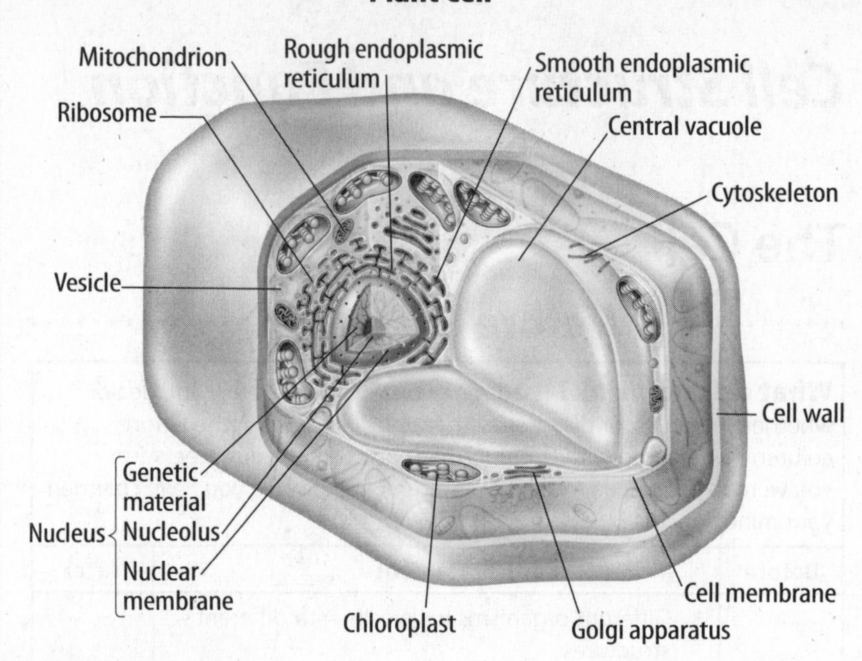

Cell Membrane

All cells have some parts, or structures, in common. One of these structures is a cell membrane. *A* **cell membrane** *is a flexible covering that protects the inside of a cell from the environment outside the cell.* You can see the cell membrane in both drawings on this page. Cell membranes are made of proteins and phospholipids.

Reading Check

2. Describe What are cell membranes made of?

____

____

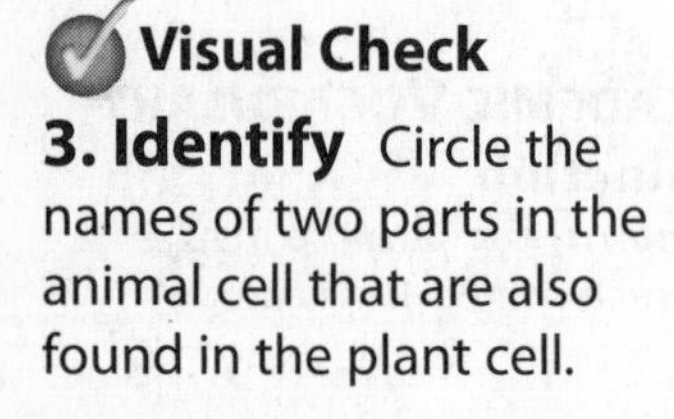

Visual Check

3. Identify Circle the names of two parts in the animal cell that are also found in the plant cell.

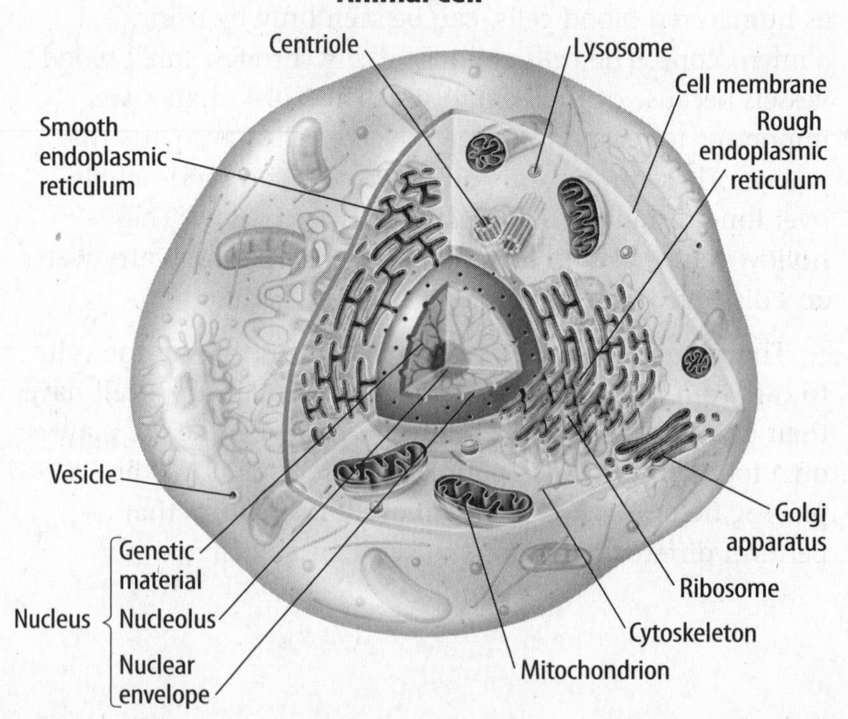

Cell Wall

Every cell has a cell membrane. But some cells also have a cell wall. Plant cells, fungal cells, bacterial cells, and some protists have cell walls. *A* **cell wall** *is a stiff structure outside the cell membrane.* A cell wall protects a cell from viruses and other harmful organisms. In some plant and fungal cells, the cell wall helps the cell keep its shape and gives it support.

Cell Appendages

If you look at a cell using a microscope, you might see structures on the outside of the cell. These appendages might look like hairs or long tails. They often help a cell move. Flagella (fluh JEH luh) (singular, flagellum) are long and tail-like. They whip back and forth to move the cell. Cilia (SIH lee uh) (singular, cilium) are short, hairlike structures. They can move a cell or move molecules away from a cell. The cilia in your windpipe move harmful particles away from your lungs.

Cytoplasm and the Cytoskeleton

The fluid inside a cell is made of water, salts, and other molecules and is called the **cytoplasm.** The cytoplasm contains a cell's cytoskeleton. *The* **cytoskeleton** *is made of threadlike proteins that are joined together.* The cytoskeleton is a framework that gives a cell its shape and helps it move. ✓

Reading Check

4. Describe the structure of the cytoskeleton.

____

____

Cell Types

Microscopes helped scientists discover that cells can be grouped into two types. There are prokaryotic (proh ka ree AH tihk) cells and eukaryotic (yew ker ee AH tihk) cells.

Prokaryotic Cells

The genetic material in a prokaryotic cell is not surrounded by a membrane. Look at the drawing below. Prokaryotic cells also do not have many of the cell parts other cells have. Most prokaryotic cells are unicellular organisms and are called prokaryotes.

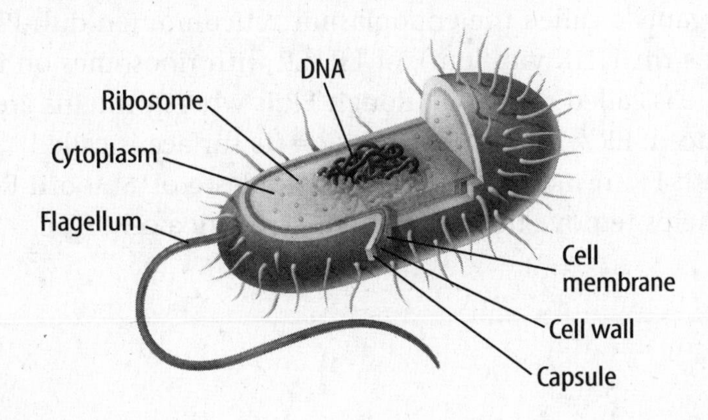

Visual Check

5. Name three parts of a prokaryotic cell.

____

____

____

Eukaryotic Cells

The cells of plants, animals, fungi, and protists are eukaryotic cells. The genetic material of eukaryotic cells is surrounded by a membrane. Every eukaryotic cell also has **organelles**—*other parts that are surrounded by a membrane and have specialized functions*. Eukaryotic cells are usually larger than prokaryotic cells.

Key Concept Check
6. Compare and Contrast How are prokaryotic cells and eukaryotic cells the same? How are they different?

Cell Organelles

The organelles of eukaryotic cells have different functions in the cell. Organelles help a cell carry out different functions at the same time. These functions include getting energy from food, storing information, and getting rid of waste material.

The Nucleus

The largest organelle inside most eukaryotic cells is the nucleus. *The* **nucleus** *is the part of a eukaryotic cell that directs cell activities and contains genetic information stored in DNA.*

DNA is in structures called chromosomes. The number of chromosomes in a nucleus is different for different species of organisms.

The nucleus also contains proteins and an organelle called the nucleolus (new KLEE uh lus). The nucleolus makes ribosomes, organelles that help produce proteins. Two membranes form the nuclear envelope that surrounds the nucleus. The nuclear envelope has many pores. Certain molecules, such as ribosomes and RNA, move into and out of the nucleus through these pores.

SCIENCE USE V. COMMON USE
envelope
Science Use *an outer covering*
Common Use a flat paper container for a letter

Manufacturing Molecules

You learned that proteins are important molecules in cells. Proteins are made of small organelles called ribosomes. A ribosome is not surrounded by a membrane. Ribosomes are in the cytoplasm of a cell. Ribosomes can be attached to an organelle called the endoplasmic reticulum (en duh PLAZ mihk • rih TIHK yuh lum), or ER. ER with ribosomes on its surface is called rough ER. Rough ER is where proteins are produced. ER without ribosomes on its surface is called smooth ER. It makes lipids such as cholesterol. Smooth ER also helps remove harmful substances from a cell.

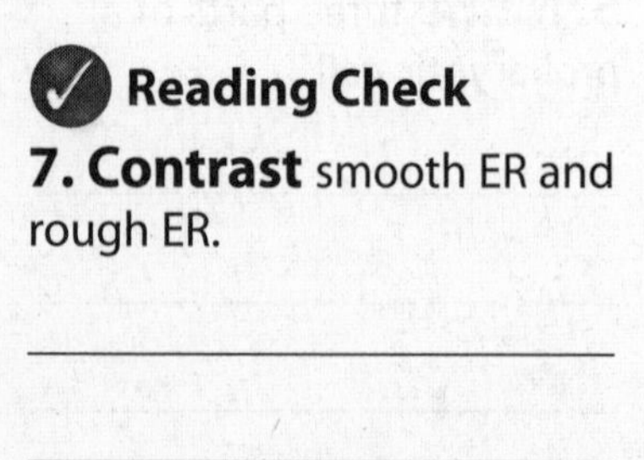

Reading Check
7. Contrast smooth ER and rough ER.

Processing Energy

All living things must have energy to survive. Cells process some energy in specialized organelles called mitochondria (mi tuh KAHN dree uh) (singular, mitochondrion). Most eukaryotic cells contain hundreds of mitochondria. Some cells in a human heart can contain 1,000 mitochondria.

ATP A mitochondrion is surrounded by two membranes. Chemical reactions within mitochondria release energy. This energy is stored in high-energy molecules called ATP—adenosine triphosphate (uh DEH nuh seen • tri FAHS fayt). The energy in ATP molecules is used by the cell for growth, cell division, and transporting materials.

Chloroplasts The cells of some organisms, such as plants and algae, contain organelles called chloroplasts (KLOR uh plasts). **Chloroplasts** *are membrane-bound organelles that use light energy and make food, a sugar called glucose, from water and carbon dioxide in a process called photosynthesis* (foh toh SIHN thuh sus). The sugar has stored energy that can be used when the cells need it. ✓

Processing, Transporting, and Storing Molecules

The Golgi (GAWL jee) apparatus is an organelle that looks like a stack of pancakes. It gets proteins ready for their specific jobs. It then packages the proteins into tiny membrane-bound, ball-like structures called vesicles. Vesicles are organelles that transport substances to other parts of the cell. Some vesicles in an animal cell are called lysosomes. Lysosomes help break down and recycle different parts of the cell.

Some cells also have structures called vacuoles (VA kyuh wohlz). Vacuoles are organelles that store food, water, and waste materials for a cell. A plant cell usually has one large vacuole. Some animal cells have many small vacuoles.

FOLDABLES®

Make a half-book to record information about cell organelles and their functions.

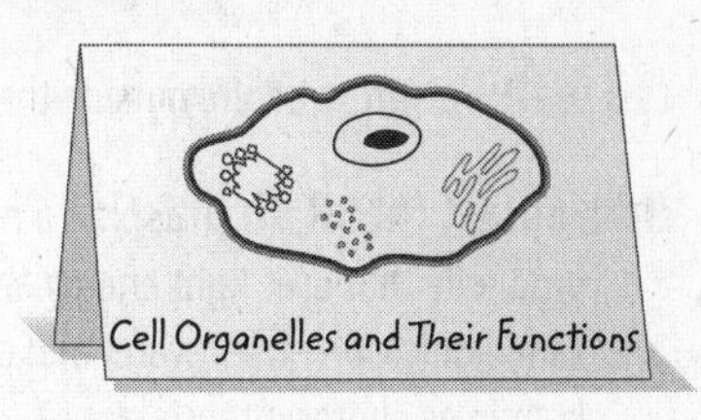

Reading Check

8. Identify the types of cells that contain chloroplasts.

____

____

____

Key Concept Check

9. Explain the function of the Golgi apparatus.

____

____

____

After You Read

Mini Glossary

cell membrane: a flexible covering that protects the inside of a cell from the environment outside a cell

cell wall: a stiff structure outside the cell membrane

chloroplast (KLOR uh plast): a membrane-bound organelle that uses light energy and makes food—a sugar called glucose—from water and carbon dioxide in a process known as photosynthesis

cytoplasm: a fluid inside a cell that contains salts and other molecules

cytoskeleton: a network of threadlike proteins that are joined together

nucleus: the part of a eukaryotic cell that directs cell activities and contains genetic information stored in DNA

organelle: a membrane-surrounded component of a cell that has specialized functions

1. Review the terms and their definitions in the Mini Glossary. Write a sentence that lists two functions of the nucleus of a eukaryotic cell.

____

____

2. Fill in the table below to identify the functions of each organelle.

Organelle	Function
Chloroplast	
Golgi apparatus	
Smooth ER	
Nucleus	

3. Name three tasks carried out by the organelles of eukaryotic cells.

____

____

What do you think NOW?

Reread the statements at the beginning of the lesson. Fill in the After column with an A if you agree with the statement or a D if you disagree. Did you change your mind?

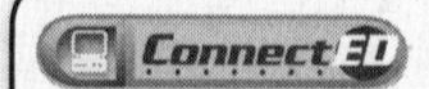

Log on to ConnectED.mcgraw-hill.com and access your textbook to find this lesson's resources.

Cell Structure and Function

Moving Cellular Material

Before You Read

What do you think? Read the two statements below and decide whether you agree or disagree with them. Place an A in the Before column if you agree with the statement or a D if you disagree. After you've read this lesson, reread the statements to see if you have changed your mind.

Before	Statement	After
	5. Diffusion and osmosis are the same process.	
	6. Cells with large surface areas can transport more than cells with smaller surface areas.	

Key Concepts 

- How do materials enter and leave cells?
- How does cell size affect the transport of materials?

Read to Learn

Passive Transport

The membranes of cells and organelles perform different functions. They form boundaries between cells. They also control the movement of substances into and out of cells.

Cell membranes are semipermeable. This means that only certain materials can enter or leave a cell. Substances can pass through a cell membrane by one of several different processes. The type of process depends on the physical and chemical properties of the substance that is passing through the membrane.

Small molecules, such as oxygen and carbon dioxide, pass through a cell's membrane by a process called passive transport. **Passive transport** *is the movement of substances through a cell membrane without using the cell's energy.* Passive transport depends on the amount of a substance on each side of the membrane. If there are more oxygen molecules outside a cell than there are inside a cell, oxygen molecules will move into the cell by passive transport. Oxygen molecules will move into a cell until the amount of oxygen outside the cell equals the amount of oxygen inside the cell. There are different types of passive transport.

Study Coach

Asking Questions Before you read the lesson, preview all the headings. Make a chart and write a *What* or *How* question for each heading. As you read, write the answers to your questions.

FOLDABLES

Make a two-tab book to organize information about the different types of passive and active transport.

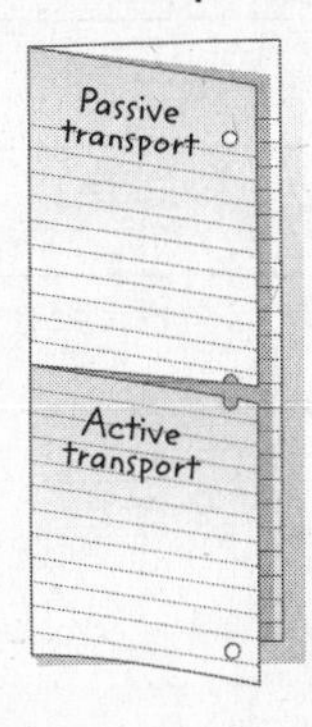

ACADEMIC VOCABULARY
concentration
(noun) the amount of a given substance in a certain area.

Visual Check
1. Predict What would the water in the beaker on the right look like if the membrane did not let anything through?

Diffusion

When the concentration, or amount per volume, of a substance is unequal on each side of a membrane, molecules will move from the side with a higher concentration of the substance to the side with the lower concentration. **Diffusion** *is the movement of substances from an area of higher concentration to an area of lower concentration.*

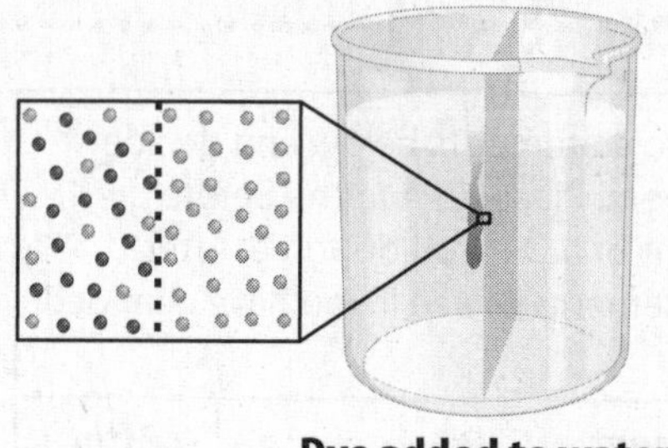

Dye added to water

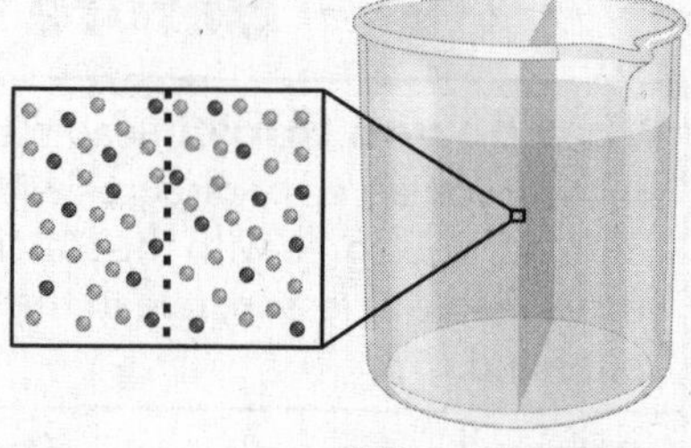

After 30 minutes

Diffusion will continue until the concentration on each side of the cell membrane is equal. The figure above shows how dye passed through the membrane into the clear water until there were equal concentrations of water and dye on both sides of the membrane.

Osmosis—The Diffusion of Water

Diffusion is the movement of any small molecules from areas of higher concentrations to areas of lower concentrations. **Osmosis** *is the diffusion of water molecules only through a membrane.* Water molecules pass through a semipermeable membrane from an area of high concentration to an area of low concentration. For example, plant cells lose water because of osmosis. The concentration of water in the air around a plant is less than the concentration of water in the cells of the plant. Water will leave plant cells and diffuse into the air. If the plant is not watered to replace the water lost by its cells, the plant will wilt and might die.

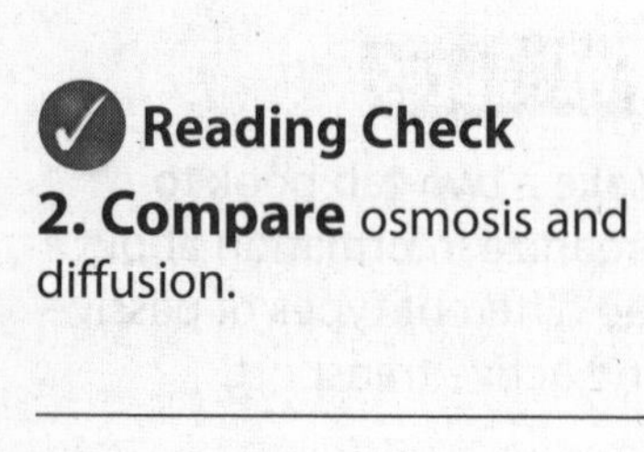

Reading Check
2. Compare osmosis and diffusion.

Facilitated Diffusion

Some molecules are too large or are chemically unable to move through a membrane by diffusion. **Facilitated diffusion** *is the movement of molecules through a cell membrane using special proteins called transport proteins.* Facilitated diffusion does not use the cell's energy to move the molecules. The transport proteins do the work. There are two types of transport proteins.

Carrier Proteins Carrier proteins are transport proteins. They carry large molecules, such as the sugar molecule glucose, through the cell membrane.

Channel Proteins Channel proteins are also transport proteins. They form pores through the cell membrane. Ions, such as sodium and potassium, pass through the cell membrane by channel proteins. Transport proteins are shown below. ✓

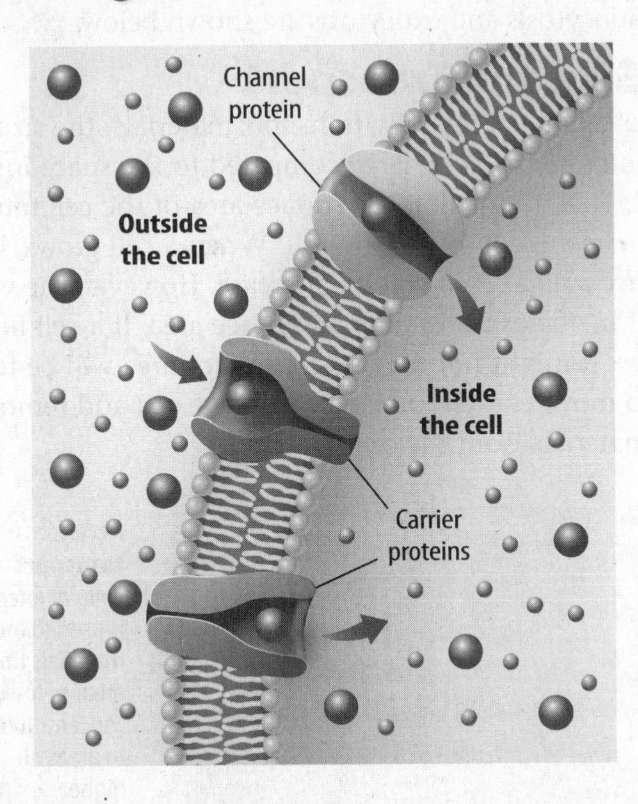

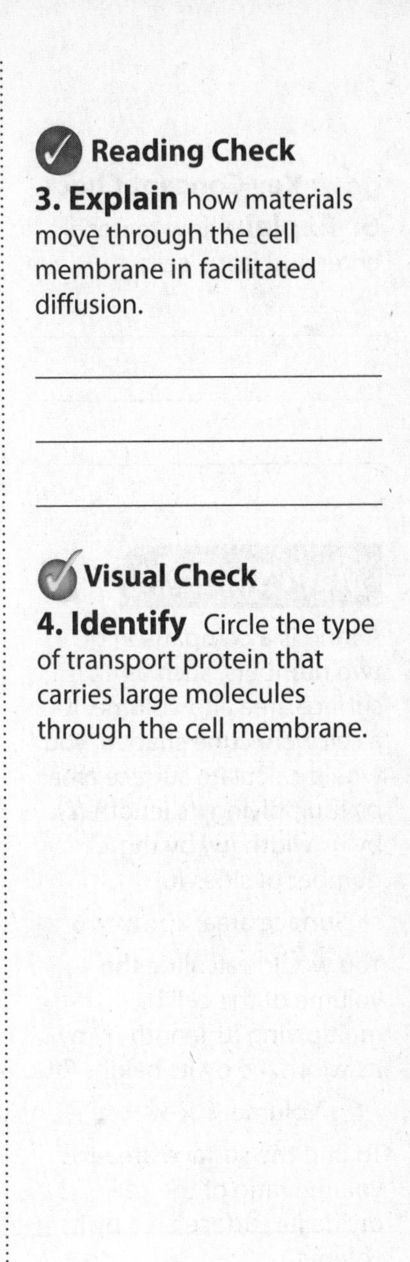

✓ **Reading Check**

3. Explain how materials move through the cell membrane in facilitated diffusion.

✓ **Visual Check**

4. Identify Circle the type of transport protein that carries large molecules through the cell membrane.

Active Transport

Sometimes a cell uses energy when a substance passes through its membrane. **Active transport** *is the movement of substances through a cell membrane only by using the cell's energy.*

Substances moving by active transport move from areas of lower concentration to areas of higher concentration. Active transport is important for cells and organelles. Cells can take in nutrients from the environment through carrier proteins by using active transport. Some molecules and waste materials leave cells by active transport. ✓

✓ **Reading Check**

5. Summarize how a cell uses active transport.

Endocytosis and Exocytosis

Some substances are too large to enter a cell membrane by diffusion or by using a transport protein. There are other ways that substances can enter a cell.

Endocytosis *The process during which a cell takes in a substance by surrounding it with the cell membrane is called* **endocytosis** (en duh si TOH sus). Some cells take in bacteria and viruses using endocytosis.

Exocytosis Some substances are too large to leave a cell by diffusion or by using a transport protein. They can leave using exocytosis (ek soh si TOH sus). **Exocytosis** *is the process during which a cell's vesicles release their contents outside the cell.* Proteins and other substances are removed from a cell through exocytosis. Both endocytosis and exocytosis are shown below.

Cell Size and Transport

For a cell to successfully transport materials, the size of the cell membrane must be large compared to the space inside of the cell. This means that the surface area of the cell must be larger than the volume of the cell. When a cell grows, both its surface area and its volume increase. However, the volume of a cell increases faster than its surface area. If a cell becomes too large, it might not survive. Its surface area will be too small to move enough nutrients into the cell and remove waste materials from the cell.

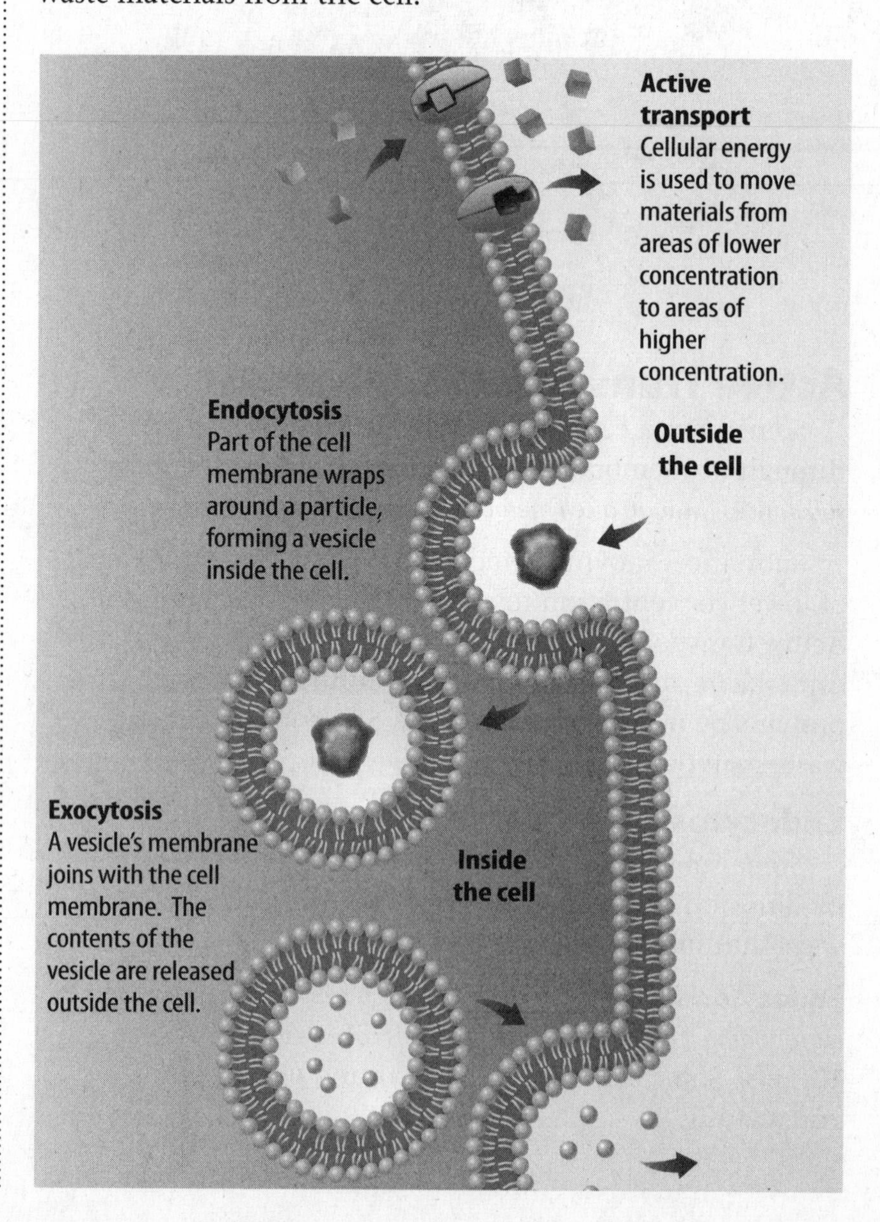

Key Concept Check

6. Explain how materials enter and leave cells.

____

____

____

Math Skills

A ratio is a comparison of two numbers, such as surface area and volume. If a cell were cube shaped, you would calculate surface area by multiplying its length (*ℓ*) by its width (*w*) by the number of sides (6).

Surface area: $\ell \times w \times 6$

You would calculate the volume of the cell by multiplying its length (*ℓ*) by its width (*w*) by its height (*h*).

Volume: $\ell \times w \times h$

To find the surface-area-to-volume ratio of the cell, divide its surface area by its volume.

$$\frac{\text{Surface area}}{\text{Volume}}$$

7. Use Ratios What is the surface-area-to-volume ratio of a cube-shaped cell whose sides are 6 mm long?

____

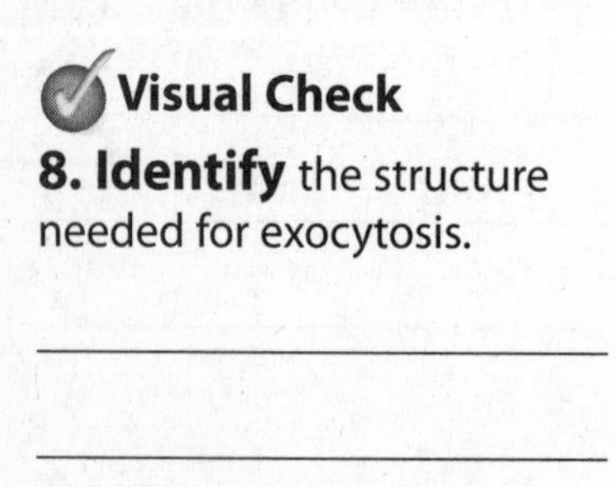

Visual Check

8. Identify the structure needed for exocytosis.

____

____

____

After You Read

Mini Glossary

active transport: the movement of substances through a cell membrane only by using the cell's energy

diffusion: the movement of substances from an area of higher concentration to an area of lower concentration

endocytosis (en duh si TOH sus): the process during which a cell takes in a substance by surrounding it with the cell membrane

exocytosis (ek soh si TOH sus): the process during which a cell's vesicles release their contents outside the cell

facilitated diffusion: when molecules pass through a cell membrane using special proteins called transport proteins

osmosis: the diffusion of water molecules only

passive transport: the movement of substances through a cell membrane without using the cell's energy

1. Review the terms and their definitions in the Mini Glossary. Write a sentence that compares passive and active transport.

2. Fill in the table below to compare active and passive transport.

	Energy needed?	Structures Involved	Examples
Active transport	yes/no		
Passive transport	yes/no		

What do you think NOW?

Reread the statements at the beginning of the lesson. Fill in the After column with an A if you agree with the statement or a D if you disagree. Did you change your mind?

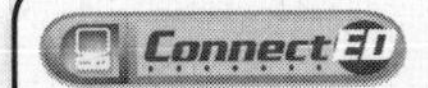

Log on to ConnectED.mcgraw-hill.com and access your textbook to find this lesson's resources.

Cell Structure and Function

Cells and Energy

Key Concepts 
- How does a cell obtain energy?
- How do some cells make food molecules?

Before You Read

What do you think? Read the two statements below and decide whether you agree or disagree with them. Place an A in the Before column if you agree with the statement or a D if you disagree. After you've read this lesson, reread the statements to see if you have changed your mind.

Before	Statement	After
	7. ATP is the only form of energy found in cells.	
	8. Cellular respiration occurs only in lung cells.	

Study Coach

Use an Outline As you read, make an outline to summarize the information in the lesson. Use the main headings in the lesson as the main headings in the outline. Complete the outline with the information under each heading.

Read to Learn

Cellular Respiration

All living organisms need energy to survive. Cells use energy from food and make an energy-storing compound, ATP. **Cellular respiration** *is a series of chemical reactions that convert the energy in food into a usable form of energy called ATP.* Cellular respiration takes place in the cytoplasm and in the mitochondria of a cell.

Reactions in the Cytoplasm

The first step of cellular respiration is called glycolysis. It takes place in the cytoplasm of all cells. **Glycolysis** *is a process by which a sugar called glucose is broken down into smaller molecules.* Glycolysis produces some ATP molecules. It also uses energy from other ATP molecules.

More ATP is made during the second step of cellular respiration than during glycolysis.

Visual Check

1. Locate Circle where sugar breaks down in the cell during glycolysis.

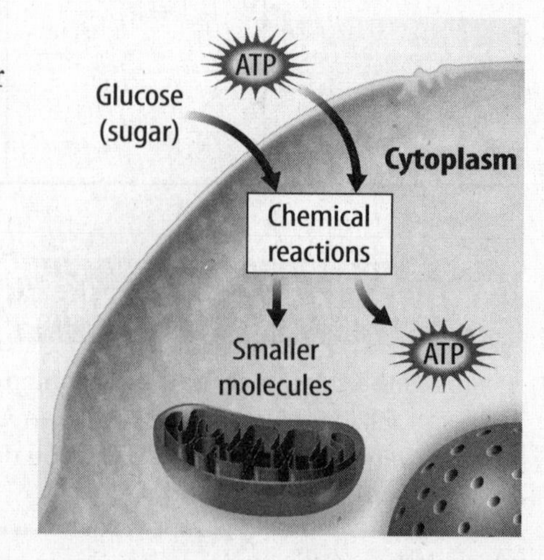

Reactions in the Mitochondria

The second step in cellular respiration, shown below, takes place in the mitochondria of eukaryotic cells. This step uses oxygen. The smaller molecules made during glycolysis are broken down. Many ATP molecules are made. Cells use ATP molecules to power all cellular processes. Two waste products, water and carbon dioxide (CO_2), are given off during this step of cellular respiration. The CO_2 released by cells as a waste product is used by plants and some unicellular organisms in a process called photosynthesis.

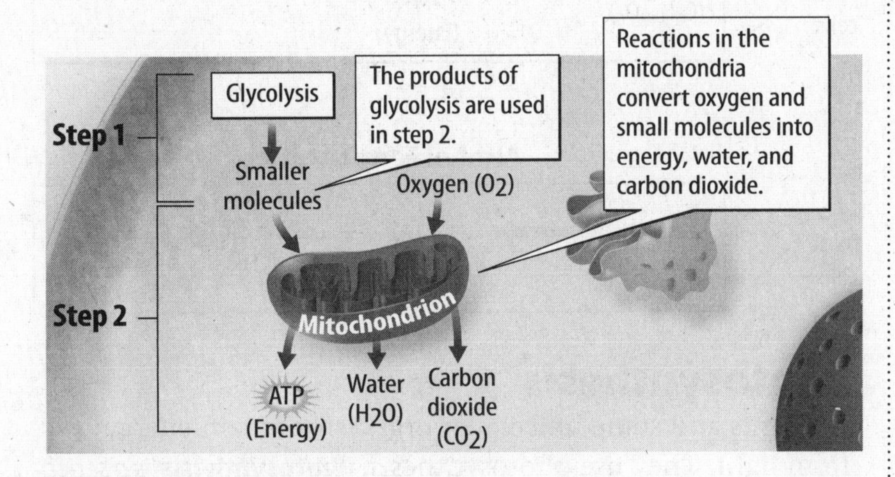

Fermentation

Sometimes, as you exercise, there is not enough oxygen in your cells to make ATP molecules through cellular respiration. When this happens, cells use a process called fermentation to obtain chemical energy. **Fermentation** *is a reaction that eukaryotic and prokaryotic cells use to obtain energy from food when oxygen levels are low.* Because no oxygen is used, fermentation makes less ATP than cellular respiration does. Fermentation takes place in a cell's cytoplasm, not in mitochondria.

Types of Fermentation

There are several types of fermentation. One type occurs when glucose is changed into ATP and a waste product called lactic acid.

Lactic-Acid Fermentation Some bacteria and fungi help produce cheese, yogurt, and sour cream using lactic-acid fermentation. The muscle cells in animals, including humans, can release energy during exercise using lactic-acid fermentation.

FOLDABLES®

Make a half-book to record information about the different types of energy production.

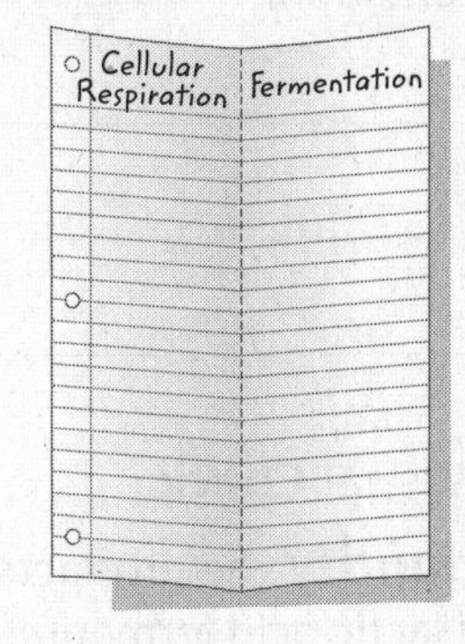

Visual Check

2. Compare the reactions in mitochondria with glycolysis.

Key Concept Check

3. Explain how a cell obtains energy.

Reading Check
4. Compare lactic-acid fermentation and alcohol fermentation.

Visual Check
5. Identify the products of both lactic-acid fermentation and alcohol fermentation.

Key Concept Check
6. Explain how some cells make food molecules.

Alcohol Fermentation Some types of bacteria and yeast make ATP through a process called alcohol fermentation. Alcohol fermentation produces an alcohol, called ethanol, and carbon dioxide. Many types of bread are made using yeast. The carbon dioxide produced by yeast during alcohol fermentation makes bread dough rise. Lactic-acid fermentation and alcohol fermentation are shown below.

Lactic-Acid Fermentation

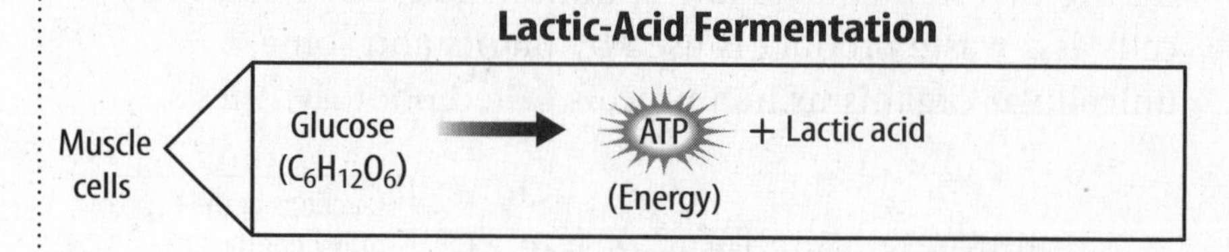

Alcohol Fermentation

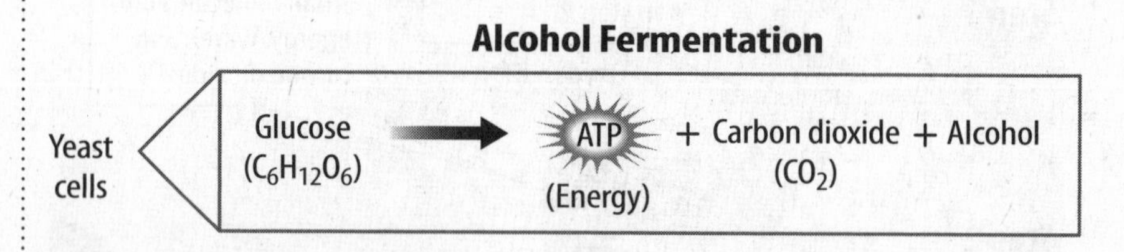

Photosynthesis

Plants and some unicellular organisms obtain energy from light. They use photosynthesis. **Photosynthesis** *is a series of chemical reactions that convert light energy, water, and carbon dioxide into the food-energy molecule glucose and the waste product oxygen.*

Light and Pigments

Photosynthesis uses light energy. In plants, pigments such as chlorophyll absorb light energy. As chlorophyll absorbs light, it absorbs all the colors in it except green.

The green light is reflected as the green color that you see in leaves and stems. Plants might also contain pigments that reflect other colors, such as red, yellow, or orange light.

Reactions in Chloroplasts

The chlorophyll that absorbs light energy for photosynthesis is in chloroplasts. Chloroplasts are organelles in plant cells that convert light energy to chemical energy in food. During photosynthesis, light energy, water, and carbon dioxide combine and make sugars. Photosynthesis also produces oxygen, which is released into the atmosphere.

Importance of Photosynthesis

Photosynthesis uses light energy and carbon dioxide to make food energy. Oxygen is released during this process. This food energy is stored as glucose. When an organism eats plant material, such as fruit, it takes in food energy. The cells of the organism will then go through cellular respiration. They will use the oxygen released during photosynthesis and convert the food energy into ATP. These organisms then release carbon dioxide into the atmosphere. The relationship between cellular respiration and photosynthesis is shown in the diagram below.

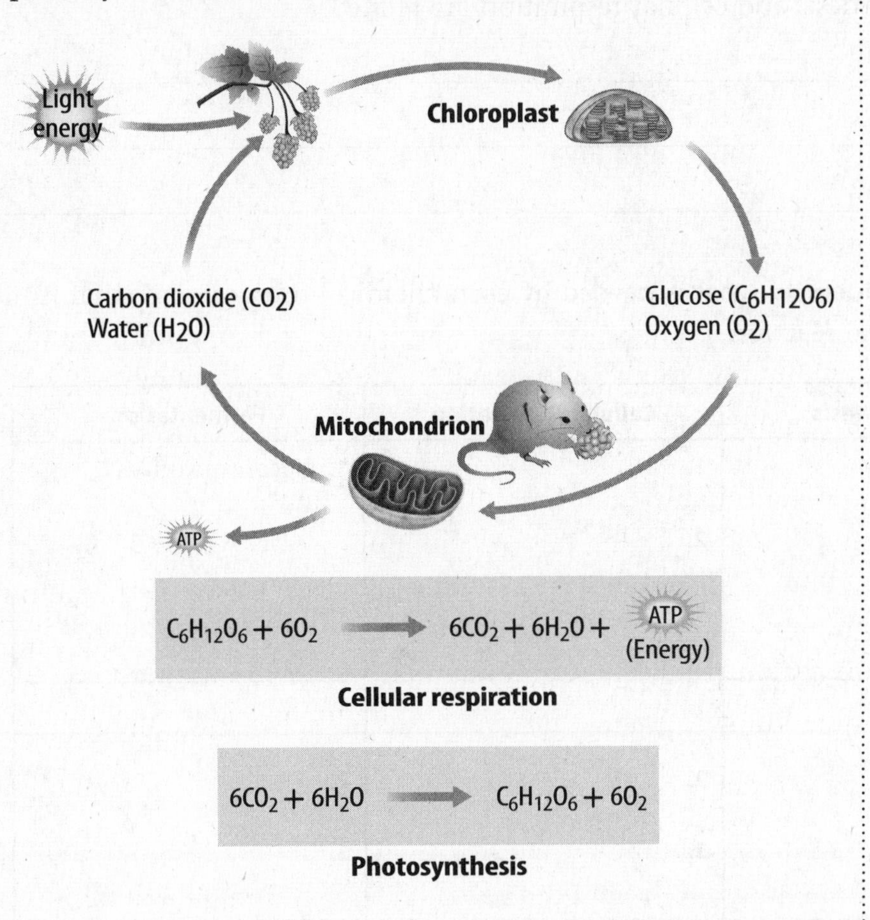

WORD ORIGIN

photosynthesis
from Greek *photo,* means "light," and *synthesis,* means "composition'

Visual Check

7. Explain the relationship between cellular respiration and photosynthesis.

____

____

____

After You Read

Mini Glossary

cellular respiration: a series of chemical reactions that convert the energy in food molecules into a usable form of energy called ATP

fermentation: a reaction that eukaryotic and prokaryotic cells use to obtain energy from food when oxygen levels are low

glycolosis: a process by which glucose, a sugar, is broken down into smaller molecules

photosynthesis: a series of chemical reactions that converts light energy, water, and carbon dioxide into the food-energy molecule glucose and gives off oxygen

1. Review the terms and their definitions in the Mini Glossary. Explain, using complete sentences, how photosynthesis and cellular respiration are related.

2. Fill in the table below to identify what is needed by each chemical reaction and what is produced by each chemical reaction.

	Photosynthesis	Cellular Respiration	Fermentation
What is needed?	1. 2. 3.	1. 2.	1. glucose molecules
What is produced?	1. 2.	1. 2. 3.	1. 2. 3.

3. As chlorophyll in plants absorbs light, it absorbs all the colors except one color. Which color is that?

What do you think NOW?

Reread the statements at the beginning of the lesson. Fill in the After column with an A if you agree with the statement or a D if you disagree. Did you change your mind?

Log on to ConnectED.mcgraw-hill.com and access your textbook to find this lesson's resources.

From a Cell to an Organism

The Cell Cycle and Cell Division

Before You Read

What do you think? Read the three statements below and decide whether you agree or disagree with them. Place an A in the Before column if you agree with the statement or a D if you disagree. After you've read this lesson, reread the statements and see if you have changed your mind.

Before	Statement	After
	1. Cell division produces two identical cells.	
	2. Cell division is important for growth.	
	3. At the end of the cell cycle, the original cell no longer exists.	

Key Concepts 

- What are the phases of the cell cycle?
- Why is the result of the cell cycle important?

Study Coach

Create a Quiz Write a question about the main idea under each heading. Exchange quizzes with another student. Together, discuss the answers to the quizzes.

Read to Learn

The Cell Cycle

No matter where you live, you have probably noticed that the weather changes in a regular pattern each year. Some areas have four seasons—winter, spring, summer, and fall. As seasons change, temperature, precipitation, and the number of hours of sunlight change in a regular cycle.

Cells also go through cycles, just like the seasons. *Most cells in an organism go through a cycle of growth, development, and division called the* **cell cycle**. The cell cycle makes it possible for organisms

- to grow and develop,
- to replace cells that are old or damaged, and
- to produce new cells.

Phases of the Cell Cycle

There are two main phases in the cell cycle. These phases are interphase and the mitotic (mi TAH tihk) phase. **Interphase** *is the period of a cell's growth and development*. A cell spends most of its life in interphase.

Key Concept Check

1. Name What are the two main phases of the cell cycle?

FOLDABLES®

Make a folded book to illustrate the cell cycle.

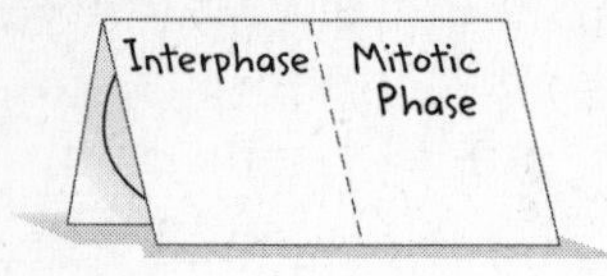

During Interphase Most cells go through three stages during interphase:

- rapid growth and replication, or copying, of the membrane-bound structures called organelles,
- copying of DNA, the genetic information in a cell, and
- preparation for cell division.

Interphase is followed by a shorter phase of the cell cycle called the mitotic phase.

During the Mitotic Phase A cell reproduces during the mitotic phase. The mitotic phase has two stages, as shown in the figure below. During the first stage, the contents of the nucleus divide. During the second stage, the cell's fluid, or cytoplasm, divides. The mitotic phase creates two new identical cells. The original cell no longer exists.

Visual Check

2. Identify Which stage of interphase is the longest?

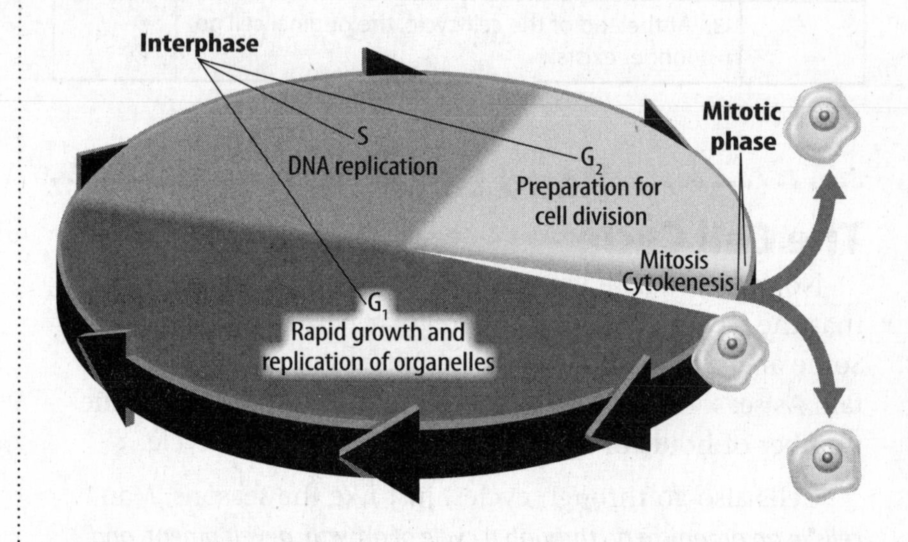

Length of a Cell Cycle

The time it takes a cell to complete the cell cycle depends on the type of cell that is dividing. Recall that a eukaryotic cell has membrane-bound organelles, including a nucleus. The cell cycle for some eukaryotic cells might only take eight minutes. The cell cycle for other eukaryotic cells might take up to one year. Most of the cells in the human body can complete the cell cycle in about 24 hours. The cells of some organisms divide very quickly. For example, the fertilized egg of the zebra fish divides into 256 cells in 2.5 hours.

REVIEW VOCABULARY

eukaryotic
related to a cell with membrane-bound structures

Interphase

A new cell begins interphase with a period of rapid growth, in which the cell gets bigger. Cellular activities, such as making proteins, follow. Each cell that is actively dividing copies its DNA and prepares for cell division. A cell's DNA is called chromatin (KROH muh tun) during interphase. Chromatin is long, thin strands of DNA in the nucleus. If scientists add dye to a cell in interphase, the nucleus looks like a plate of spaghetti. This is because the nucleus contains strands of chromatin tangled together. ✓

Phases of Interphase

Interphase can be divided into three different stages, as shown in the table below.

The G_1 Stage The first stage of interphase is the G_1 stage. This is a period of rapid growth. G_1 is the longest stage of the cell cycle. During G_1, a cell grows and carries out its normal cell functions. For example, during G_1 the cells that line your stomach make enzymes that help you digest your food. Most cells continue the cell cycle. However, some cells stop the cell cycle at the G_1 stage. Mature nerve cells in your brain remain in G_1 and do not divide again.

Phases of the Cell Cycle

Phase	Stage	Description
Interphase	G_1	growth and cellular functions; organelle replication
	S	growth and chromosome replication; organelle replication
	G_2	growth and cellular functions; organelle replication
Mitotic Phase	mitosis	division of nucleus
	cytokinesis	division of cytoplasm

The S Stage The second stage of interphase is the S stage. During the S stage, a cell grows and copies its DNA. Strands of chromatin are copied, so there are now two identical strands of DNA. This is necessary because each new cell gets a copy of the genetic information. The new strands coil up and form chromosomes. A cell's DNA is arranged as pairs. Each pair is called a duplicated chromosome. *Two identical chromosomes called* **sister chromatids** *make up a duplicated chromosome. The sister chromatids are held together by a structure called a* **centromere.** ✓

✓ **Reading Check**

3. Identify What is the location of the chromatin in the cell?

______

______

______

Interpreting Tables

4. Name What are the stages of interphase?

______

______

______

✓ **Reading Check**

5. Explain What happens during the S stage?

______

______

______

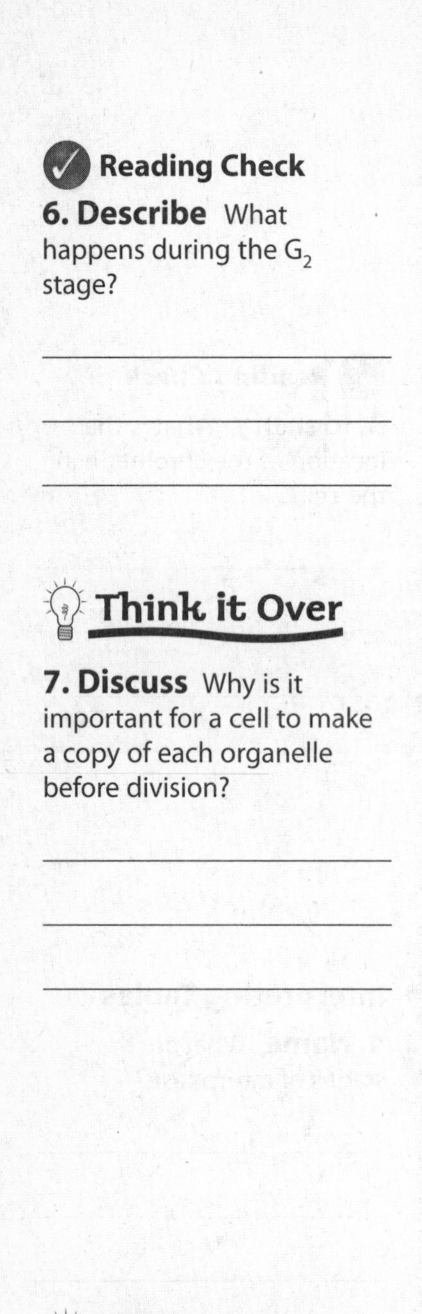

The G_2 Stage The last stage of interphase is the G_2 stage. This is another period of growth and the final preparation for mitosis. A cell uses energy to copy DNA during the S stage. During G_2, the cell stores energy that will be used during the mitotic phase of the cell cycle.

Organelle Replication

During cell division, the organelles are distributed between the two new cells. Before a cell divides, it makes a copy of each organelle. This way, the two new cells can function properly. Some organelles, such as the energy-processing mitochondria and chloroplasts, have their own DNA. These organelles can make copies of themselves on their own. A cell produces other organelles from materials such as proteins and lipids. A cell makes these materials using the information in the DNA inside the nucleus. Organelles are copied during all stages of interphase.

The Mitotic Phase

The mitotic phase of the cell cycle follows interphase. There are two stages of the mitotic phase: mitosis (mi TOH sus) and cytokinesis (si toh kuh NEE sus). *In* ***mitosis,*** *the nucleus and its contents divide. In* ***cytokinesis,*** *the cytoplasm and its contents divide.* ***Daughter cells*** *are the two new cells that result from mitosis and cytokinesis.*

During mitosis, the contents of the nucleus divide, forming two identical nuclei. The sister chromatids of the duplicated chromosomes separate from each other. This gives each daughter cell the same genetic information. For example, a cell that has ten duplicated chromosomes actually has 20 chromatids. When the cell divides, each daughter cell will have ten different chromatids. Chromatids are now called chromosomes.

During cytokinesis, the cytoplasm divides and two new daughter cells form. The organelles that were made during interphase are divided between the daughter cells.

Phases of Mitosis

Mitosis, like interphase, is a process that can be divided into different phases. Follow along with the diagrams on the next page as you read the descriptions in this section.

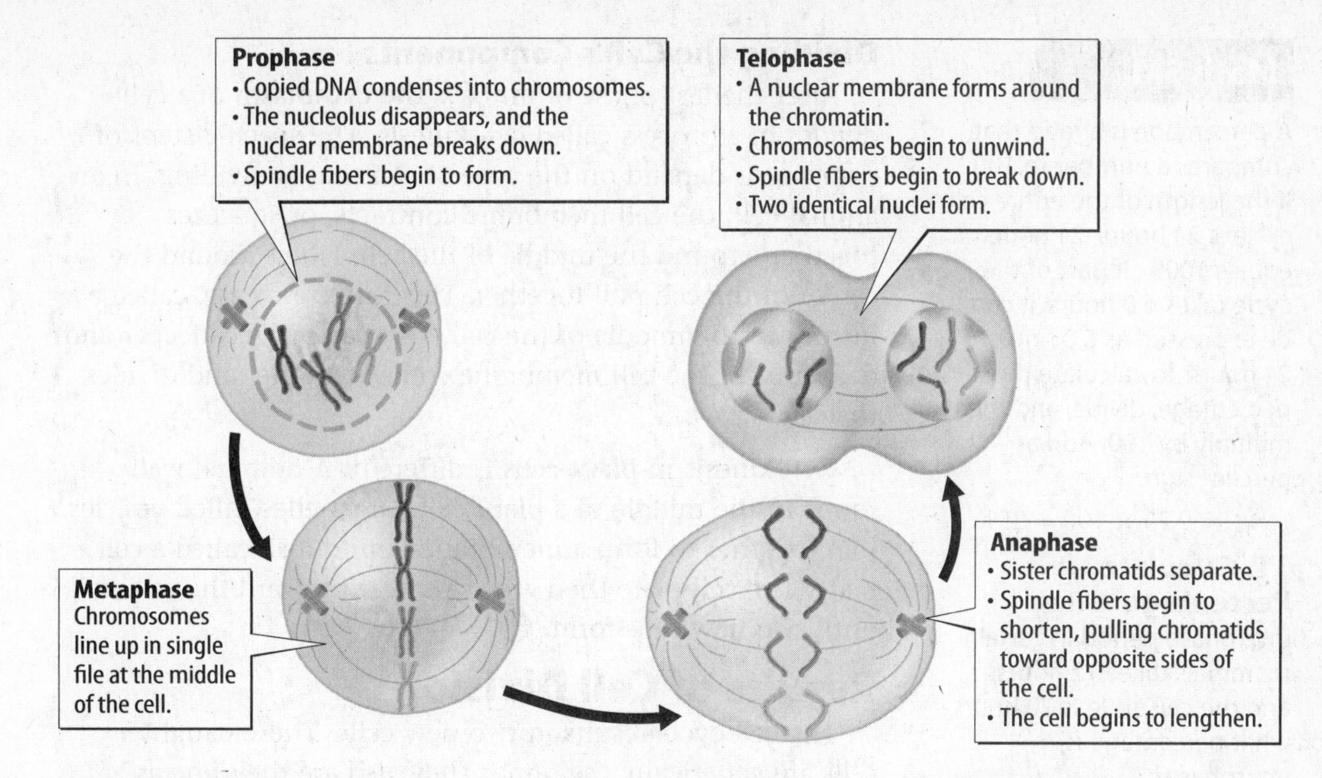

Prophase During the first phase of mitosis, called prophase, the copied chromatin coils together tightly. The coils form duplicated chromosomes that can be seen with a microscope. The nucleolus disappears and the nuclear membrane breaks down. Structures called spindle fibers form in the cytoplasm.

Metaphase The spindle fibers pull and push the duplicated chromosomes to the middle of the cell during metaphase. Notice in the figure above that the chromosomes line up along the middle of the cell. This makes sure that each new cell will receive one copy of each chromosome. Metaphase is the shortest phase in mitosis. It is an important phase because it makes the new cells the same.

Anaphase In anaphase, the two sister chromatids separate from each other and are pulled in opposite directions. Once they are separated, the chromatids are now two identical single-stranded chromosomes. As the single-stranded chromosomes move to opposite sides of the cell, the cell begins to get longer. Anaphase ends when the two sets of identical chromosomes reach opposite ends of the cell.

Telophase During telophase, the spindle fibers that helped divide chromosomes begin to disappear. The chromosomes begin to uncoil. A nuclear membrane grows around each set of chromosomes at either end of the cell. Two new identical nuclei form.

Visual Check

9. Identify During which phase of mitosis do the chromosomes line up in the center of a cell?

____

____

Reading Check

10. State What are the phases of mitosis?

____

____

____

____

Math Skills

A percentage is a ratio that compares a number to 100. If the length of the entire cell cycle is 24 hours, 24 hours equals 100%. If part of the cycle takes 6.0 hours, it can be expressed as 6.0 hours/24 hours. To calculate the percentage, divide, and then multiply by 100. Add a percent sign.

$\frac{6.0}{24} = 0.25 \times 100 = 25\%$

11. Calculate Percentage If the interphase period in certain mammals takes 12 hours, and the cell cycle is 20 hours, what percentage is interphase?

____

Reading Check

12. Compare cytokinesis in plant cells and animal cells.

____

____

____

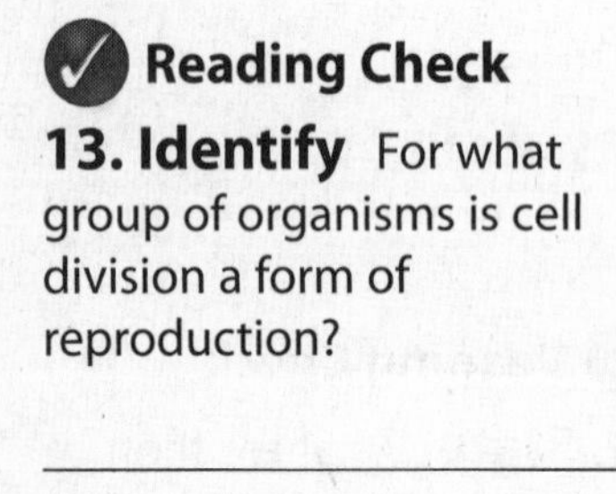

Reading Check

13. Identify For what group of organisms is cell division a form of reproduction?

____

____

Dividing the Cell's Components

After the last phase of mitosis, the cytoplasm of a cell divides in a process called cytokinesis. The specific steps of cytokinesis depend on the type of cell that is dividing. In an animal cell, the cell membrane contracts, or squeezes together, around the middle of the cell. Fibers around the center of the cell pull together. This forms a crease, called a furrow, in the middle of the cell. This furrow gets deeper and deeper until the cell membrane comes together and divides the cell.

Cytokinesis in plant cells is different. A new cell wall forms in the middle of a plant cell. Organelles called vesicles join together to form a membrane-bound disk called a cell plate. The cell plate then grows outward toward the cell wall until two new cells form.

Results of Cell Division

The cell cycle results in two new cells. These daughter cells are genetically the same. They also are the same as the original cell that no longer exists. A human cell has 46 chromosomes. When that cell divides, it produces two new cells, each with 46 chromosomes. The cell cycle is important for reproduction in some organisms. It is important for growth in multicellular organisms. The cell cycle also helps replace worn-out or damaged cells and repair damaged tissues.

Reproduction

Cell division is a form of reproduction for some unicellular organisms. For example, an organism called a paramecium reproduces by dividing into two new daughter cells, or two new paramecia. Cell division is also important in other methods of reproduction in which the offspring are identical to the parent organism.

Growth

Cell division allows multicellular organisms, such as humans, to grow and develop from one cell (a fertilized egg). In humans, cell division begins about 24 hours after fertilization. Cell division continues quickly for the first few years of life. During the next few years, you will probably go through a period of rapid growth and development. This happens because cells divide and increase in number as you grow and develop.

Replacement

Cell division continues even after an organism is fully grown. Cell division replaces cells that wear out or are damaged. The outermost layer of your skin is always rubbing or flaking off. A layer of cells below the skin's surface is constantly dividing. This produces millions of new cells each day to replace the ones that rub off.

Repair

Cell division is also important for repairing damage. When a bone breaks, cell division produces new bone cells. These new cells patch the broken pieces of bone back together.

Not all damage can be repaired. This is because not all cells continue to divide. Some nerve cells stop the cell cycle in interphase. Injuries to nerve cells often cause permanent damage.

Key Concept Check

14. Discuss Why is the result of the cell cycle important?

After You Read

Mini Glossary

cell cycle: a cycle of growth, development, and division that most cells in an organism go through

centromere: the structure that holds together two sister chromatids

cytokinesis (si toh kuh NEE sus): the stage in which the cytoplasm and its contents divide

daughter cells: the two new cells that result from mitosis and cytokinesis

interphase: the period of a cell's growth and development

mitosis (mi TOH sus): the phase in which the nucleus and its contents divide

sister chromatids: two identical chromosomes that make up a duplicated chromosome

1. Review the terms and their definitions in the Mini Glossary. Write a sentence that describes the cell cycle.

2. Complete the table below to explain what happens in each phase of mitosis.

Phases of Mitosis	Prophase	Metaphase	Anaphase	Telophase
What happens within the cell?		duplicated chromosomes line up in the middle of the cell		chromosomes uncoil, spindle fibers disappear, nuclear membrane grows around each set of chromosomes, and two new nuclei form

3. Choose one of your quiz questions and write it on the first line below. Then write your answer to that question on the line that follows.

What do you think NOW?

Reread the statements at the beginning of the lesson. Fill in the After column with an A if you agree with the statement or a D if you disagree. Did you change your mind?

Log on to ConnectED.mcgraw-hill.com and access your textbook to find this lesson's resources.

From a Cell to an Organism

Levels of Organization

Before You Read

What do you think? Read the three statements below and decide whether you agree or disagree with them. Place an A in the Before column if you agree with the statement or a D if you disagree. After you've read this lesson, reread the statements and see if you have changed your mind.

Before	Statement	After
	4. Unicellular organisms do not have all the characteristics of life.	
	5. All the cells in a multicellular organism are the same.	
	6. Some organs work together as part of an organ system.	

Key Concepts 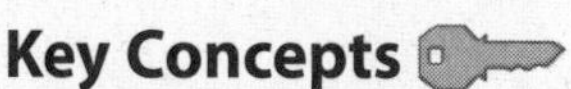

- How do unicellular and multicellular organisms differ?
- How does cell differentiation lead to the organization within a multicellular organism?

Read to Learn

Life's Organization

All matter is made of atoms. Atoms combine and form molecules. Molecules make up cells. A large animal, such as a Komodo dragon, is not made of one cell. Instead, it is made of trillions of cells working together. The skin of the Komodo dragon is made of many cells that are specialized for protection. The Komodo dragon has other types of cells, such as blood cells and nerve cells, which perform other functions. Cells work together in the Komodo dragon and enable the whole organism to function. This is the same way that cells work together in you and in other multicellular organisms.

Recall that some organisms are made of only one cell. These unicellular organisms carry out all the activities necessary to survive, such as absorbing nutrients and getting rid of wastes. No matter their sizes, all organisms are made of cells.

Study Coach

Make Flash Cards As you read, write each vocabulary word and key term from the text on one side of a flash card and its definition on the other side. Use your cards to review the material later.

Reading Check

1. Identify What are all organisms made from?

Unicellular Organisms

Unicellular organisms have only one cell. These organisms do all the things needed for their survival within that one cell. An amoeba is a unicellular organism. It takes in, or ingests, other unicellular organisms for food to get energy. Unicellular organisms also respond to their environment, get rid of waste, grow, and reproduce. Unicellular organisms include both prokaryotes and some eukaryotes.

Prokaryotes

A cell without a membrane-bound nucleus is a prokaryotic cell. In general, prokaryotic cells are smaller than eukaryotic cells. As shown below on the left, prokaryotic cells also have fewer cell structures. A unicellular organism made of one prokaryotic cell is called a prokaryote. Some prokaryotes live in groups called colonies. Some can also live in extreme environments. The heat-loving bacteria that live in hot springs get their energy from sulfur instead of light.

Visual Check

2. Highlight each area where the hereditary material is located.

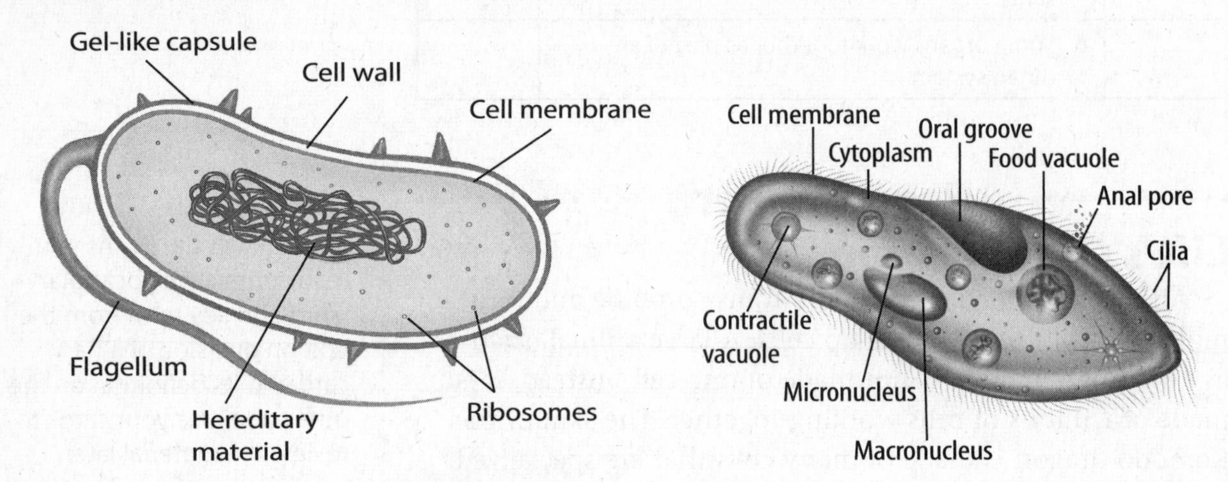

Eukaryotes

A eukaryotic cell has a nucleus surrounded by a membrane and many specialized organelles as shown above on the right. This paramecium has an organelle called a contractile vacuole. The contractile vacuole collects extra water from the paramecium's cytoplasm and pumps it out. The contractile vacuole keeps the paramecium from swelling and bursting.

A unicellular organism that is made of one eukaryotic cell is called a eukaryote. There are thousands of different unicellular eukaryotes. The alga that grows on the inside of an aquarium and the fungus that causes athlete's foot are unicellular eukaryotes.

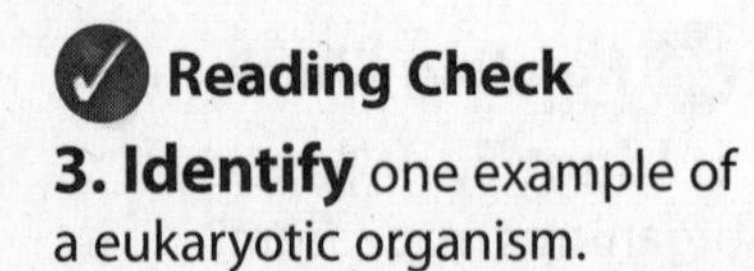

Reading Check

3. Identify one example of a eukaryotic organism.

Multicellular Organisms

A multicellular organism is made of many eukaryotic cells working together. Each type of cell in a multicellular organism has a specific job that is important to the survival of the organism.

Cell Differentiation

Remember that all cells in a multicellular organism come from one cell, a fertilized egg. Cell division starts quickly after fertilization. The first cells made can become any type of cell, such as a muscle cell, a nerve cell, or a blood cell. *The process by which cells become different types of cells is called* **cell differentiation** (dihf uh ren shee AY shun).

A cell's instructions are contained in its chromosomes. Nearly all the cells in an organism have identical sets of chromosomes. If an organism's cells have identical sets of instructions, how can the cells be different? Different cell types use different parts of the instructions on the chromosomes. A few of the many different types of cells that can result from cell differentiation are shown in the figure below.

Animal Stem Cells Not all cells in a developing animal differentiate. **Stem cells** *are unspecified cells that are able to develop into many different cell types.* There are many stem cells in embryos but fewer in adult organisms. Adult stem cells are important for cell repair and replacement. For example, stem cells in your blood marrow can produce more than a dozen different types of blood cells. These replace the cells that are damaged or worn out. Stem cells in your muscles can produce new muscle cells. These can replace torn muscle fibers.

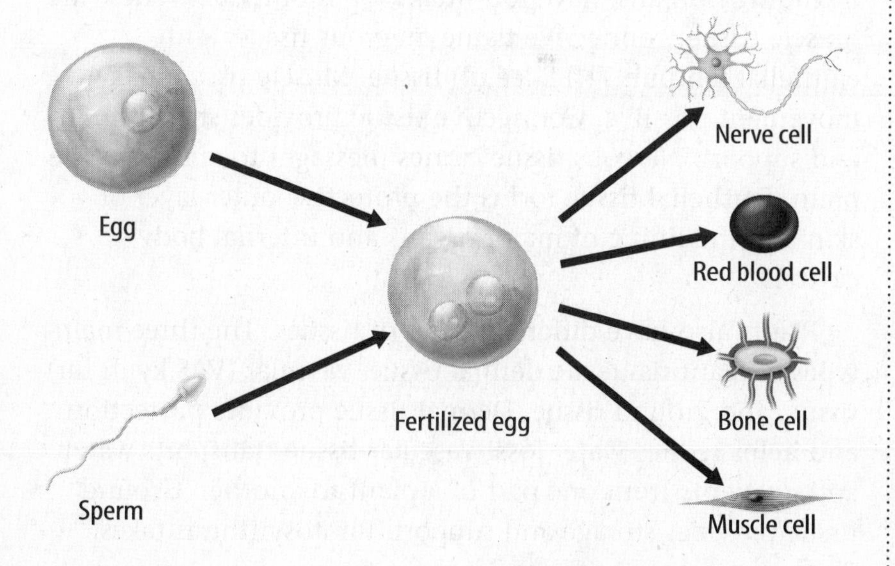

Key Concept Check

4. Describe How do unicellular and multicellular organisms differ?

FOLDABLES

Use a layered book to describe the levels of organization that make up organisms.

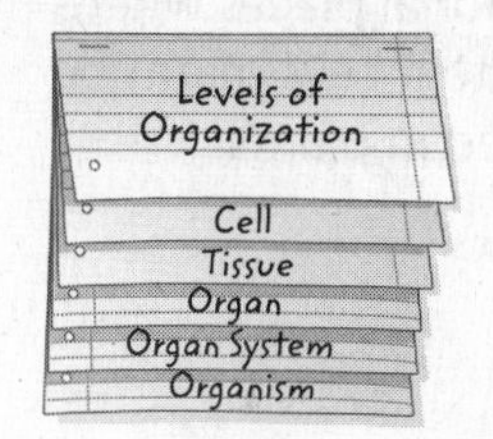

Visual Check

5. Name two types of cells that can result from cell differentiation.

Plant Cells Plants also have unspecialized cells, similar to the stem cells of animals. These cells are grouped in areas called meristems (MER uh stemz). Meristems are in different areas of a plant, including the tips of roots and stems. Cell division in meristems produces different types of plant cells with specialized structures and functions. These functions include transporting materials, making and storing food, or protecting the plant. Meristem cells might become part of stems, leaves, flowers, or roots. Meristems are shown in the figure below. ✓

Reading Check
6. Identify the three possible functions of meristems.

____

____

____

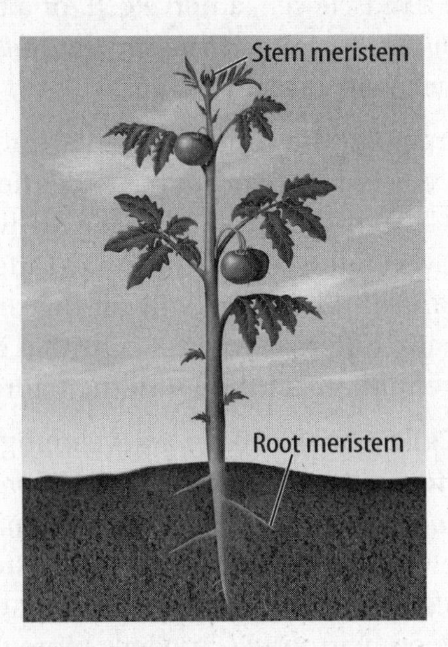

Visual Check
7. Identify Circle two additional places where meristem cells might be located in plants.

Tissues

In multicellular organisms, similar types of cells are organized into groups. **Tissues** *are groups of similar types of cells that work together to carry out specific tasks.* Most animals, including humans, have four main types of tissues. These are muscle tissue, connective tissue, nervous tissue, and epithelial (eh puh THEE lee ul) tissue. Muscle tissue makes movement possible. Connective tissue provides structure and support. Nervous tissue carries messages to and from the brain. Epithelial tissue forms the protective outer layer of skin and the lining of major organs and internal body cavities.

Plants also have different types of tissues. The three main types of plant tissue are dermal tissue, vascular (VAS kyuh lur) tissue, and ground tissue. Dermal tissue provides protection and helps reduce water loss. Vascular tissue transports water and nutrients from one part of a plant to another. Ground tissue provides storage and support. Photosynthesis takes place in ground tissue. ✓

Reading Check
8. Compare animal and plant tissues.

____

____

____

Organs

Complex jobs in organisms require more than one type of tissue. **Organs** *are groups of different tissues working together to perform a particular job.* Your stomach is an organ that breaks down food. It is made of all four types of tissue: muscle, epithelial, nervous, and connective. Each type of tissue performs a specific function necessary for the stomach to work properly and break down food. Muscle tissue contracts and breaks up food. Epithelial tissue lines the stomach. Nervous tissue signals when the stomach is full. Connective tissue supports the stomach wall.

Plants also have organs. A leaf is an organ specialized for photosynthesis. Each leaf is made of dermal tissue, ground tissue, and vascular tissue. Dermal tissue covers the outer surface of a leaf. The leaf is an important organ because it contains ground tissue that produces food for the rest of the plant. Ground tissue is where photosynthesis takes place. The ground tissue is tightly packed on the top half of the leaf. The vascular tissue moves both the food produced by photosynthesis and water throughout the leaf and plant.

ACADEMIC VOCABULARY
complex
(adjective) made of two or more parts

Organ Systems

Most organs do not function alone. Instead, **organ systems** *are groups of different organs that work together to complete a series of tasks.* Human organ systems can be made of many different organs working together. For example, the digestive system is made of the stomach, the small intestine, the liver, and the large intestine. These organs all work together to break down food. Blood absorbs and transports nutrients from food to cells throughout the body.

Plants have two main organ systems—the shoot system and the root system. The shoot system includes leaves, stems, and flowers. The shoot system transports food and water throughout the plant. The root system anchors the plant and takes in water and nutrients.

Reading Check
9. Identify the major organ systems in plants.

Organisms

Multicellular organisms usually have many organ systems. The cells of these systems work together and carry out all the jobs needed for the organism to survive. There are many organ systems in the human body. Each organ system depends on the others and cannot work alone. For example, the respiratory system and circulatory system carry oxygen to the cells of the muscle tissue of the stomach. The oxygen aids in the survival of muscle tissue cells.

Key Concept Check
10. Explain How does cell differentiation lead to the organization within a multicellular organism?

After You Read

Mini Glossary

cell differentiation (dihf uh ren shee AY shun): the process by which cells become different types of cells

organ: a group of tissues working together to perform a particular job

organ system: a group of organs that work together to complete a series of tasks

stem cell: an unspecified cell that is able to develop into many different cell types

tissue: a group of similar types of cells that work together to carry out specific tasks

1. Review the terms and their definitions in the Mini Glossary. Write two sentences describing some of the different types of cells within an organism.

____

____

____

____

2. Fill in the chart below to show the different levels of organization in a multicellular organism.

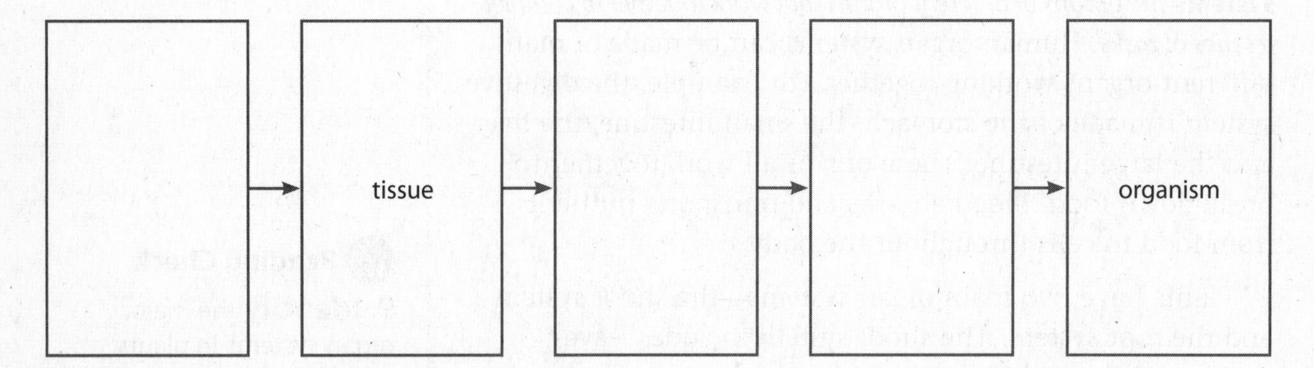

3. How did making flash cards of the important terms in the text help you review the material?

____

____

What do you think NOW?

Reread the statements at the beginning of the lesson. Fill in the After column with an A if you agree with the statement or a D if you disagree. Did you change your mind?

Log on to ConnectED.mcgraw-hill.com and access your textbook to find this lesson's resources.

Reproduction of Organisms

Sexual Reproduction and Meiosis

Before You Read

What do you think? Read the three statements below and decide whether you agree or disagree with them. Place an A in the Before column if you agree with the statement or a D if you disagree. After you've read this lesson, reread the statements to see if you have changed your mind.

Before	Statement	After
	1. Humans produce two types of cells: body cells and sex cells.	
	2. Environmental factors can cause variation among individuals.	
	3. Two parents always produce the best offspring.	

Key Concepts 

- What is sexual reproduction, and why is it beneficial?
- What is the order of the phases of meiosis, and what happens during each phase?
- Why is meiosis important?

Read to Learn

What is sexual reproduction?

Have you ever seen a litter of kittens? One kitten might have orange fur like its mother. A second kitten might have gray fur like its father. A third kitten might look like a combination of both parents. How does this happen?

The kittens look different because of sexual reproduction. **Sexual reproduction** *is a type of reproduction in which the genetic materials from two different cells combine, producing an offspring.* The cells that combine are called sex cells. Sex cells form in reproductive organs. There are two types of sex cells—eggs and sperm. *An* **egg** *is the female sex cell, which forms in an ovary. A* **sperm** *is the male sex cell, which forms in a testis.* **Fertilization** (fur tuh luh ZAY shun) *occurs when an egg cell and a sperm cell join together.* When an egg and a sperm join together, a new cell is formed. *The new cell that forms from fertilization is called a* **zygote.** ✓

Study Coach

Vocabulary Quiz Write a question about each vocabulary term in this lesson. Exchange questions with another student. Together, discuss the answers to the questions.

✓ **Reading Check**

1. Describe What is sexual reproduction?

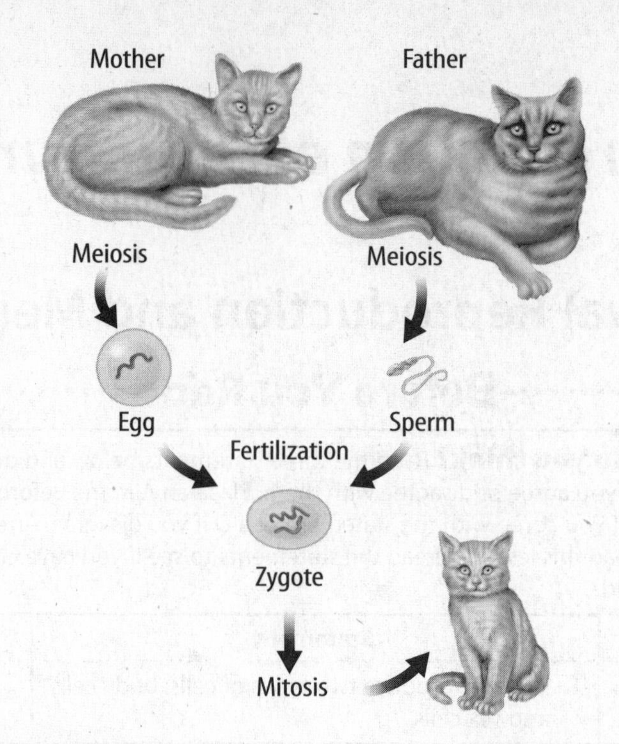

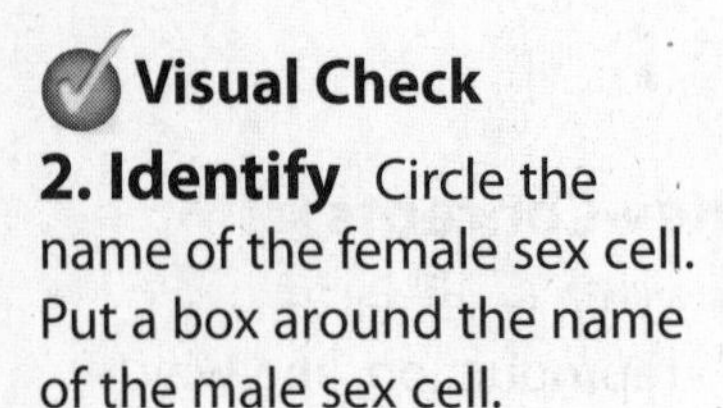
Visual Check

2. Identify Circle the name of the female sex cell. Put a box around the name of the male sex cell.

Diploid Cells

After fertilization, a zygote goes through mitosis and cell division, as shown above. Mitosis and cell division produce nearly all of the cells in a multicellular organism. The kitten in the picture above is a multicellular organism. Organisms that reproduce sexually form two kinds of cells—body cells and sex cells. In the body cells of most organisms, chromosomes occur in pairs. *Cells that have pairs of chromosomes are called* **diploid** *cells.*

Chromosomes

Pairs of chromosomes that have genes for the same traits arranged in the same order are called **homologous** (huh MAH luh gus) **chromosomes.** Because one chromosome is inherited from each parent, the chromosomes are not always identical. For example, the kittens you read about earlier inherited a gene for orange fur color from their mother. They also inherited a gene for gray fur color from their father. Some kittens might be orange, and some kittens might be gray. No matter what the color of a kitten's fur, both genes for fur color are found at the same place on homologous chromosomes. In this case, each gene codes for a different color.

FOLDABLES

Make the following shutter-fold book, then use it to describe and illustrate the phases of meiosis.

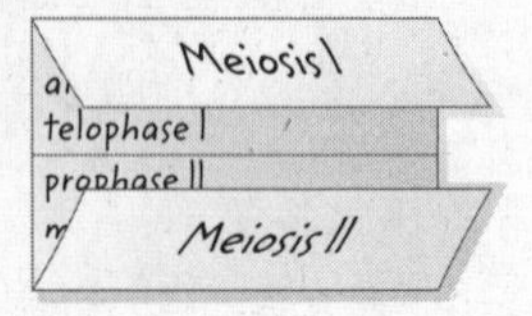

Chromosomes of a Human Cell

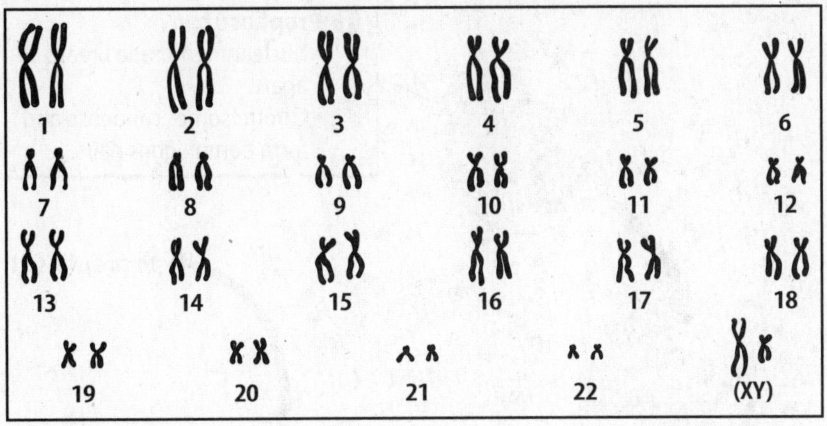

Different organisms have different numbers of chromosomes. Recall that diploid cells have pairs of chromosomes. Human diploid cells have 23 pairs of chromosomes, as shown in the picture above. This means that human diploid cells have a total of 46 chromosomes.

It is important to have the correct number of chromosomes. If a zygote has too many or too few chromosomes, it will not develop properly. The process of meiosis helps maintain the correct number of chromosomes.

Haploid Cells

Organisms that reproduce sexually also form egg and sperm cells, or sex cells. Sex cells have only one chromosome from each pair of chromosomes. **Haploid** *cells are cells that have only one chromosome from each pair.*

Organisms produce sex cells using a special type of cell division called meiosis. *In* **meiosis,** *one diploid cell divides and makes four haploid sex cells.* Meiosis occurs only during the formation of sex cells.

The Phases of Meiosis

Recall that mitosis and cytokinesis involve one division of the nucleus and cytoplasm. Meiosis involves two divisions of the nucleus and the cytoplasm. These two divisions are phases called meiosis I and meiosis II. Meiosis results in four haploid cells, each with half the number of chromosomes as the original cell.

Phases of Meiosis I

A reproductive cell goes through interphase before beginning meiosis I. During interphase, the reproductive cell grows and copies, or duplicates, its chromosomes. Each duplicated chromosome consists of two sister chromatids joined by a centromere.

Visual Check

3. Identify How many chromosomes do human diploid cells have?

Reading Check

4. Explain Why is it important for an organism to have the correct number of chromosomes?

Reading Check

5. Contrast How do diploid cells differ from haploid cells?

Meiosis I

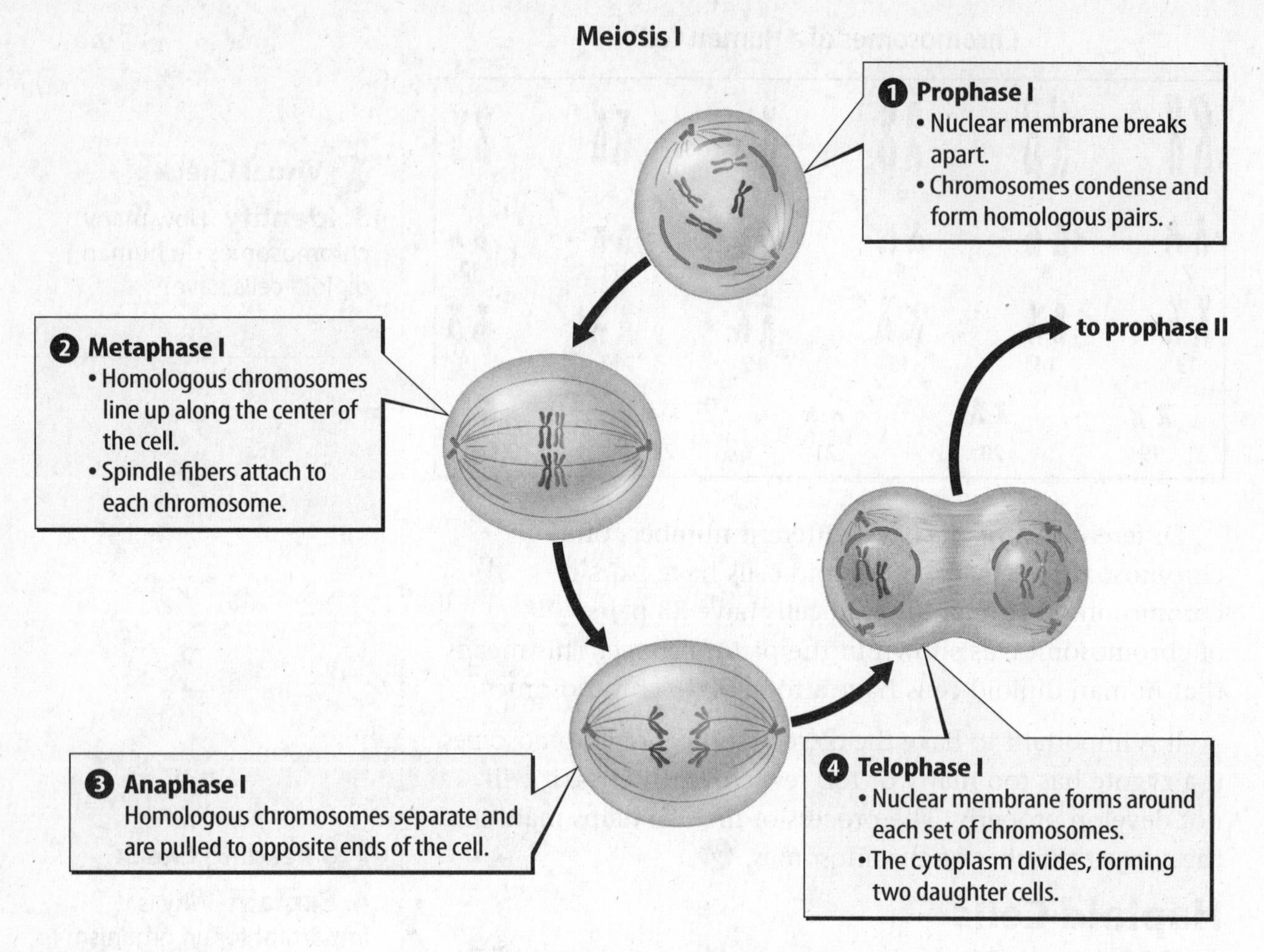

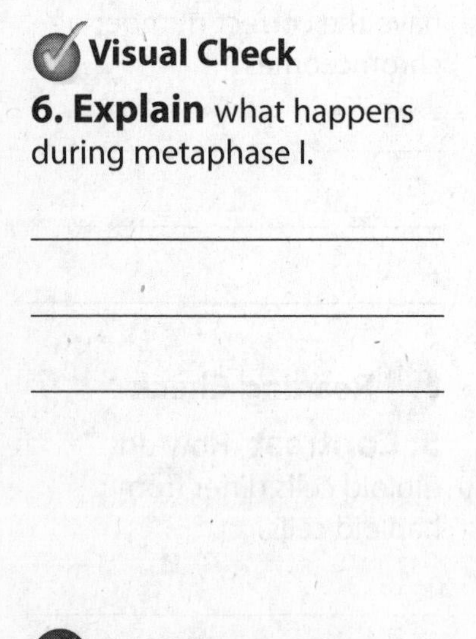

Visual Check

6. Explain what happens during metaphase I.

____

____

____

Reading Check

7. Describe what happens to the sister chromatids at the end of anaphase I.

____

____

____

As you read about the phases of meiosis I, refer to the figure above. Think about the process that produces cells with a reduced number of chromosomes.

1. Prophase I During prophase I, duplicated chromosomes condense, or shorten, and thicken. Homologous chromosomes come together and form pairs. The membrane around the nucleus breaks apart and the nucleolus disappears.

2. Metaphase I During metaphase I, homologous chromosome pairs line up along the middle of the cell, as shown in the figure above. A spindle fiber attaches to each chromosome.

3. Anaphase I During anaphase I, chromosome pairs separate and are pulled toward opposite ends of the cell. Notice in the figure above that the sister chromatids stay together.

4. Telophase I During telophase I, a membrane forms around each group of duplicated chromosomes. The cytoplasm divides through cytokenesis, and two daughter cells form. Sister chromatids remain together.

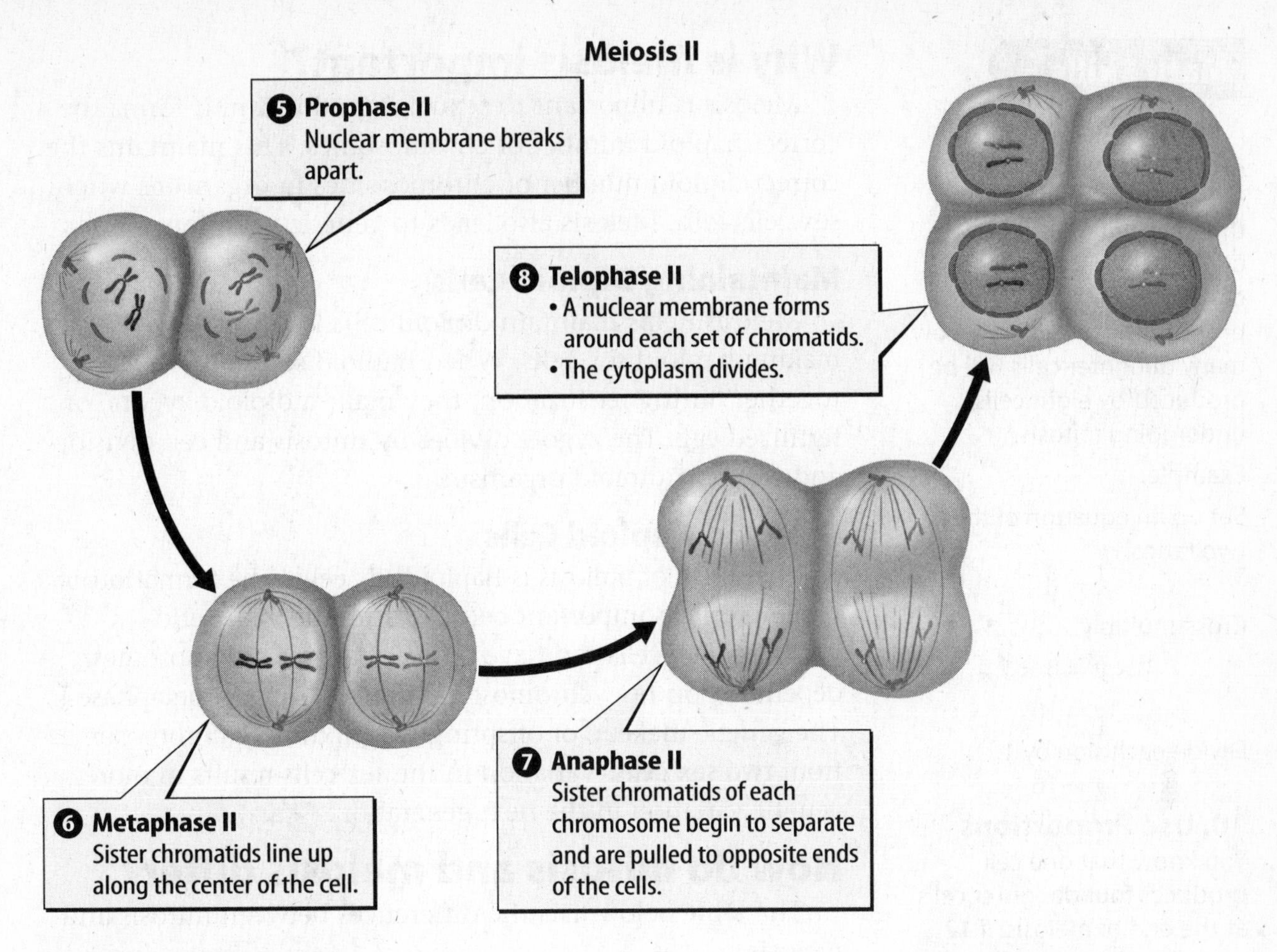

Phases of Meiosis II

After meiosis I, the two cells formed during this stage go through a second division of the nucleus and the cytoplasm. This process is called meiosis II. Meiosis II is shown in the figure above.

1. Prophase II Unlike prophase I, chromosomes are not copied again before prophase II. They remain short and thick sister chromatids. During prophase II, the membrane around the nucleus breaks apart, and the nucleolus disappears in each cell.

2. Metaphase II During metaphase II, the pairs of sister chromatids line up along the middle of the cell in single file.

3. Anaphase II During anaphase II, the sister chromatids of each duplicated chromosome are pulled apart. They then move toward opposite ends of the cells.

4. Telophase II The final phase of meiosis is telophase II. During telophase II, a nuclear membrane forms around each set of chromatids. The chromatids are again called chromosomes. The cytoplasm divides through cytokinesis, and four haploid cells form.

Visual Check

8. Differentiate How does telophase II differ from telophase I?

Key Concept Check

9. Name the phases of meiosis in order.

A proportion is an equation that shows that two ratios are equivalent. If you know that one cell produces two daughter cells at the end of mitosis, you can use proportions to calculate how many daughter cells will be produced by eight cells undergoing mitosis.

Example:

Set up an equation of the two ratios.

$$\frac{1}{2} = \frac{8}{y}$$

Cross-multiply.

$$1 \times y = 8 \times 2$$

$$1y = 16$$

Divide each side by 1.

$$y = 16$$

10. Use Proportions You know that one cell produces four daughter cells at the end of meiosis. If 12 sex cells undergo meiosis, how many daughter cells will be produced?

Key Concept Check

11. State why meiosis is important.

Visual Check

12. Compare How many cells are produced during mitosis? During meiosis?

Why is meiosis important?

Meiosis is important to sexual reproduction. It forms the correct haploid number of chromosomes. This maintains the correct diploid number of chromosomes in organisms when sex cells join. Meiosis also leads to genetic variation.

Maintaining Diploid Cells

Meiosis helps maintain diploid cells in offspring by making haploid sex cells. When haploid sex cells join together during fertilization, they make a diploid zygote, or fertilized egg. The zygote divides by mitosis and cell division and creates a diploid organism.

Creating Haploid Cells

The result of meiosis is haploid sex cells. The formation of haploid cells is important because it results in genetic variation. Sex cells can have different sets of chromosomes, depending on how chromosomes line up during metaphase I. The genetic makeup of offspring is a mixture of chromosomes from two sex cells. Variation in the sex cells results in more genetic variation in the next generation.

How do mitosis and meiosis differ?

The table below lists the differences between mitosis and meiosis.

Characteristic	Meiosis	Mitosis
Number of chromosomes in parent cell	diploid	diploid
Type of parent cell	reproductive	body
Number of divisions of the nucleus	2	1
Number of daughter cells produced	4	2
Chromosome number in daughter cells	haploid	diploid
Function in organism	forms sperm and egg cells	growth, cell repair, some types of reproduction

Advantages of Sexual Reproduction

The main advantage of sexual reproduction is that it results in genetic variation among offspring. Offspring inherit half their DNA from each parent. Inheriting different DNA means that each offspring has a different set of traits.

REVIEW VOCABULARY
DNA
the genetic information in a cell

Genetic Variation

Genetic variation exists among humans. You can look at your friends to see genetic variations. Some people have blue eyes; others have brown eyes. Some people have blonde hair; others have red hair. Genetic variation occurs in all organisms that reproduce sexually.

Because of genetic variation, individuals within a population have slight differences. These differences might be an advantage if the environment changes. Some individuals might have traits that make them able to survive harsh conditions. For example, some plants within a population might be able to survive long periods of dry weather. Sometimes the traits might help keep an organism from getting infected by a disease.

Key Concept Check
13. Identify Why is sexual reproduction beneficial?

Selective Breeding

Selective breeding is a process that involves breeding certain individuals within a population because of the traits they have. For example, a farmer might choose plants with the biggest flowers and stems. These plants would be allowed to reproduce and grow. Over time, the offspring of the plants would all have big flowers and stems. Selective breeding has been used to produce many types of plants and animals with certain traits.

Disadvantages of Sexual Reproduction

Sexual reproduction takes time and energy. Organisms have to grow and develop until they are mature enough to produce sex cells. Before they can reproduce, organisms have to find mates. Searching for a mate takes time and energy. The search might also expose individuals to predators, diseases, or harsh environmental conditions. Sexual reproduction can be limited by certain factors. For example, fertilization cannot take place during pregnancy, which can last as long as two years in some mammals.

Reading Check
14. State the disadvantages of sexual reproduction.

After You Read

Mini Glossary

diploid: cells that have pairs of chromosomes

egg: the female sex cell, which forms in an ovary

fertilization (fur tuh luh ZAY shun): the process in which an egg cell and a sperm cell join together

haploid: cells that have only one chromosome from each pair

homologous (huh MAH luh gus) chromosomes: pairs of chromosomes that have genes for the same traits arranged in the same order

meiosis: the process in which one diploid cell divides and makes four haploid cells

sexual reproduction: a type of reproduction in which the genetic materials from two different cells combine, producing an offspring

sperm: the male sex cell, which forms in a testis

zygote: the new cell that forms from fertilization

1. Review the terms and their definitions in the Mini Glossary. Use at least two words from the Mini Glossary in a sentence to describe the difference between the female and male sex cells.

2. In the table below, list the advantages and disadvantages of sexual reproduction.

Advantages	Disadvantages

3. How is genetic variation related to meiosis?

What do you think NOW?

Reread the statements at the beginning of the lesson. Fill in the After column with an A if you agree with the statement or a D if you disagree. Did you change your mind?

Log on to ConnectED.mcgraw-hill.com and access your textbook to find this lesson's resources.

Reproduction of Organisms

Asexual Reproduction

Before You Read

What do you think? Read the three statements below and decide whether you agree or disagree with them. Place an A in the Before column if you agree with the statement or a D if you disagree. After you've read this lesson, reread the statements to see if you have changed your mind.

Before	Statement	After
	4. Cloning produces identical individuals from one cell.	
	5. All organisms have two parents.	
	6. Asexual reproduction occurs only in microorganisms.	

Key Concepts 

- What is asexual reproduction, and why is it beneficial?
- How do the types of asexual reproduction differ?

Read to Learn

What is asexual reproduction?

In **asexual reproduction,** *one parent organism produces offspring without meiosis and fertilization.* Offspring produced by asexual reproduction inherit all of their DNA from one parent. Therefore, they are genetically the same as each other and their parent.

You have seen the results of asexual reproduction if you have ever seen mold on bread or fruit. Mold is a type of fungus (FUN gus) that can reproduce either sexually or asexually. Asexual reproduction is different from sexual reproduction.

Recall that sexual reproduction involves two parent organisms and the processes of meiosis and fertilization. Offspring inherit half of their DNA from each parent, resulting in genetic variation among the offspring.

Study Coach

Discuss Read the first two paragraphs about asexual reproduction. Then take turns with a partner saying something about what you learned. Repeat this process with the other paragraphs in this lesson.

Key Concept Check

1. Describe What is asexual reproduction?

______

______

______

Types Of Asexual Reproduction

There are many different types of organisms that reproduce asexually. Not only fungi, but also bacteria, protists, plants, and animals can reproduce asexually.

Make the following six-celled chart, then use it to compare types of asexual reproduction.

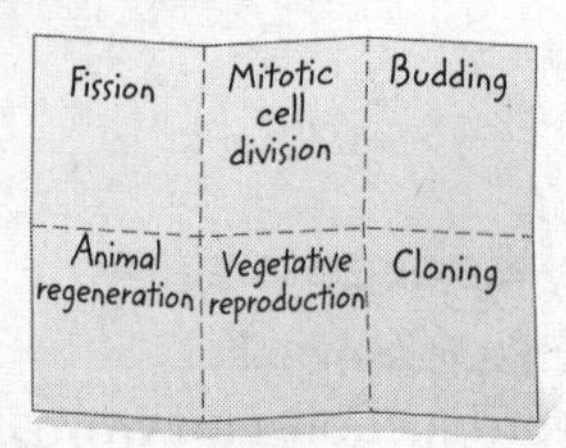

Visual Check

2. Recognize What happens to the original cell's chromosome during fission?

Reading Check

3. Evaluate What advantage might asexual reproduction by fission have compared to sexual reproduction?

Fission

Recall that a prokaryotic cell, such as a bacterial cell, has a simpler cell structure than a eukaryotic cell. A prokaryote's DNA is not contained in a nucleus. For this reason, mitosis does not occur. Cell division in a prokaryote is a simpler process than in a eukaryote. *Cell division in prokaryotes that forms two genetically identical cells is known as* ***fission.***

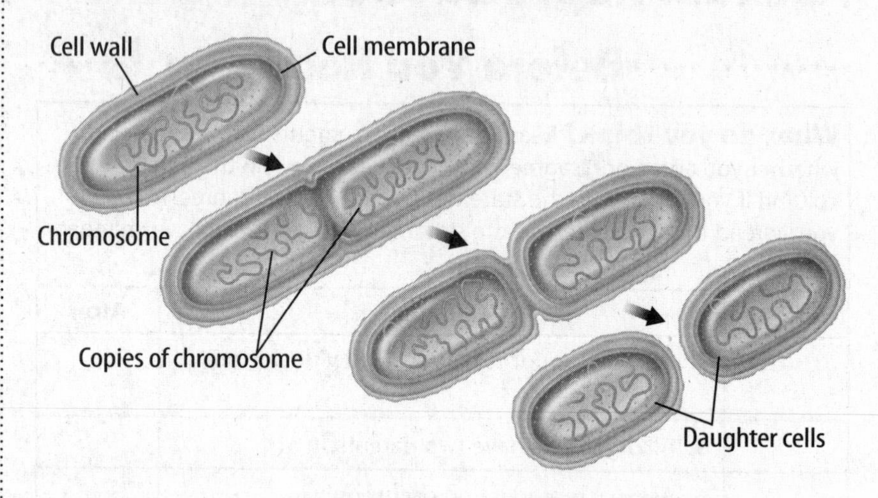

Fission begins when a prokaryote's DNA is copied, as shown in the figure above. Each copy attaches to the cell membrane. Then the cell begins to grow longer. The two copies of DNA are pulled apart. At the same time, the cell membrane starts to pinch inward along the middle of the cell. Finally the cell splits and forms two new identical offspring. The original cell no longer exists. Fission makes it possible for prokaryotes to divide rapidly.

Mitotic Cell Division

Many unicellular eukaryotes, such as amoebas, reproduce by mitotic cell division. In this type of asexual reproduction, an organism forms two offspring through mitosis and cell division. The nucleus of the cell divides by mitosis. Next, the cytoplasm and its contents divide through cytokinesis. Two new amoebas form.

Budding

In **budding,** *a new organism grows by mitosis and cell division on the body of its parent.* The bud, or offspring, is genetically identical to its parent. When the bud is large enough, it can break from the parent and live on its own. Organisms such as yeasts, which are fungi, reproduce through budding. Sometimes the bud stays attached to the parent and starts to form a colony. Corals are animals that form colonies through budding.

Original planarian is divided into two pieces.

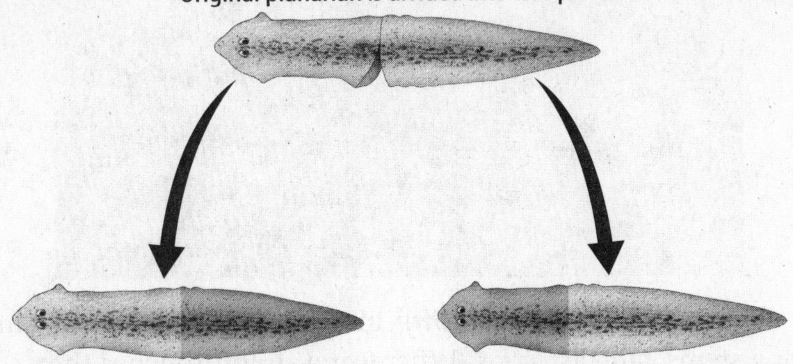

The head end regenerates a new tail.

The tail end regenerates a new head.

Animal Regeneration

Another type of asexual reproduction, **regeneration,** *occurs when an offspring grows from a piece of its parent.* Animals that can reproduce asexually through regeneration include sponges, sea stars, and planarians.

Producing New Organisms The figure above shows how a planarian reproduces through regeneration. If the planarian is cut into two pieces, each piece of the original planarian becomes a new organism.

If the arms are separated from the parent sea star, each of these arms has the potential to grow into a new organism. To regenerate a new sea star, the arm must have a part of the central disk of the parent. If conditions are right, one five-armed sea star can produce five new organisms. As with all types of asexual reproduction, the offspring are genetically the same as the parent.

Producing New Parts Some animals, such as newts, tadpoles, crabs, hydras, zebra fish, and salamanders, can regenerate a lost or damaged body part. Even humans are able to regenerate some damaged body parts, such as the skin and the liver. This type of regeneration is not considered asexual reproduction. It does not produce a new organism.

Vegetative Reproduction

Plants can also reproduce asexually in a process similar to regeneration. **Vegetative reproduction** *is a form of asexual reproduction in which offspring grow from part of a parent plant.* Strawberries, raspberries, potatoes, and geraniums are other plants that can reproduce this way

Visual Check

4. Describe What happens to a planarian when it is cut into two pieces?

ACADEMIC VOCABULARY
potential
(noun) possibility

Reading Check

5. Specify What is true of all cases of asexual reproduction?

Visual Check

6. Locate Circle the offspring of the strawberry plant in the figure.

The strawberry plant shown in the figure above sends out long stems called stolons. Wherever a stolon touches the ground, it can produce roots. Once a stolon grows roots, a new plant can grow, even if the stolon breaks off from the parent plant. Each new plant grown from a stolen is genetically identical to the parent plant. Roots, leaves, and stems are the structures that usually produce new plants.

Cloning

Cloning *is a type of asexual reproduction performed in laboratories. It produces identical individuals from a cell or from a cluster of cells taken from a multicellular organism.* Farmers and scientists often clone cells or organisms that have desirable traits.

Plant Cloning Some plants can be cloned from just a few cells using a method called a tissue culture. Tissue cultures make it possible for plant growers and scientists to make many copies of a plant with desirable traits. The new plants are genetically the same as the parent plant. Also, cloning produces plants more quickly than vegetative reproduction does.

Science Use v. Common Use

culture

Science Use the process of growing living tissue in a laboratory

Common Use the social customs of a group of people

A plant might be infected with a disease. To clone such a plant, a scientist can use cells from the meristem of the plant. Cells in meristems are disease-free. Therefore, if a plant becomes diseased, it can be cloned using meristem cells.

Think it Over

7. Synthesize Do all cloned plants have the same genetic makeup? Why or why not?

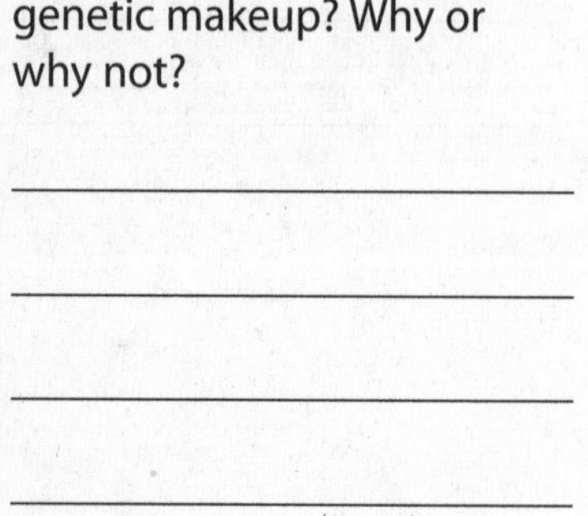

Animal Cloning In addition to cloning plants, scientists have been able to clone many animals. All of a clone's chromosomes come from one parent, the donor of the nucleus. This means that the clone is genetically the same as its parent. The first mammal cloned was a sheep named Dolly.

Steps in Cloning Dolly The first step in cloning Dolly was to remove cells from a sheep, as shown in the figure on the next page. DNA was then removed from an unfertilized egg of a second sheep. In a laboratory, the cells were fused, or combined, and the new cell had the DNA from the first sheep. The cell developed into an embryo. The embryo was then placed in a third sheep. The cloned sheep developed inside the third sheep and was later born.

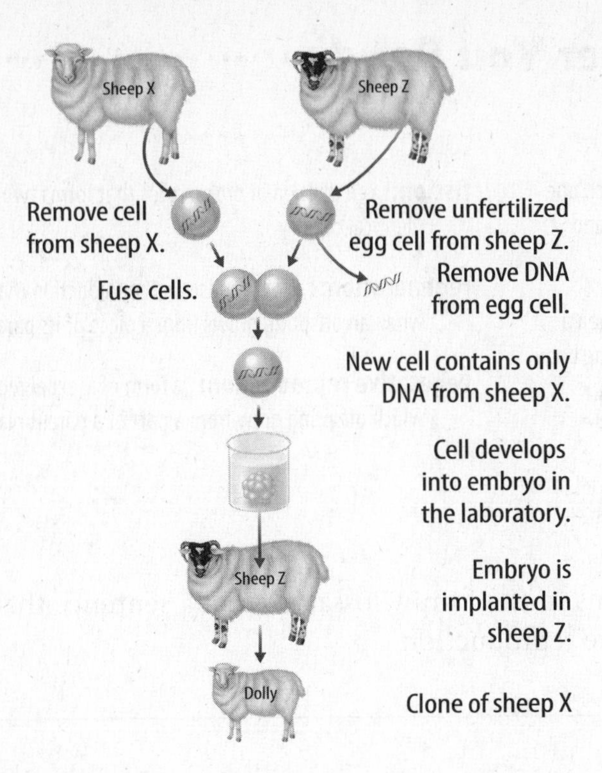

Visual Check

8. Identify Circle the two sheep that are genetically identical.

Cloning Issues Scientists are working to save some endangered species from extinction by cloning. Some people are concerned about the cost and ethical issues of cloning. Ethical issues include the possibility of human cloning.

Key Concept Check

9. Compare and contrast the different types of asexual reproduction.

Advantages of Asexual Reproduction

One advantage of asexual reproduction is that an organism can reproduce without a mate. Recall that finding a mate takes time and energy. Another advantage is that some organisms can quickly produce a large number of offspring. For example, crabgrass reproduces by underground stolons. This enables one plant to spread and colonize an area in a short period of time.

Key Concept Check

10. State the advantages of asexual reproduction.

Disadvantages of Asexual Reproduction

Asexual reproduction produces offspring that are genetically the same as the parent. This results in little genetic variation within a population. Genetic variation can give organisms a better chance of surviving if the environment changes. Imagine that all of the crabgrass plants in a lawn are genetically the same. If a weed killer can kill the parent plant, then it can kill all of the crabgrass plants in the lawn. This might be good for the lawn, but it is a disadvantage for the crabgrass. Another disadvantage involves genetic changes called mutations. A harmful mutation passed to asexually reproduced offspring could affect the offspring's ability to survive.

After You Read

Mini Glossary

asexual reproduction: a form of reproduction in which one parent organism produces offspring without meiosis and fertilization

budding: a form of asexual reproduction that occurs when a new organism grows by mitosis and cell division on the body of its parent

cloning: a type of asexual reproduction performed in a laboratory that produces identical individuals from a cell or from a cluster of cells taken from a multicellular organism

fission: cell division in prokaryotes that forms two genetically identical cells

regeneration: a form of asexual reproduction that occurs when an offspring grows from a piece of its parent

vegetative reproduction: a form of asexual reproduction in which offspring grow from a part of a parent plant

1. Review the terms and their definitions in the Mini Glossary. Write a sentence that compares regeneration and vegetative reproduction.

____

____

2. Fill in the spider map below with the different types of asexual reproduction. Use terms from the Mini Glossary.

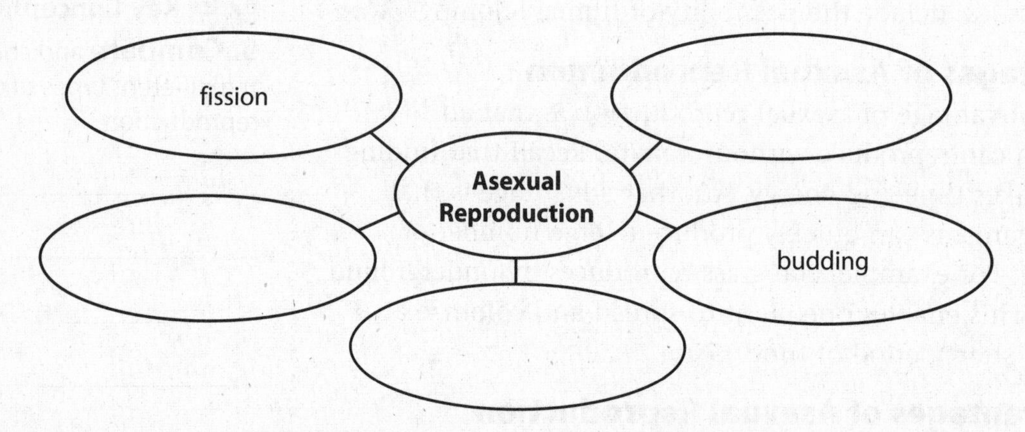

3. How did discussing what you learned from each paragraph with another student help you learn about asexual reproduction?

____

____

What do you think NOW?

Reread the statements at the beginning of the lesson. Fill in the After column with an A if you agree with the statement or a D if you disagree. Did you change your mind?

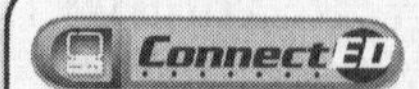

Log on to ConnectED.mcgraw-hill.com and access your textbook to find this lesson's resources.

Genetics

Mendel and His Peas

Before You Read

What do you think? Read the two statements below and decide whether you agree or disagree with them. Place an A in the Before column if you agree with the statement or a D if you disagree. After you've read this lesson, reread the statements to see if you have changed your mind.

Before	Statement	After
	1. Like mixing paints, parents' traits always blend in their offspring.	
	2. If you look more like your mother than you look like your father, then you received more traits from your mother.	

Key Concepts 

- Why did Mendel perform cross-pollination experiments?
- What did Mendel conclude about inherited traits?
- How do dominant and recessive factors interact?

Read to Learn

Early Ideas About Heredity

Have you ever mixed two different colors of paint to make a new color? Long ago, people thought that an organism's characteristics, or traits, were determined in the same way that paint colors can be mixed. People assumed this because offspring often resemble both parents. This is known as blending inheritance.

Today, scientists know that heredity (huh REH duh tee) is more complex. **Heredity** *is the passing of traits from parents to offspring.* For example, you and your brother might have blue eyes but both of your parents have brown eyes. How does this happen?

More than 150 years ago, Gregor Mendel, an Austrian monk, performed experiments that helped answer many questions about heredity. The results of his experiments also disproved the idea of blending inheritance.

Mendel's research into the questions of heredity gave scientists a basic understanding of genetics. **Genetics** (juh NE tihks) *is the study of how traits are passed from parents to offspring.* Because of his research, Mendel is known as the father of genetics. ✓

Study Coach

Vocabulary Quiz Write a question about each vocabulary term in this lesson. Exchange quizzes with a partner. After completing the quizzes, discuss the answers with your partner.

✓ **Reading Check**

1. Define What is genetics?

Mendel's Experimental Methods

During the 1850s, Mendel studied genetics by doing controlled breeding experiments with pea plants. Pea plants were ideal for genetics studies because

- they reproduce quickly. Mendel was able to grow many plants and collect a lot of data.
- they have easily observed traits, such as flower color and pea shape. Mendel was able to observe whether or not a trait was passed from one generation to the next.
- Mendel could control which pairs of plants reproduced. He was able to find out which traits came from which plant pairs.

Think it Over

2. Explain In his breeding experiments, how did Mendel know which traits came from which pair of plants?

Pollination in Pea Plants

To observe how a trait was inherited, Mendel controlled which plants pollinated other plants. Pollination occurs when pollen lands on the pistil of a flower. Sperm cells from the pollen then fertilize egg cells in the pistil.

REVIEW VOCABULARY

sperm
a haploid sex cell formed in the male reproductive organs

egg
a haploid sex cell formed in the female reproductive organs

Self-pollination occurs when pollen from one plant lands on the pistil of a flower on the same plant. Cross-pollination occurs when pollen from one plant reaches the pistil of a flower on a different plant. Mendel allowed one group of flowers to self-pollinate. With another group, he cross-pollinated the plants himself.

True-Breeding Plants

Mendel began his experiments with plants that were true-breeding for the trait that he would test. When a true-breeding plant self-pollinates, it always produces offspring with traits that match the parent. For example, when a true-breeding pea plant with wrinkled seeds self-pollinates, it produces only plants with wrinkled seeds. In fact, it will produce wrinkled seeds generation after generation.

Mendel's Cross-Pollination

By cross-pollinating plants himself, Mendel was able to select which plants pollinated other plants. Mendel cross-pollinated hundreds of plants for each set of traits he wanted to learn more about. The traits included flower color (purple or white), seed color (green or yellow), and seed shape (round or wrinkled).

Key Concept Check

3. Explain Why did Mendel perform cross-pollination experiments?

With each cross-pollination, Mendel recorded the traits that appeared in the offspring. By testing such a large number of plants, Mendel was able to predict which crosses would produce which traits.

Mendel's Results

Once Mendel had enough true-breeding plants for a trait that he wanted to test, he cross-pollinated selected plants. His results are described below.

First-Generation Crosses

Crosses between true-breeding plants with purple flowers produced true-breeding plants with only purple flowers. Crosses between true-breeding plants with white flowers produced true-breeding plants with only white flowers. However, when Mendel crossed true-breeding plants with purple flowers and true-breeding plants with white flowers, all of the offspring had purple flowers.

New Questions Raised

Why did crossing plants with purple flowers and plants with white flowers always produce offspring with purple flowers? Why were there no white flowers? Why didn't the cross produce offspring with pink flowers—a combination of white and purple? Mendel carried out more experiments to answer these questions. ✓

Second-Generation (Hybrid) Crosses

Mendel's first-generation purple-flowering plants are called hybrid plants. They came from true-breeding parent plants with different forms of the same trait. When Mendel cross-pollinated two purple-flowering hybrid plants, some of the offspring had white flowers. The trait that had disappeared in the first-generation always reappeared in the second-generation.

Mendel got similar results each time he cross-pollinated hybrid plants. For example, a true-breeding yellow-seeded pea plant crossed with a true-breeding green-seeded pea plant always produced yellow-seeded hybrids. A second-generation cross of two yellow-seeded hybrids always produced plants with yellow seeds and plants with green seeds.

More Hybrid Crosses

Mendel cross-pollinated many hybrid plants. He counted and recorded the traits of offspring. He analyzed these data and noticed patterns. In crosses between hybrid plants with purple flowers, the ratio of purple flowers to white flowers was about 3:1. This means that purple-flowering pea plants grew from this cross three times more often than white-flowering pea plants grew from the cross. Mendel calculated similar ratios for all seven traits that he tested.

FOLDABLES®

Make a two-tab book and organize your notes on dominant and recessive factors.

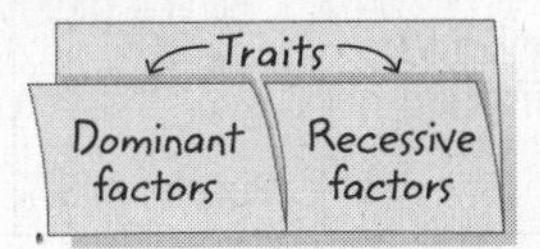

✓ Reading Check

4. Predict the offspring of a cross between two true-breeding pea plants with smooth seeds.

____

____

____

Math Skills

A ratio is a comparison of two numbers or quantities by division. For example, the ratio comparing 6,022 yellow seeds to 2,001 green seeds can be written as follows:

6,022 to 2,001 or
6,022:2,001 or
$\frac{6,022}{2,001}$

To simplify the ratio, divide the first number by the second number.

$$\frac{6,022}{2,001} = \frac{3}{1} = 3:1$$

5. Use Ratios A science class has 14 girls and 7 boys. Simplify the ratio.

____

Results of Hybrid Crosses

Characteristics	Flower Color	Flower Position	Seed Color	Seed Shape	Pod Shape	Pod Color	Stem Length
Dominant Trait; # of Offspring	Purple; 705	Axial; 651	Yellow; 6022	Round; 5474	Smooth; 882	Green; 428	Long; 781
Recessive Trait: # of Offspring	White; 224	Terminal; 207	Green; 2001	Wrinkled; 1850	Bumpy; 299	Yellow; 152	Short; 277
Ratio	3.15 : 1	3.14 : 1	3.01 : 1	2.96 : 1	2.95 : 1	2.82 : 1	2.84 : 1

Visual Check

6. Predict If a cross between two hybrid plants with purple flowers produced 12 offspring, how many offspring would you expect to have purple flowers?

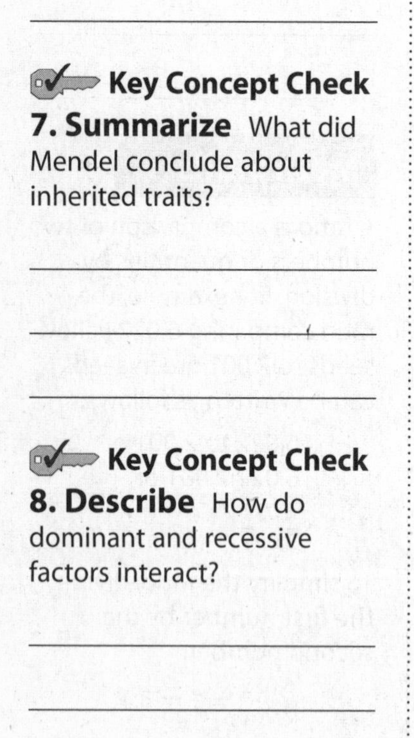

Key Concept Check

7. Summarize What did Mendel conclude about inherited traits?

Key Concept Check

8. Describe How do dominant and recessive factors interact?

Mendel's Conclusions

After analyzing the results of his experiments, Mendel concluded that two factors control each inherited trait. He also proposed that when organisms reproduce, the sperm and the egg each contribute one factor for each trait. Mendel's results are shown in the table above.

Dominant and Recessive Traits

Recall that when Mendel cross-pollinated a true-breeding plant with purple flowers and a true-breeding plant with white flowers, the hybrid offspring had only purple flowers. He hypothesized that the hybrid offspring had one genetic factor for purple flowers and one genetic factor for white flowers. But why were there no white flowers? Mendel also hypothesized that the purple factor was dominant, blocking the white factor. *A genetic factor that blocks another genetic factor is called a* **dominant** (DAH muh nunt) **trait.** A dominant trait, such as purple pea flowers, is seen when offspring have either one or two dominant factors. *A genetic factor that is blocked by the presence of a dominant factor is called a* **recessive** (rih SE sihv) **trait.** A recessive trait, such as white pea flowers, is seen only when two recessive genetic factors are present in offspring.

From Parents to Second Generation

For the second generation, Mendel cross-pollinated two hybrids that had purple flowers. About 75 percent of the second-generation plants had purple flowers. These plants had at least one dominant factor. Twenty-five percent of the second-generation plants had white flowers. These plants had the same two recessive factors.

After You Read

Mini Glossary

dominant (DAH muh nunt) trait: a genetic factor that blocks another genetic factor

genetics (juh NE tihks): the study of how traits are passed from parents to offspring

heredity (huh REH duh tee): the passing of traits from parents to offspring

recessive (rih SE sihv) trait: a genetic factor that is blocked by the presence of a dominant factor

1. Review the terms and their definitions in the Mini Glossary. Write one or two sentences that compare and contrast dominant traits and recessive traits.

____

____

____

2. The tables below show a sequence of crosses for the trait of pod color in a type of plant. Study the tables and fill in the trait or traits that the second-generation cross would produce in the offspring.

First-Generation Cross	
Plants Crossed	**Offspring**
true-breeding green-pod × true-breeding yellow-pod	all green-pod hybrids

Second-Generation (Hybrid) Cross	
Plants Crossed	**Offspring**
green-pod hybrid × green-pod hybrid	

3. Why were the pea plants that Mendel used in his experiments a good choice for genetics studies?

____

____

What do you think NOW?

Reread the statements at the beginning of the lesson. Fill in the After column with an A if you agree with the statement or a D if you disagree. Did you change your mind?

Log on to ConnectED.mcgraw-hill.com and access your textbook to find this lesson's resources.

Genetics

Understanding Inheritance

Key Concepts 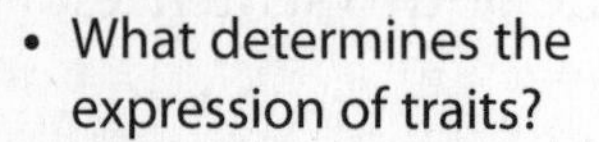

- What determines the expression of traits?
- How can inheritance be modeled?
- How do some patterns of inheritance differ from Mendel's model?

Before You Read

What do you think? Read the two statements below and decide whether you agree or disagree with them. Place an A in the Before column if you agree with the statement or a D if you disagree. After you've read this lesson, reread the statements to see if you have changed your mind.

Before	Statement	After
	3. All inherited traits follow Mendel's patterns of inheritance.	
	4. Scientists have tools to predict the form of a trait an offspring might inherit.	

Study Coach

Build Vocabulary Skim this lesson and circle any words you do not know. After you've read the lesson, review the circled words. Look up the definitions in the dictionary for any words you cannot define.

Read to Learn

What controls traits?

Mendel concluded that two factors control each trait. One factor comes from the egg cell and one factor comes from the sperm cell. What are these factors? How are they passed from parents to offspring?

Chromosomes

Inside each cell is a nucleus that has threadlike structures called chromosomes. Chromosomes contain genetic information that controls traits. What Mendel called "factors" are parts of chromosomes. Each cell in an offspring contains chromosomes from both parents. These chromosomes exist in pairs—one chromosome from each parent.

Genes and Alleles

Each chromosome can have information about hundreds or thousands of traits. *A* **gene** (JEEN) *is a section on a chromosome that has genetic information for one trait.* For example, a gene of a pea plant might have information about flower color. An offspring inherits two genes (factors) for each trait, one from each parent. The genes can be the same or different, such as purple or white for pea flower color. *The different forms of a gene are called* **alleles** (uh LEELS). Pea plants can have two purple alleles, two white alleles, or one of each allele.

1. Identify How many alleles controlled flower color in Mendel's experiments?

Genotype and Phenotype

Geneticists call how a trait appears, or is expressed, the trait's **phenotype** (FEE nuh tipe). A person's eye color is an example of phenotype. The trait of eye color can be expressed as blue, brown, green, or other colors.

Mendel concluded that two alleles control the expression or phenotype of each trait. *The two alleles that control the phenotype of a trait are called the trait's* **genotype** (JEE nuh tipe). You cannot see an organism's genotype. But you can make guesses about a genotype based on its phenotype. For example, you have already learned that a pea plant with white flowers has two recessive alleles for that trait. These two alleles are its genotype. The white flower is its phenotype.

Symbols for Genotype Scientists use symbols to represent the alleles in a genotype. The table below shows the possible genotypes for both round and wrinkled seed phenotypes. Uppercase letters represent dominant alleles and lowercase letters represent recessive alleles. The dominant allele, if present, is written first.

The round pea seed can have either of these two genotypes—*RR* or *Rr*. Both genotypes have a round phenotype. *Rr* results in round seeds because the round allele *(R)* is dominant to the wrinkled allele *(r)*.

The wrinkled pea can have only one genotype—*rr*. The wrinkled phenotype is possible only when two recessive alleles *(rr)* are present in the genotype.

Homozygous and Heterozygous *When the two alleles of a gene are the same, its genotype is* **homozygous** (hoh muh ZI gus). Both *RR* and *rr* are homozygous genotypes. The *RR* genotype has two dominant alleles. The *rr* genotype has two recessive alleles.

If the two alleles of a gene are different, its genotype is **heterozygous** (he tuh roh ZI gus). *Rr* is a heterozygous genotype. It has one dominant and one recessive allele.

Phenotype and Genotype	
Phenotype (observed traits)	**Genotype (alleles of a gene)**
Round	Homozygous dominant (*RR*)
	Heterozygous (*Rr*)
Wrinkled	Homozygous recessive (*rr*)

Think it Over

2. Draw Conclusions In a pea plant, an allele for a tall stem is dominant to an allele for a short stem. You see a pea plant with a short stem. What can you conclude about the genotype of this plant?

Key Concept Check

3. Explain How do alleles determine the expression of traits?

Interpreting Tables

4. Identify the genotype of a plant that produces a wrinkled pea.

Modeling Inheritance

Plant breeders and animal breeders use two tools to help them predict how often traits will appear in offspring. These tools, Punnett squares and pedigrees, can be used to predict and identify traits among genetically related individuals.

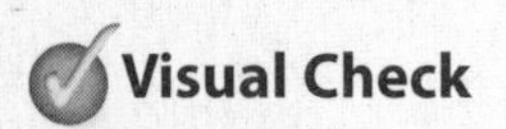

5. Specify What phenotypes are possible for pea offspring of this cross?

Punnett Squares

If the genotypes of the parents are known, then the different genotypes and phenotypes of the offspring can be predicted. *A* **Punnett square** *is a model used to predict possible genotypes and phenotypes of offspring.* Follow the steps shown below to learn how to make a Punnett square.

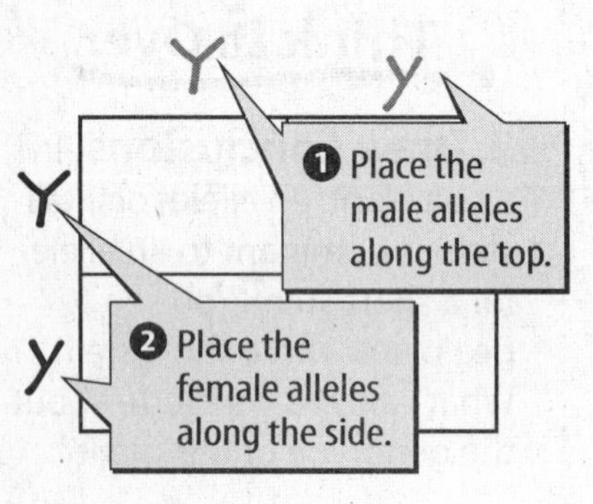

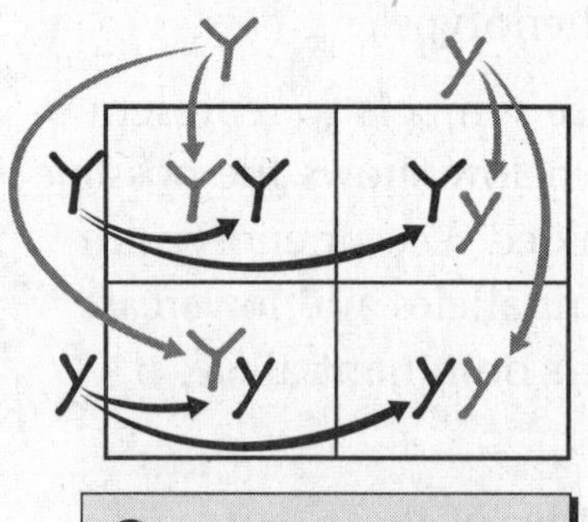

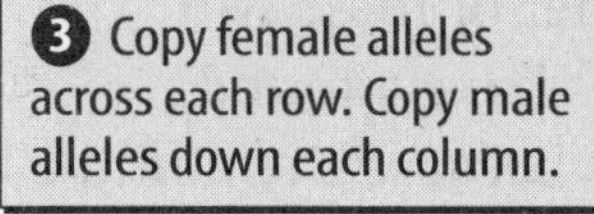

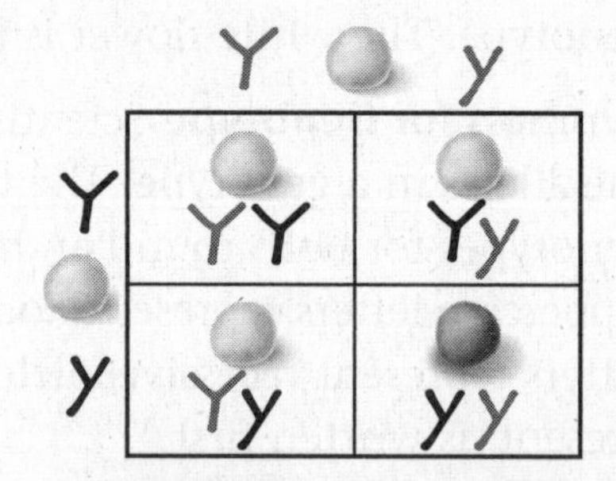

Analysis The ratio of phenotypes is 3:1, yellow:green. The ratio of genotypes is 1:2:1, *YY:Yy:yy.*

Analyzing a Punnett Square

The figure shows a cross between two pea plants that are heterozygous for seed color—*Yy* and *Yy*. Yellow is the dominant allele—*Y*. Green is the recessive allele—*y*. The offspring can have one of three genotypes—*YY, Yy,* or *yy*. The ratio of genotypes is written as 1:2:1.

6. Apply If the cross shown in the figure produced 100 offspring, about how many offspring would have green peas?

Because *YY* and *Yy* represent the same phenotype (yellow), the offspring can have one of only two phenotypes—yellow or green. The ratio of phenotypes is written 3:1. About 75 percent of the offspring of the cross between two heterozygous plants will produce yellow seeds. About 25 percent of the plants will produce green seeds.

Using Ratios to Predict

A 3:1 ratio means that an offspring of heterozygous parents has a 3:1 chance of having yellow seeds. It does not mean that any group of four offspring will have three plants with yellow seeds and one with green seeds. This is because one offspring does not affect the phenotype of other offspring. But if you examine large numbers of offspring from a particular cross, as Mendel did, the overall ratio will be close to the ratio predicted by a Punnett square.

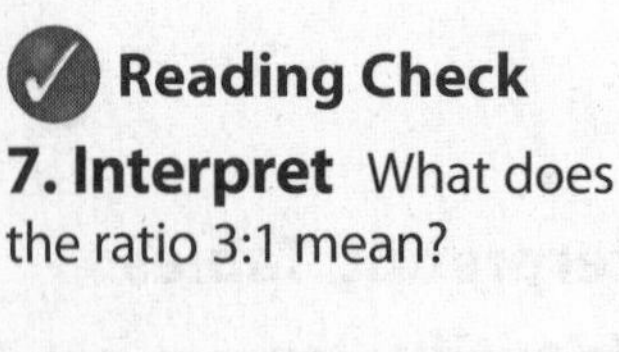

7. Interpret What does the ratio 3:1 mean?

Pedigrees

Another tool that can show inherited traits is a pedigree. A pedigree shows phenotypes of genetically related family members. It can also help determine genotypes. In the pedigree shown below, three offspring have a trait—attached earlobes—that the parents do not have. If these offspring received one allele for this trait from each parent but neither parent displays this trait, the offspring must have received two recessive alleles. If either allele was dominant, the offspring would have the dominant phenotype—unattached earlobes.

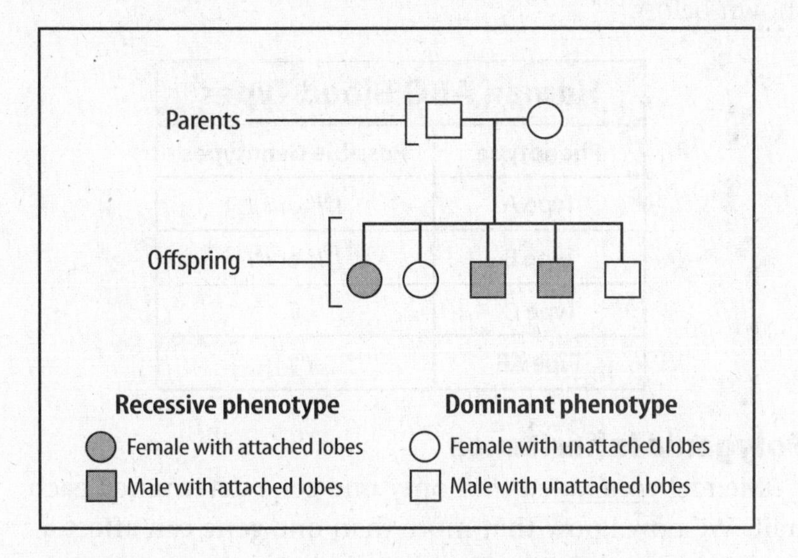

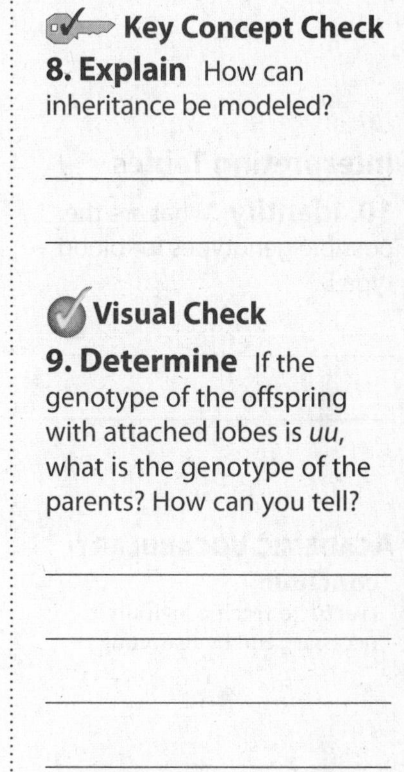

Key Concept Check

8. Explain How can inheritance be modeled?

Visual Check

9. Determine If the genotype of the offspring with attached lobes is *uu*, what is the genotype of the parents? How can you tell?

Complex Patterns of Inheritance

Mendel studied traits influenced by only one gene with two alleles. We know now that not all traits are inherited this way. Some traits have more complex inheritance patterns.

Types of Dominance

Recall that in pea plants, the presence of one dominant allele produces a dominant phenotype. However, not all allele pairs have a dominant-recessive interaction.

Incomplete Dominance Sometimes traits appear to be blends of alleles. *Alleles show* ***incomplete dominance*** *when the offspring's phenotype is a blend of the parents' phenotypes.* For example, a pink camellia flower results from incomplete dominance. A cross between a camellia plant with white flowers and a camellia plant with red flowers produces only camellia plants with pink flowers.

Codominance *When both alleles can be observed in a phenotype, this type of interaction is called* ***codominance.*** For example, if a cow inherits the allele for white coat color from one parent and the allele for red coat color from the other parent, the cow will have both red and white hairs.

FOLDABLES®

Make a layered book and organize your notes on inheritance patterns.

Inheritance Patterns
Incomplete dominance
Multiple alleles
Polygenic inheritance

Multiple Alleles

Unlike the genes in Mendel's pea plants, some genes have more than two alleles, or multiple alleles. Human ABO blood type is an example of a trait that is determined by multiple alleles. There are three alleles for the ABO blood type—I^A, I^B, and *i*. The way the alleles combine results in one of four blood types—A, B, AB, or O. The I^A and I^B alleles are codominant to each other, but they both are dominant to the *i* allele. Even though there are multiple alleles, a person inherits only two of these alleles, one from each parent, as shown below.

Interpreting Tables

10. Identify What are the possible genotypes for blood type B?

____

____

Human ABO Blood Types	
Phenotype	**Possible Genotypes**
Type A	I^AI^A or I^Ai
Type B	I^BI^B or I^Bi
Type O	*ii*
Type AB	I^AI^B

Polygenic Inheritance

ACADEMIC VOCABULARY

conclude
(verb) to reach a logically necessary end by reasoning

Mendel concluded that only one gene determined each trait. We now know that more than one gene can affect a trait. **Polygenic inheritance** *occurs when multiple genes determine the phenotype of a trait.* Because several genes determine a trait, many alleles affect the phenotype even though each gene has only two alleles. Therefore, polygenic inheritance has many possible phenotypes. Eye color in humans is an example of polygenic inheritance. Polygenic inheritance also determines the human characteristics of height, weight, and skin color.

Key Concept Check

11. Explain How does polygenic inheritance differ from Mendel's model?

____

____

____

____

Genes and the Environment

Recall that an organism's genotype determines its phenotype. However, genes are not the only factors that can affect phenotypes. An organism's environment can also affect its phenotype. For example, the flower color of one type of hydrangea is determined by the soil in which the hydrangea plant grows. Acidic soil produces blue flowers. Basic, or alkaline, soil produces pink flowers.

For humans, healthful choices can also affect phenotype. Many genes affect a person's chances of having heart disease. However, what a person eats and the amount of exercise he or she gets can influence whether heart disease will develop.

After You Read

Mini Glossary

allele (uh LEEL): any of the different forms of a gene

codominance: occurs when both alleles can be observed in a phenotype

gene (JEEN): a section on a chromosome that has genetic information for one trait

genotype (JEE nuh tipe): the two alleles that control the phenotype of a trait

heterozygous (he tuh roh ZI gus): an organism's genotype when the two alleles of a gene are different

homozygous (hoh muh ZI gus): an organism's genotype when the two alleles of a gene are the same

incomplete dominance: occurs when an offspring's phenotype is a blend of the parents' phenotypes

phenotype (FEE nuh tipe): how a trait appears or is expressed

polygenic inheritance: occurs when multiple genes determine the phenotype of a trait

Punnett square: a model used to predict possible genotypes and phenotypes of offspring

1. Review the terms and their definitions in the Mini Glossary. Compare and contrast genotype and phenotype.

2. Complete the Punnett square below. Predict the genotypes of offspring produced by crossing a heterozygous pea plant with round seeds and a homozygous pea plant with wrinkled seeds. Round (*R*) is dominant to wrinkled (*r*).

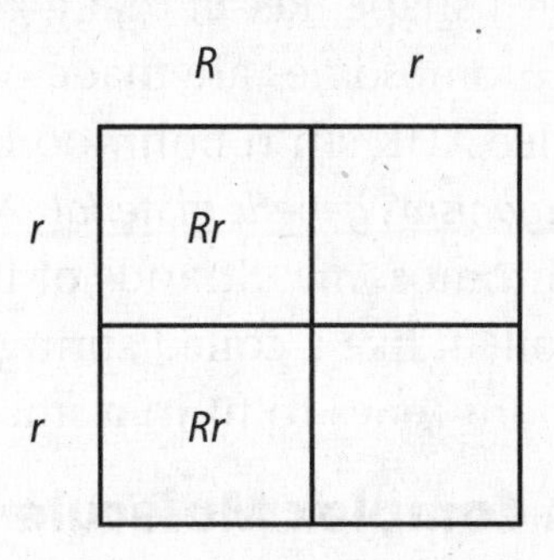

	R	*r*
r	*Rr*	
r	*Rr*	

3. Predict the ratio of round phenotype to wrinkled phenotype in the offspring from the cross in Question 2.

What do you think NOW?

Reread the statements at the beginning of the lesson. Fill in the After column with an A if you agree with the statement or a D if you disagree. Did you change your mind?

Log on to ConnectED.mcgraw-hill.com and access your textbook to find this lesson's resources.

CHAPTER 5
LESSON 3

Genetics

DNA and Genetics

Key Concepts
- What is DNA?
- What is the role of RNA in protein production?
- How do changes in the sequence of DNA affect traits?

Before You Read

What do you think? Read the two statements below and decide whether you agree or disagree with them. Place an A in the Before column if you agree with the statement or a D if you disagree. After you've read this lesson, reread the statements to see if you have changed your mind.

Before	Statement	After
	5. Any condition present at birth is genetic.	
	6. A change in the sequence of an organism's DNA always changes the organism's traits.	

Study Coach

Asking Questions As you read the lesson, write a question for each paragraph. Answer the question using information from the paragraph.

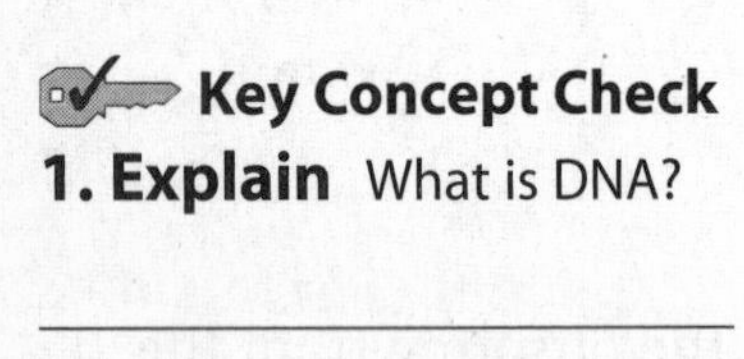

Key Concept Check
1. Explain What is DNA?

Read to Learn

The Structure of DNA

Cells put molecules together by following a set of directions. Genes provide the directions for a cell to put together molecules that express traits, such as eye color or seed shape. Recall that a gene is a section of a chromosome. Chromosomes are made of proteins and deoxyribonucleic (dee AHK sih ri boh noo klee ihk) acid, or DNA. **DNA** *is an organism's genetic material.* A gene is a segment of DNA on a chromosome. Strands of DNA in a chromosome are tightly coiled, like a coiled spring. This coiling makes it possible for more genes to fit in a small space.

A Complex Molecule

The shape of DNA is like a twisted ladder. It is called a double helix. You can see a double helix in the figure on the next page. How did scientists discover the shape of DNA? Rosalind Franklin and Maurice Wilkins used X-rays to study DNA. Some of the X-rays showed that DNA has a helix shape. Another scientist, James Watson, saw one of the DNA X-rays. Watson worked with Francis Crick to build a model of DNA. They used information from the X-rays and chemical information about DNA discovered by another scientist, Erwin Chargaff. Eventually, Watson and Crick were able to build a model that showed how smaller molecules of DNA bond together and form a double helix.

Four Nucleotides Shape DNA

DNA has a twisted-ladder shape that is caused by molecules called nucleotides. *A* **nucleotide** *is a molecule made of a nitrogen base, a sugar, and a phosphate group.* Sugar-phosphate groups form the sides of the DNA ladder. The nitrogen bases bond and form the rungs of the ladder. There are four nitrogen bases: adenine (A), cytosine (C), thymine (T), and guanine (G). A and T always bond together, and C and G always bond together. The figure below shows how the sugar-phosphate groups and the nitrogen bases form the twisted DNA shape. ✓

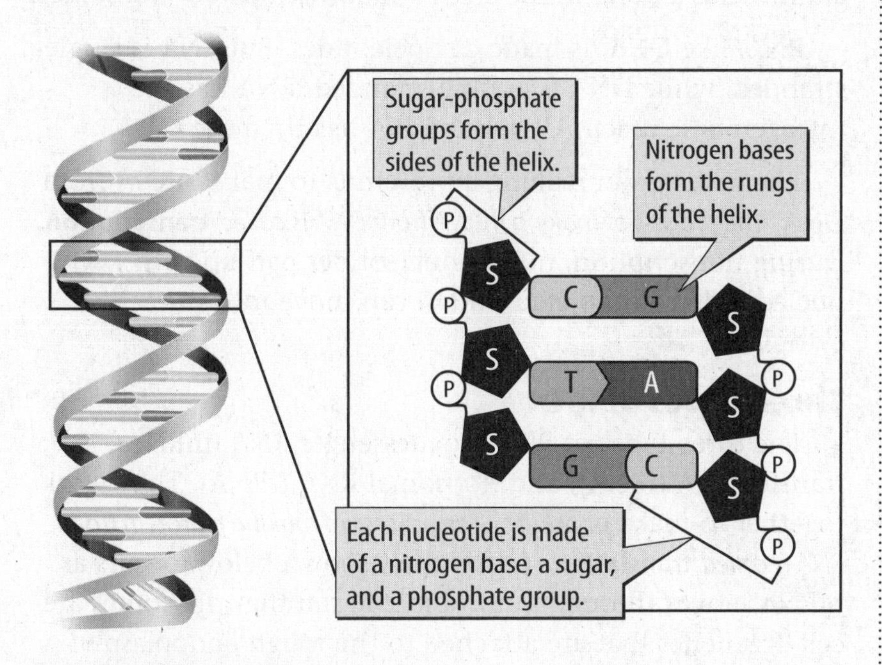

✓ **Reading Check**

2. Define What is a nucleotide?

✓ **Visual Check**

3. Identify What forms the rungs of the DNA double helix?

How DNA Replicates

Cells contain DNA in chromosomes. So, every time a cell divides, all chromosomes must be copied for the new cell. The new DNA is identical to existing DNA. **Replication** *is the process of copying a DNA molecule to make another DNA molecule.*

In the first part of replication, the strands separate in many places and the nitrogen bases are exposed. Nucleotides move into place and form new nitrogen base pairs. This produces two identical strands of DNA. ✓

✓ **Reading Check**

4. Describe What is replication?

Making Proteins

Proteins are important for every cellular process. The DNA of each cell carries a complete set of genes that provides instructions for making all the proteins a cell needs. Most genes contain instructions for making proteins. Some genes contain instructions for when and how quickly proteins are made.

FOLDABLES

Make a vertical three-tab book and record information about the three types of RNA and their functions.

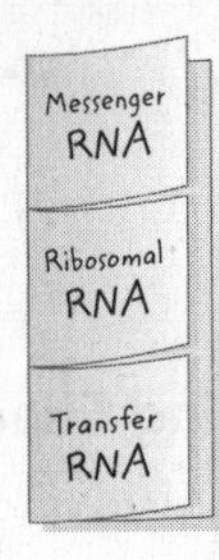

Junk DNA

All genes are segments of DNA on a chromosome. However, about 97 percent of DNA on human chromosomes is not part of any gene. Segments of DNA that are not parts of genes are often called junk DNA. It is not known whether junk DNA has functions that are important to cells.

The Role of RNA in Making Proteins

Proteins are made with the help of ribonucleic acid. **Ribonucleic acid (RNA)** *is a type of nucleic acid that carries the code for making proteins from the nucleus to the cytoplasm.* RNA also carries amino acids around inside a cell and forms a part of ribosomes.

RNA, like DNA, is made of nucleotides. But RNA is single-stranded, while DNA is double-stranded. RNA has the nitrogen base uracil (U), while DNA has thymine (T).

The first step in making a protein is to make mRNA from DNA. *The process of making mRNA from DNA is called* **transcription.** During transcription, mRNA nucleotides pair up with DNA nucleodtides. Completed mRNA can move into the cytoplasm.

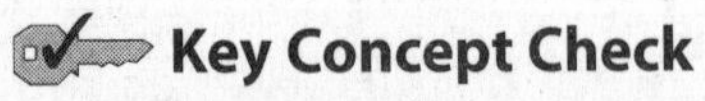
Key Concept Check

5. Explain What is the role of RNA in protein production?

____

____

Three Types of RNA

The three types of RNA are messenger RNA (mRNA), transfer RNA (tRNA), and ribosomal RNA (rRNA). They work together to make proteins. *The process of making a protein from RNA is called* **translation.** Translation, shown below, occurs as mRNA moves through a ribosome. Recall that ribosomes are cell organelles that are attached to the rough endoplasmic reticulum (rough ER).

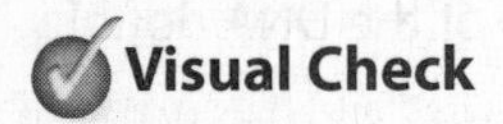
Visual Check

6. Explain What is the role of rRNA during translation?

____

____

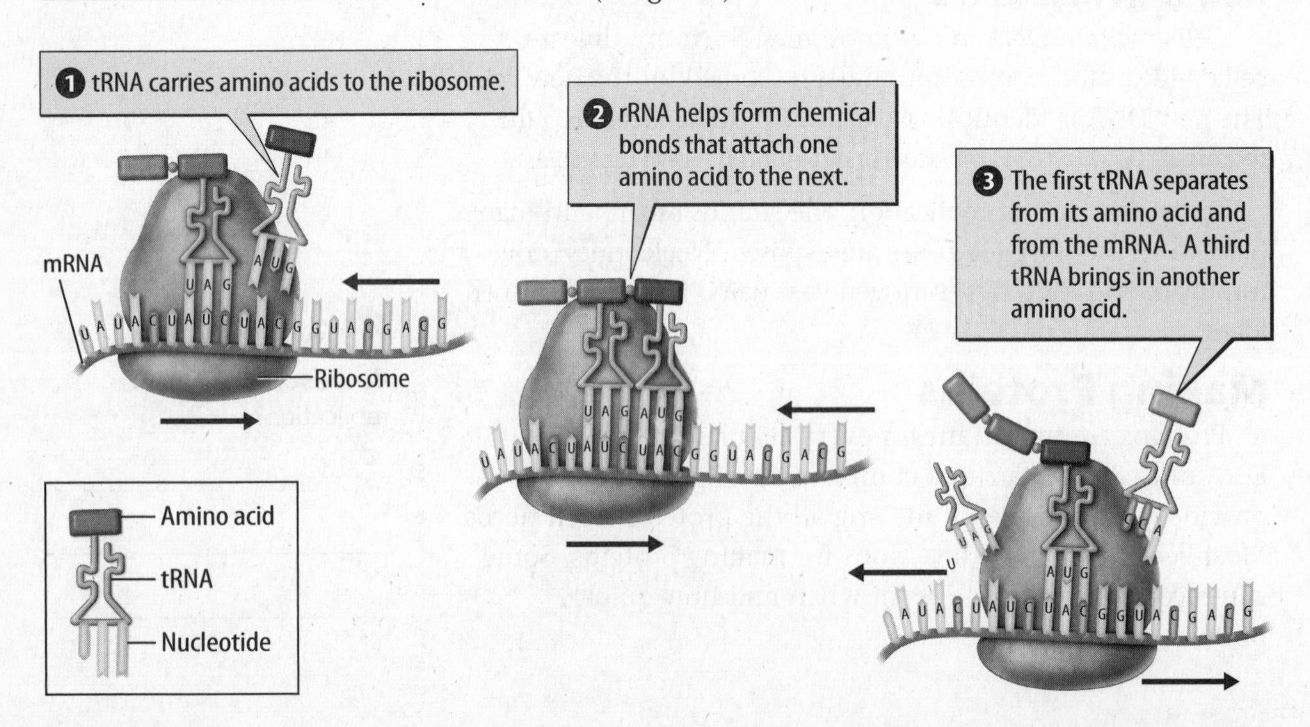

Translating the RNA Code

Making a protein from mRNA is like using a secret code. Proteins are made of amino acids. The order of the nitrogen bases in mRNA determines the order of the amino acids in a protein. Three nitrogen bases on mRNA form the code for one amino acid.

Each series of three nitrogen bases on mRNA is called a codon. There are 64 codons, but only 20 amino acids. Some of the codons code for the same amino acid. One of the codons codes for an amino acid that is the beginning of a protein. This codon signals that translation should start. Three of the codons do not code for any amino acid. Instead, they code for the end of a protein. They signal that translation should stop.

Reading Check
7. Define What is a codon?

Mutations

A change in the nucleotide sequence of a gene is called a **mutation.** Sometimes, mistakes happen during replication. Most mistakes are corrected before replication is finished. An uncorrected mistake can result in a mutation. Mutations can be caused by exposure to X-rays, ultraviolet light, radioactive materials, and some kinds of chemicals.

Think it Over
8. Specify What are some causes of mutations?

Types of Mutations

There are several types of DNA mutations. In a deletion mutation, one or more nitrogen bases are left out of the DNA sequence. In an insertion mutation, one or more nitrogen bases are added to the DNA. In a substitution mutation, one nitrogen base is replaced by a different nitrogen base.

Each type of mutation changes the sequence of nitrogen base pairs. A change can cause a mutated gene to code for a protein that is different from a normal gene. Some mutated genes do not code for any protein. For example, a cell might lose the ability to make one of the proteins it needs.

Results of a Mutation

The effects of a mutation depend on where in the DNA sequence the mutation happens and the type of mutation. Proteins express traits. Because mutations can change proteins, they can cause traits to change. Some mutations in human DNA cause genetic disorders. With more research, scientists hope to find cures and treatments for genetic disorders.

Not all mutations have negative effects. Some mutations do not change proteins, so they do not affect traits. Other mutations can cause a trait to change in a way that benefits an organism.

Key Concept Check
9. Explain How do changes in the sequence of DNA affect traits?

After You Read

Mini Glossary

DNA: an organism's genetic material

mutation: a change in the nucleotide sequence of a gene

nucleotide: a molecule made of a nitrogen base, a sugar, and a phosphate group

replication: the process of copying a DNA molecule to make another DNA molecule

RNA: a type of nucleic acid that carries the code for making proteins from the nucleus to the cytoplasm

transcription: the process of making mRNA from DNA

translation: the process of making a protein from RNA

1. Review the terms and their definitions in the Mini Glossary. Write as many sentences as you need to explain how DNA and RNA are related.

2. Under each heading in the chart below, explain what happens during each type of mutation.

Deletion	Insertion	Substitution
	one or more nitrogen bases are added to the DNA sequence	

3. How did writing a question for each paragraph and then finding the answer in the paragraph help you learn about DNA and genetics?

What do you think NOW?

Reread the statements at the beginning of the lesson. Fill in the After column with an A if you agree with the statement or a D if you disagree. Did you change your mind?

Log on to ConnectED.mcgraw-hill.com and access your textbook to find this lesson's resources.

The Environment and Change Over Time

Fossil Evidence of Evolution

Before You Read

What do you think? Read the two statements below and decide whether you agree or disagree with them. Place an A in the Before column if you agree with the statement or a D if you disagree. After you've read this lesson, reread the statements to see if you have changed your mind.

Before	Statement	After
	1. Original tissues can be preserved as fossils.	
	2. Organisms become extinct only in mass extinction events.	

Key Concepts 

- How do fossils form?
- How do scientists date fossils?
- How are fossils evidence of biological evolution?

Read to Learn

The Fossil Record

An oak tree changes a little when it loses its leaves. A robin changes when it loses some of its feathers. Yet, these living organisms change little from day to day. It might seem as if oak trees and robins have been on Earth forever. If you were to go back a few million years in time, you would not see oak trees or robins. You would see different species of trees and birds. That is because species change over time.

You might already know that fossils are the remains or evidence of once-living organisms. *The* **fossil record** *is made up of all the fossils ever discovered on Earth.* It has millions of fossils that come from many thousands of species. Most of these species are no longer alive on Earth.

The fossil record provides evidence that species have changed over time. Fossils help scientists picture what species looked like. Based on fossil evidence, scientists can re-create the physical appearance of species that are no longer alive.

The fossil record is huge, but it still has many missing parts. Scientists are still looking for more fossils to fill these missing parts. Scientists hypothesize that the fossil record represents only a small fraction of all the organisms that ever lived on Earth.

Study Coach

Vocabulary Quiz Write a question about each vocabulary term in this lesson. Exchange quizzes with a partner. After completing the quizzes, discuss the answers with your partner.

Think it Over

1. Explain Why don't scientists know the exact number of species that have lived on Earth?

SCIENCE USE V. COMMON USE

tissue

Science Use similar cells that work together and perform a function

Common Use a piece of soft, absorbent paper

Reading Check

2. Analyze Why is it rare for soft tissue to become a fossil?

______

______

______

______

Fossil Formation

When animals eat a dead animal, they usually leave little behind. Any soft tissues that are not eaten are broken down by bacteria. Only the hard parts—bones, shells, and teeth—remain. Usually, these parts also break down over time. Sometimes they become fossils. Very rarely, the soft tissues of animals or plants—skin, muscles, or leaves—can also become fossils.

Mineralization

After an organism dies, its body might be buried under mud or sand in a river. Minerals in the water can take the place of the organism's original material and harden into rock. When this happens, a fossil forms. This process is called mineralization. Minerals in water also can fill the small spaces of a dead organism's tissues and become rock. Shells and bones are the most common mineralized fossils. Wood can also become a fossil in this way.

Carbonization

In carbonization, a fossil forms when a dead plant or animal is subjected to pressure over time. Pressure drives off the organism's liquids and gases. Only the carbon outline, or film, of the organism is left behind.

Molds and Casts

Sometimes the shell or bone of a dead animal makes an impression, the outline of its shape, in mud or sand. When the mud or sand hardens, so does the impression. *The impression of an organism in a rock is called a* ***mold.*** Sand or mud can later fill in the mold and harden to form a cast. *A* ***cast*** *is a fossil copy of an organism in a rock.* Molds and casts show only the outside parts of living organisms.

Trace Fossils

Fossils can give clues about the movement or behavior of once-living organisms. *A* ***trace fossil*** *is the preserved evidence of the activity of an organism.* For example, an organism might walk across mud and leave tracks. The tracks can become trace fossils if they fill with mud or sand that later hardens.

Key Concept Check

3. List the different ways fossils can form.

______

______

______

______

______

Original Material

In rare cases, the actual parts of an organism can be preserved. Some original-material fossils include mammoths frozen in ice and saber-toothed cats preserved in tar pits. Even the bodies of ancient humans have been found in bogs. Some insects were preserved when they got stuck in sap that hardened into amber.

Determining a Fossil's Age

Scientists cannot date most fossils directly. Instead, they find the age of the rocks in which the fossils are found.

Relative-Age Dating

You might be younger than a brother but older than a sister. This is your relative age. In the same way, a rock is either older or younger than rocks near it. In relative-age dating, scientists find the order in which rock layers formed. Some layers of rock have not moved since they formed. In these layers, scientists know that the bottom layers are older than the top layers, as shown in the figure below. Knowing the order in which the rocks formed helps scientists date the fossils in them. In this way, they can find the relative order in which species have appeared on Earth over time.

Dating Fossils

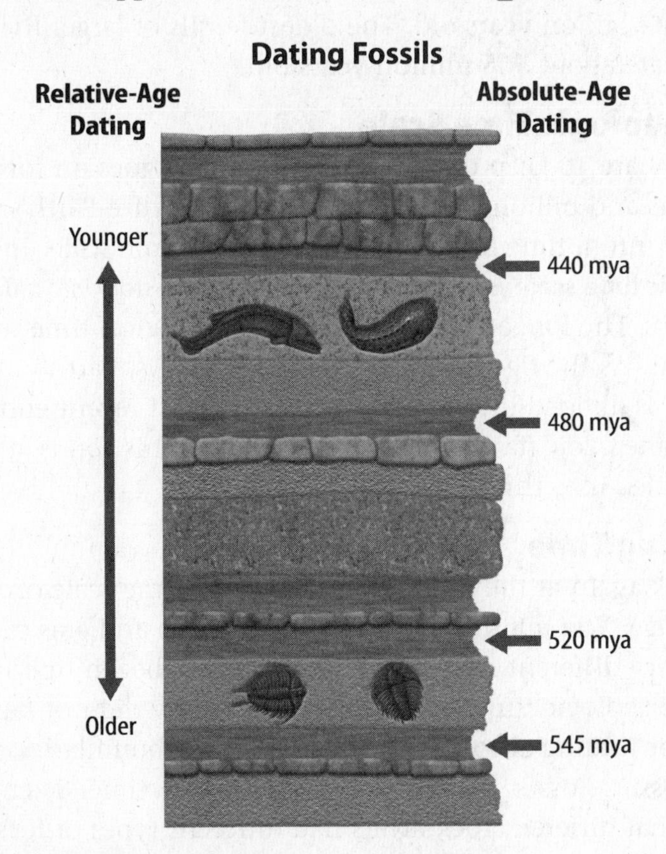

Absolute-Age Dating

Absolute-age dating is more exact than relative-age dating. A rock's absolute age is its age in years. To find absolute age, scientists use radioactive decay, which is a natural process that happens at a known rate. In radioactive decay, unstable isotopes in rocks change into stable isotopes over time. Scientists measure the ratio of unstable isotopes to stable isotopes to find the age of a rock. This ratio is best measured in igneous rocks, as shown above.

FOLDABLES

Make a small shutterfold book. Label it as shown. Under the left tab, describe relative-age dating. Under the right tab, describe absolute-age dating.

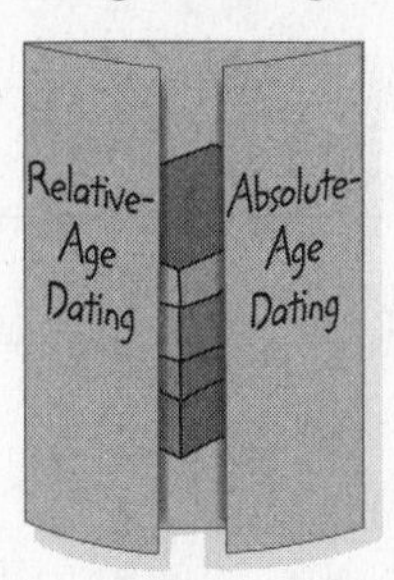

Key Concept Check

4. Explain How does relative-age dating help scientists learn about fossils?

____

____

____

____

Visual Check

5. Estimate What is the estimated age of the trilobite fossils (bottom layer of fossils)?

____

REVIEW VOCABULARY

isotopes
atoms of the same element that have different numbers of neutrons

6. Analyze Why is it difficult for scientists to absolute-age date sedimentary rock formations using isotopes?

Dating Igneous Rock Absolute age is easiest to determine in igneous rocks. Igneous rocks form from volcanic magma. Magma is so hot that it is rare for parts of organisms in it to remain and form fossils.

Dating Sedimentary Rock Most fossils form in mud and sand, which become sedimentary rock. To measure the age of a sedimentary rock layer, scientists find the ages of igneous layers above and below it. They can estimate an age in between these ages for the fossils found in the sedimentary layer. Absolute-age dating is illustrated in the figure on the previous page.

Fossils over Time

How old do you think Earth's oldest fossils are? Evidence of microscopic, unicellular organisms has been found in rocks 3.4 billion years old. The oldest fossils of larger living things are about 565 million years old.

The Geologic Time Scale

It is hard to keep track of something that goes on for millions and billions of years. Scientists organize Earth's history into a time line called the geologic time scale. *The* ***geologic time scale*** *is a chart that divides Earth's history into different time units.* The longest time units in the geological time scale are eons. As the figure on the next page shows, Earth's history is divided into four eons. Earth's most recent eon is the Phanerozoic (fa nuh ruh ZOH ihk) eon. This eon is subdivided into three eras.

Reading Check

7. Describe What is the geologic time scale?

Dividing Time

Look again at the figure of the geologic time scale on the next page. You might have noticed that eons and eras can have very different lengths. When scientists began figuring out the geologic time scale in the 1800s, they did not have ways for finding absolute age. To mark time boundaries, they used fossils. Fossils were an easy way to mark time. Scientists knew that different rock layers had different types of fossils. Some of the fossils scientists use to mark the time boundaries are shown in the figure.

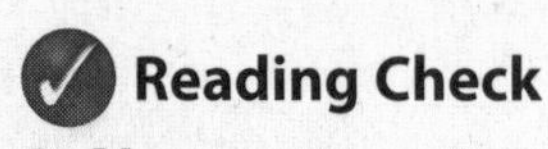

8. Name What do scientists use to mark boundaries in the geologic time scale?

Often, a type of fossil found in one rock layer was not in layers above it. Even more surprising, entire groups of fossils found in one layer were sometimes missing from layers above them. It seemed as if whole communities of living organisms had suddenly disappeared. What could have caused them to disappear?

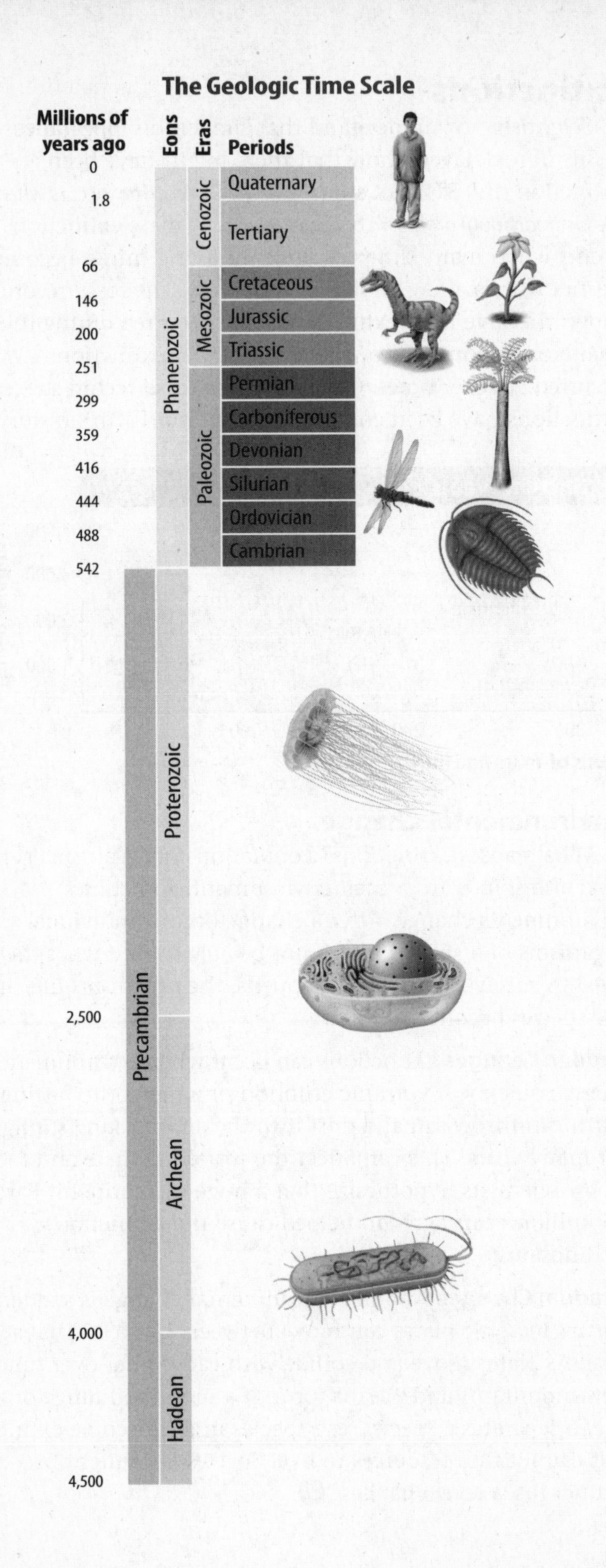

Visual Check

9. Sequence List the following living organisms in the order in which they first appeared on Earth: humans, insects, single cells, dinosaurs.

____

____

____

____

Math Skills

Numbers that refer to the ages of Earth's fossils are very large, so scientists use scientific notation to work with them. For example, mammals appeared on Earth about 200 mya or 200,000,000 years ago. Change this number to scientific notation using the following process.

Move the decimal point until only one nonzero digit remains on the left.

$200{,}000{,}000 = 2.00000000$

Count the number of places you moved the decimal point (8) and use that number as a power of ten.

$200{,}000{,}000 = 2.0 \times 10^8$ years.

10. Use Scientific Notation The first vertebrates appeared on Earth about 490,000,000 years ago. Express this time in scientific notation.

____

Visual Check

11. Analyze When did the most recent mass extinction happen? Circle the name of the time period on the graph. About how long ago did it happen?

____

Extinctions

Scientists now understand that sudden disappearances of fossils in rock layers show that there might have been an extinction (ihk STINGK shun) event. **Extinction** *occurs when the last individual organism of a species dies.* A mass extinction occurs when many different kinds of living things become extinct within a few million years or less. The fossil record shows that five mass extinctions have occurred during the Phanerozoic eon, as shown below. Smaller extinctions occurred at other times. Clues from the fossil record suggest extinctions have been common throughout Earth's history.

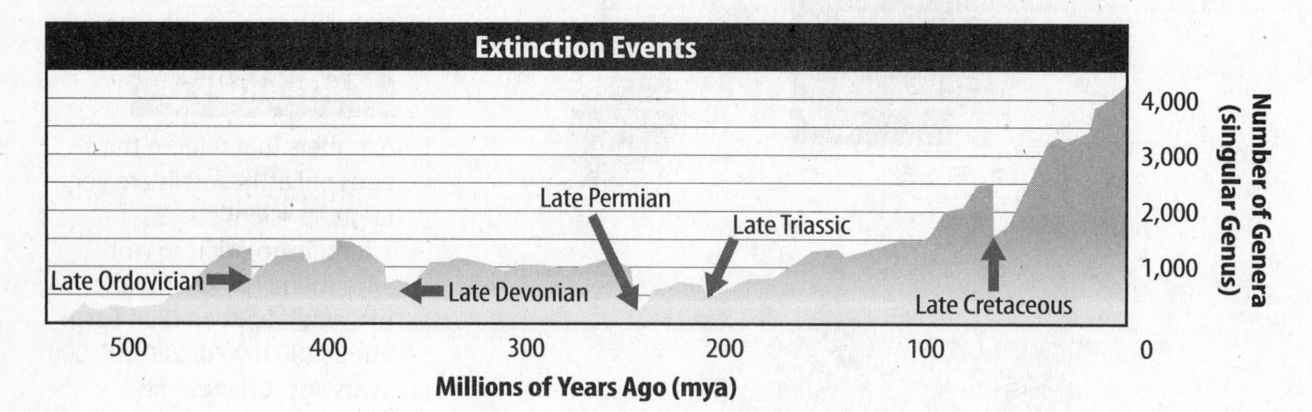

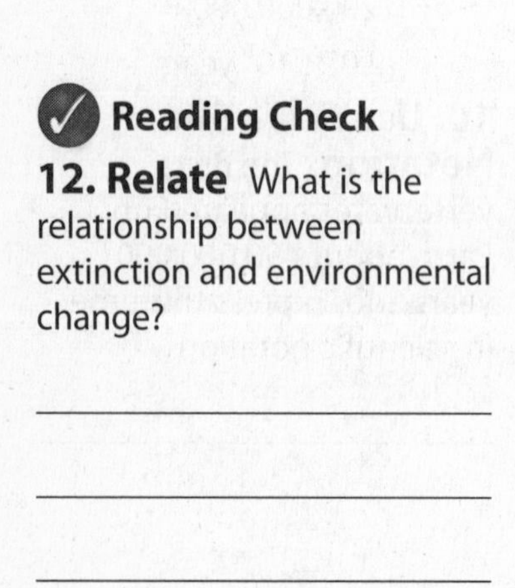

Reading Check

12. Relate What is the relationship between extinction and environmental change?

____

____

____

____

____

Environmental Change

What causes extinctions? Populations of organisms get food and shelter from their environment. Sometimes environments change. After a change occurs, individual organisms of a species might not be able to find what they need to survive. When this happens, the organisms die, and the species becomes extinct.

Sudden Changes Extinctions can occur when environments change quickly. A volcanic eruption or a meteorite hitting Earth can throw ash and dust into the air, blocking sunlight for many years. This can affect the world's climate and food webs. Scientists hypothesize that a huge meteorite hit Earth 65 million years ago and helped cause the extinction of dinosaurs.

Gradual Changes Not all environmental change is sudden. Earth's tectonic plates can move between 1 and 15 cm each year. As plates move and collide with each other over time, new mountains and oceans form. If a mountain range or an ocean separates a species, the species might become extinct if it cannot find resources to live. Species also might become extinct if sea level changes. ✓

Extinctions and Evolution

The fossil record has obvious clues about the extinction of species over time. But it also has clues about the appearance of many new species. How do new species form?

Many early scientists thought that each species appeared on Earth independently of every other species. However, as scientists found more fossils, they began to see patterns in the fossil record. Many fossil species in nearby rock layers had similar body plans and similar body parts. These similar species seemed to be related to each other. For example, the series of horse fossils in the figure below suggests that the modern horse is related to other extinct species.

These species changed over time in what appeared to be a sequence. Change over time is evolution. **Biological evolution** *is the change over time in populations of related organisms.* Charles Darwin developed a theory about how species evolve from other species. You will read about Darwin's theory in the next lesson.

Horse Fossils

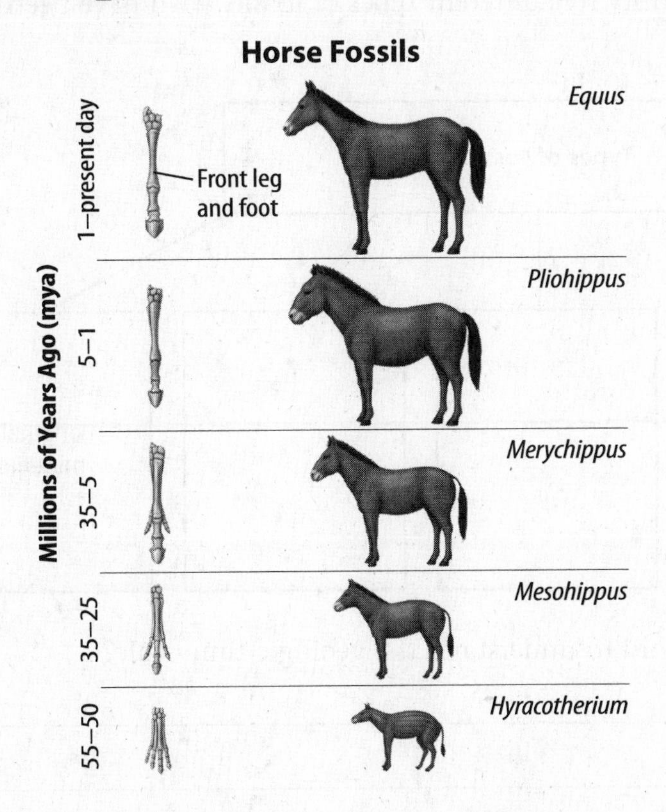

Reading Check

13. State What patterns did scientists find in the fossil record?

Key Concept Check

14. Analyze How are fossils evidence of biological evolution?

Visual Check

15. Evaluate In what ways did the horse change over time?

After You Read

Mini Glossary

biological evolution: the change over time in populations of related organisms

cast: a fossil copy of an organism in a rock

extinction (ihk STINGK shun): when the last individual organism of a species dies

fossil record: all the fossils ever discovered on Earth

geologic time scale: a chart that divides Earth's history into different time units

mold: the impression of an organism in a rock

trace fossil: the preserved evidence of the activity of an organism

1. Review the terms and their definitions in the Mini Glossary. Write a sentence that explains the difference between a cast and a mold.

2. Fill in the graphic organizer to identify five different types of fossils. Two have been done for you.

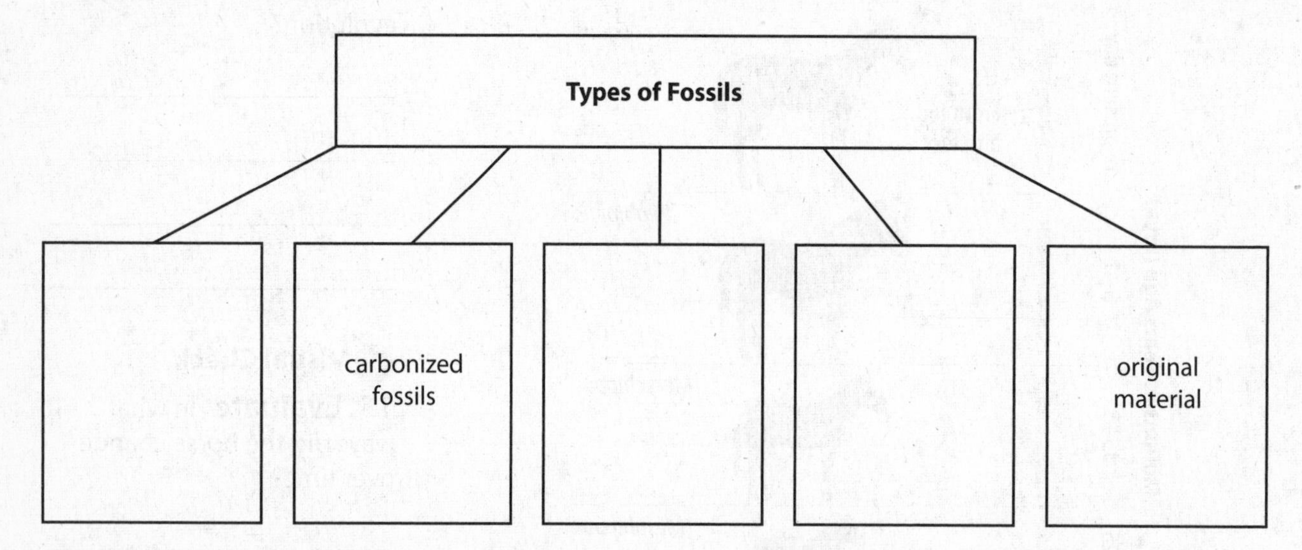

3. How do scientists use the fossil record to understand the geologic time scale?

What do you think NOW?

Reread the statements at the beginning of the lesson. Fill in the After column with an A if you agree with the statement or a D if you disagree. Did you change your mind?

Log on to ConnectED.mcgraw-hill.com and access your textbook to find this lesson's resources.

CHAPTER 6
LESSON 2

The Environment and Change Over Time

Theory of Evolution by Natural Selection

Before You Read

What do you think? Read the two statements below and decide whether you agree or disagree with them. Place an A in the Before column if you agree with the statement or a D if you disagree. After you've read this lesson, reread the statements to see if you have changed your mind.

Before	Statement	After
	3. Environmental change causes variations in populations.	
	4. Variations can lead to adaptations.	

Key Concepts 

- Who was Charles Darwin?
- How does Darwin's theory of evolution by natural selection explain how species change over time?
- How are adaptations evidence of natural selection?

Read to Learn

Charles Darwin

How many species of birds can you name? Robins, penguins, and chickens are a few. There are about 10,000 species of birds on Earth today. Each species has wings, feathers, and beaks. Scientists hypothesize that all birds evolved from an earlier, or ancestral, group of birdlike organisms. As this group evolved into different species, birds developed different sizes, colors, songs, and ways of eating. Yet, they kept their key bird traits.

How do species evolve? Charles Darwin, a scientist, worked to answer this question. Darwin was an English naturalist who, in the mid-1800s, developed a theory of how evolution works. *A* **naturalist** *is a person who studies plants and animals by observing them.* Darwin spent years studying plants and animals in nature before developing his theory. Recall that a theory is an explanation of the natural world that is well supported by evidence. Darwin's theory of evolution was not the first, but his theory is the one best supported by evidence today.

Study Coach

Make Flash Cards Think of a quiz question for each paragraph. Write the question on one side of a flash card. Write the answer on the other side. Work with a partner to quiz each other using your flash cards.

Key Concept Check

1. Describe Who was Charles Darwin?

Voyage of the *Beagle*

Darwin worked as a naturalist on the HMS *Beagle,* a ship of the British navy. During his trip around the world, Darwin observed and collected many plants and animals.

FOLDABLES

Make a small four-door shutterfold book. Use it to investigate the who, what, when, and where of Charles Darwin, the Galápagos Islands, and the theory of evolution by natural selection.

 **Reading Check**

2. Explain What made Darwin become curious about the organisms that lived on the Galápagos Islands?

____

____

____

____

ACADEMIC VOCABULARY
convince
(verb) to overcome by argument

The Galápagos Islands

Darwin was interested in the organisms he saw on the Galápagos (guh LAH puh gus) Islands. These islands are 1,000 km off the South American coast in the Pacific Ocean. Darwin saw that each island had a slightly different environment. Some were dry. Some were more humid. Others had mixed environments.

Tortoises Darwin saw that the giant tortoises on each island looked different. On one island, tortoises had shells that came close to their necks. They could eat only short plants. On other islands, tortoises had more space between the shell and neck. They could eat taller plants.

Mockingbirds and Finches Darwin was also curious about the different mockingbirds and finches he saw. Like the tortoises, different types of mockingbirds and finches lived in different island environments. Later, he was surprised to find that many were different enough to be separate species.

Darwin's Theory

Darwin discovered a relationship between each species and the food found on the island where it lived. Tortoises with long necks lived on islands that had tall cacti. Their long necks made it possible for them to reach high to eat the cacti. The tortoises with short necks lived on islands that had plenty of short grass.

Common Ancestors

Darwin became convinced that all the tortoise species were related. He thought they all shared a common ancestor. He suggested that millions of years before, a storm had carried a group of tortoises to one of the islands from South America. In time, the tortoises spread to the other islands. Their neck lengths and shell shapes changed to match their islands' food sources. How did this happen?

Variations

Darwin knew that individual members of a species have slight differences, or variations. *A* ***variation*** *is a slight difference in the appearance of individual members of a species.* Variations arise naturally in populations. They occur in the offspring as a result of sexual reproduction. You might recall that variations are caused by random mutations, or changes, in genes. Mutations can lead to changes in phenotype. Recall that an organism's phenotype is all of the observable traits and characteristics of the organism. Genetic changes to phenotype can be passed on to future generations.

Natural Selection

Darwin did not know about genes. But he saw that variations were the key to how evolution worked. He knew that there was not enough food on each island to feed every tortoise that was born. Tortoises had to compete for food. As the tortoises spread to the different islands, some were born with random variations in neck length. If a variation helped a tortoise compete for food, the tortoise lived longer than other tortoises without the variation. Because it lived longer, it reproduced more. It passed on the helpful variation to its offspring.

This is Darwin's theory of evolution by natural selection. **Natural selection** *is the process by which populations of organisms with variations that help them survive in their environments live longer, compete better, and reproduce more than those that do not have the variations.* Natural selection explains how Galápagos tortoises became matched to their food sources, as shown below. It also explains why there were so many different kinds of Galápagos finches and mockingbirds. Birds with beak variations that helped them compete for food lived longer and reproduced more.

Key Concept Check

3. Analyze What role do variations have in the theory of evolution by natural selection?

____

____

____

Visual Check

4. Illustrate Mark all the tortoises in the figure that have short necks with the letter *S*. Mark those that have long necks with the letter *L*. What trend do you see over time?

____

____

Natural Selection

1 Reproduction A population of tortoises produces many offspring that inherit its characteristics.

2 Variation A tortoise is born with a variation that makes its neck slightly longer.

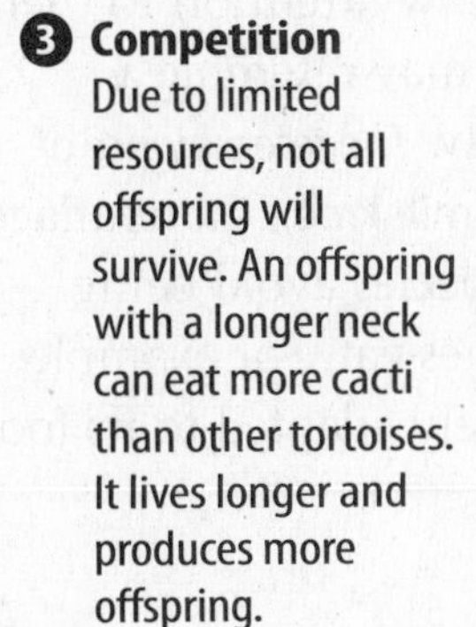

3 Competition Due to limited resources, not all offspring will survive. An offspring with a longer neck can eat more cacti than other tortoises. It lives longer and produces more offspring.

4 Selection Over time, the variation is inherited by more and more offspring. Eventually, all tortoises have longer necks.

Adaptations

Natural selection explains how all species change over time as their environments change. Through natural selection, a helpful variation in one individual can eventually pass to future members of a population.

As time passes, more variations come about. The buildup of many similar variations can lead to an adaptation (a dap TAY shun). *An* **adaptation** *is a characteristic of a species that enables the species to survive in its environment.* The long neck of certain species of tortoises is an adaptation to an environment with tall cacti.

Key Concept Check

5. Explain How do variations lead to adaptations?

____

____

____

____

Types of Adaptations

Every species has many adaptations. Scientists classify adaptations into three categories: structural, behavioral, and functional.

Structural Adaptations These adaptations involve color, shape, and other physical characteristics. The shape of a tortoise's neck is a structural adaptation.

Behavioral Adaptations The way an organism behaves or acts is a behavioral adaptation. Hunting at night and moving in herds are behavioral adaptations.

Think it Over

6. Apply An opossum will play dead when a predator frightens it. That way the predator might think it is not good food and will leave it alone. What kind of adaptation is this? (Circle the correct answer.)

a. structural

b. behavioral

c. functional

Functional Adaptations The last category is functional adaptations. These adaptations involve chemical changes in body systems. A drop in body temperature during hibernation is a functional adaptation.

Environmental Interactions

Many species have evolved adaptations that make them nearly invisible. For example, a seahorse may be the same color as and similar in texture to the coral it rests on. This is a structural adaptation called camouflage (KAM uh flahj). **Camouflage** *is an adaptation that enables species to blend in with their environments.*

Some species have adaptations that draw attention to them or make them more visible. A caterpillar may resemble a snake. Predators see it and are scared away. *The resemblance of one species to another species is* **mimicry** (MIH mih kree). Camouflage and mimicry are adaptations that help species avoid being eaten. Many other adaptations help species eat. For example, the pelican has a beak and mouth uniquely adapted to its food source—fish.

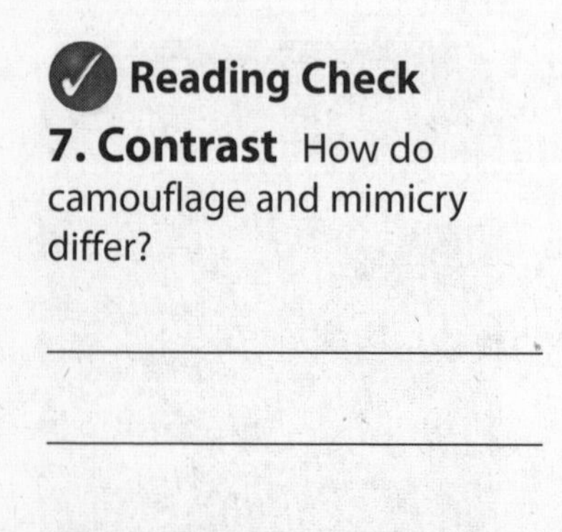

Reading Check

7. Contrast How do camouflage and mimicry differ?

____

____

____

____

Role of Environment Environments are complex. Species must adapt to an environment's living parts as well as to its nonliving parts. Some nonliving things are temperature, water, nutrients in soil, and climate. Deciduous trees shed their leaves due to changes in climate. Camouflage, mimicry, and mouth shape are adaptations mostly to an environment's living parts.

Extinct Species Living and nonliving factors are always changing. Even slight environmental changes affect how species adapt. If a species is unable to adapt, it becomes extinct. The fossil record contains many fossils of species unable to adapt to change.

Artificial Selection

Adaptations show how closely Earth's species match their environments. This is exactly what Darwin's theory of evolution by natural selection predicted. Darwin gave many examples of adaptation in *On the Origin of Species,* the book he wrote to explain his theory. Darwin wrote his book 20 years after he developed his theory. He spent those years collecting more evidence for his theory.

Darwin also had a hobby of breeding pigeons. He bred pigeons of different colors and shapes. In this way, he produced new, fancy varieties. *The breeding of organisms for desired characteristics is called* ***selective breeding.*** Like many plants and animals produced from selective breeding, pigeons look different from their ancestors.

Darwin saw that changes caused by selective breeding were much like changes caused by natural selection. Instead of nature selecting variations, humans selected them. Darwin called this process artificial selection. ✓

Artificial selection explains and supports Darwin's theory. In Lesson 3, you will read about other evidence that supports the idea that species evolve from other species.

Think it Over

8. Synthesize How do you think some fur-bearing species might adapt to a gradual change in climate in which global temperature increased?

____

____

____

✓ Reading Check

9. Compare How are artificial selection and natural selection alike?

____

____

____

After You Read

Mini Glossary

adaptation (a dap TAY shun): a characteristic of a species that enables the species to survive in its environment

camouflage (KAM uh flahj): an adaptation that enables species to blend in with their environments

mimicry (MIH mih kree): the resemblance of one species to another species

naturalist: a person who studies plants and animals by observing them

natural selection: the process by which populations of organisms with variations that help them survive in their environments live longer, compete better, and reproduce more than those that do not have the variations

selective breeding: the breeding of organisms for desired characteristics

variation: a slight difference in the appearance of individual members of a species

1. Review the terms and their definitions in the Mini Glossary. Describe a living organism that depends on camouflage or mimicry to survive.

2. Write a letter in each box to show the correct sequence that demonstrates the process of natural selection.
   a. Birds eat more light green beetles. Dark green beetles live longer and reproduce more.
   b. A beetle is born with a variation in its color: It is dark green.
   c. Over time, all beetles in the environment are dark green.
   d. A population of beetles is light green. They stand out against dark green leaves.

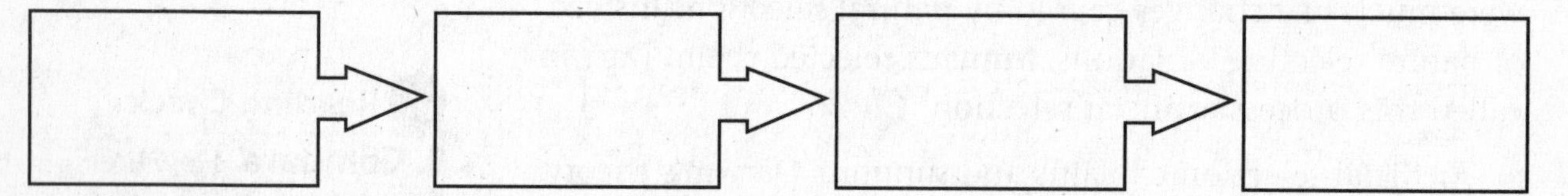

3. Compare selective breeding and evolution.

What do you think NOW?

Reread the statements at the beginning of the lesson. Fill in the After column with an A if you agree with the statement or a D if you disagree. Did you change your mind?

Log on to ConnectED.mcgraw-hill.com and access your textbook to find this lesson's resources.

The Environment and Change Over Time

Biological Evidence of Evolution

Before You Read

What do you think? Read the two statements below and decide whether you agree or disagree with them. Place an A in the Before column if you agree with the statement or a D if you disagree. After you've read this lesson, reread the statements to see if you have changed your mind.

Before	Statement	After
	5. Living species contain no evidence that they are related to each other.	
	6. Plants and animals share similar genes.	

Key Concepts 

- What evidence from living species supports the theory that species descended from other species over time?
- How are Earth's organisms related?

Read to Learn

Evidence for Evolution

The pictures of horse fossils in Lesson 1 seem to show that horses evolved in a straight line. That is, one species replaced another in a series of orderly steps. Evolution does not occur this way. Different horse species were sometimes alive at the same time. They are related to one another because each descended from a common ancestor.

Living species that are closely related share a close common ancestor. How closely they are related depends on how closely in time they diverged, or split, from that ancestor. Evidence of common ancestors can be found in the fossil record and in living organisms.

Comparative Anatomy

It is easy to see that some species evolved from a common ancestor. For example, robins, finches, and hawks have similar body parts. They all have feathers, wings, and beaks. The same is true for tigers, leopards, and house cats. But how are hawks related to cats?

Studying the structural and functional similarities and differences in species that do not look alike can show the relationships. *The study of similarities and differences among structures of living species is called* **comparative anatomy.**

Mark the Text

Identify Main Ideas Highlight the main idea of each paragraph. Highlight two details that support each main idea with a different color. Use your highlighted copy to review what you studied in this lesson.

FOLDABLES®

Make a table with five rows and three columns. Label the rows and columns of the table as shown below. Give your table a title.

	Explanation	Example
Comparative Anatomy		
Vestigial Structures		
Developmental Biology		
Molecular Biology		

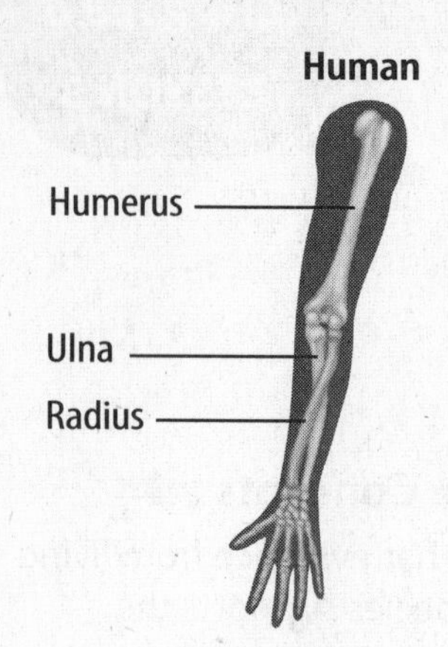

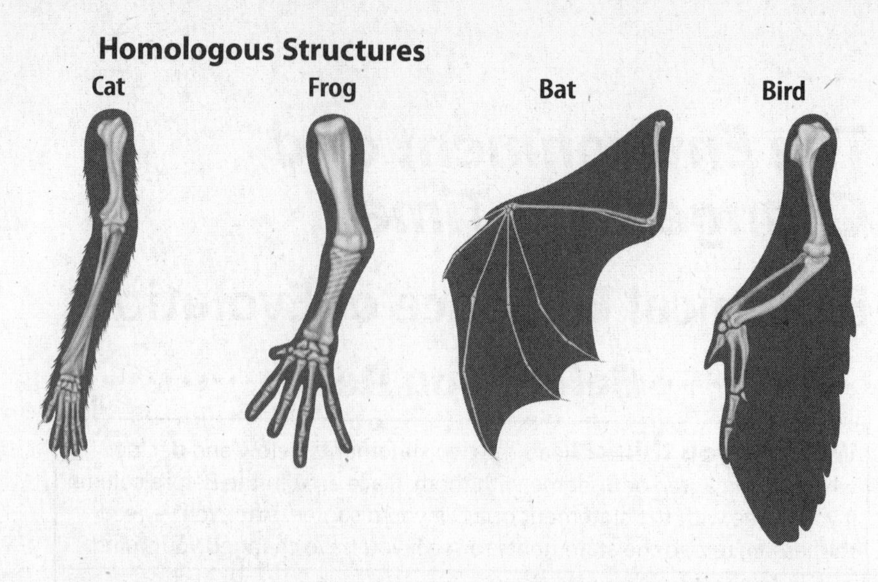

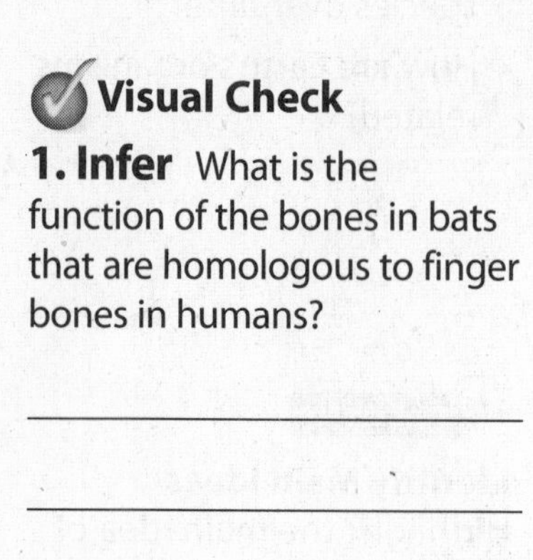

Visual Check

1. Infer What is the function of the bones in bats that are homologous to finger bones in humans?

Key Concept Check

2. Explain How do homologous structures provide evidence for evolution?

Homologous Structures Humans, cats, frogs, bats, and birds look different and move in different ways. Humans use their arms for balance and their hands to grasp objects. Cats use their forelimbs to walk, run, and jump. Frogs use their forelimbs to jump. The forelimbs of bats and birds are wings and are used for flying. However, the forelimb bones of all these species show similar patterns, as shown in the figure above. The forelimbs of the species in the figure are different sizes, but their placement and structure suggest common ancestry.

Homologous (huh MAH luh gus) **structures** *are body parts of organisms that are similar in structure and position but different in function.* Homologous structures, such as the forelimbs of humans, cats, frogs, bats, and birds, suggest that these species are related. The more alike two structures are, the more likely it is that the species have evolved from a recent common ancestor.

Analogous Structures Can you think of a body part in two species that does the same job but differs in structure? How about the wings of birds and flies? The wings in both species are used for flight. But bird wings are covered with feathers. Fly wings are covered with tiny hairs. Though used for the same function—flight—the wings of birds and insects are too different in structure to suggest close common ancestry.

Bird wings and fly wings are analogous (uh NAH luh gus) structures. **Analogous structures** *are body parts that perform a similar function but differ in structure.* The differences in wing structure show that birds and flies are not closely related.

Vestigial Structures

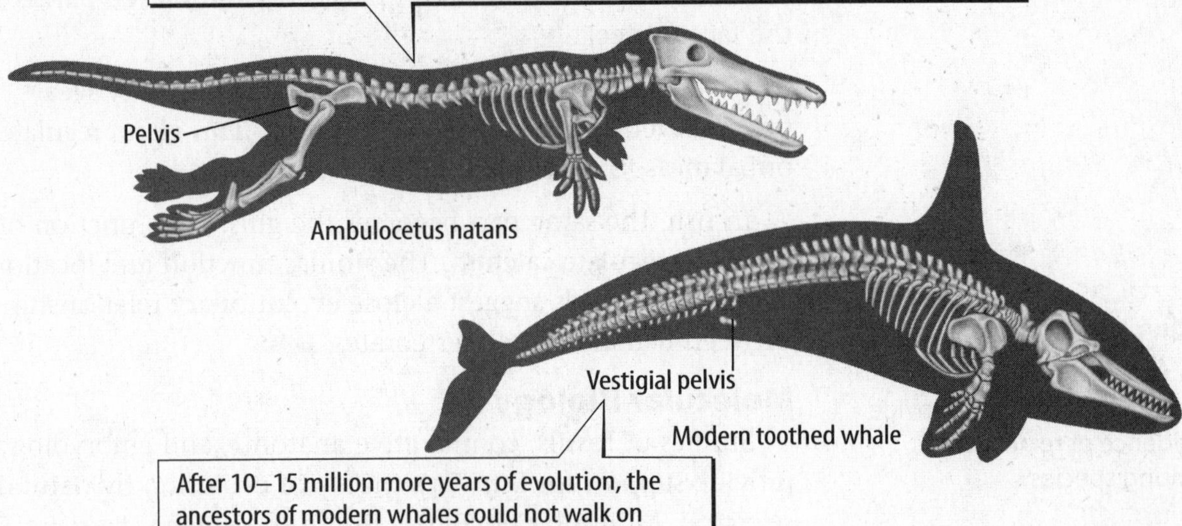

Vestigial Structures

Ostriches have wings. Yet they cannot fly. An ostrich's wings are an example of vestigial structures. **Vestigial** (veh STIH jee ul) **structures** *are body parts that have lost their original function through evolution.* The best explanation for vestigial structures is that the species with a vestigial structure is related to an ancestral species that used the structure for a specific purpose.

The whale shown in the figure above has tiny pelvic bones inside its body. Pelvic bones are hip bones, which in many species attach the leg bones to the body. Modern whales do not have legs. The pelvic bones in whales suggest that whales came from ancestors that used legs for walking on land. The fossil evidence supports this conclusion. Many fossils of whale ancestors show a slow loss of legs over millions of years. They also show, at the same time, that whale ancestors became better adapted to their watery environments.

Developmental Biology

Studying the internal structures of living organisms is not the only way that scientists learn about common ancestors. Studying how embryos develop can also show how species are related. *The science of the development of embryos from fertilization to birth is called* **embryology** (em bree AH luh jee).

Visual Check

3. Infer Why does a vestigial pelvis show that the ancestors of the modern whale once had legs?

Key Concept Check

4. Explain How are vestigial structures evidence of descent from ancestral species?

Pharyngeal Pouches Embryos of different species often look like each other at different stages of their growth. For example, all vertebrate embryos have pharyngeal (fuh rihn JEE ul) pouches at one stage. These pouches become different body parts in each vertebrate. Yet, in all vertebrates, each part is in the face or neck.

In reptiles, birds, and humans, part of the pharyngeal pouch develops into a gland in the neck. This gland regulates, or balances, the body's calcium levels.

In fish, the same part becomes the gills. One function of gills is to regulate calcium. The similar function and location of gills and glands suggest a close evolutionary relationship between fish and other vertebrates.

Key Concept Check
5. Analyze How do pharyngeal pouches provide evidence of relationships among species?

Molecular Biology

Studies of fossils, comparative anatomy, and embryology provide support for Darwin's theory of evolution by natural selection. Molecular biology is the study of gene structure and function.

Discoveries in molecular biology have confirmed and extended much of the data already collected about the theory of evolution. Darwin did not know about genes, but scientists today know that mutations in genes are the source of variations upon which natural selection acts. Genes provide powerful support for evolution.

Reading Check
6. Describe What is molecular biology?

Comparing Sequences All living organisms have genes. All genes are made of DNA, and all genes work in similar ways. This supports the idea that all living organisms are related.

Scientists can study how living organisms are related by comparing their genes. For example, nearly all organisms have a gene for cytochrome *c*, a protein required for cellular respiration. Some species, such as humans and rhesus monkeys, have nearly identical cytochrome *c*. The more closely related two species are, the more similar their genes and proteins are.

Key Concept Check
7. Explain How is molecular biology used to determine relationships among species?

Divergence Scientists have found that some stretches of shared DNA mutate at regular, predictable rates. Scientists use this "molecular clock" to estimate when in the past living species split from common ancestors. This is how scientists have shown that whales and porpoises are more closely related to hippopotamuses than they are to other living things. Whales and hippopotamuses share an ancestor that lived 50–60 million years ago.

The Study of Evolution Today

The theory of evolution by natural selection is the cornerstone of modern biology. Since Darwin published his theory, scientists have confirmed, refined, and extended his work. They have observed natural selection in hundreds of living species. Their studies of fossils, anatomy, embryology, and molecular biology have shown relationships among living and extinct species. ✓

How New Species Form

New evidence supporting the theory of evolution by natural selection is discovered nearly every day. But scientists debate some of the details. The figure below shows how scientists have different ideas about the rate at which natural selection produces new species. Some say it works slowly and gradually. Others say it works quickly, in bursts. How different species first came about is difficult to study on human time scales. It is also difficult to study with the incomplete fossil record. Yet, new fossils that fill in the holes are discovered all the time. Further fossil discoveries will help scientists study more details about the origin of new species. ✓

Diversity

Evolution has produced Earth's wide diversity of living things using the same basic building blocks called genes. This is an active area of study in evolutionary biology. Scientists are finding that genes can be reorganized in simple ways and give rise to dramatic changes in organisms. Scientists now study evolution by looking at molecules. Yet, they still use the same basic ideas that Darwin came up with over 150 years ago.

Rates of Evolution

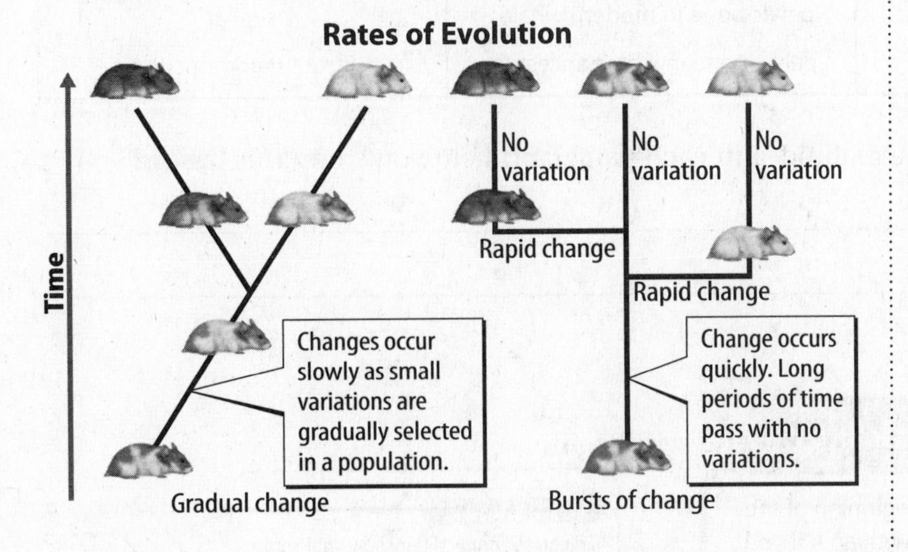

✓ **Reading Check**

8. Connect What is the connection between modern biology and the theory of evolution by natural selection?

✓ **Reading Check**

9. Describe a difference of opinion in regard to how scientists interpret the theory of evolution by natural selection.

✓ **Visual Check**

10. Analyze What does a flat (horizontal) line mean in the figure? (Circle the correct answer.)

a. gradual change

b. no variation

c. rapid change

After You Read

Mini Glossary

analogous (uh NAH luh gus) structure: a body part that performs a similar function to the body part of another organism, though it differs in structure

comparative anatomy: the study of similarities and differences among structures of living species

embryology (em bree AH luh jee): the science of the development of embryos from fertilization to birth

homologous (huh MAH luh gus) structure: a body part that is similar in structure and position to the body part of another organism, though it has a different function

vestigial (veh STIH jee ul) structure: a body part that has lost its original function through evolution

1. Review the terms and their definitions in the Mini Glossary. Use one of the terms to write your own sentence.

____

____

2. Use what you have learned about analogous, homologous, and vestigial structures to complete the table. The last row has been completed for you.

Structures	Example Pair of Structures	Similar Structure or Function (circle one)
Analogous		similar structure or function
Homologous		similar structure or function
Vestigial	pelvic bone in modern whales pelvic bone in whale ancestors	similar (structure) or function

3. How did highlighting the main idea in each paragraph help you study this lesson?

____

____

What do you think NOW?

Reread the statements at the beginning of the lesson. Fill in the After column with an A if you agree with the statement or a D if you disagree. Did you change your mind?

Log on to ConnectED.mcgraw-hill.com and access your textbook to find this lesson's resources.

Human Body Systems

Transport and Defense

Before You Read

What do you think? Read the two statements below and decide whether you agree or disagree with them. Place an A in the Before column if you agree with the statement or a D if you disagree. After you've read this lesson, reread the statements to see if you have changed your mind.

Before	Statement	After
	1. A human body has organ systems that carry out specific functions.	
	2. The body protects itself from disease.	

Key Concepts 

- How does food enter and leave the body?
- How do nutrients travel through the body?
- How does the body defend itself from harmful invaders?

Read to Learn

The Body's Organization

Libraries have thousands of books grouped together by subject. Grouping books by subject helps keep them organized and easier to find. Your body's organization helps it function.

All organisms have different parts with special functions. Cells are the basic unit of all living organisms. Organized groups of cells that work together are tissues. Groups of tissues that perform a specific function are organs. *Groups of organs that work together and perform a specific task are* **organ systems.** Organ systems provide movement, transport (carry) substances, and perform many other functions.

Organ systems work together and maintain **homeostasis** (hoh mee oh STAY sus), *or steady internal conditions when external conditions change.* When you exercise, your body uses stored energy. Your body releases excess energy as thermal energy. Sweat, also called perspiration (pur spuh RAY shun), helps the body release thermal energy and maintain homeostasis.

Study Coach

Building Vocabulary Work with another student to write a question about each vocabulary term in this lesson. Answer the questions and compare your answers. Reread the text to clarify the meaning of the terms.

Digestion and Excretion

Humans need food, water, and oxygen to survive. Food contains energy that is processed by the body. The process by which food is broken down is called digestion. After digestion, substances the body does not use are removed through excretion (ihk SKREE shun).

Reading Check

1. Define What is digestion?

____

____

____

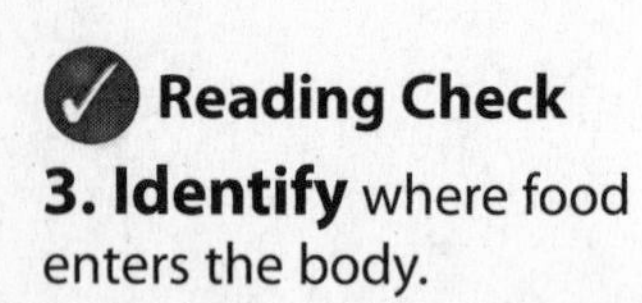

FOLDABLES®

Make three two-tab books, label them with the body systems in this lesson, and use them to organize information about each body system.

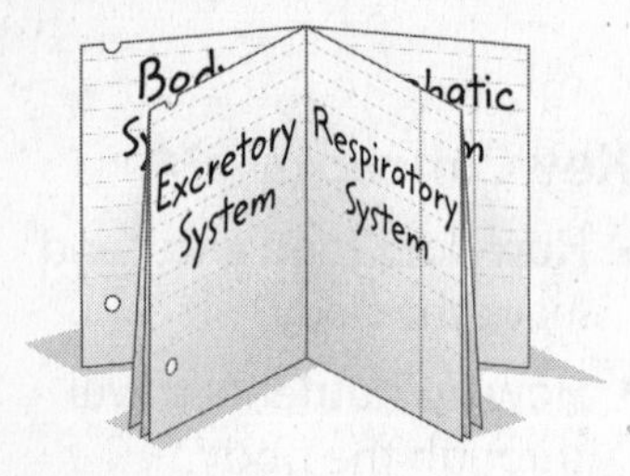

Visual Check

2. Name the organs in the digestive system.

Reading Check

3. Identify where food enters the body.

The Digestive System

The digestive system is made up of several organs. The organs are shown in the figure below. Food and water enter the digestive system through the mouth.

The Digestive System

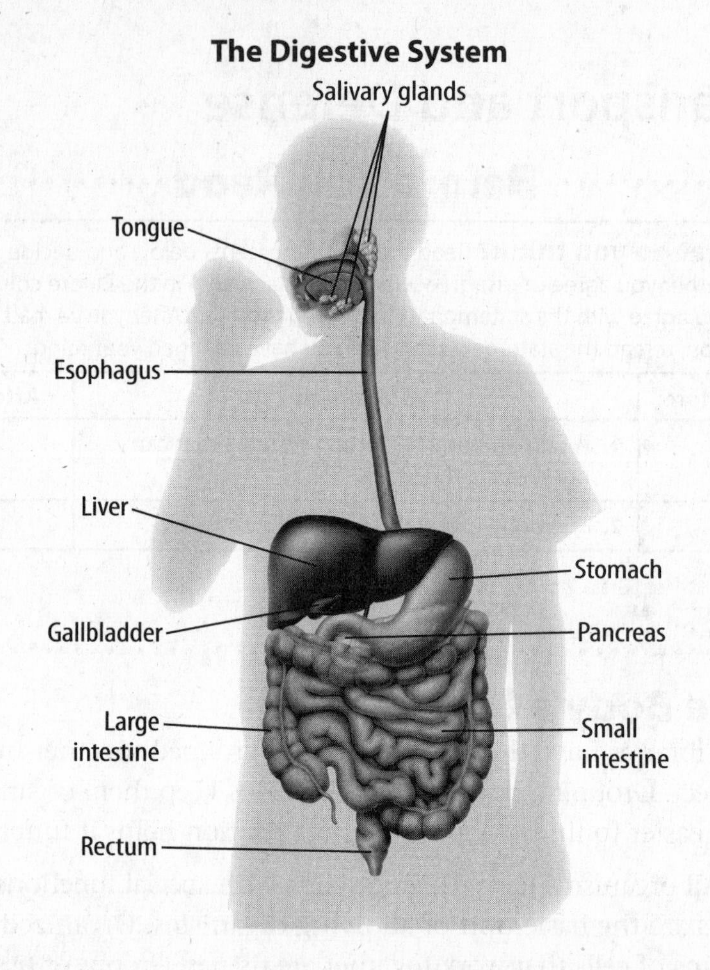

Digestion After food enters the mouth, chewing breaks food into smaller parts. Saliva also helps the mouth break down food. Saliva contains enzymes, which are proteins that speed up chemical reactions.

When you swallow, food, water, and other liquids move into a hollow tube called the esophagus (ih SAH fuh gus). The esophagus connects the mouth to the stomach. Digestion continues as food moves into the stomach.

The stomach is a flexible, baglike organ. It contains other enzymes that break down food into smaller parts. After the food is broken down by enzymes, it can be used by the body.

Absorption After food leaves the stomach, it moves into the small intestine. By the time food gets to the small intestine, it is a soupy mixture.

The small intestine is a tube that has two functions—digestion and absorption. The liver makes a substance called bile. The pancreas makes enzymes. Bile and enzymes are used in the small intestine to break down food even more. Because the small intestine is very long, it takes hours for food to move through it. During that time, particles of food and water are absorbed into the blood.

Excretion The large intestine, or colon (KOH lun), receives digested food that the small intestine did not absorb. The large intestine also absorbs water from the remaining waste material.

Most foods are completely digested into smaller parts. These small parts can be absorbed easily by the small intestine. However, some foods travel through the entire digestive system without being digested or absorbed. For example, some types of fiber, called insoluble fiber, in vegetables and whole grains are not digested and leave the body through the rectum.

Nutrition

As you have read, one of the functions of the small intestine is absorption. **Nutrients** *are the parts of food used by the body to grow and survive.* There are several types of nutrients. Proteins, fats, carbohydrates, vitamins, and minerals are nutrients. Nutrition labels on food packages show the amount of each nutrient in that food. Studying food labels can help you make sure you get all the nutrients you need.

Different people need different amounts of nutrients. For example, football players, swimmers, and other athletes need a lot of nutrients for energy. Pregnant women also need lots of nutrients to provide for their developing babies.

Digestion helps release energy from food. *A* **Calorie** *is the amount of energy it takes to raise the temperature of 1 kg of water by 1°C.* The body uses Calories from proteins, fats, and carbohydrates. Each of these nutrients contains a different amount of energy. ✓

The Excretory System

The excretory system removes solid, liquid, and gas waste materials from the body. The lungs, skin, liver, kidneys, bladder, and rectum are parts of the excretory system.

Math Skills

A proportion is an equation of two equal ratios. You can solve a proportion for an unknown value. For example, a 50-g egg provides 70 Calories (C) of energy. How many Calories would you get from 125 g of scrambled eggs?

Write a proportion.

$$\frac{50\text{ g}}{70\text{ C}} = \frac{125\text{ g}}{x}$$

Find the cross products.

$$50\text{ g }(x) = 70\text{ C} \times 125\text{ g}$$

$$50\text{ g }(x) = 8{,}750\text{ C g}$$

Divide both sides by 50.

$$\frac{50\text{ g}(x)}{50\text{ g}} = \frac{8{,}750\text{ C g}}{50\text{ g}}$$

Simplify the equation.

$$x = 175\text{ C}$$

4. Use Proportions The serving size of a large fast-food hamburger with cheese is 316 g. It contains 790 C of energy. How many Calories would you consume if you ate 100 g of the burger?

____

✓ **Reading Check**

5. Identify Name five types of nutrients.

____

____

____

The Lungs When you breathe out, or exhale, the lungs remove carbon dioxide (CO_2) and excess water as water vapor. The skin removes water and salt when you sweat.

The Liver The organ that removes wastes from the blood is the liver. As you have read, the liver is also a part of the digestive system. The digestive and excretory systems work together to break down, absorb, and remove food.

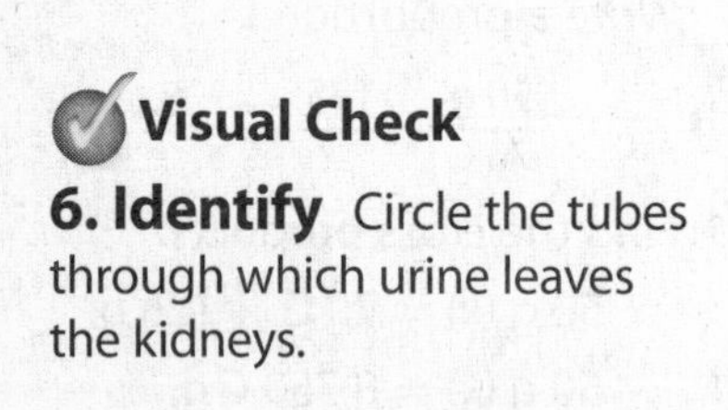

Visual Check

6. Identify Circle the tubes through which urine leaves the kidneys.

The Kidneys and Bladder When the liver breaks down proteins, urea forms. Urea is deadly if it stays in the body. See the figure to the right. The kidneys remove urea by making urine. Urine contains water, urea, and other waste. Urine leaves each kidney through a tube called the ureter (YOO ruh tur) and is stored in a flexible sac called the bladder. Urine is removed from the body through a tube called the urethra (yoo REE thruh).

The Kidneys

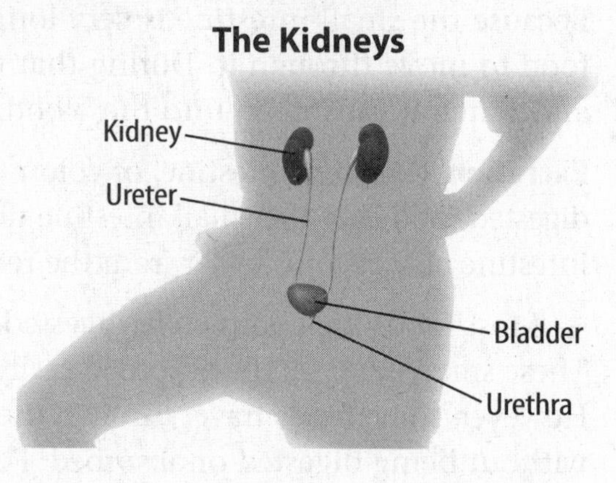

The Rectum Like the liver, the rectum is part of the excretory system and the digestive system. Food substances that are not absorbed by the small intestine are mixed with other wastes and form feces. The rectum stores feces until it moves out of the body.

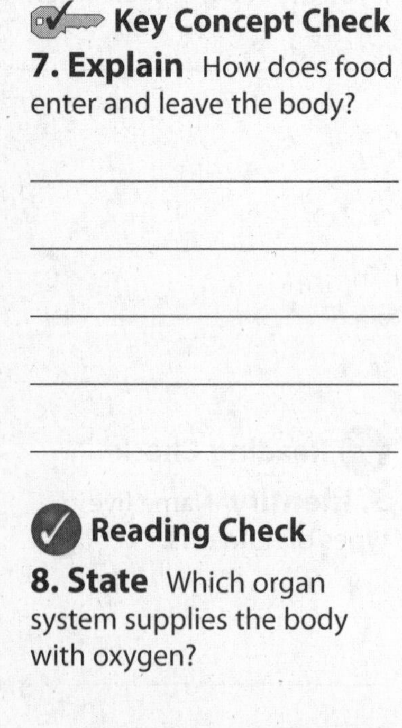

Key Concept Check

7. Explain How does food enter and leave the body?

Reading Check

8. State Which organ system supplies the body with oxygen?

Respiration and Circulation

You have read about how the body converts food into nutrients and absorbs them. But how do the oxygen you breathe in and the nutrients you absorb get to the rest of your body? And how do waste products leave the body?

The Respiratory System

The respiratory system exchanges gases between the body and the environment. As air flows through the respiratory system, it passes through the nose and mouth, pharynx (FER ingks), trachea (TRAY kee uh), bronchi (BRAHN ki; singular, bronchus), and lungs. The parts of the respiratory system work together to supply the body with oxygen. They also rid the body of wastes, such as carbon dioxide. The respiratory system is shown on the next page.

The Respiratory System

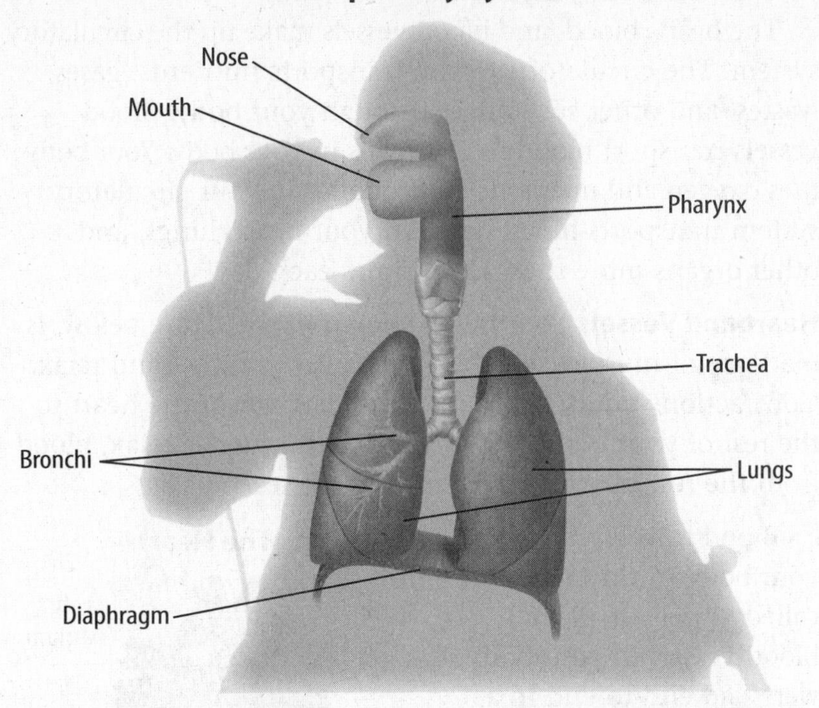

Pharynx and Trachea Oxygen enters the body when you inhale, or breathe in. Carbon dioxide leaves the body when you exhale. When you inhale, air enters the nostrils and passes through the pharynx. Because the pharynx is part of the throat, it is a part of the digestive and respiratory systems. Food goes through the pharynx to the esophagus. Air travels through the pharynx to the trachea. The trachea is also called the windpipe because it is a long, tubelike organ that connects the pharynx to the bronchi. 

Bronchi and Alveoli The trachea branches into two bronchi. One enters the left lung, and one enters the right lung. The bronchi divide into smaller tubes that end in tiny groups of cells that look like bunches of grapes. These groups of cells are called alveoli (al VEE uh li).

More than 100 million alveoli are inside each lung. They are surrounded by blood vessels called capillaries. Oxygen in the alveoli enters the capillaries. The blood inside capillaries transports oxygen to the rest of the body. Carbon dioxide enters the alveoli from the capillaries. 

Inhaling and exhaling require the movement of a thin muscle under your lungs called the diaphragm (DI uh fram). The diaphragm contracts when you inhale and air enters your lungs. The diaphragm relaxes when you exhale.

Visual Check

9. Illustrate Trace the path of airflow through the respiratory system.

Reading Check

10. Name Which organ is part of the digestive system and the respiratory system?

____

Reading Check

11. Recognize What are alveoli, and what do they do?

____

____

____

The Circulatory System

The heart, blood, and blood vessels make up the circulatory system. The circulatory system transports nutrients, gases, wastes, and other substances through your body. Blood vessels transport blood to all organs in your body. Your body uses oxygen and nutrients continually. So your circulatory system transports blood between your heart, lungs, and other organs more than 1,000 times each day!

Heart and Vessels Your heart, shown in the figure below, is made up of muscle cells that constantly contract and relax. Contractions pump blood in your heart out of the heart to the rest of your body. When your heart muscles relax, blood from the rest of your body enters your heart.

SCIENCE USE V. COMMON USE
vessel
Science Use a tube in the body that carries fluid such as blood

Common Use a ship

Blood travels through your body in tiny tubes called vessels. If all the blood vessels in your body were laid end-to-end in a single line, it would be more than 95,000 km long.

The Heart

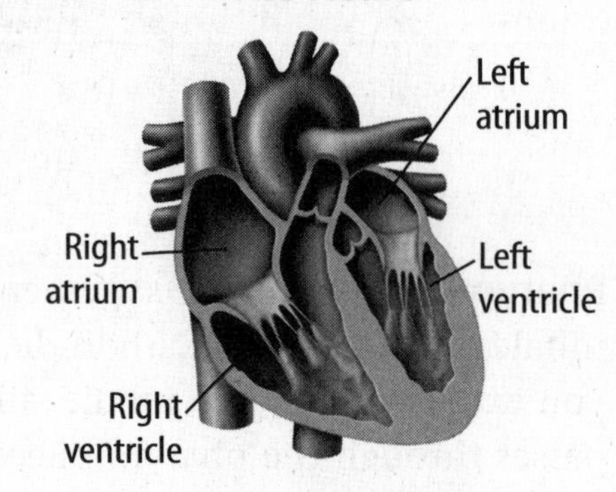

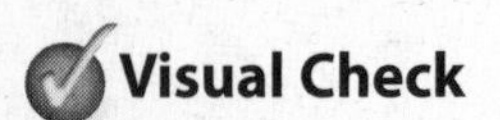

Visual Check
12. Classify What are the names of the two main sections of the heart?

The three main types of blood vessels are arteries, veins, and capillaries.
Arteries carry blood away from your heart. This blood is usually full of oxygen and contains nutrients. Blood in the pulmonary arteries contains CO_2. Arteries are large. They are surrounded by muscle cells that help blood move through the vessels faster. Veins transport blood that contains CO_2 back to your heart, except for the blood in the pulmonary veins, which is full of oxygen. Capillaries are tiny vessels that enable oxygen, CO_2, and nutrients to move between your circulatory system and your entire body.

Capillaries surround the alveoli in your lungs. Capillaries also surround your small intestine, where they absorb nutrients and transport them to the rest of your body.

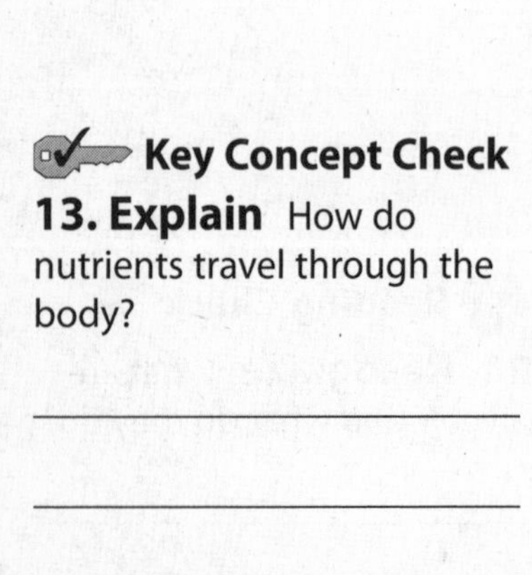

Key Concept Check
13. Explain How do nutrients travel through the body?

Blood The blood that circulates through vessels has several parts. The liquid part of blood is called plasma. Plasma contains nutrients, water, and CO_2. Blood also contains red blood cells, platelets, and white blood cells. Red blood cells carry oxygen. Platelets help your body heal when you get a cut. White blood cells help your body defend itself from toxins and diseases.

Human Blood Types				
Blood Type	**Type A**	**Type B**	**Type AB**	**Type O**
Percentage of U.S. population with this blood type	42	10	4	44
Clumping proteins in plasma	anti-B	anti-A	none	anti-A and anti-B
Blood type(s) that can be RECEIVED in a transfusion	A or O	B or O	A, B, AB, or O	O only
This blood type can DONATE TO these blood types	A or AB	B or AB	AB only	A, B, AB, O

Everyone has red blood cells. However, different people have different proteins on the surfaces of their red blood cells. Scientists classify these different red-blood-cell proteins into groups called blood types. These different human blood types are shown in the table above.

People with A proteins on their red blood cells have type A blood. People with B proteins on their red blood cells have type B blood. Some people have both A and B proteins on their red blood cells. They have type AB blood. People with type O blood have neither A nor B proteins on the surfaces of their red blood cells. ✓

Medical professionals use blood types to determine which type of blood a person can receive from a blood donor. For example, people with type O blood can receive blood only from a donor who also has type O blood. This is because people with type O blood have no proteins on the surfaces of their red blood cells.

The Lymphatic System

Have you ever had a cold and found it painful to swallow? This can happen when your tonsils swell. Tonsils are small organs on both sides of your throat. They are part of the lymphatic (lihm FA tihk) system. ✓

Parts of the Lymphatic System The spleen, the thymus, bone marrow, and lymph nodes also are parts of the lymphatic system. The spleen stores blood for use in an emergency. The thymus, the spleen, and bone marrow make white blood cells.

Interpreting Tables

14. Name To which blood group can type A donate?

____

✓ Reading Check

15. Identify People with no proteins on the surfaces of their red blood cells have __________. (Circle the correct answer.)

a. type AB blood
b. type O blood
c. type A blood

✓ Reading Check

16. Locate Where are your tonsils?

____

____

____

Functions of the Lymphatic System Your lymphatic system has three main functions: removing excess fluid around organs, producing white blood cells, and absorbing and transporting fats. The lymphatic system helps your body maintain fluid homeostasis. About 65 percent of the human body is water. Most of this water is inside cells.

Sometimes, when water, wastes, and nutrients move between capillaries and organs, not all of the fluid is taken up by the organs. When fluid builds up around organs, swelling can occur. The lymphatic system removes the fluid and keeps swelling from occurring. ✓

Reading Check
17. Identify a function of the lymphatic system.

Lymph vessels are all over your body. Fluid that travels through the lymph vessels flows into organs called lymph nodes. Humans have more than 500 lymph nodes. The lymph nodes work together and protect the body by removing toxins, wastes, and other harmful substances.

The lymphatic system makes white blood cells. White blood cells help the body defend against infection. There are many different types of white blood cells. *A* **lymphocyte** (LIHM fuh site) *is a type of white blood cell that is made in the thymus, the spleen, or the bone marrow.* Lymphocytes protect the body by traveling through the circulatory system, defending against infection. ✓

Reading Check
18. Explain why your spleen is important.

Immunity

The lymphatic system protects your body from harmful substances and infection. *The resistance to specific pathogens or disease-causing agents, is called* **immunity.** The skeletal system produces immune cells. The circulatory system transports immune cells throughout the body.

Immune cells include lymphocytes and other white blood cells. These cells detect viruses, bacteria, and other foreign substances that are not normally made in the body. The immune cells attack and destroy them.

ACADEMIC VOCABULARY
detect
(verb) to discover the presence of

If the body is exposed to the same bacteria, virus, or substance later, some immune cells remember and make proteins called antibodies. These antibodies recognize specific proteins on the harmful agent and help the body fight infection faster. There are many different types of bacteria and viruses, so humans make billions of different types of antibodies. Each type responds to a different harmful agent.

Types of Diseases

There are two main groups of diseases—infectious and noninfectious. Infectious diseases are caused by pathogens, such as bacteria and viruses. Infectious diseases are usually contagious. This means that they can be spread from one person to another. The flu is an example of an infectious disease. Viruses that invade organ systems of the body, such as the respiratory system, cause infectious diseases.

A noninfectious disease is caused by the environment or a genetic disorder, not a pathogen. Skin cancer, diabetes, and allergies are noninfectious diseases. Noninfectious diseases are not contagious and cannot be spread from one person to another. The table below lists some examples of infectious and noninfectious diseases.

Examples of Diseases

Infectious Disease		Noninfectious Disease
Disease	**Pathogen**	
colds	virus	cancer
AIDS	virus	diabetes
strep throat	bacteria	heart disease
chicken pox	virus	allergy

Interpreting Tables

19. Classify Circle all the diseases in the table that are caused by viruses. Underline the diseases that are caused by bacteria.

Lines of Defense

The human body has many ways of protecting itself from viruses, bacteria, and harmful substances. Skin and mucus (MYEW kus) are parts of the first line of defense. They prevent toxins and other substances from entering the body. Mucus is a thick, gel-like substance in the nostrils, trachea, and lungs. Mucus traps harmful substances and prevents them from entering your body.

The second line of defense is the immune response. In the immune response, white blood cells attack and destroy harmful substances.

The third line of defense protects your body against substances that have infected the body before. As you have read, immune cells make antibodies that destroy the harmful substances. Vaccines are used to help the body develop antibodies against infectious diseases. For example, many people get an influenza vaccine annually to protect them against the flu.

Key Concept Check

20. Describe How does the body defend itself from harmful invaders?

____

____

____

____

After You Read

Mini Glossary

Calorie: the amount of energy it takes to raise the temperature of 1 kg of water by 1°C.

homeostasis (hoh mee oh STAY sus): steady internal conditions when external conditions change

immunity: the resistance to specific pathogens, or disease-causing agents

lymphocyte (LIHM fuh site): a type of white blood cell that is made in the thymus, the spleen, or the bone marrow

nutrient: the part of food used by the body to grow and survive

organ system: a group of organs that work together and perform a specific task

1. Review the terms and their definitions in the Mini Glossary. Write a sentence explaining how lymphocytes aid in immunity.

___

___

2. Use the graphic organizer to identify the organ that removes the type of waste material indicated.

Name of Organ	Type of Waste Material Removed
	removes carbon dioxide and excess water vapor when you exhale
	removes urea from the body
	removes wastes from the blood
	removes feces from the body

3. Explain what is unique about people with type AB blood.

___

___

___

What do you think NOW?

Reread the statements at the beginning of the lesson. Fill in the After column with an A if you agree with the statement or a D if you disagree. Did you change your mind?

Log on to ConnectED.mcgraw-hill.com and access your textbook to find this lesson's resources.

Human Body Systems

Structure, Movement, and Control

Before You Read

What do you think? Read the two statements below and decide whether you agree or disagree with them. Place an A in the Before column if you agree with the statement or a D if you disagree. After you've read this lesson, reread the statements to see if you have changed your mind.

Before	Statement	After
	3. All bones in the skeletal system are hollow.	
	4. The endocrine system makes hormones.	

Key Concepts 
- How does the body move?
- How does the body respond to changes in its environment?

Read to Learn

Structure and Movement

The human body can move in many different directions and do many different things. It can do things that require many parts of the body to move, such as swimming or shooting a basketball. It also can remain very still, such as when posing for a picture or balancing on one leg.

In this lesson, you will read more about two organ systems that give the body structure, help the body move, and protect other organ systems. These organ systems are the skeletal system and the muscular system.

Mark the Text

Sticky Notes As you read, use sticky notes to mark information that you do not understand. Read the text carefully a second time. If you still need help, write a list of questions to ask your teacher.

The Skeletal System

The skeletal system has four major jobs. It protects internal organs, provides support, helps the body move, and stores minerals. The skeletal system is mostly bones. Adults have 206 bones. Ligaments, tendons, and cartilage are also parts of the skeletal system.

Storage The skeletal system stores important minerals such as calcium. Your body uses calcium in many ways. Muscles need calcium for contractions. The nervous system needs calcium for communication. Most of the calcium in the body is stored in bone. Calcium helps build stronger compact bone. Cheese and milk are good sources of calcium. 

Reading Check

1. Identify Which mineral is stored by the skeletal system?

FOLDABLES®

Make two horizontal two-tab books, label them as shown, and use them to organize information about each body system.

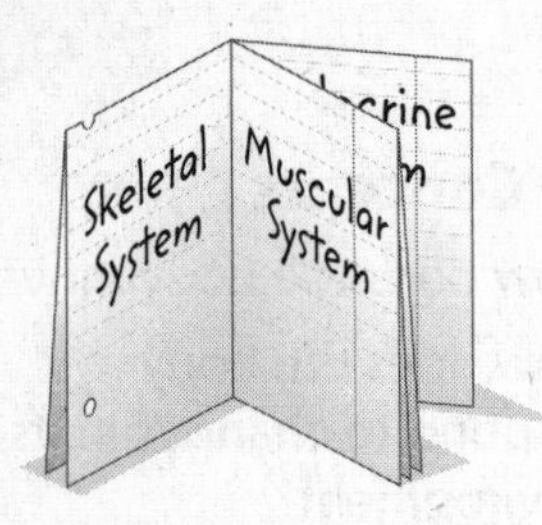

Support Without a skeleton, your body would look like a beanbag. Your skeleton gives your body structure and support. Your bones help you stand, sit up, and raise your arms.

Protection Many of the bones in the body protect organs that are made of softer tissue. For example, the skull protects the soft tissue of the brain. The rib cage protects the soft tissue of the lungs and heart.

Movement The skeletal system helps the body move by working with the muscular system. Bones can move because they are attached to muscles.

Bone Types Bones are organs that contain two types of tissue. **Compact bone** *is the hard outer layer of bone.* **Spongy bone** *is the interior region of bone that contains many tiny holes.*

Spongy bone is inside compact bone. Some bones also contain bone marrow. Remember that bone marrow is a part of the lymphatic system and makes white blood cells.

Reading Check

2. Differentiate How do the two types of bone tissue differ?

______

______

______

______

______

The Muscular System

You might already know that your arms and legs have muscle cells. But did you know that your eyes, heart, and blood vessels have them too? Without muscle cells, you would not be able to talk, write, or run.

As shown in the figure on the next page, muscle cells are everywhere in the body. Almost one-half of your body mass is muscle cells. Muscle cells make up the muscular system. By working together, they help the body move.

The muscular system is made of three different types of muscle tissue—skeletal muscle, cardiac muscle, and smooth muscle. Skeletal muscle works with the skeletal system and helps you move. Tendons connect skeletal muscles to bones. Skeletal muscle also gives you the strength to lift heavy objects.

Another type of muscle tissue is cardiac muscle. Cardiac muscle is only in the heart. It continually contracts and relaxes and moves blood throughout your body.

Smooth muscle tissue is another type of muscle tissue. Smooth muscle tissue is in organs such as the stomach and the bladder. Blood vessels also have smooth muscle tissue.

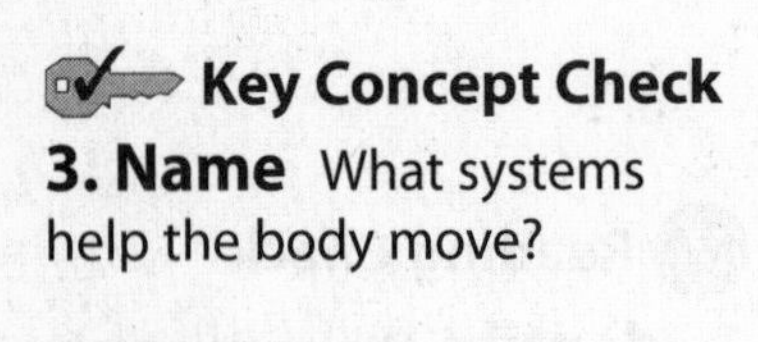

Key Concept Check

3. Name What systems help the body move?

______

______

______

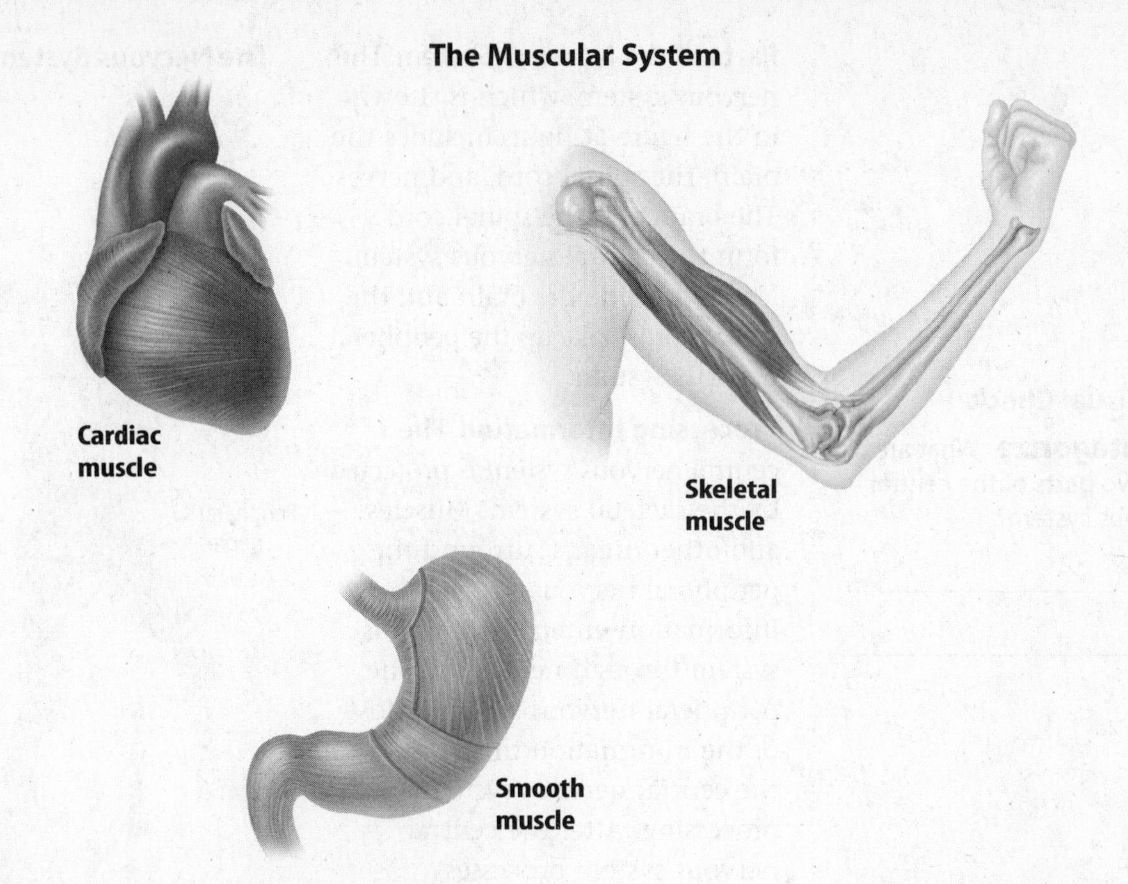

Control and Coordination

The nervous system receives and processes information about your internal and external environments. The nervous system works with the endocrine system, which you will read about later. These two systems control many functions, including movement, communication, and growth, by working with other systems in the body. They also help your body maintain homeostasis.

The Nervous System

The nervous system is a group of organs and specialized cells that detect, process, and respond to information. The nervous system constantly receives information from your external environment and from inside your body. It can receive information, process it, and respond in less than 1 second.

Nerve Cells *The basic units of the nervous system are* **neurons,** or nerve cells. Neurons can be many different lengths. In adults, some neurons are more than 1 m long. This is about as long as the distance between a toe and the spinal cord.

Visual Check

4. Recognize Which type of muscle is in your arms?

Key Concept Check

5. Explain How does the body respond to changes in its environment?

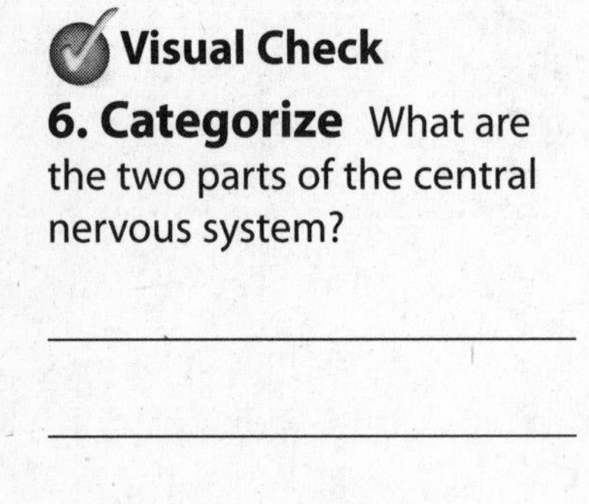

Visual Check

6. Categorize What are the two parts of the central nervous system?

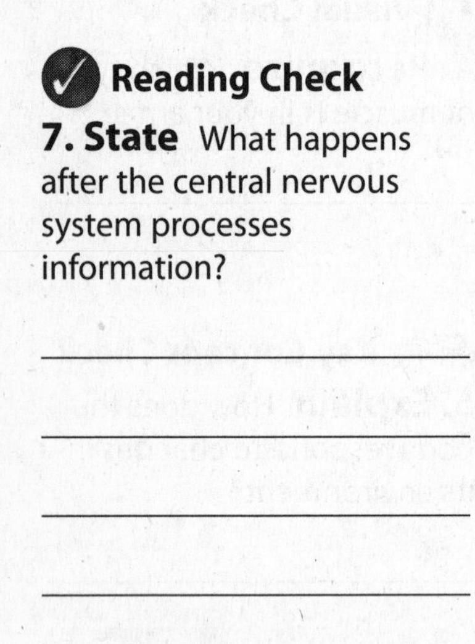

Reading Check

7. State What happens after the central nervous system processes information?

Think it Over

8. Consider Is frying an egg a voluntary or an involuntary function? Explain.

Parts of the Nervous System The nervous system, which is shown in the figure at right, includes the brain, the spinal cord, and nerves. The brain and the spinal cord form the central nervous system. Nerves outside the brain and the spinal cord make up the peripheral nervous system.

The Nervous System

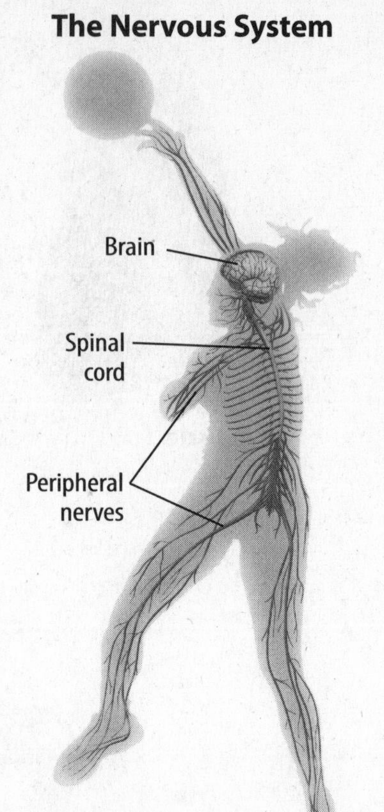

Processing Information The central nervous system is protected by the skeletal system. Muscles and other organs surround the peripheral nervous system. Information enters the nervous system through neurons in the peripheral nervous system. Most of the information then is sent to the central nervous system for processing. After the central nervous system processes information, it signals the peripheral nervous system to respond.

Voluntary and Involuntary Control The body carries out many functions that depend on the nervous system. Some of these functions, such as breathing and digestion, are automatic, or involuntary. You do not have to think about them to make them happen. The nervous system automatically controls these functions and maintains homeostasis.

Most of the other functions of the nervous system are not automatic. You must think about them to make them happen. Reading, talking, and walking are voluntary functions. These tasks require input, processing, and a response.

Reflexes Touching a hot object with your hand sends a rapid signal that your hand is in pain. The signal is so fast that you do not think about moving your hand. It just happens automatically. *Automatic movements in response to a signal are called* **reflexes.** The table at the top of the next page lists how the spinal cord receives and processes reflex signals. Processing information in the spinal cord instead of the brain helps the body respond faster.

Receiving and Processing Reflex Signals	
Step	**Action**
1	When you touch something hot, receptors in your hand detect stimuli (the hot temperature). They send signals through nerves to your spinal cord.
2	Responding nerve signals travel directly from the spinal cord to muscles in your arm. This causes you to pull your hand away.
3	After you respond to the stimuli, nerve signals travel from the spinal cord to the brain. This causes you to feel pain.

The Senses Humans detect their external environment with five senses—vision, hearing, smell, touch, and taste—as shown in the following table. Each of the five senses has specific neurons that receive signals from the environment.

The Senses				
Vision	**Hearing**	**Smell**	**Touch**	**Taste**
The visual system receives light signals.	The auditory system detects sound.	The olfactory system receives odor signals.	There are many sensory receptors for touch. Some receive signals that detect temperature.	Taste buds receive chemical signals.

Information detected by the senses is sent to the spinal cord and then to the brain for processing and a response. Responses depend on the specific signal detected. Some responses cause muscles to contract and move, such as when you touch a hot surface. The aroma of baking cookies might cause your mouth to water. It is producing saliva.

The Endocrine System

How tall were you in first grade? How tall are you now? From the time you were born until now, your body has changed. These changes are controlled by the endocrine system.

Like the nervous system, the endocrine system sends signals to the body. *Chemical signals released by the organs of the endocrine system are called* **hormones.** Hormones cause organ systems to carry out specific functions.

Interpreting Tables

9. State What detects heat when you touch something hot?

Interpreting Tables

10. Infer Name a part of the body that is included in the olfactory system.

Think it Over

11. Identify at least three senses you use every time you eat breakfast.

The Endocrine System

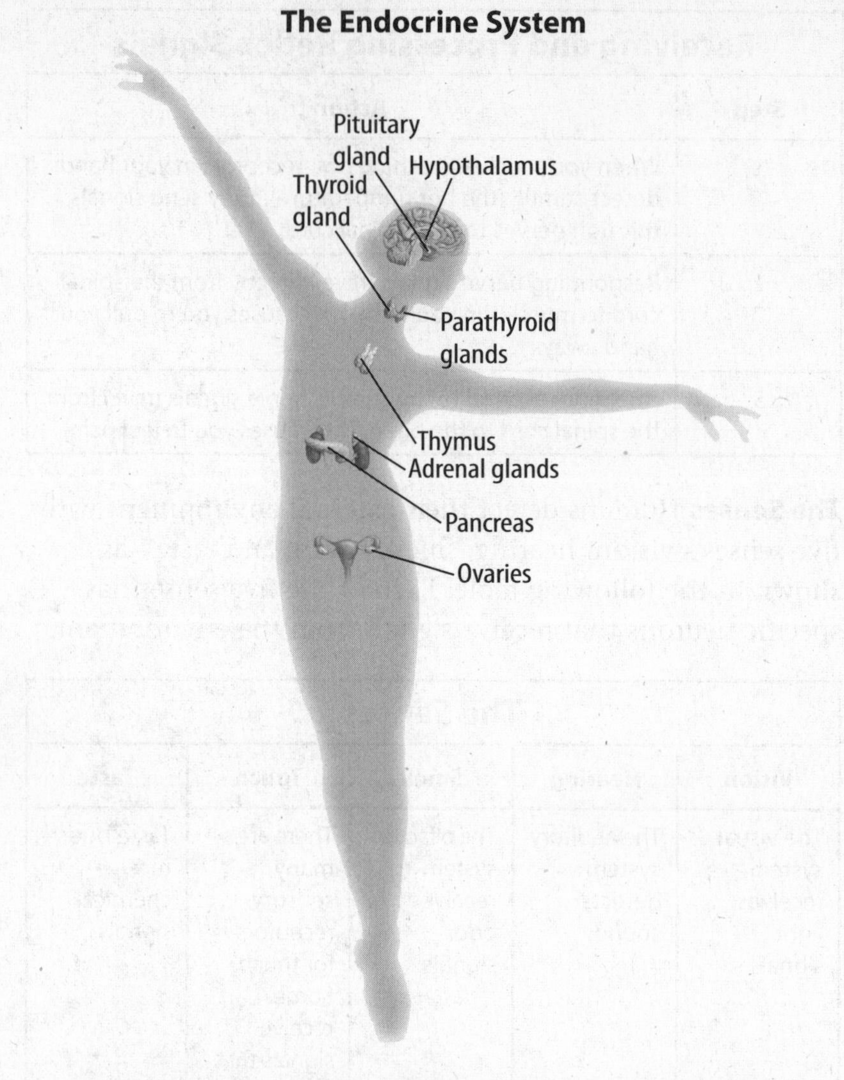

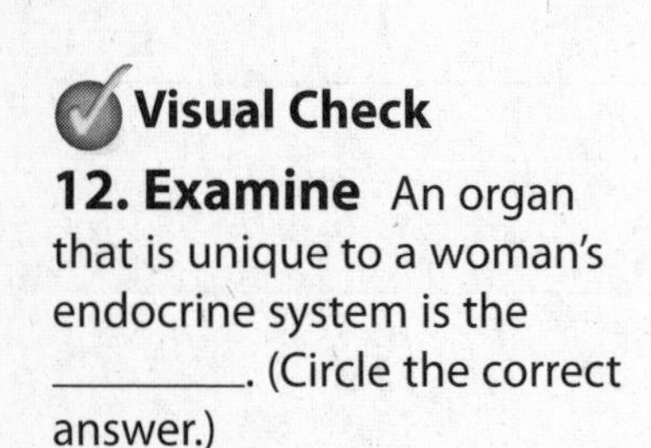

Visual Check

12. Examine An organ that is unique to a woman's endocrine system is the ________. (Circle the correct answer.)

a. thymus

b. pancreas

c. ovaries

Why does your body need two organ systems to process information? Signals sent by the nervous system travel quickly through neurons. Hormones travel in blood through blood vessels in the circulatory system. These messages travel more slowly than nerve messages. A signal sent by the nervous system can travel from your head to your toes in less than 1 second. A hormone needs about 20 seconds to make the trip. Although hormones take longer to reach their target organ system, their effects usually last longer.

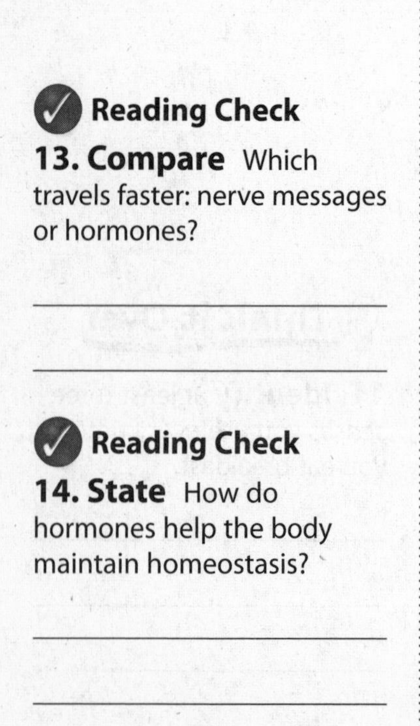

Reading Check

13. Compare Which travels faster: nerve messages or hormones?

____

____

The endocrine system is shown in the figure above. Many of the hormones made by the endocrine system work with other organ systems and maintain homeostasis. Parathyroid hormone works with the skeletal system and controls calcium storage. Insulin is a hormone released from the pancreas. It signals the digestive system to control nutrient homeostasis. Other hormones, such as growth hormone, work with many organ systems to help you grow. In the next lesson, you will read about another system that the endocrine system works with.

Reading Check

14. State How do hormones help the body maintain homeostasis?

____

____

____

After You Read

Mini Glossary

compact bone: the hard outer layer of bone

hormone: a chemical signal released by the organs of the endocrine system

neuron: the basic unit of the nervous system

reflex: an automatic movement in response to a signal

spongy bone: the interior region of bone that contains many tiny holes

1. Review the terms and their definitions in the Mini Glossary. Write a sentence explaining the relationship between neurons and reflexes.

2. Fill in the graphic organizer below to show four major jobs of the skeletal system.

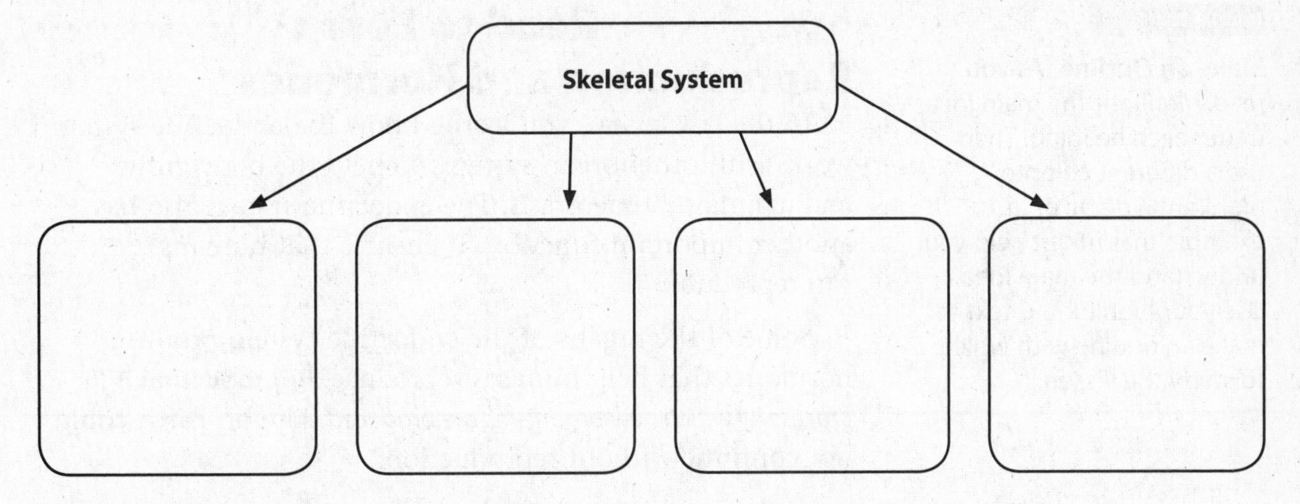

3. Review the information you marked with sticky notes as you read the lesson. In the space below, write the answer to one question from the list that you asked your teacher.

What do you think NOW?

Reread the statements at the beginning of the lesson. Fill in the After column with an A if you agree with the statement or a D if you disagree. Did you change your mind?

Log on to ConnectED.mcgraw-hill.com and access your textbook to find this lesson's resources.

Human Body Systems

Reproduction and Development

Key Concepts

- What do the male and female reproductive systems do?
- How do humans grow and change?

Before You Read

What do you think? Read the two statements below and decide whether you agree or disagree with them. Place an A in the Before column if you agree with the statement or a D if you disagree. After you've read this lesson, reread the statements to see if you have changed your mind.

Before	Statement	After
	5. The testes produce sperm.	
	6. Puberty occurs during infancy.	

Mark the Text

Make an Outline As you read, highlight the main idea under each heading. Then use a different color to highlight a detail or an example that might help you understand the main idea. Use your highlighted text to make an outline with which to study the lesson.

Read to Learn

Reproduction and Hormones

In the last lesson, you learned how the endocrine system works with other organ systems to help the body grow and maintain homeostasis. The endocrine system also has another important function. It ensures that humans can reproduce.

Some of the organs of the endocrine system produce hormones that help humans reproduce. **Reproduction** *is the process by which new organisms are produced.* Life on Earth could not continue without reproduction.

A male and a female each have special organs for reproduction. Organs in the male reproductive system are different from the organs in the female reproductive system.

Human reproductive cells, called **gametes** (GA meets), are made by the male and female reproductive systems. *Male gametes are called* **sperm.** *Female gametes are called* **ova** (OH vah; singular, ovum), *or eggs.*

A sperm joins with an egg in a reproductive process called **fertilization.** *The cell that forms when a sperm cell fertilizes an egg cell is called a* **zygote** (ZI goht). A zygote is the first cell of a new human. It contains genetic information from the sperm and the ovum. The zygote will grow and develop in the female's reproductive system.

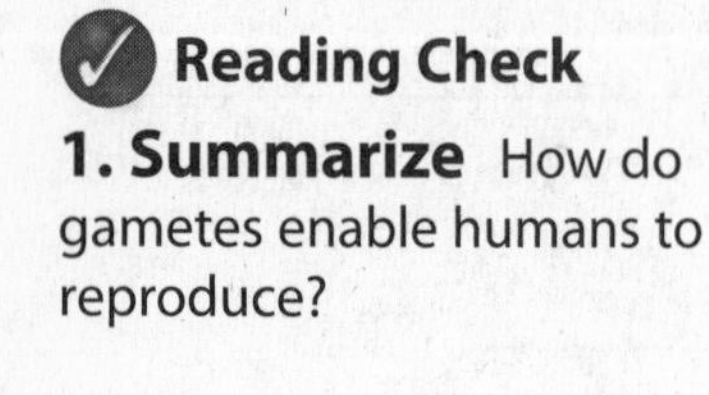

Reading Check

1. Summarize How do gametes enable humans to reproduce?

The Male Reproductive System

The male reproductive system is shown at left in the figure below. It produces sperm and delivers it to the female reproductive system. Sperm are produced in the testes (TES teez; singular, testis). Sperm develop inside each testis and then are stored in tubes called sperm ducts. Sperm mature in the sperm ducts.

The testes also produce a hormone called testosterone. Testosterone helps sperm change from round cells to long, slender cells that can swim. When sperm are fully developed, they can travel to the penis. The penis is a tubelike structure that delivers sperm to the female reproductive system. Sperm are transported in a fluid called semen (SEE mun). Semen contains millions of sperm and nutrients. The nutrients provide the sperm with energy.

The Female Reproductive System

The female reproductive system is shown at right in the figure below. It contains two ovaries, where eggs grow and mature. Two hormones are made by the ovaries—estrogen (ES truh jun) and progesterone (proh JES tuh rohn). These hormones help eggs mature. When eggs are mature, they are released from the ovaries and enter the fallopian tubes. The fallopian tubes connect the ovaries to the uterus.

FOLDABLES

Make a horizontal two-tab book, label it as shown, and use it to organize information about the male and female reproductive systems.

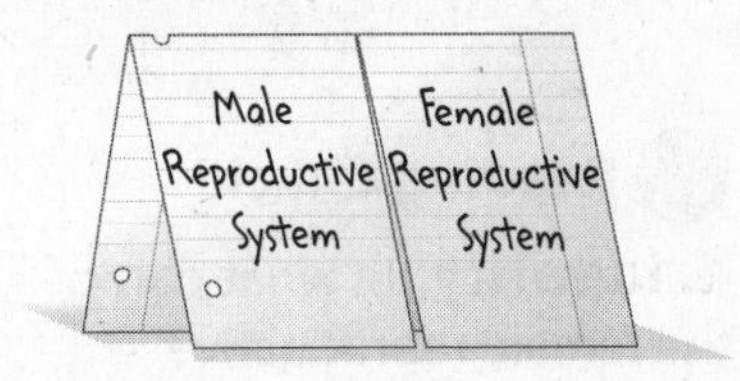

Key Concept Check

2. Explain What does the male reproductive system do?

______

______

Visual Check

3. Locate Circle the names of the organs that produce male and female gametes.

Male and Female Reproductive Systems

Male Reproductive System

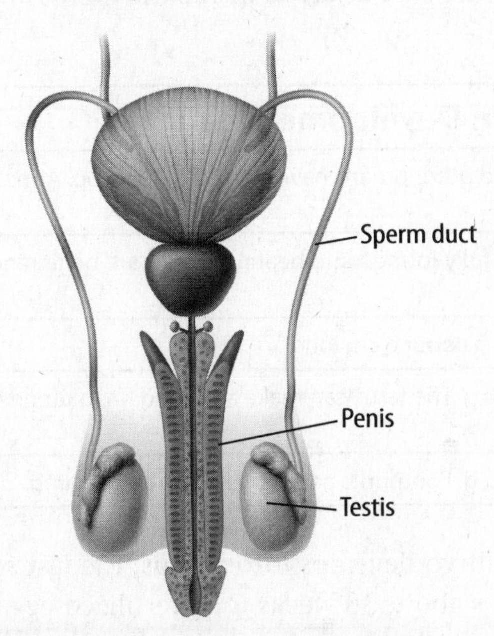

The organs of the male reproductive system produce sperm and deliver it to the female reproductive system.

Female Reproductive System

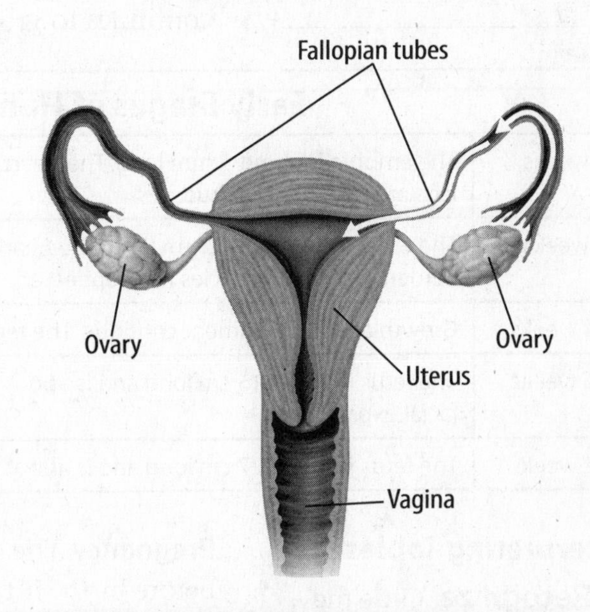

The female reproductive system produces eggs and provides a place for a new human to grow and develop before birth.

Fertilization If sperm are also present in the fallopian tube, fertilization can occur as the egg enters the fallopian tube. Sperm enter the female reproductive system through the vagina. The vagina is a tube-shaped organ that leads to the uterus. A fertilized egg, or zygote, can move through the fallopian tube and attach inside the uterus. If there are no sperm in the fallopian tube, the egg will not be fertilized. But the egg will still travel through the fallopian tube and uterus. Then it will break down.

Reading Check

4. Identify In what part of the female reproductive system does fertilization take place?

____

The Menstrual Cycle The endocrine system controls egg maturation and release and thickening of the lining of the uterus in a process called the menstrual (MEN stroo ul) cycle. The menstrual cycle takes about 28 days and has three parts:

1. Eggs grow and mature. The thickened lining of the uterus leaves the body.
2. Mature eggs are released from the ovaries. The lining of the uterus thickens.
3. Unfertilized eggs and the thickened lining break down. The lining leaves the body in the first part of the next cycle.

Key Concept Check

5. State What does the female reproductive system do?

____

____

____

____

Human Development

Humans develop in many stages, as shown in the table below. When a sperm fertilizes an egg, a zygote forms. The zygote develops into an embryo (EM bree oh). An embryo is a ball-shaped structure that attaches inside the uterus and continues to grow.

Early Stages of Human Development	
5 weeks	The embryo is about 7 mm long. The heart and other organs have started to develop. Arms and legs are beginning to bud.
8 weeks	The embryo is about 2.5 cm long. The heart is fully formed and beating. Bones are beginning to harden. Nearly all muscles have appeared.
14 weeks	Growth and development continue. The fetus is about 6 cm long.
16 weeks	The fetus is about 15 cm long and is about 140 g. The fetus can make a fist and has a range of facial expressions.
22 weeks	The fetus is about 27 cm long and is about 430 g. Footprints and fingerprints are forming.

Interpreting Tables

6. Recognize Circle the stage when the heart is fully formed in the embryo.

Pregnancy The embryo develops into a fetus, the last stage before birth. It takes about 38 weeks for a fertilized egg to fully develop. This developmental period is called pregnancy. During pregnancy, the organ systems of the fetus will develop. The fetus will get larger.

Birth Pregnancy ends with birth. During birth, the endocrine system releases hormones that help the uterus push the fetus through the vagina and out of the body.

From Birth Through Childhood

The first life stage after birth is infancy, the first 2 years of life. The figure below shows several stages of development in infancy. During infancy, the muscular and nervous systems develop. An infant begins walking. Growth and development continue in childhood, which is from about 2 years to about 12 years of age. Bones in the skeletal system grow longer and stronger. The lymphatic system matures.

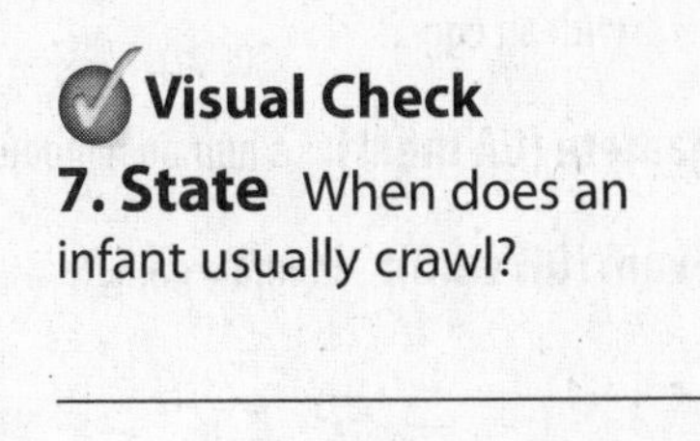

Visual Check

7. State When does an infant usually crawl?

____

____

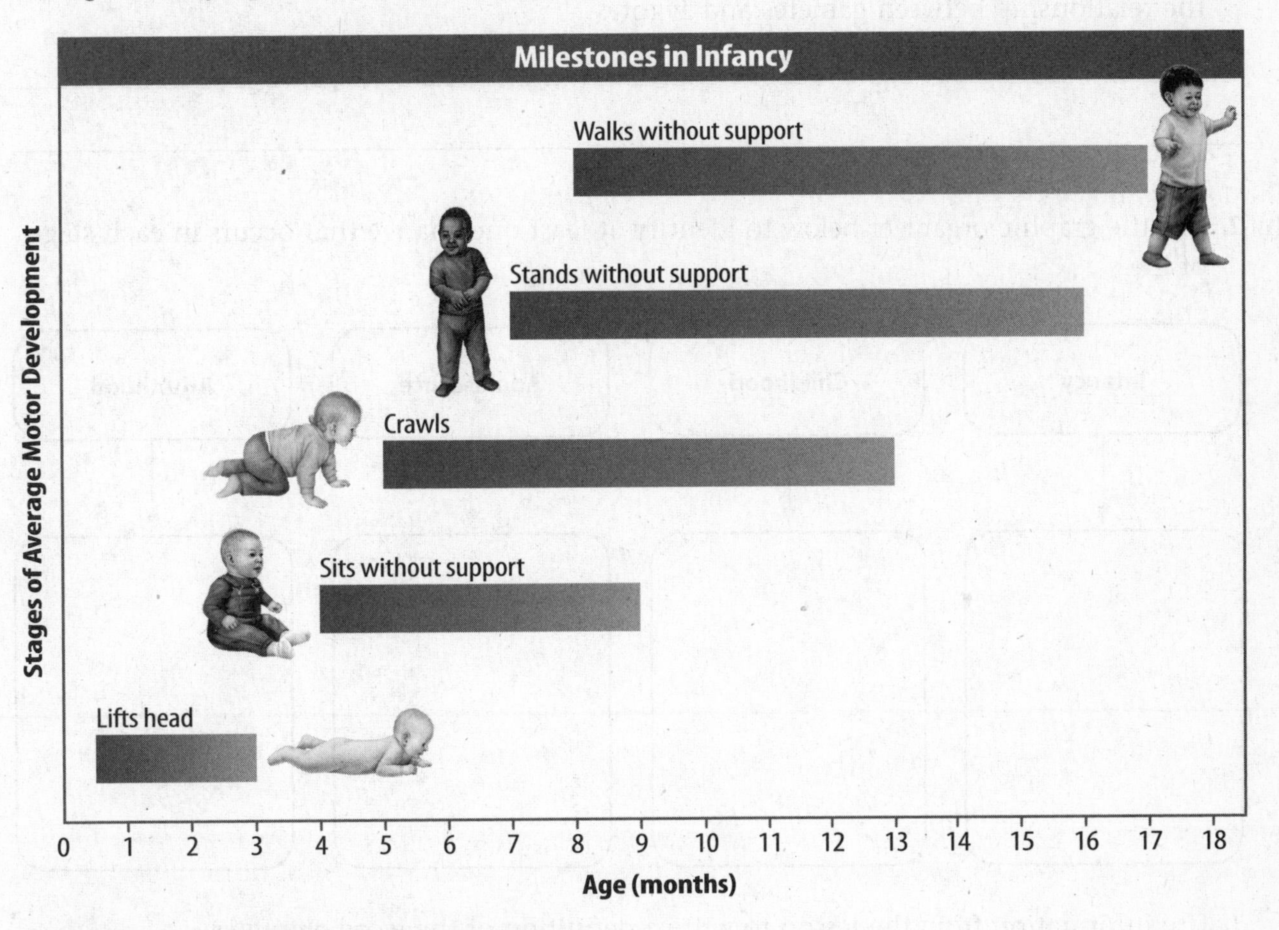

Adolescence Through Adulthood

Adolescence follows childhood. During adolescence, growth of the skeletal and muscular systems continues. Organs get larger. As the endocrine system develops, the male and female reproductive systems mature. The time during which the reproductive system matures is called puberty.

After adolescence is adulthood. During adulthood, humans continue to change. In later adulthood, hair turns gray, wrinkles might form in the skin, and bones become weaker. This process is called aging. Aging is a slow process that can last for decades.

Key Concept Check

8. Explain How do humans change during adulthood?

____

____

____

After You Read

Mini Glossary

fertilization: a reproductive process in which a sperm joins with an egg

gamete (GA meet): a human reproductive cell

ovum (OH vum): a female gamete

reproduction: the process by which new organisms are produced

sperm: a male gamete

zygote (ZI goht): the cell that forms when a sperm cell fertilizes an egg cell

1. Review the terms and their definitions in the Mini Glossary. Write a sentence explaining the relationship between gametes and zygotes.

2. Use the graphic organizer below to identify at least one change that occurs in each stage of life.

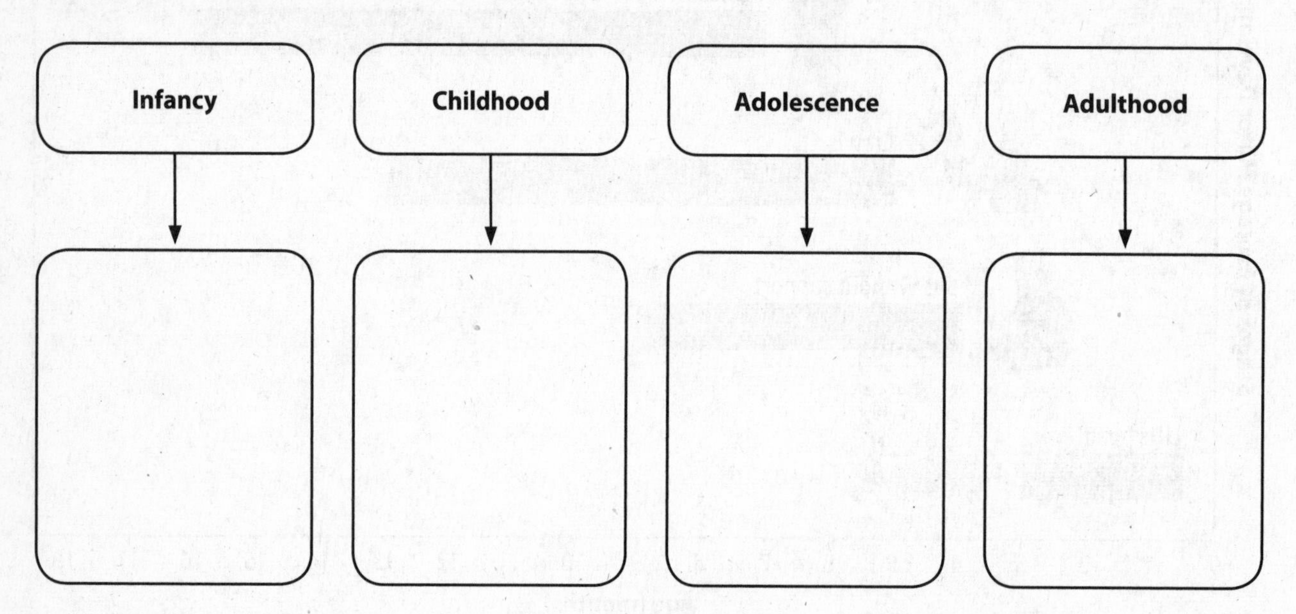

3. Use information from the lesson to write a definition of the word *pregnancy*.

What do you think NOW?

Reread the statements at the beginning of the lesson. Fill in the After column with an A if you agree with the statement or a D if you disagree. Did you change your mind?

Log on to ConnectED.mcgraw-hill.com and access your textbook to find this lesson's resources.

Plant Processes and Reproduction

Energy Processing in Plants

Before You Read

What do you think? Read the three statements below and decide whether you agree or disagree with them. Place an A in the Before column if you agree with the statement or a D if you disagree. After you've read this lesson, reread the statements to see if you have changed your mind.

Before	Statement	After
	1. Plants do not carry on cellular respiration.	
	2. Plants are the only organisms that carry on photosynthesis.	
	3. Plants make food in their underground roots.	

Key Concepts 

- How do materials move inside plants?
- How do plants perform photosynthesis?
- What is cellular respiration?
- How are photosynthesis and cellular respiration alike, and how are they different?

Read to Learn

Materials for Plant Processes

Food, water, and oxygen are three things you need to survive. Some of your organ systems process these materials. Other systems transport them throughout your body. Like you, plants need food, water, and oxygen to survive. Unlike you, plants do not take in food. Most plants make their own food.

Study Coach

Make an Outline Summarize the information in the lesson in an outline. Use the main headings in the lesson as the main headings in your outline. Use your outline to review the lesson.

Moving Materials Inside Plants

Xylem (ZI lum) and phloem (FLOH em) are the vascular tissue in most plants. These tissues transport materials throughout a plant.

After water enters a plant's roots, it moves into xylem. Water then flows inside xylem to all parts of a plant. Without enough water, plant cells wilt.

Most plants make their own food—a liquid sugar. The liquid sugar moves out of food-making cells, enters phloem, and flows to all plant cells. Cells break down the sugar and release energy. Some plant cells can store food.

Plants require oxygen and carbon dioxide to make food. Like you, plants produce water vapor as a waste product. Carbon dioxide, oxygen, and water vapor pass into and out of a plant through tiny openings in leaves.

Key Concept Check

1. Determine How do materials move through plants?

____

____

____

____

Photosynthesis

Plants make their own food through a process called photosynthesis (foh toh SIHN thuh sus). **Photosynthesis** *is a series of chemical reactions that convert light energy, water, and carbon dioxide into the food-energy molecule glucose and give off oxygen.*

WORD ORIGIN

photosynthesis
from Greek *photo–*, means "light"; and *synthesis*, means "composition"

Leaves are the major food-producing organs of plants. This means that photosynthesis takes place in the plant's leaves. The structure of a leaf is well-suited to its role in photosynthesis.

Leaves and Photosynthesis

The figure below shows the many types of cells in a leaf. The epidermal (eh puh DUR mul) cells make up the upper and lower layers of the leaf. Epidermal cells are flat and irregularly shaped. The bottom epidermal layer of most leaves has small openings called stomata (STOH muh tuh). Carbon dioxide, water vapor, and oxygen pass through stomata. Epidermal cells can produce a waxy covering called the cuticle.

FOLDABLES®

Make a shutterfold book and label it as shown to use as a diagram of leaf structure.

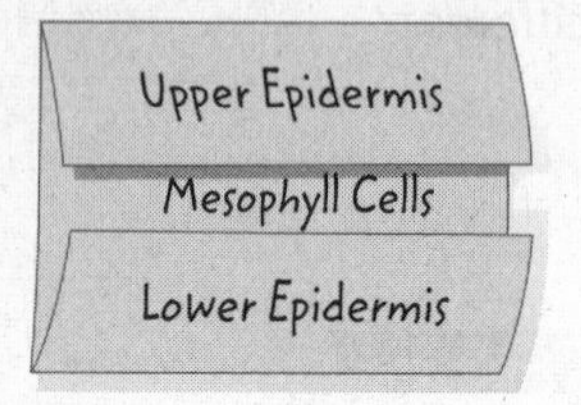

Most photosynthesis occurs in two types of mesophyll (ME zuh fil) cells inside a leaf. Mesophyll cells contain chloroplasts, which are the organelles where photosynthesis occurs. The palisade mesophyll cells are near the top surface of the leaf. They are packed close together. This arrangement exposes the most cells to light. Spongy mesophyll cells are below the palisade mesophyll cells. They have open spaces between them. Gases needed for photosynthesis flow through the spaces between the spongy mesophyll cells.

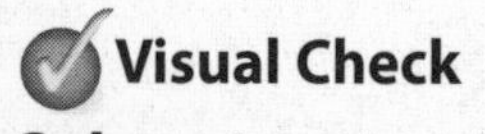

Visual Check

2. Locate Which layer of cells contains vascular tissue?

____

____

Cross Section of a Leaf

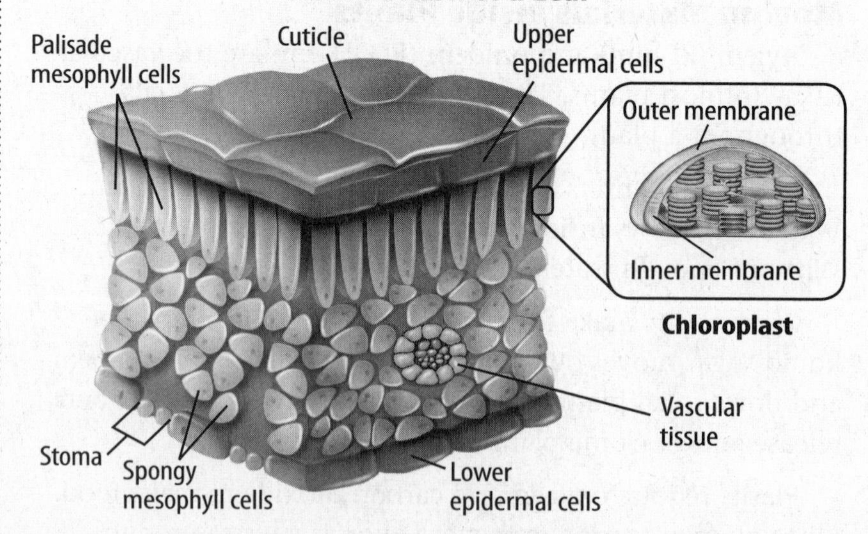

Capturing Light Energy

Photosynthesis is a complex chemical process. It consists of two basic steps: capturing light energy and using that energy to make sugars. Look at the figure below as you read about these steps to help you understand the process.

In the first step, chloroplasts capture the energy in light. Chloroplasts contain plant pigments. Pigments are chemicals that can absorb and reflect light. Chlorophyll is the most common plant pigment. It is necessary for photosynthesis. Most plants appear green because chlorophyll reflects green light. Chlorophyll absorbs other colors of light. This light energy is used during photosynthesis.

Chlorophyll traps and stores light energy. Then this energy can be transferred to other molecules. During photosynthesis, water molecules are split apart. The oxygen from the water molecules is released into the atmosphere, as shown below. The hydrogen atoms in the water are used to make sugars in the second step of photosynthesis.

Making Sugars

In the second step of photosynthesis, sugars are made from the light energy. In chloroplasts, carbon dioxide from the air is converted into sugars by using the energy stored and trapped by chlorophyll. Carbon dioxide combines with hydrogen atoms from the splitting of water molecules and forms sugar molecules. Plants can use this sugar as an energy source. Plants can also store the sugar for later use. Potatoes and carrots are examples of plant structures where plants store excess sugar.

Photosynthesis

Carbon dioxide CO_2

Light energy

Oxygen O_2

Sugar $C_6H_{12}O_6$

Water H_2O

Photosynthesis

Carbon dioxide + Water → Sugar + Oxygen

$$6CO_2 + 6H_2O \xrightarrow[\text{Chlorophyll}]{\text{Light energy}} C_6H_{12}O_6 + 6O_2$$

Reading Check

3. Identify How do plants capture light energy?

Key Concept Check

4. Name What are the two steps of photosynthesis?

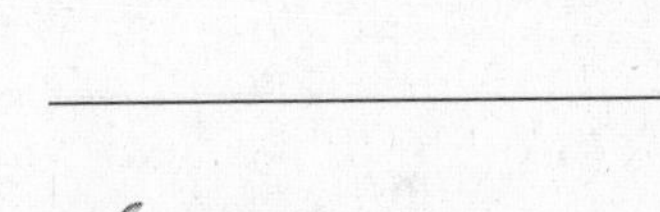

Visual Check

5. Describe Use the figure to explain to a partner the first step of photosynthesis. Then have your partner use the figure to explain the second step of photosynthesis.

Why is photosynthesis important?

Try to imagine a world without plants. The world would certainly look different, and its atmosphere would also be different. How would humans or other animals get the oxygen they need?

Plants help maintain the atmosphere you breathe. Photosynthesis produces as much as 90 percent of the oxygen in the atmosphere. Without green plants, humans would not have enough oxygen to breathe.

ACADEMIC VOCABULARY
energy
(noun) usable power

Cellular Respiration

All organisms require energy to survive. Energy is in the chemical bonds in food molecules. A process called cellular respiration releases energy. ***Cellular respiration*** *is a series of chemical reactions that convert the energy in food molecules into a usable form of energy called ATP.* ✓

✓ Reading Check
6. Name Which cellular process converts food energy into usable energy?

____

____

Releasing Energy from Sugars

Glucose is the sugar produced by photosynthesis. Glucose molecules break down during cellular respiration. Much of the energy released during this process is used to make ATP. ATP is an energy storage molecule.

Cellular respiration occurs in the cytoplasm and mitochondria of cells. This process requires oxygen and produces water and carbon dioxide as waste products.

Why is cellular respiration important?

If your body did not break down the food you eat, you would not have energy to do anything. All organisms must break down their food to produce energy. Plants produce their own food—glucose—during photosynthesis. Without cellular respiration, plants could not grow, reproduce, or repair tissues.

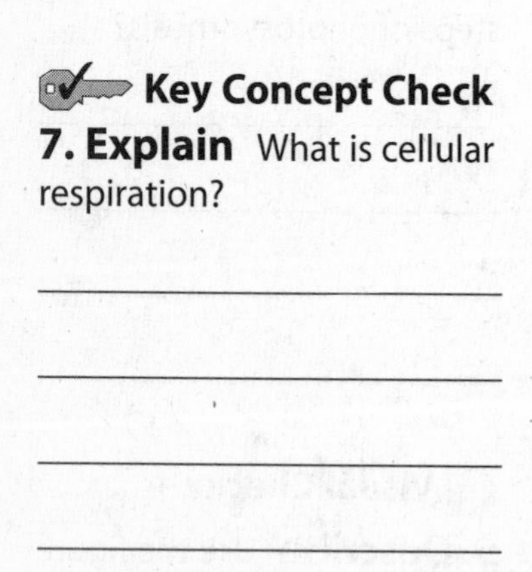

Key Concept Check
7. Explain What is cellular respiration?

____

____

____

____

Comparing Photosynthesis and Cellular Respiration

Photosynthesis requires light energy and the reactants carbon dioxide and water. Reactants are substances that react with one another during the process.

Oxygen and glucose are the products, or end substances, of photosynthesis. Glucose is a molecule of stored energy. Most plants, some protists, and some bacteria carry on photosynthesis.

Photosynthesis and Cellular Respiration Work Together

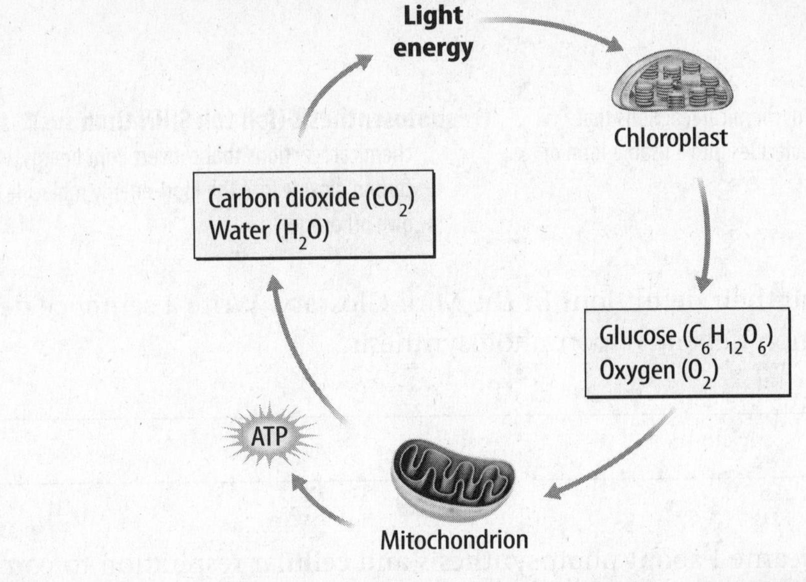

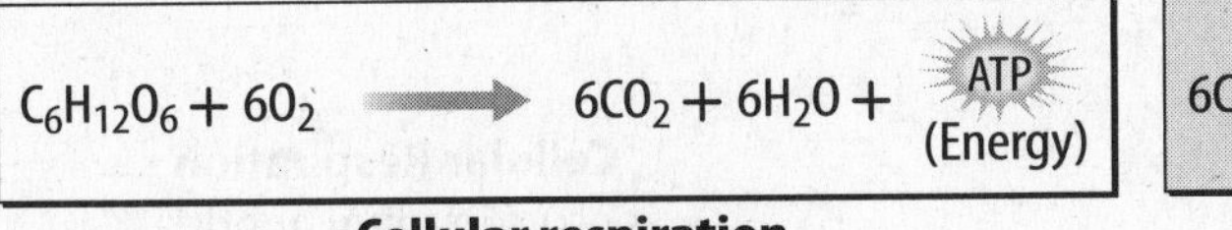

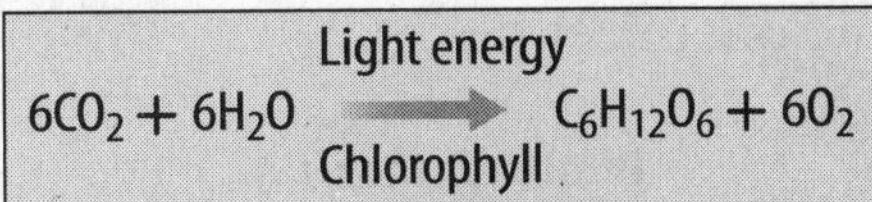

Chemical Equation for Photosynthesis Look at the chemical equation for photosynthesis on the right in the figure above. Notice that photosynthesis requires carbon dioxide (CO_2) and water (H_2O) molecules. These molecules react with light energy and produce glucose ($C_6H_{12}O_6$) and oxygen (O_2).

Chemical Equation for Cellular Respiration The chemical equation for cellular respiration is on the left in the figure above. The reactants are glucose ($C_6H_{12}O_6$) and oxygen (O_2). It produces carbon dioxide (CO_2) and water (H_2O) molecules. Cellular respiration releases energy in the form of ATP.

Most organisms carry on cellular respiration. The connection between photosynthesis and cellular respiration is shown in the table below. Life on Earth depends on a balance of these two processes.

Comparing Photosynthesis and Cellular Respiration

Process	Photosynthesis	Cellular Respiration
Reactants	light energy, CO_2, H_2O	glucose (sugar), O_2
Products	glucose, O_2	CO_2, H_2O, ATP
Organelle in which it occurs	chloroplasts	mitochondria
Type of organism	photosynthetic organisms including plants and algae	most organisms, including plants and animals

Visual Check

8. Distinguish Refer to the figure above and the table below. What are the reactants of cellular respiration? What are the products?

____

____

____

Key Concept Check

9. Compare How are photosynthesis and cellular respiration alike, and how are they different?

____

____

____

____

____

After You Read

Mini Glossary

cellular respiration: a series of chemical reactions that convert the energy in food molecules into a usable form of energy called ATP

photosynthesis (foh toh SIHN thuh sus): a series of chemical reactions that convert light energy, water, and carbon dioxide into the food-energy molecule glucose and give off oxygen

1. Review the terms and their definitions in the Mini Glossary. Write a sentence describing how cellular respiration depends upon photosynthesis.

___

___

2. Use what you have learned about photosynthesis and cellular respiration to complete the graphic organizer below.

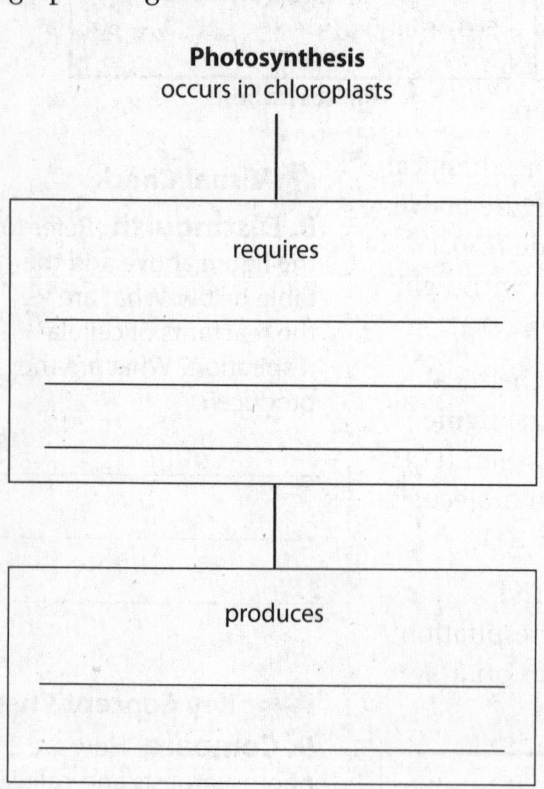

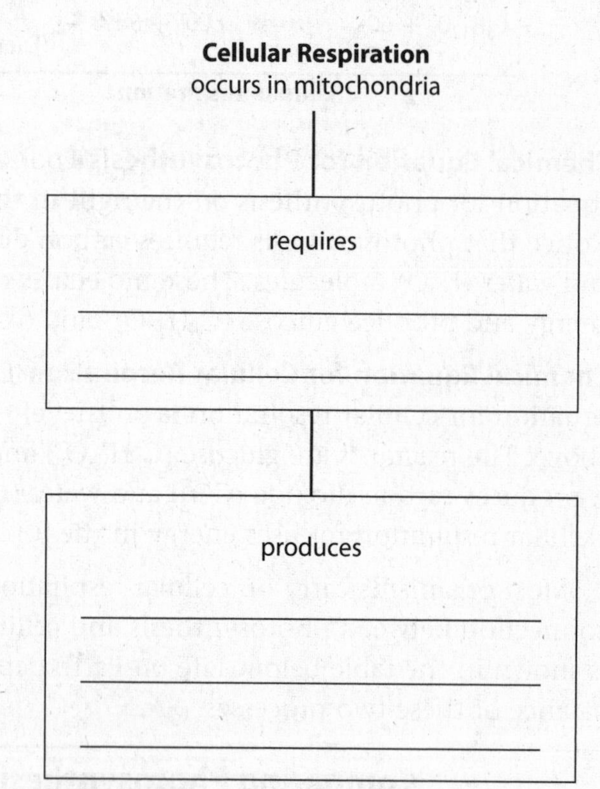

What do you think NOW?

Reread the statements at the beginning of the lesson. Fill in the After column with an A if you agree with the statement or a D if you disagree. Did you change your mind?

Log on to ConnectED.mcgraw-hill.com and access your textbook to find this lesson's resources.

Plant Processes and Reproduction

Plant Responses

Before You Read

What do you think? Read the three statements below and decide whether you agree or disagree with them. Place an A in the Before column if you agree with the statement or a D if you disagree. After you've read this lesson, reread the statements to see if you have changed your mind.

Before	Statement	After
	4. Plants do not produce hormones.	
	5. Plants can respond to their environments.	
	6. All plants flower when nights are 10–12 hours long.	

Key Concepts

- How do plants respond to environmental stimuli?
- How do plants respond to chemical stimuli?

Read to Learn

Stimuli and Plant Responses

Have you ever been in a dark room when someone suddenly turned on a light? How did you react when the light suddenly came on? You might have shut your eyes or covered them. Organisms can respond to changes in their environments in many different ways. In this lesson, you will learn how plants respond to environmental and chemical stimuli.

Stimuli (STIM yuh li; singular, stimulus) *are any changes in an organism's environment that cause a response.* Many plant responses to stimuli occur slowly. In fact, they are so slow that it is hard to see them happen. The response might occur gradually over a period of hours or days. Light is a stimulus. The stems and leaves of many houseplants grow toward a window. The plants are responding to the light stimulus that comes through the window. This response occurs gradually over many hours.

The response to a stimulus can be quick. For example, a Venus flytrap is a plant with unusual leaves that close when a stimulus, such as a fly, brushes against hairs on the leaf. The trap snaps shut, like jaws, when stimulated by an insect touching the leaf. The insect is trapped inside the plant.

Mark the Text

Identify the Main Ideas Write a phrase beside each paragraph that summarizes the main point of the paragraph. Use the phrases to review the lesson.

Reading Check

1. Determine Why is it sometimes hard to see a plant's response to a stimulus?

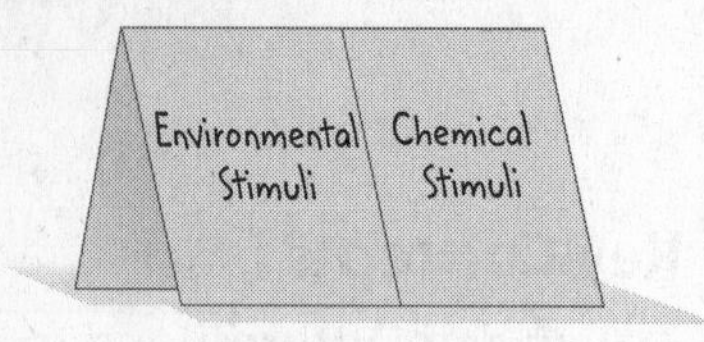

FOLDABLES®

Make a two-tab book to record what you learn about the two types of stimuli that affect plant growth.

Environmental Stimuli

Plants respond to their environments in a variety of ways. In the spring, some trees flower and new, green leaves sprout. In the fall, the same trees drop their leaves. Both are plant responses to environmental stimuli.

Growth Responses

Plants respond to different environmental stimuli. These include light, touch, and gravity. *A* **tropism** (TROH pih zum) *is a response that results in plant growth toward or away from a stimulus.* Positive tropism is growth toward a stimulus. Negative tropism is growth away from a stimulus.

Light Look at the figure below. The plant's growth toward or away from light is a tropism called phototropism. A light-sensing chemical in a plant helps it detect light. Leaves and stems tend to grow in the direction of light.

Recall that photosynthesis occurs in a plant's leaves, and photosynthesis requires light. By growing in the direction of light, the plant's leaves are exposed to more light. Roots generally grow away from light. This usually means that the roots of the plant grow down into the soil and help anchor the plant. ✓

Reading Check

2. Evaluate How is phototropism beneficial to a plant?

____

____

____

Touch Thigmotropism (thihg MAH truh pih zum) is the name given to a plant's response to touch. You might have seen vines growing up the side of a building. Some plants have special structures that respond to touch. These structures are called tendrils. The tendrils wrap around or cling to objects, such as when vine tendrils coil around a blade of grass. This is positive thigmotropism.

Roots display negative thigmotropism. They grow away from objects in soil, enabling them to follow the easiest path through the soil.

Response to Light

Visual Check

3. Explain Use the figure to explain to a partner what happened to the plant.

Gravity A plant's response to gravity is called gravitropism. Stems grow away from gravity, so they show a negative gravitropism. Roots grow toward gravity, showing a positive gravitropism.

When a seed lands in the soil and starts to grow, its roots will always grow down into the soil. The stem grows up. This will happen even when a seed is grown in a dark chamber. This shows that the response of the root and the stem can occur independently of light.

Flowering Responses

Flowering is a plant response to environmental stimuli. Some plants flower in response to the amount of darkness they are exposed to. **Photoperiodism** *is a plant's response to the number of hours of darkness in its environment.* Scientists once hypothesized that photoperiodism was a response to light. For that reason, the flowering responses are called long-day, short-day, and day-neutral. The names relate to the number of hours of daylight in a plant's environment. Scientists now know that plants respond to the number of hours of darkness.

Long-Day Plants Plants that flower when exposed to less than 10 to 12 hours of darkness are called long-day plants. Long-day plants usually produce flowers in summer. During the summer, the number of hours of daylight is greater than the number of hours of darkness.

Short-Day Plants Short-day plants begin to flower when there are 12 or more hours of darkness. A poinsettia is an example of a short-day plant. Poinsettias tend to flower in the late summer or early fall when the number of hours of darkness is increasing.

Day-Neutral Plants The number of hours of darkness doesn't seem to affect the flowering of some plants. These plants are called day-neutral plants. These plants flower when they reach maturity and the other environmental conditions are right. Roses are day-neutral plants.

Chemical Stimuli

Plants respond to chemical stimuli as well as environmental stimuli. **Plant hormones** *are substances that act as chemical messengers within plants.* These chemicals are produced in tiny amounts. They are called messengers because the chemicals are usually produced in one part of a plant but affect another part of that plant.

Key Concept Check

4. Identify What types of environmental stimuli do plants respond to? Give three examples.

Reading Check

5. State Why are plant responses named according to length of day?

Reading Check

6. Identify How is the flowering of day-neutral plants affected by exposure to hours of darkness?

Auxins

One plant hormone is auxin (AWK sun). There are many different kinds of auxins. Plant cells respond to auxins with increased growth. The growth of leaves toward light is a response to auxin. Auxins concentrate on the dark side of a plant's stem, and these cells grow longer. This causes the stem to grow toward the light. The figure below shows auxin on the left side of the seedling. It causes more growth on the left side, leading the seedling to bend to the right.

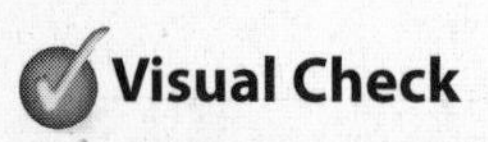

Visual Check

7. Explain Use the figure to explain to a partner how auxins have affected the growth of this seedling.

Response to Auxins

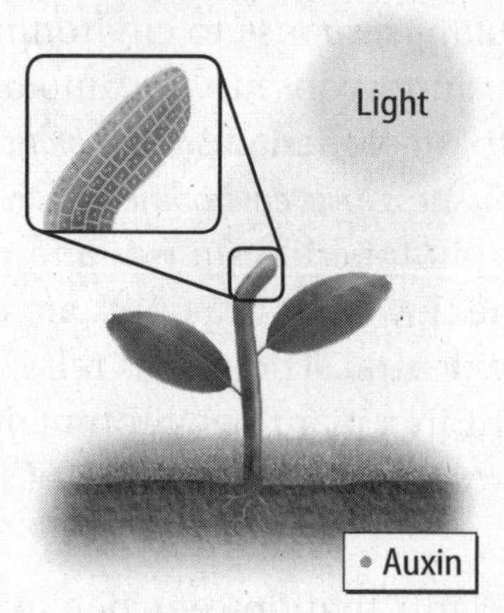

Ethylene

The plant hormone ethylene helps fruit ripen. Ethylene is a gas that can be produced by fruits, seeds, flowers, and leaves. Have you heard the expression "one rotten apple spoils the whole barrel"? This is based on the fact that rotting fruits release ethylene. This can cause other fruits nearby to ripen and possibly rot. Ethylene also can cause plants to drop their leaves.

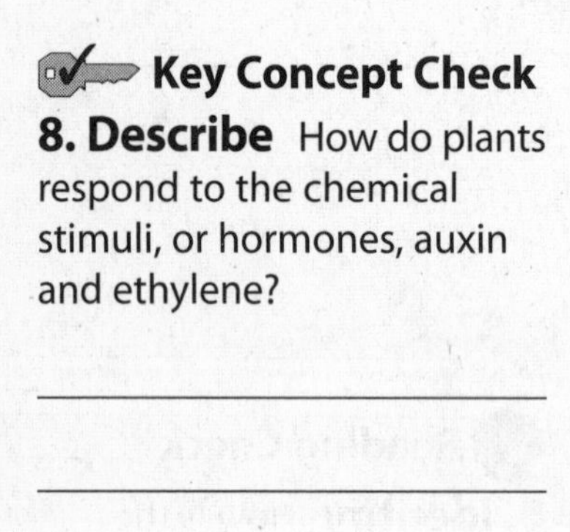

Key Concept Check

8. Describe How do plants respond to the chemical stimuli, or hormones, auxin and ethylene?

Gibberellins and Cytokinins

Rapidly growing areas of a plant, such as roots and stems, produce gibberellins (jih buh REL unz). These hormones increase the rate of cell division and cell elongation. This results in the increased growth of stems and leaves. Sometimes gibberellins are applied to the outside of plants to encourage plant growth. Fruit-producing plants can be treated with gibberellins to produce more fruit and larger fruit.

Cytokinins (si tuh KI nunz) are another type of hormone. They are produced mostly in root tips. Xylem carries cytokinins to other parts of a plant. Cytokinins increase the rate of cell division. In some plants, cytokinins slow the aging process of flowers and fruits.

Summary of Plant Hormones

Plants produce many different hormones. The hormones discussed in this lesson are groups of similar compounds. Often, two or more hormones interact and produce a plant response. Scientists are still discovering new information about plant hormones.

Humans and Plant Responses

Humans depend on plants for food, fuel, shelter, and clothing. Humans use plant hormones to make plants more productive. Some crops have become easier to grow because humans understand how the plants respond to hormones. For example, bananas and tomatoes can be picked and shipped while they are still green. They can then be treated with ethylene to make them ripen. ✓

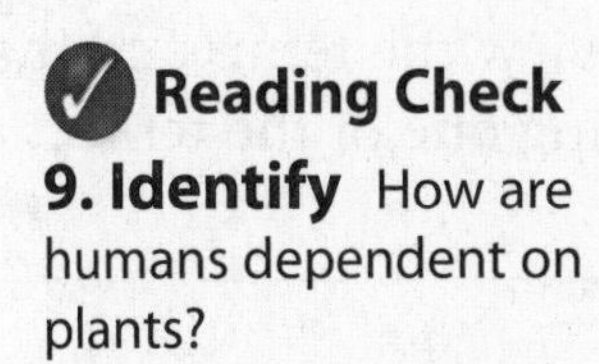

Math Skills

A percentage is a ratio that compares a number to 100. For example, if a plant grows 2 cm per day with no chemical stimulus and 3 cm per day with a chemical stimulus, what is the percentage increase in growth?

Subtract the original value from the final value.

$$3\text{ cm} - 2\text{ cm} = 1\text{ cm}$$

Set up a ratio between the difference and the original value. Find the decimal equivalent.

$$\frac{1\text{ cm}}{2\text{ cm}} = 0.5\text{ cm}$$

Multiply by 100 and add a percent sign.

$$0.5 \times 100 = 50\%$$

10. Use Percentages Without gibberellins, pea seedlings grew to 2 cm in 3 days. With gibberellins, the seedlings grew to 4 cm in 3 days. What was the percentage increase in growth?

After You Read

Mini Glossary

photoperiodism: a plant's response to the number of hours of darkness in its environment

plant hormone: a substance that acts as a chemical messenger within plants

stimulus: a change in an organism's environment that causes a response

tropism (TROH pih zum): a response that results in plant growth toward or away from a stimulus

1. Review the terms and their definitions in the Mini Glossary. Write your own sentence using one of the terms.

2. Complete the chart below to summarize what you have learned about environmental stimuli.

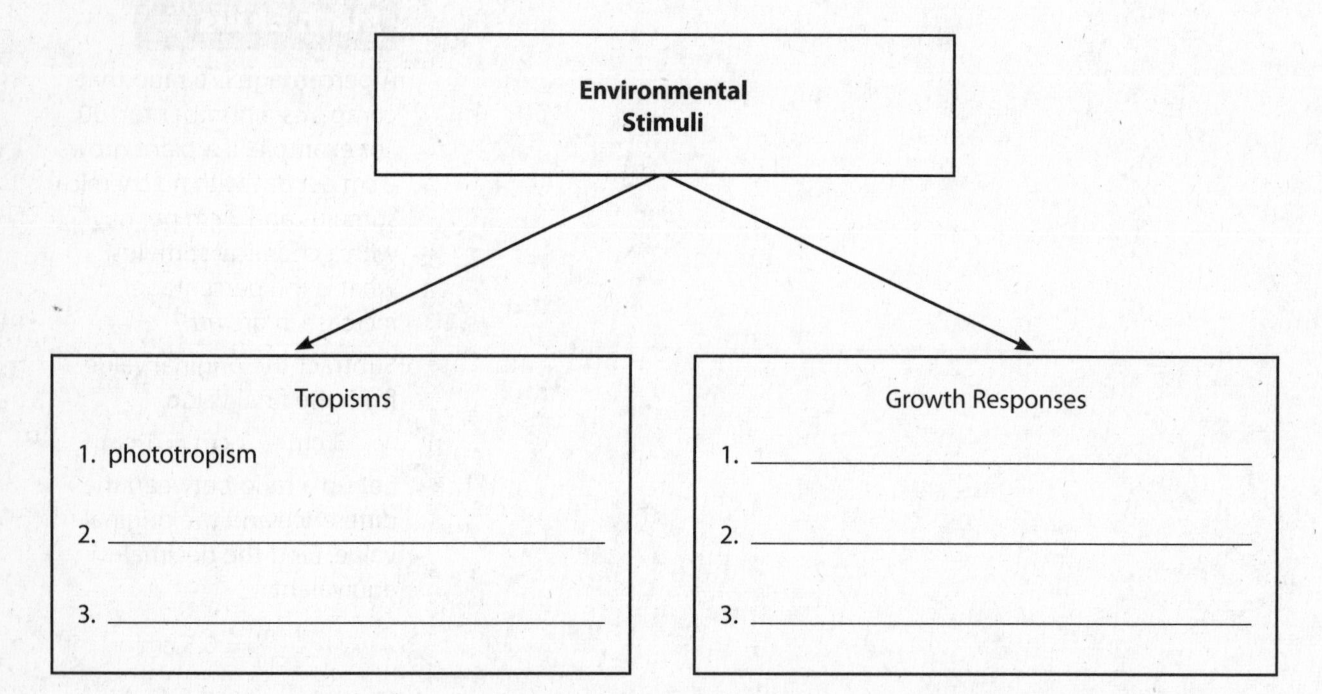

3. How did writing a phrase next to each paragraph help you identify the main ideas?

What do you think NOW?

Reread the statements at the beginning of the lesson. Fill in the After column with an A if you agree with the statement or a D if you disagree. Did you change your mind?

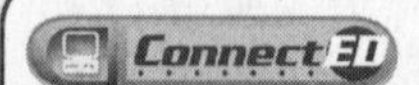

Log on to ConnectED.mcgraw-hill.com and access your textbook to find this lesson's resources.

Plant Processes and Reproduction

Plant Reproduction

Before You Read

What do you think? Read the two statements below and decide whether you agree or disagree with them. Place an A in the Before column if you agree with the statement or a D if you disagree. After you've read this lesson, reread the statements to see if you have changed your mind.

Before	Statement	After
	5. Seeds contain tiny plant embryos.	
	6. Flowers are needed for plant reproduction.	

Key Concepts 

- What is the alternation of generations in plants?
- How do seedless plants reproduce?
- How do seed plants reproduce?

Read to Learn

Asexual Reproduction Versus Sexual Reproduction

Plants can reproduce either asexually, sexually, or both ways. Asexual reproduction occurs when part of a plant develops into a separate new plant. The new plant is genetically the same as the original, or parent, plant. Irises and daylilies are plants that reproduce asexually through their underground stems. Houseleeks also reproduce asexually. Horizontal stems called stolons grow from the main plant. New plants grow at the ends of the stolons.

Asexual reproduction requires just one parent organism to produce offspring. Sexual reproduction in plants usually requires two parent plants. Sexual reproduction occurs when a plant's sperm combines with a plant's egg. The new plant that results is a genetic combination of its parents.

Study Coach

Building Vocabulary Make a vocabulary card for each bold term in this lesson. Write each term on one side of the card. On the other side, write the definition. Use these terms to review the vocabulary for the lesson.

Alternation of Generations

Most human cells are diploid. Sperm and eggs are the only human haploid cells. The human life cycle includes only a diploid stage. This isn't true for all organisms. Plants, for example, have two life stages called generations. One generation is almost all diploid cells. The other generation has only haploid cells. **Alternation of generations** *occurs when the life cycle of an organism alternates between diploid and haploid generations.*

Key Concept Check

1. Define What is alternation of generations in plants?

Alternation of Generations

Fertilization → Diploid zygote → Diploid plant → Meiosis → Haploid spores → Haploid plant → Sperm, Egg → Fertilization

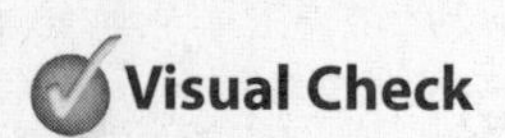

Visual Check

2. Identify Highlight in one color the diploid generation. Highlight in another color the haploid generation.

The Diploid Generation

Look at the figure above. A tree or a flower is part of a plant's diploid generation. Meiosis occurs in certain cells in the reproductive structures of a diploid plant. *The daughter cells produced from haploid structures are called* **spores.** Spores grow by mitosis and cell division. They form the haploid generation of a plant.

Reading Check

3. Explain How do spores grow?

____

____

____

____

The Haploid Generation

In most plants, the haploid generation is tiny and lives surrounded by tissues of the diploid plant. In other plants the haploid generation lives on its own. Certain reproductive cells in the haploid generation produce haploid sperm or eggs by mitosis and cell division. Fertilization takes place when a sperm and an egg fuse and form a diploid zygote. The zygote grows into the diploid generation of a plant through mitosis and cell division.

Reproduction in Seedless Plants

Some plants grow from haploid spores, not from seeds. They are known as seedless plants. The first land plants to inhabit Earth were probably seedless plants. Mosses and ferns are examples of seedless plants found on Earth today.

REVIEW VOCABULARY

mitosis
the process during which a nucleus and its contents divide

Life Cycle of a Moss

The tiny, green moss plants that grow in moist areas are haploid plants. These plants grow by mitosis and cell division from haploid spores produced by the diploid generation. They have male structures that produce sperm and female structures that produce eggs. After fertilization, the diploid zygote grows by mitosis and cell division into the diploid generation of moss. A diploid moss is tiny and hard to see.

Life Cycle of a Fern

An alternation of generations is also seen in the life cycle of a fern. The leafy, green plants that grow in many forests are the diploid generations. These plants produce haploid spores. The spores grow into tiny haploid plants. The haploid plants produce eggs and sperm that can unite and form the diploid generations.

How do seed plants reproduce?

Most land plants on Earth grow from seeds. There are two groups of seed plants—flowerless seed plants and flowering seed plants.

The haploid generation of all seed plants is within diploid tissue. Separate diploid male reproductive structures produce haploid sperm. Separate diploid female reproductive structures produce haploid eggs. The haploid sperm and the haploid egg join during fertilization.

The Role of Pollen Grains

A **pollen** (PAH lun) **grain** *forms from tissue in a male reproductive structure of a seed plant.* Each pollen grain has a hard outer covering that protects it. All the nutrients the pollen grain needs are contained inside the covering. Pollen grains produce sperm cells. Wind, animals, gravity, or water currents can carry pollen grains to female reproductive structures.

Plants can't move on their own. They do not find mates as most animals do. The male reproductive structures of plants produce a large number of pollen grains. **Pollination** (pah luh NAY shun) *occurs when pollen grains land on a female reproductive structure of a plant that is the same species as the pollen grains.*

The Role of Ovules and Seeds

The female reproductive structure of a seed plant where the haploid egg develops is called the **ovule.** After pollination, sperm enter the ovule and fertilization occurs. A zygote forms and develops into an embryo. *An* **embryo** *is an immature diploid plant that develops from the zygote. An embryo, its food supply, and a protective covering make up a* **seed.** A seed's food supply provides the embryo with the nourishment it needs for its early growth.

Key Concept Check

4. Describe How do seedless plants such as mosses and ferns reproduce?

Reading Check

5. Identify What occurs during fertilization?

Key Concept Check

6. Describe How do seed plants reproduce?

Visual Check

7. Name three things the seeds have in common.

FOLDABLES

Make a two-tab book to record information about reproduction in flowerless and flowering plants.

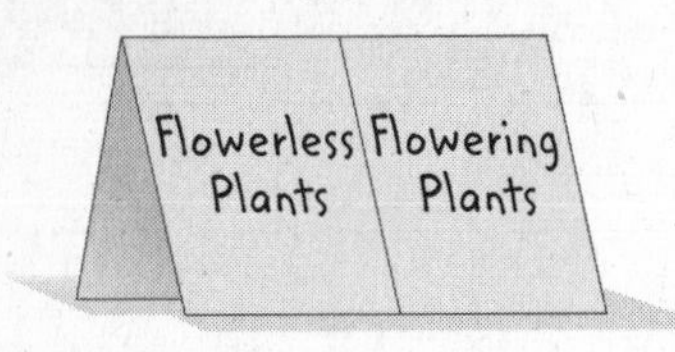

Reading Check

8. Describe Where is the haploid generation of conifers contained?

Seed Structures

Food supply
Covering
Embryo
Corn
Embryo
Food supply
Covering
Food supply
Bean
Food supply
Embryo
Covering
Pine

Corn, bean, and pine seeds are shown in the figure above. A seed contains a diploid plant embryo and a food supply protected by a hard outer covering.

Reproduction in Flowerless Seed Plants

Flowerless seed plants are also known as gymnosperms (JIHM nuh spurmz). The word *gymnosperm* means "naked seed." Gymnosperm seeds are not surrounded by a fruit.

The most common gymnosperms are conifers. Conifers, such as pines, firs, cypresses, redwoods, and yews, are trees and shrubs with needlelike or scalelike leaves. Most conifers are evergreens, meaning they keep their leaves all year.

Life Cycle of a Gymnosperm The life cycle of a gymnosperm includes an alternation of generations. The life cycle is shown in the figure at the top of the next page.

Cones are the male and female reproductive structures of conifers. They contain the haploid generation. Male cones are small structures that produce pollen grains. Female cones can be woody, berrylike, or soft. They produce eggs.

Male cones release clouds of pollen grains containing the sperm. A zygote forms when a sperm fertilizes an egg. The zygote is the beginning of the diploid generation. Seeds form as part of the female cone.

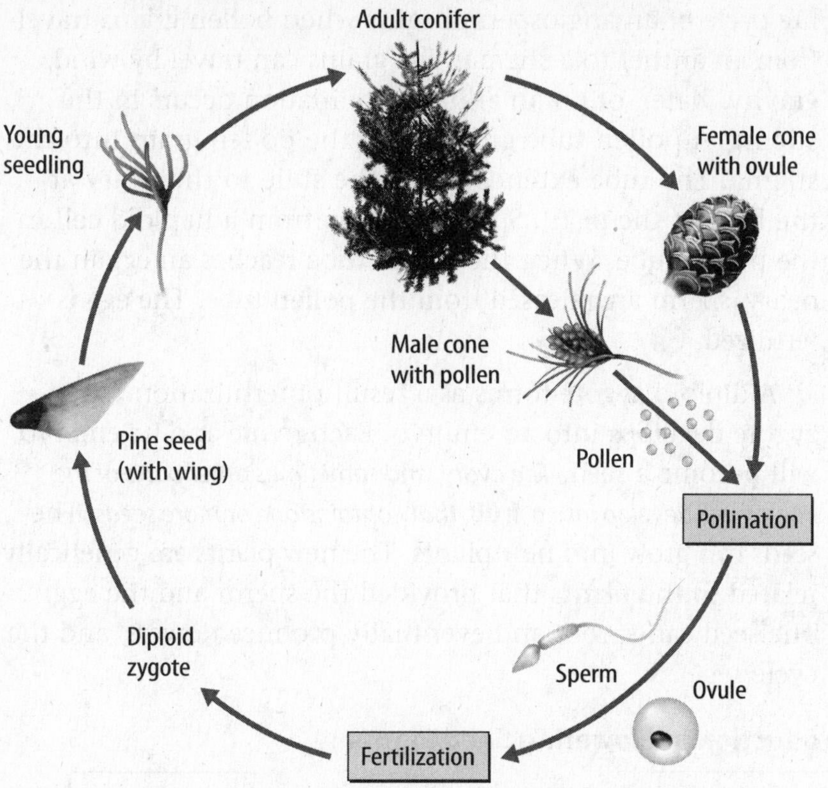

Reproduction in Flowering Seed Plants

Flowering seed plants are called angiosperms. Fruits and vegetables come from angiosperms. Many animals depend on angiosperms for food.

The Flower Reproduction of an angiosperm begins in a flower. Most flowers have male and female reproductive structures. See the figure below.

The male reproductive organ of a flower is the **stamen.** Pollen grains form at the tip of the stamen in the anther. The anther is connected to the base of the flower by the filament. *The female reproductive organ of a flower is the* **pistil.** Pollen can land on the stigma at the tip of the pistil. The stigma is at the top of a long tube called the style. *At the base of the style is the* **ovary,** *which contains one or more ovules.* Each ovule eventually will contain a haploid egg and might become a seed if fertilized.

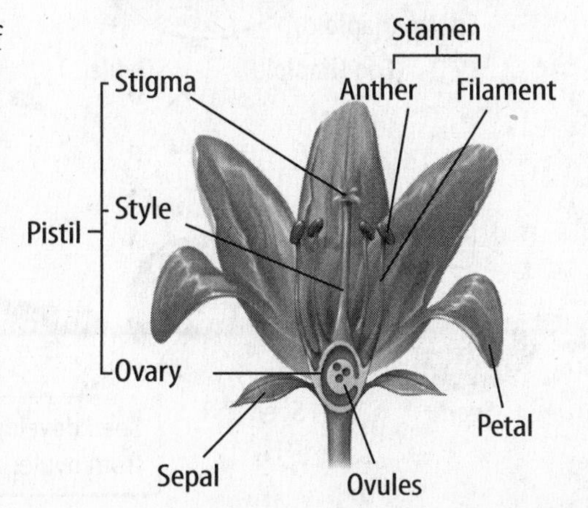

Visual Check

9. Identify Circle each structure of the haploid generation.

Reading Check

10. Name What is another name for flowering seed plants?.

____

Visual Check

11. Identify In which structure do pollen grains form?

____

____

Reading Check

12. Determine Do sperm develop before or after pollination?

Visual Check

13. Determine How does sperm in a pollen grain reach an egg in the ovule?

Life Cycle of an Angiosperm Look at the figure below. The life cycle of an angiosperm begins when pollen grains travel from an anther to a stigma. The grains can travel by wind, gravity, water, or on an animal. Pollination occurs in the stigma. A pollen tube grows from the pollen grain into the stigma. The tube extends down the style to the ovary at the base of the pistil. Sperm develop from a haploid cell in the pollen tube. When the pollen tube reaches an egg in the ovary, sperm are released from the pollen tube. The egg is fertilized.

A diploid zygote forms as a result of fertilization. The zygote develops into an embryo. Each ovule and its embryo will become a seed. *The ovary, and sometimes other parts of the flower, will develop into a* **fruit** *that contains one or more seeds.* The seeds can grow into new plants. The new plants are genetically related to the plants that provided the sperm and the egg. The seed can sprout and eventually produce flowers, and the cycle repeats.

Reproduction in Flowering Seed Plants

Haploid pollen grain in anther of stamen
Stigma
Pollen grains land on stigma.
Pollen grain
Pollen tube
Stamen
Anther
Filament
Style
Ovary
Ovule
Mature diploid plant produces flowers.
Pollen tube grows toward an egg in ovary. Sperm are released from the pollen tube and fertilize the egg.
FERTILIZATION
Ovule
Sperm (haploid)
Egg (haploid)
Ovule
Zygote (diploid)
Embryo (diploid)
Fruit develops from ovary.
Seed
Seed
Seed develops from ovule.
Diploid seedling
Sprouting seed

Fruit and Seed Dispersal Fruits and seeds are important sources of food for people and animals. In most cases, seeds of flowering plants are inside fruits. For example, pods are the fruits of pea plants. The peas inside the pods are the seeds of a pea plant. Each ear of corn is made up of many fruits, or kernels. The main part of each kernel is the seed. Strawberries have tiny seeds on the outside of the fruit.

Many fruits are juicy and good to eat, such as an orange or a watermelon. However, some fruits are hard and dry and not good to eat. Each parachute-like structure of a dandelion is a dry fruit.

Fruits help protect seeds and help scatter, or disperse, them. The fruits of a dandelion are light and float on air currents. When an animal eats a fruit, the fruit's seeds can pass through the animal's digestive system with little or no damage to the seed. Imagine a mouse eating a blackberry. The animal digests the fruit but deposits the seeds on the soil with its wastes. The mouse might have traveled some distance before depositing the seeds. This means the mouse helped move the seeds to a location away from the blackberry bush.

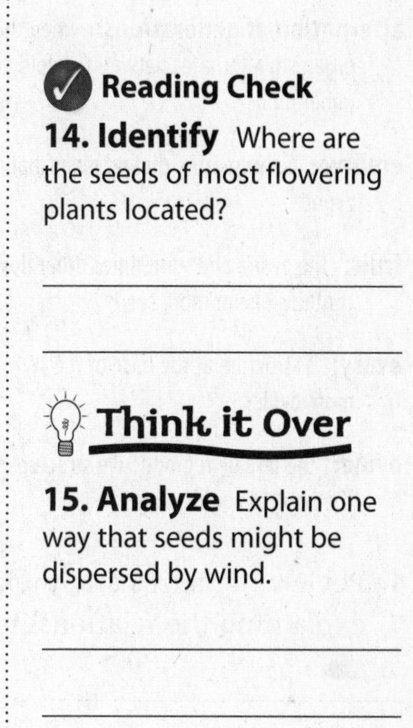

After You Read

Mini Glossary

alternation of generations: when the life cycle of an organism alternates between diploid and haploid generations

embryo: an immature diploid plant that develops from the zygote

fruit: the ovary, and sometimes other flower parts, that contains one or more seeds

ovary: a structure at the base of the style that contains one or more ovules

ovule: the female reproductive structure of a seed plant where the haploid egg develops

pistil: the female reproductive organ of a flower

pollen (PAH lun) grain: forms from tissue in a male reproductive structure of a seed plant

pollination (pah luh NAY shun): occurs when pollen grains land on a female reproductive structure of a plant that is the same species as the pollen grains

seed: an embryo, its food supply, and a protective covering

spore: a daughter cell produced from a haploid structure

stamen: the male reproductive organ of a flower

1. Review the terms and their definitions in the Mini Glossary. Write one sentence explaining the relationship between an embryo and a seed.

2. Write the letter of each statement below in the correct box to show whether the statement relates to the reproduction of seedless plants, the reproduction of seed plants, or both. The first one has been done for you as an example.

   a. Has alternation of generations
   b. The haploid egg develops in the ovule.
   c. Can be gymnosperms or angiosperms
   d. Produces spores
   e. Both pollination and fertilization are part of reproduction.
   f. Produces fruit
   g. Grows by mitosis and cell division

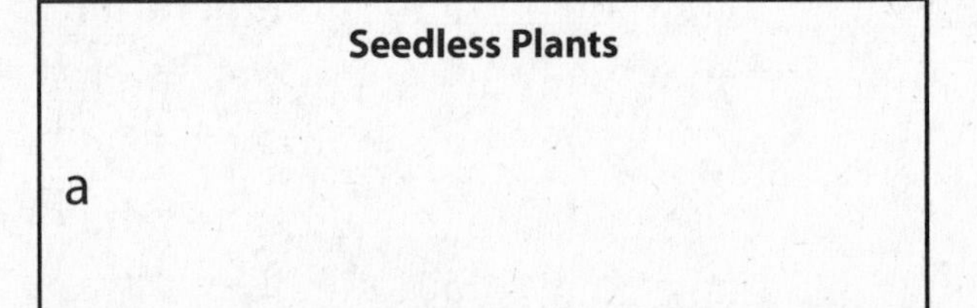

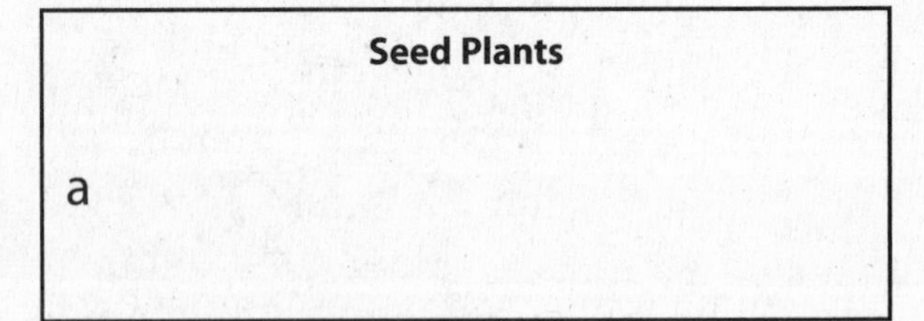

What do you think NOW?

Reread the statements at the beginning of the lesson. Fill in the After column with an A if you agree with the statement or a D if you disagree. Did you change your mind?

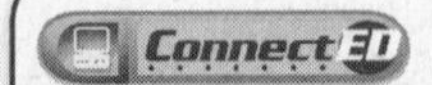

Log on to ConnectED.mcgraw-hill.com and access your textbook to find this lesson's resources.

Interactions of Living Things

Ecosystems and Biomes

Before You Read

What do you think? Read the two statements below and decide whether you agree or disagree with them. Place an A in the Before column if you agree with the statement or a D if you disagree. After you've read this lesson, reread the statements to see if you have changed your mind.		
Before	**Statement**	**After**
	1. An ecosystem contains both living and nonliving things.	
	2. All changes in an ecosystem occur over a long period of time.	

Key Concepts
- What are ecosystems?
- What are biomes?
- What happens when environments change?

Read to Learn

What are ecosystems?

You, a wolf, and a pine tree are all living things. Living things are also called organisms. All organisms use energy and do certain things to survive. Organisms interact with parts of the environment around them. Ecology is the study of how organisms interact with each other and with their environments.

Every organism lives in an ecosystem. *An* **ecosystem** *is all the living and nonliving things in one place*. Different organisms depend on different parts of an ecosystem to survive. For example, a deer eats the plants and drinks the water available in its woodland environment. The plants that the deer eats are alive; the water it drinks is not alive. A deer needs both the plants and the water to survive. A fish in the stream needs water to survive. But it interacts differently with the water than the deer does.

Mark the Text

Building Vocabulary Read all the headings in this section and circle any word that you cannot define. Then underline the part of the text that helps you define each circled word.

Key Concept Check

1. Define What is an ecosystem?

Abiotic Factors

Water is an example of a part of an ecosystem that was never alive. **Abiotic factors** *are the nonliving parts of an ecosystem*. Abiotic factors include water, light, temperature, atmosphere, and soil. Ecosystems have different types and amounts of abiotic factors. The types and amounts of these factors in an ecosystem help determine which organisms can live there.

Water

All organisms need water to live, but some need more water than others. A cactus grows in a desert, where it does not rain often. Ferns and vines live in rain forests, where it rains often. The limited water in a desert means that ferns and vines could not live successfully there.

The type of water in an ecosystem also helps determine which organisms can live there. Some organisms need saltwater environments, such as oceans. Humans and other organisms must have freshwater to survive.

FOLDABLES®

Choose an ecosystem. Make a two-tab book and use it to describe the abiotic and biotic factors that might be found in that ecosystem.

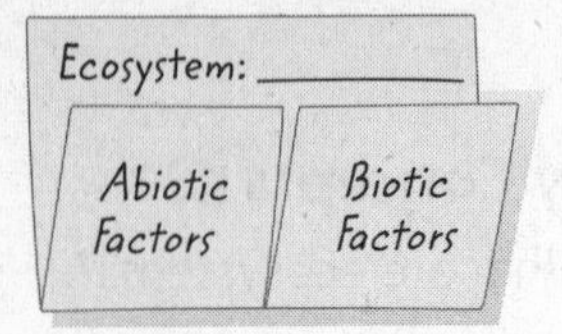

Light and Temperature

The amount of light an ecosystem receives and the temperature of an ecosystem can also determine which organisms can live there. Many organisms, such as plants, require light energy for making food.

Temperatures in ecosystems vary. Ecosystems that receive more sunlight generally have higher temperatures. Some organisms need higher temperatures to survive. A fern, for example, grows in a warm rain forest. Other organisms can survive a wide range of temperatures. For example, a cactus survives in the hot days and cold nights of a desert.

Atmosphere

Very few organisms can live in an ecosystem without oxygen. Earth's atmosphere contains oxygen gas as well as other gases that organisms need. Some of these gases are water vapor, carbon dioxide, and nitrogen.

SCIENCE USE V. COMMON USE

atmosphere

Science Use the mix of gases surrounding a planet

Common Use a surrounding influence or feeling

Soil

Different ecosystems have soil that contains different amounts and types of nutrients, minerals, and rocks. The texture and amount of water soil can hold also varies. Soil is deeper in some ecosystems than in others. All of these factors determine which organisms can live in an ecosystem.

Biotic Factors

You read that nonliving, or abiotic, parts of an ecosystem are important to living things. **Biotic factors** *are all of the living or once-living things in an ecosystem.* A parrot and a fallen tree are both biotic factors in a rain-forest ecosystem.

2. Apply Suppose an ecosystem is made up of 20 elk of the same species and 10 bears of the same species. How many populations does the ecosystem contain?

____

____

Populations

A **population** *is made up of all the members of one species that live in an area.* For example, all the gray squirrels in a neighborhood are a population. Organisms in a population interact and compete for food, shelter, and mates.

Communities

Most ecosystems have many populations. These populations form a community, as shown below. *A* **community** *is all the populations living in an ecosystem at the same time*. For example, populations of trees, worms, insects, and toads are part of a forest community.

Populations interact with each other in some way. Trees lose their leaves in the fall and the leaves become food for worms and insects. Toads might use the leaves as they hide from predators. Waste from these animals provides nutrients to the trees and insects.

Biomes

The populations and communities that interact in a desert are different from those that interact in an ocean. Deserts and oceans are different biomes. *A* **biome** *is a geographic area on Earth that contains ecosystems with similar biotic and abiotic features*. Biomes contain ecosystems, populations, and communities. Biomes also have specific biotic and abiotic factors. As a result, biomes can be very different from each other.

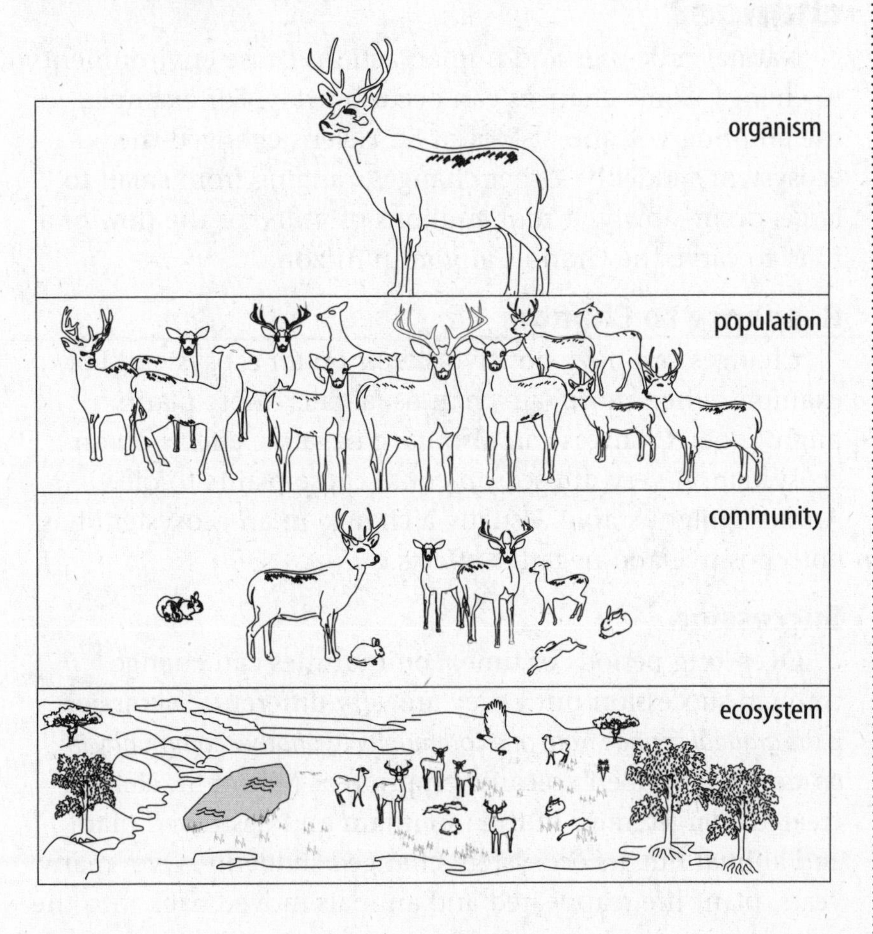

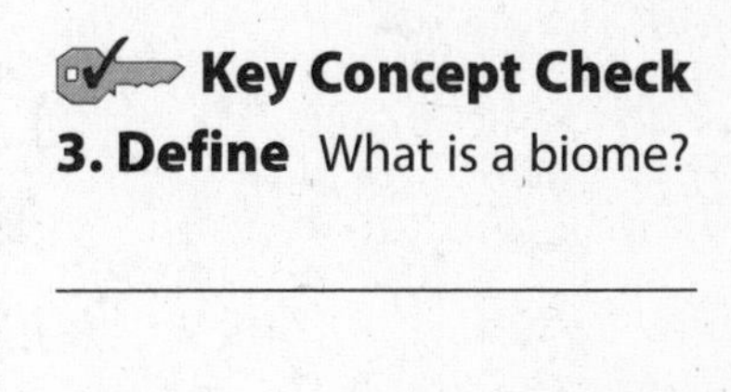

Key Concept Check

3. Define What is a biome?

Visual Check

4. Identify The figure shows two different populations in the community. Circle the members of one population in one color. Circle the members of the other population in another color.

Earth's Biosphere	
Terrestrial Biomes	**Aquatic Biomes**
forests	salt water
deserts	freshwater
tundra	
grasslands	

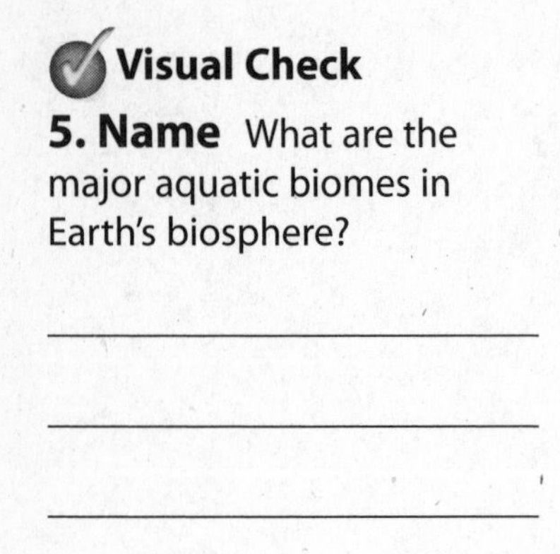

Visual Check

5. Name What are the major aquatic biomes in Earth's biosphere?

The part of Earth that supports life is the biosphere. As shown above, both terrestrial and aquatic biomes are part of the biosphere. *Terrestrial* means "related to land," and *aquatic* means "related to water." Terrestrial biomes include forests, deserts, tundra, and grasslands. Aquatic biomes include saltwater areas and freshwater areas. Biomes can affect each other. For example, a beach ecosystem is part of both a terrestrial and an aquatic biome. Some organisms from the terrestrial biome interact with organisms in the beach ecosystem.

What happens when environments change?

Natural processes and human actions cause environments to change. Some changes can occur quickly. For example, the erupting volcano at Mount St. Helens changed the ecosystem suddenly. Other changes, ranging from small to large, occur slowly. It took millions of years for the flow of a river to carve the Grand Canyon in Arizona.

Response to Change

Changes can have positive effects on an ecosystem. For example, when more rain than usual falls, more plants might grow. Changes can also have negative effects on an ecosystem. A very dry season could cause plants to die. Animals might starve. Usually a change in an ecosystem has both positive and negative effects.

Succession

Over long periods of time, communities can change through succession until they are very different. **Succession** *is the gradual change from one community to another community in an area.* The sudden volcanic eruption at Mount St. Helen's created a large crater in the mountain and destroyed plant and animal life. Its ecosystem changed suddenly. Over many years, plant life reappeared and animals moved back into the area. The area looks very different today than it did immediately after the eruption in 1980.

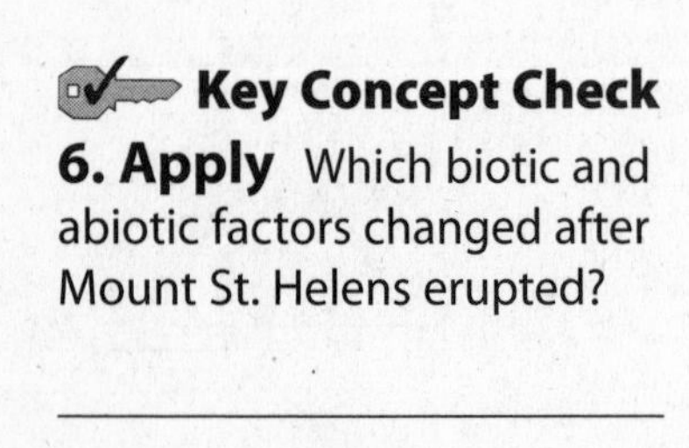

Key Concept Check

6. Apply Which biotic and abiotic factors changed after Mount St. Helens erupted?

After You Read

Mini Glossary

abiotic factors: the nonliving parts of an ecosystem

biome: a geographic area on Earth that contains ecosystems with similar biotic and abiotic features

biotic factors: all of the living or once-living things in an ecosystem

community: the populations living in an ecosystem at one time

ecosystem: all the living and nonliving things in one place

population: all the members of one species that live in an area

succession: the gradual change from one community to another community in an area

1. Review the terms and their definitions in the Mini Glossary. Write a sentence that explains the difference between a population and a community.

____

____

2. Write each example below in the correct part of the diagram.

woodpecker	**oak tree**	**rocks**
sunlight	**dead grasshopper**	**freshwater in a pond**

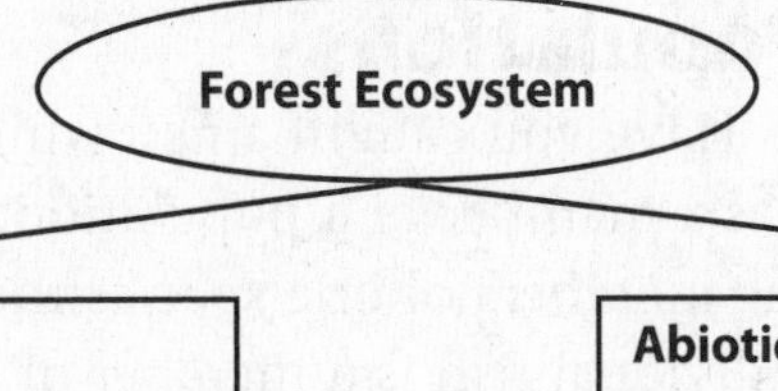

Biotic factors	Abiotic factors
•	•
•	•
•	•

3. How did finding the meanings of words help you learn about the biotic factors that make up an ecosystem?

____

____

What do you think 

Reread the statements at the beginning of the lesson. Fill in the After column with an A if you agree with the statement or a D if you disagree. Did you change your mind?

Log on to ConnectED.mcgraw-hill.com and access your textbook to find this lesson's resources.

Interactions of Living Things

Populations and Communities

Key Concepts

- How do individuals and groups of organisms interact?
- What are some examples of symbiotic relationships?

Before You Read

What do you think? Read the two statements below and decide whether you agree or disagree with them. Place an A in the Before column if you agree with the statement or a D if you disagree. After you've read this lesson, reread the statements to see if you have changed your mind.

Before	Statement	After
	3. Changes that occur in an ecosystem can cause populations to become larger or smaller.	
	4. Some organisms have relationships with other types of organisms that help them to survive.	

Study Coach

Create a Quiz After you read this lesson, write five questions about what you learned. Exchange quizzes with another student. Together, discuss the answers to your quizzes.

Read to Learn

Populations

Have you caught a fish while fishing in a lake? That fish was a member of a population. Recall that a population is all the members of one species that live in an area. An individual fish is a member of a population. That population of fish might live in a small area, such as a lake, or in a large area, such as an ocean.

You learned that abiotic factors and biotic factors in an ecosystem affect the organisms in the ecosystem. Sunlight, temperature, and water quality are some of the abiotic factors that affect the fish in a lake and in the ocean. Biotic factors that affect the fish include plants that they eat and other organisms that hunt them. If any of these factors change, the fish population can also change.

Think it Over

1. Draw a Conclusion Can a trout and a catfish be members of the same population? Explain why or why not.

Population Size

Think about the fish in the lake. What might happen to the fish population if a large number of fish eggs hatched? Now consider what might happen to the fish population if hundreds of people caught fish from the lake. The figure on the next page shows possible changes that can happen to the fish population of a lake over time.

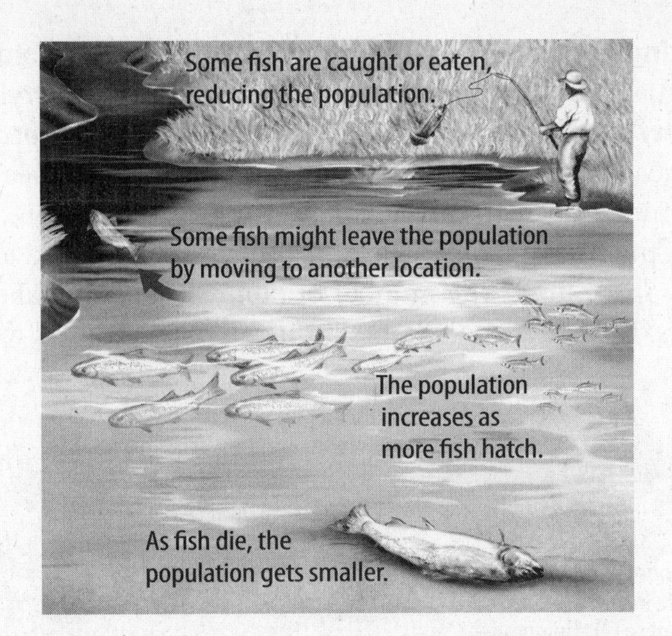

Population size can increase, or it can decrease. If more fish move into the area or more fish hatch, the fish population increases. The fish population can decrease if more people catch fish or if some fish move to another location. Sometimes the size of a population changes because the ecosystem changes. For example, when there is little rainfall, a pond might become smaller and some fish might die.

Population Density

Population density describes the number of organisms in the population compared to the amount of space available. A dense population might have one fish for every few cubic meters of water. If a population is dense, organisms might not find enough resources to live. They might not grow as large as individuals in less-crowded conditions.

Limiting Factors

What keeps a population from becoming too large? **Limiting factors** *are factors that can limit the growth of a population.* The amount of water, space, shelter, and food affects a population's size. When there are not enough resources, some individuals will die. Other factors, such as predators, competition, and disease, can also limit how many individuals in a population survive.

Biotic Potential Imagine a population of fish with unlimited food and water and no predators. The population would keep growing until it reached its biotic potential. **Biotic potential** *is the potential growth of a population if it could grow in perfect conditions with no limiting factors.* The population's rate of birth is the highest it can be. Its rate of death is the lowest it can be.

Visual Check

2. Infer How do you think fish hatching affects the other populations in the community?

Think it Over

3. Predict What would happen to a population of fish if one hundred people caught fish in the lake?

Reading Check

4. Name four limiting factors.

Carrying Capacity Almost no population reaches its biotic potential. Instead, it reaches its carrying capacity. **Carrying capacity** *is the largest number of individuals of one species that an ecosystem can support over time.* Limiting factors in an area determine the area's carrying capacity, as shown below. The rabbit population competes for resources such as food and water. Space limits the number of rabbits who can make homes. Disease and predators limit the population as well.

Visual Check

5. Identify List the limiting factors described in the figure.

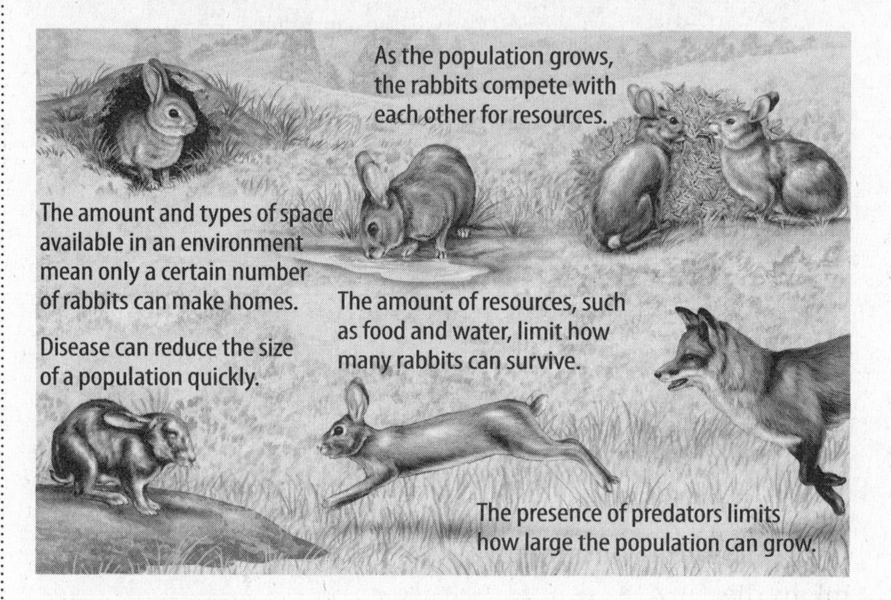

Overpopulation An area is overpopulated when a population grows larger than its carrying capacity. In other words, the population has grown beyond the area's ability to support it. Overpopulation often results in overcrowding, a lack of resources, and an unhealthful environment. A trout, for example, might not grow very large in overcrowded conditions. Waste from a large trout population might build up faster than it can be broken down. The waste buildup might make the population sick.

Reading Check

6. Explain Why is overpopulation harmful to organisms?

Communities

Recall that populations in the same area interact as a community. Think about the fish in a lake. Many other populations of organisms live in and around the lake. These include different species of fish, frogs, algae, bacteria, insects, plants, raccoons, and other organisms. All the populations of the lake community interact with each other in different ways. Populations might compete with each other for some of the resources available in the lake. The organisms of each population must have a certain amount of space in which to live. Some populations hunt each other for food or compete for places to hide from predators.

Key Concept Check

7. Discuss How do the different populations in a lake interact with each other?

Symbiotic Relationships

Populations affect their community by the ways they interact with each other. Each population has different ways to stay alive and reproduce. All populations in a community share a habitat. *A **habitat** is the physical place where a population or organism lives.* Each organism also has a niche in the community. *A **niche** is the unique ways an organism survives, obtains food and shelter, and avoids danger in its habitat.*

Some organisms develop relationships with other organisms that help them survive. *A **symbiotic relationship** is one in which two different species live together and interact closely over a long period of time.* Some symbiotic relationships benefit both organisms. Other relationships benefit one organism and harm the other organism. Some relationships benefit one organism, but have no effect on the other organism.

Mutualism

Mutualism is a symbiotic relationship in which two species in a community benefit from the relationship. Leafcutter ants collect plant material and bring it back to their nest. They do not eat the leaves, but they chew the leaves into small pieces. A fungus grows on the small pieces of leaves. The ants eat the fungus. The leafcutter ants have provided a place for the fungus to grow. In turn, the fungus becomes a food source for the ants. The ants and the fungus have a mutualistic relationship.

Parasitism

Parasitism is a symbiotic relationship in which one species (the parasite) benefits while another (the host) is harmed. Mistletoe is a parasite. It grows in the branches of trees, sending its roots into the tissues of its host, the tree. The mistletoe takes food and water from the tree. This can weaken the tree and might eventually kill it.

Commensalism

Commensalism is a symbiotic relationship in which one species benefits and the other is neither helped nor harmed. For example, cocklebur plants produce spiny seeds. These spiny burs with hooks are seed pods. They stick to passing animals and humans when they touch them. The plants' seeds are spread to other areas as the animals and humans move from place to place. The plants benefit from their seeds being spread. Humans and animals are not harmed or helped by the cocklebur.

FOLDABLES®

Use a three-tab book to organize your notes about the types of symbiotic relationships.

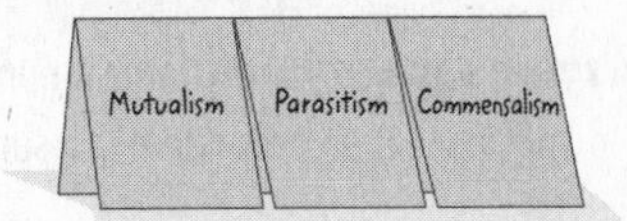

Think it Over

8. Contrast How is mutualism different from parasitism?

______

______

______

Key Concept Check

9. Identify What is one example of a symbiotic relationship?

______

______

After You Read

Mini Glossary

biotic potential: the potential growth of a population if it could grow in perfect conditions with no limiting factors

carrying capacity: the largest number of individuals of one species that an ecosystem can support over time

habitat: the physical place where a population or organism lives

limiting factor: a factor that can limit the growth of a population

niche: the unique ways an organism survives, obtains food and shelter, and avoids danger in its habitat

symbiotic relationship: one in which two different species live together and interact closely over a long period of time

1. Review the terms and their definitions in the Mini Glossary. Write a sentence that explains how limiting factors and biotic potential are related.

____

____

2. The table below describes changes in the environment where a population of deer live. Predict whether each change will increase or decrease the size of the deer population. Place a check mark in the correct box in the table.

Effect on Population of Deer

Change in the Environment	Increase	Decrease
a. A company dumped harmful waste into a stream where the deer drink.		
b. Deer hunters have been very successful this season.		
c. A disease has killed many wolves that prey on deer.		
d. Winter was especially harsh this year.		
e. Heavy rains this spring helped the grass that the deer eat grow very thick.		

What do you think NOW?

Reread the statements at the beginning of the lesson. Fill in the After column with an A if you agree with the statement or a D if you disagree. Did you change your mind?

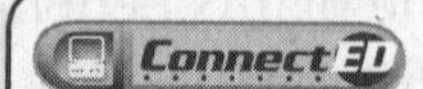

Log on to ConnectED.mcgraw-hill.com and access your textbook to find this lesson's resources.

Interactions of Living Things

Energy and Matter

Before You Read

What do you think? Read the two statements below and decide whether you agree or disagree with them. Place an A in the Before column if you agree with the statement or a D if you disagree. After you've read this lesson, reread the statements to see if you have changed your mind.

Before	Statement	After
	5. Most of the energy used by organisms on Earth comes from the Sun.	
	6. Both nature and humans affect the environment.	

Key Concepts 

- How does energy move in ecosystems?
- How is the movement of energy in an ecosystem modeled?
- How does matter move in ecosystems?

Read to Learn

Energy Flow

The food you eat is your energy source. It gives you fuel to walk, play games, read books, sit at a desk, and even sleep. All living things need energy for cell processes.

Some organisms get energy from food that they make using light or chemical energy. Other organisms get energy by eating other organisms. The energy from the organism that is eaten is transferred to the organism that eats it. In this way, energy travels through organisms, populations, communities, and ecosystems in a flow. A flow, like the one shown below, is different from a cycle. Energy that moves in a flow does not return to its source, as it does when matter cycles.

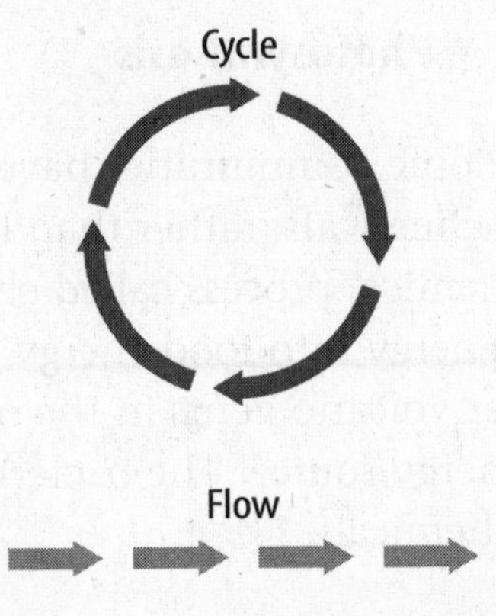

Mark the Text

Identify the Main Points Highlight the main idea of each paragraph. In another color, highlight the details that support the main idea to help you study this lesson.

Visual Check

1. Contrast Look at the figure. How is energy in a flow different from matter in a cycle?

Organisms and Energy

All organisms need energy to survive. Scientists classify organisms by the way they get the energy they need. Almost all energy on Earth comes from the Sun. Some organisms use the Sun's energy directly. Plants directly transform the Sun's energy into the energy-rich sugars that they use for food. A few organisms can make food using the energy from chemicals in the environment. Other organisms cannot capture energy from sunlight or chemicals and make food. They must get energy by eating food. Organisms that cannot make their own food using the Sun must depend on organisms that can.

ACADEMIC VOCABULARY
transform
(verb) to change in composition or structure

Producers

Recall that energy cannot be created or destroyed, but it can change form. **Producers** *change the energy available in their environment into food energy.* They then use this food energy for living and reproducing. Humans and other organisms can get this energy by eating producers.

Reading Check
2. Explain Why must producers be present in an environment?

Photosynthesis Energy from the Sun always enters a community through producers. Some producers use a chemical process called photosynthesis to transform light energy from the Sun into food energy. Producers that use light energy include most plants, algae, and some microorganisms. The figure below shows how photosynthesis converts light energy into food energy.

Visual Check
3. Summarize What happens to light energy during photosynthesis?

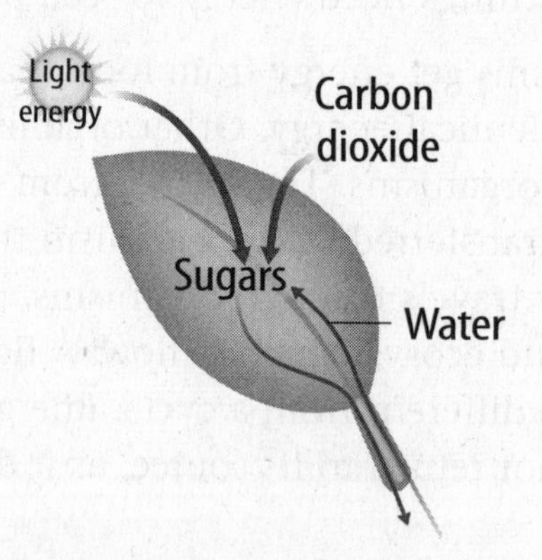

Photosynthesis

Chemosynthesis Some communities have producers that get their energy from chemicals, rather than light energy. Some producers use a chemical process called chemosynthesis to change chemical energy into food energy. For example, bacteria living near volcano vents in the ocean floor use chemicals as an energy source. The bacteria make food energy from the chemicals.

Key Concept Check
4. Describe How does energy move from a producer to other organisms?

Consumers

Organisms that cannot make food get their energy by eating other organisms. These organisms are called consumers. **Consumers** *cannot make their own food and get energy by eating other organisms.* Scientists classify consumers by what they eat. Consumers are either herbivores, omnivores, carnivores, or detritivores. Some examples are shown in the table below.

Detritivores eat dead plant and animal material. A type of detritivore called a decomposer breaks down dead material into simple molecules. These molecules can be used by other organisms, such as plants.

Type of Consumer	What They Eat	Examples
Herbivores	only producers, such as plants	cows
Omnivores	both producers and other consumers	human beings
Carnivores	only other consumers	lions
Detritivores (Decomposers)	energy and nutrients from dead plants and animals	some insects, fungi, worms, some bacteria and protists

Reading Check

5. Name What are the four types of consumers?

Modeling Energy Flow

Energy is always moving through ecosystems as organisms eat other organisms. *A* **food chain** *models how energy flows in an ecosystem through feeding relationships.* The figure below is a food chain. It shows how energy passes from a plant to a snake. Each stage of a food chain has less available food energy than the last one. Some food energy is lost to the environment as thermal energy, commonly called heat.

Visual Check

6. Predict Which type of consumer likely has the most food choice? Explain.

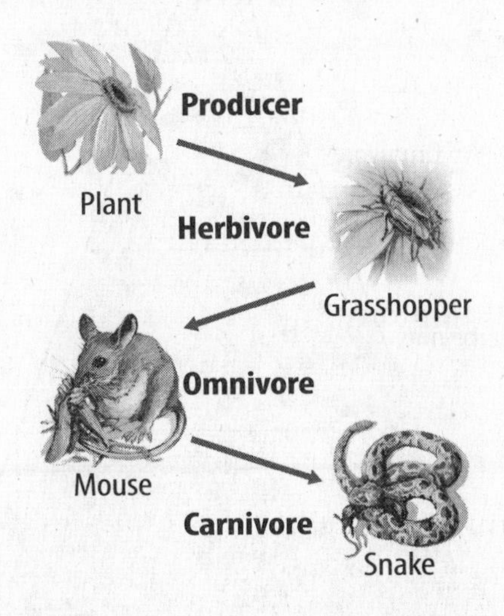

Visual Check

7. Connect How does energy move from the mouse to the snake in the food chain?

Food Webs

A **food web** *is a model that shows several connected food chains.* The figure below shows how food energy moves through a forest ecosystem. Notice that food energy can move through several different pathways.

Key Concept Check
8. Compare a food chain with a food web.

Modeling Energy Pyramids

As energy travels through different organisms, the amount of available food energy decreases. *An* **energy pyramid** *shows the amount of energy available at each step of a food chain.*

The figure below is an energy pyramid. More food energy is available at the base of the pyramid, where producers are. At each level, organisms use some of the energy and some is transformed to thermal energy. As a result, there is less food energy to pass on to the next level. Carnivores are usually at the top of the energy pyramid for an ecosystem. These are the animals that prey only on other animals.

Reading Check
9. Identify a Main Point What happens to the amount of available energy in higher levels of an energy pyramid?

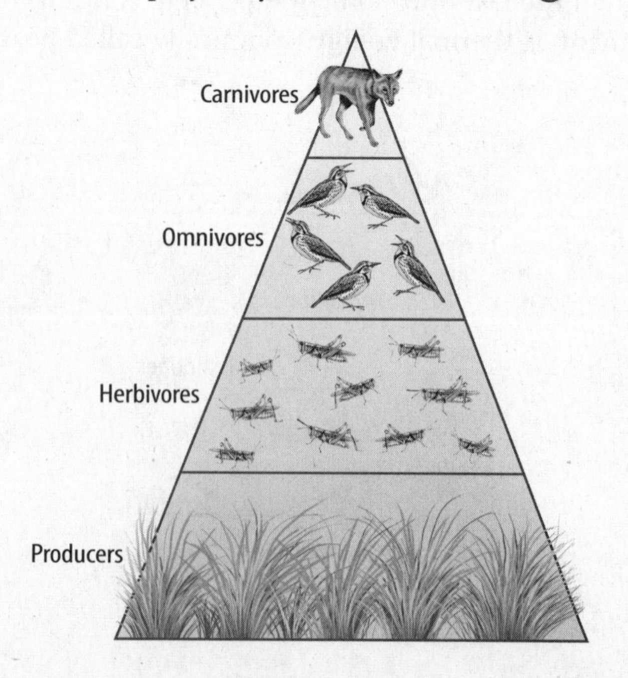

Visual Check
10. Explain Energy moves from producers to which consumer at the highest level on the energy pyramid?

Matter Cycles

Matter is the physical material that makes up Earth and everything on it. Like everything else on Earth, your body is made up of matter. Most of the matter in your body is water. Your body also contains matter in other forms, such as carbon and oxygen.

Like energy, matter is not created or destroyed. Like energy, matter is transferred through the environment. Unlike energy, matter cycles through an environment, rather than flowing through it. Matter is used again and again. The types of matter found in an environment and the amounts of that matter determine which organisms can live there.

Water Cycle

Water is important to all life. It moves through every ecosystem on Earth. Water cycles in different forms. Its forms include a liquid, a gas (water vapor), and a solid (ice). The water cycle is shown below.

- Water evaporates from oceans, rivers, and other bodies of water. Plants release water vapor during transpiration. Some organisms also release water vapor when they breathe out (exhalation).
- The water vapor then rises into the atmosphere. It condenses and falls as rain or snow (precipitation).
- Water moves across the surface of Earth in lakes, streams, and rivers. It soaks into the ground, or organisms take it in. The cycle continues when the water is released again. 

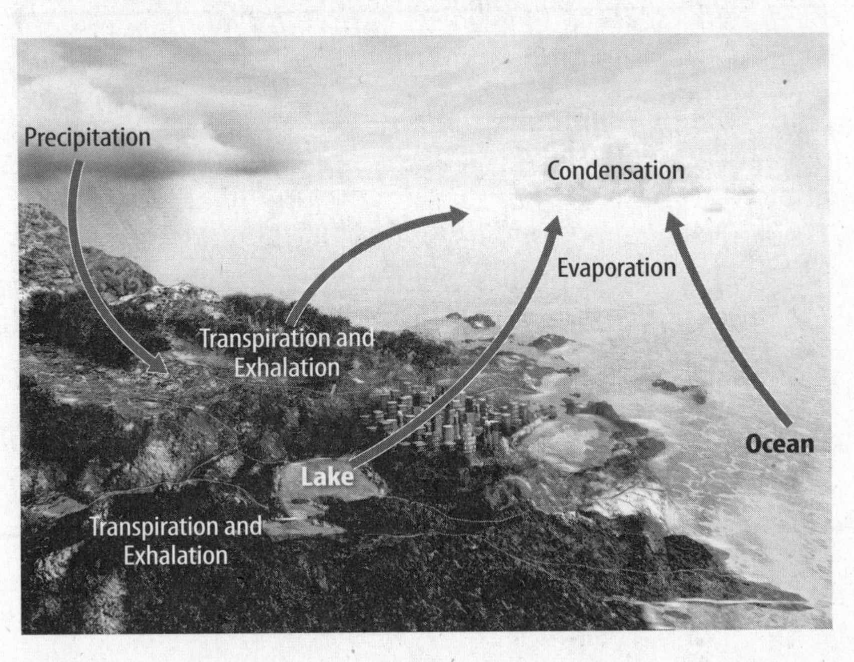

FOLDABLES®

Use a trifold book to summarize information about how matter cycles in an ecosystem.

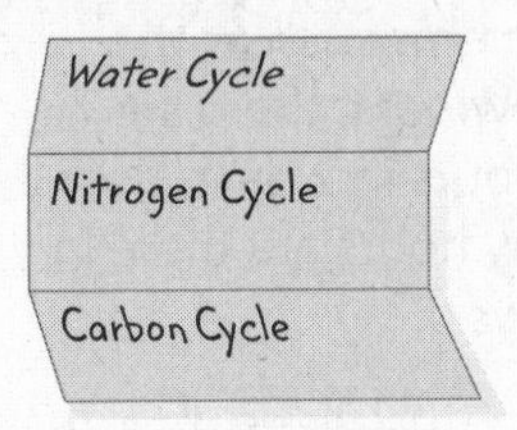

Key Concept Check

11. Specify What forms does water take in the water cycle?

____

____

____

____

Visual Check

12. Specify Circle the two processes in which water vapor condenses and falls as rain or snow.

Math Skills

When two ratios form a proportion, the cross products are equal. This method can be used to solve problems such as the following: If there are 48 hours in 2 days, how many hours are there in 5 days?

$$\frac{48 \text{ hr}}{2 \text{ days}} = \frac{n \text{ hr}}{5 \text{ days}}$$

a. Find the cross products.

$$48 \times 5 = n \times 2$$

$$240 = 2n$$

b. Solve for n by dividing both sides by 2.

$$\frac{240}{2} = \frac{2n}{2}$$

$$n = 120 \text{ hours}$$

13. Use Proportions Your heart beats an average of 84 times per minute. How many times will it beat in 5 minutes?

____

Reading Check

14. Identify What organisms are part of the oxygen cycle?

____

____

Visual Check

15. Specify What do the consumers in this image release to the atmosphere?

____

Oxygen Cycle

Like water, oxygen cycles through the environment. Oxygen is another example of matter that is important to the survival of many organisms. You take in oxygen when you breathe. Your blood carries oxygen to all parts of your body. Oxygen is also part of many molecules, such as sugars, that are important to life.

The figure below shows the oxygen cycle in a rain forest. Plants in a rain forest release large amounts of oxygen. The oxygen as a waste product of photosynthesis. The oxygen from these producers enters the atmosphere.

Consumers, such as monkeys and parrots, take in oxygen when they breathe. When these organisms exhale, they release carbon dioxide. Carbon dioxide, which is a by-product of cellular respiration, contains oxygen. Producers, such as plants, take in the carbon dioxide, and the cycle continues. ✓

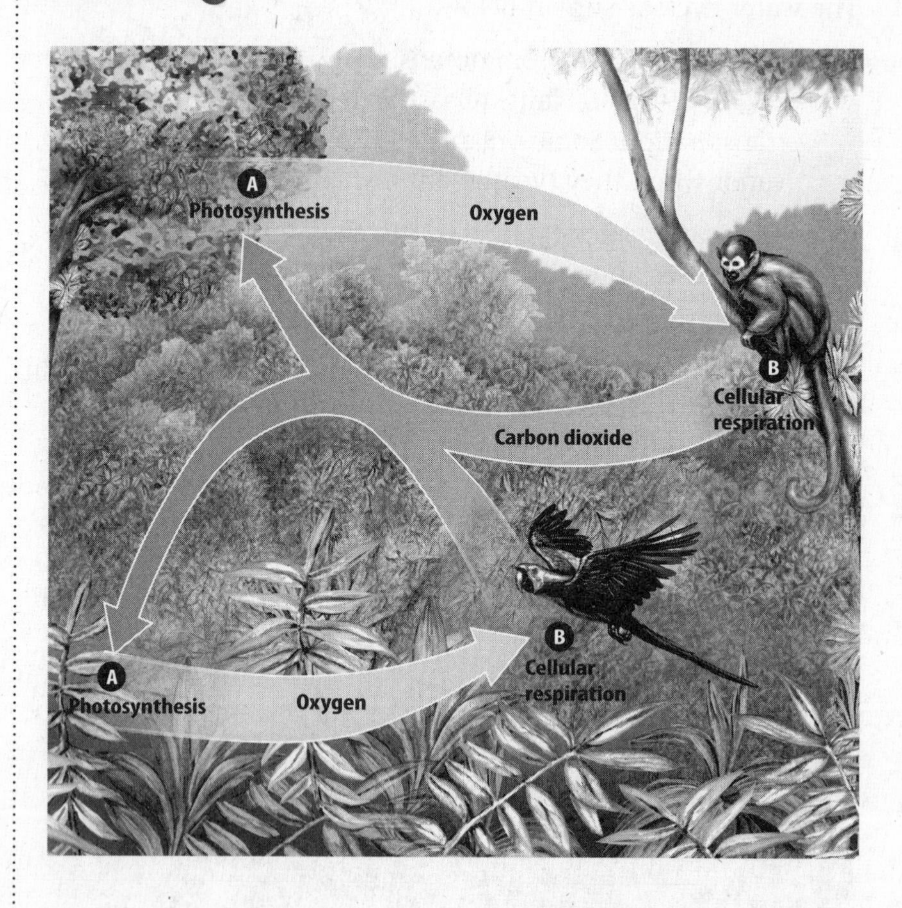

Carbon Cycle

Carbon is a basic building block of all living things. It is part of molecules such as proteins, carbohydrates, and fats. Carbon is also almost everywhere on Earth, even in nonliving things.

Producers use carbon dioxide during photosynthesis. This removes carbon from the atmosphere. Consumers eat these producers and release carbon back into the environment as a waste product. Human activities, such as the burning of fossil fuels, also add carbon to the atmosphere. Producers again remove carbon dioxide from the atmosphere as they continue making food, and the cycle continues. The carbon cycle is shown below. ✓

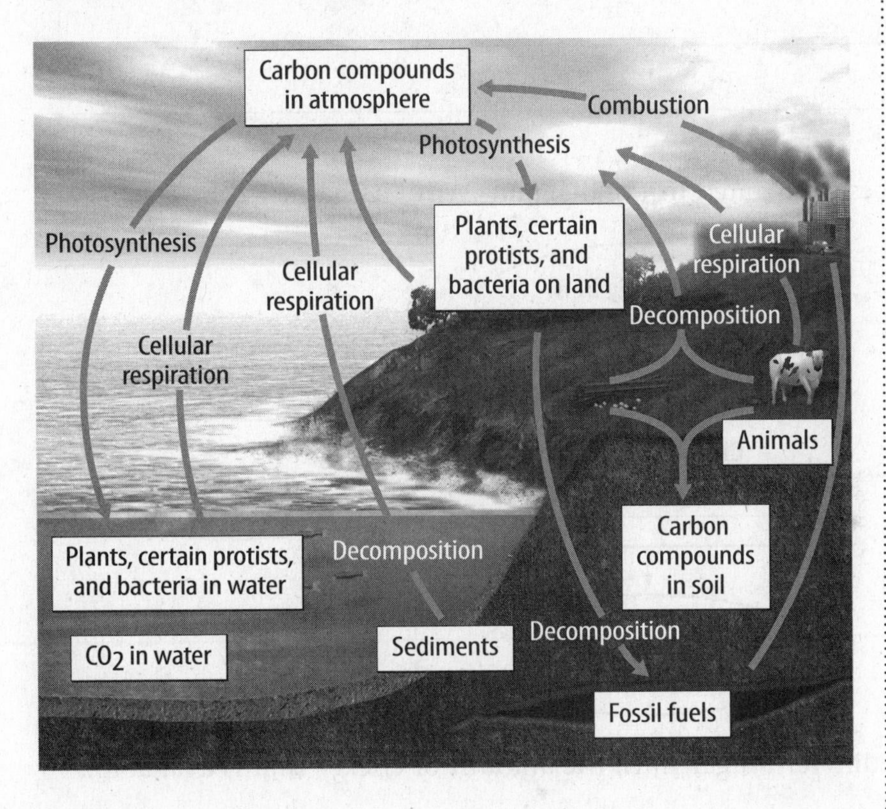

✓ **Reading Check**

16. Specify Where is carbon in living organisms?

✓ **Visual Check**

17. Distinguish What processes add carbon compounds to the atmosphere?

After You Read

Mini Glossary

consumer: an organism that cannot make its own food and gets energy by eating other organisms

energy pyramid: a model that shows the amount of energy available at each step of a food chain

food chain: a model of how energy flows in an ecosystem through feeding relationships

food web: a model that shows several connected food chains

producer: an organism that changes the energy available in its environment into food energy

1. Review the terms and their definitions in the Mini Glossary. Write a sentence that explains how consumers depend on producers.

____

____

2. Identify the chemical process illustrated in the diagram. Write its name in the box.

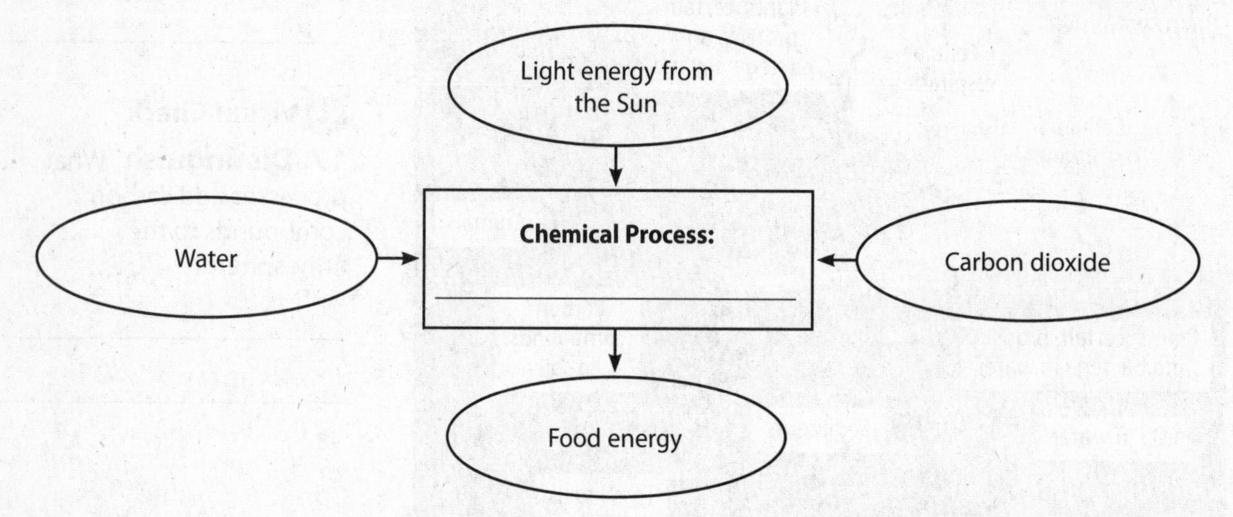

3. As energy travels through different organisms, the amount of energy that is available decreases. Why?

____

____

____

What do you think NOW?

Reread the statements at the beginning of the lesson. Fill in the After column with an A if you agree with the statement or a D if you disagree. Did you change your mind?

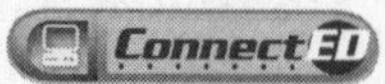

Log on to ConnectED.mcgraw-hill.com and access your textbook to find this lesson's resources.

Foundations of Chemistry

Classifying Matter

Before You Read

What do you think? Read the two statements below and decide whether you agree or disagree with them. Place an A in the Before column if you agree with the statement or a D if you disagree. After you've read this lesson, reread the statements to see if you have changed your mind.

Before	Statement	After
	1. The atoms in all objects are the same.	
	2. You cannot always tell by an object's appearance whether it is made of more than one type of atom.	

Key Concepts 

- What is a substance?
- How do atoms of different elements differ?
- How do mixtures differ from substances?
- How can you classify matter?

Read to Learn

Understanding Matter

Have you ever seen a rock that has more than one color? Why are different parts of the rock different in color? Why might some parts of the rock feel harder than other parts? The parts of the rock look and feel different because they are made of different types of matter. **Matter** *is anything that has mass and takes up space.*

Look around. Many types of matter surround you. In your classroom, you might see things made of metal, wood, or plastic. In a park, you might see trees, soil, or water in a pond. Look up at the sky. You might see clouds and the Sun. All of these things are made of matter.

Everything you can see is matter. However, some things you cannot see also are matter. Air, for example, is matter because it has mass and takes up space. Sound and light are not matter. Forces and energy also are not matter. To decide whether something is matter, ask yourself if it has mass and takes up space.

An **atom** *is a small particle that is a building block of matter.* In this lesson, you will explore the parts of an atom. You will read how atoms can differ. You also will read how different arrangements of atoms make up the many types of matter.

Study Coach

Building Vocabulary
Write each vocabulary term in this lesson on an index card. Shuffle the cards. After you have studied the lesson, take turns picking cards with a partner. Each of you should define the term using your own words.

FOLDABLES®

Make a layered Foldable to summarize the lesson.

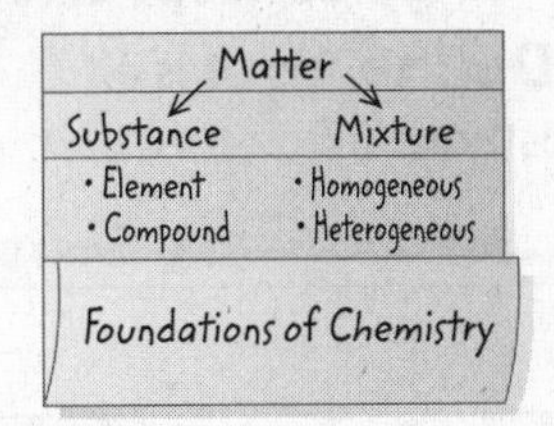

Atoms

Look at the diagram of an atom in the figure below. At the center of an atom is the nucleus. Protons and neutrons make up the nucleus. Electrons move in an area outside of the nucleus.

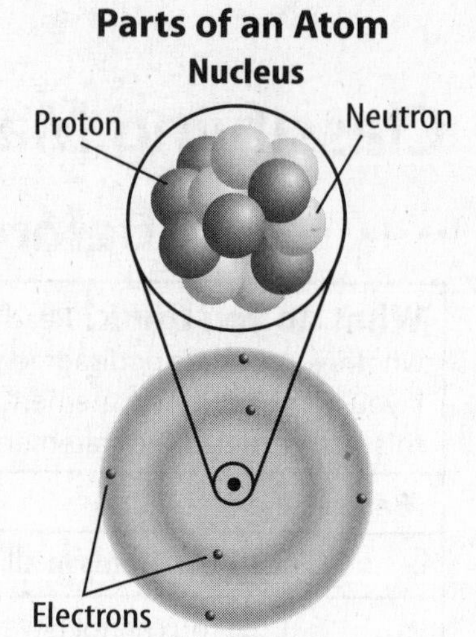

Reading Check

1. Name What are the parts of an atom?

Protons have a positive charge. Neutrons have a neutral charge. Electrons have a negative charge. Electrons move quickly throughout an area around the nucleus called the electron cloud.

Visual Check

2. Identify Which part of an atom exists outside of the nucleus?

Not all atoms have the same number of protons, neutrons, and electrons. Atoms that have different numbers of protons differ in their properties. You will read more about the differences in atoms later in this lesson.

An atom is almost too small to imagine. Think about how thin a human hair is. The diameter of a human hair is about a million times greater than the diameter of an atom. Also, an atom is about 10,000 times wider than its nucleus! Even though atoms are tiny, they determine the properties of the matter they compose.

Substances

Atoms make up most of the matter on Earth. Atoms can combine and arrange in millions of different ways. In fact, these different combinations and arrangements of atoms determine the makeup of the various types of matter. Matter can be divided into two main classifications—substances and mixtures.

A **substance** *is matter with a composition that is always the same.* This means that a given substance is always made up of one or more atoms in the same combinations. Aluminum, oxygen, water, and sugar are examples of substances. A sample of any one of these substances, or any other substance, always has the same composition, or makeup. Any sample of aluminum is always made up of the same type of atoms. A sample of oxygen, sugar, or water always has the same combination of atoms as any other sample of that substance.

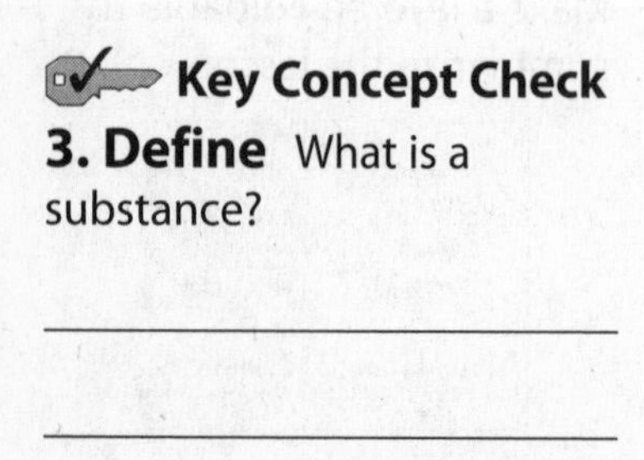

Key Concept Check

3. Define What is a substance?

There are two types of substances. Substances can be elements or compounds.

Elements

Look at the periodic table of elements on the inside back cover of this book. Find the substances oxygen and aluminum. They are elements. *An* **element** *is a substance that consists of just one type of atom*. We currently know of about 115 elements. As a result, there are about 115 different types of atoms. Each type of atom contains a different number of protons in its nucleus. For example, each aluminum atom has 13 protons. The number of protons in an atom is the atomic number of the element. As the figure below shows, the atomic number of aluminum is 13.

The atoms of most elements exist as individual atoms. A roll of pure aluminum foil consists of trillions of individual aluminum atoms. However, the atoms of some elements usually exist in groups. For example, the oxygen atoms in air exist in pairs. Whether the atoms of an element exist individually or in groups, each element contains only one type of atom. Thus, its composition is always the same.

4. Apply Find carbon (C) on the periodic table. How many protons are in the nucleus of a carbon atom? How do you know?

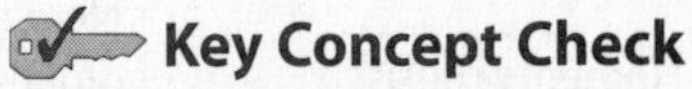

5. Describe How do atoms of different elements differ?

Elements in the Periodic Table

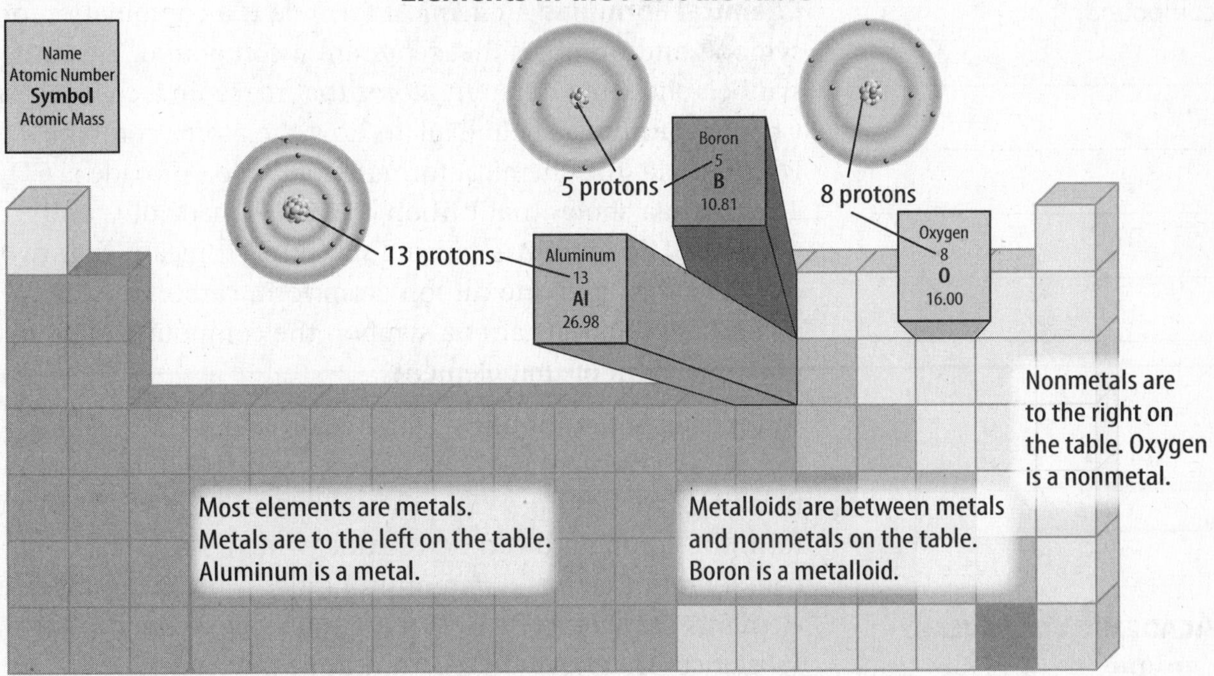

Compounds

Water is a substance, but it is not an element. It is a compound. *A* **compound** *is a type of substance containing atoms of two or more different elements chemically bonded together*. Carbon dioxide (CO_2) is also a compound. It consists of atoms of two elements, carbon (C) and oxygen (O), bonded together. Carbon dioxide is a substance because the C and the O atoms are always combined in the same way.

Visual Check

6. Contrast Which element has a higher atomic number, aluminum or oxygen?

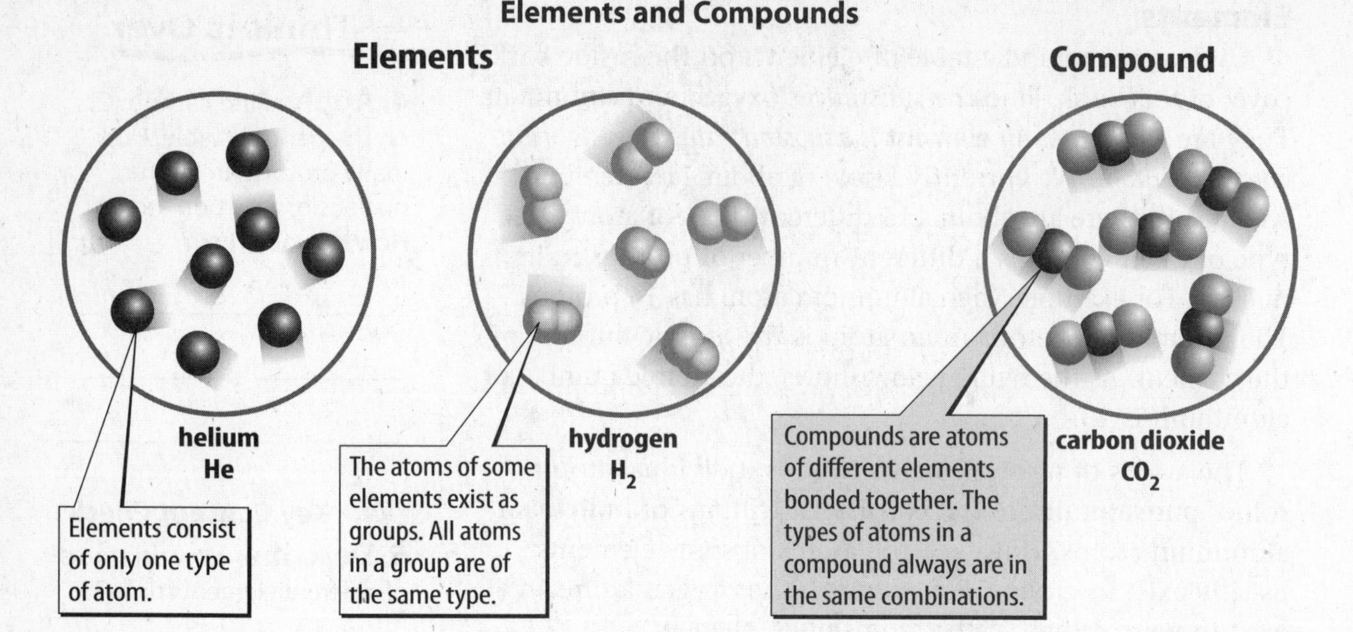

Visual Check

7. Explain Why is CO_2 a compound but H_2 is not a compound?

The figure above contrasts elements and compounds. If a substance contains only one type of atom, it is an element. If it contains more than one type of atom, it is a compound.

Chemical Formulas A chemical formula is a combination of symbols and numbers that represent a compound. Element symbols show the different atoms that make up a compound. Chemical formulas help explain how the atoms combine. For example, the chemical formula for carbon dioxide is CO_2. The formula shows that carbon dioxide is made of C and O atoms. The small 2 is called a subscript. It means that two oxygen atoms and one carbon atom form carbon dioxide. If no subscript appears after a symbol, the compound contains only one atom of that element.

Properties of Compounds Think again about the elements carbon and oxygen. Carbon is a black solid. Oxygen is a gas that helps fuels burn: When they chemically combine, they form the compound carbon dioxide, which is a gas. A compound often has different properties from the individual elements that comprise it. Compounds, like elements, are substances that have their own unique properties.

ACADEMIC VOCABULARY
unique
(adjective) having nothing else like it

Mixtures

Mixtures are another classification of matter. *A* **mixture** *is matter that can vary in composition.* Mixtures are combinations of two or more substances that are physically blended together. The amounts of the substances can vary in different parts of a mixture and from mixture to mixture. The substances in a mixture do not combine chemically. Therefore, you can separate them by physical methods.

Heterogeneous Mixtures

Mixtures can differ by how well substances mix. Sand and water at the beach form a mixture. However, the sand is not evenly mixed throughout the water. Sand and water form a heterogeneous mixture. *A* **heterogeneous mixture** *is a type of mixture in which the individual substances are not evenly mixed.*

Because the substances in a heterogeneous mixture are not evenly mixed, two samples of the same mixture can have different amounts of the substances. For example, if you fill two buckets with sand and water at the beach, one bucket might have more sand in it than the other.

Homogeneous Mixtures

Unlike a mixture of water and sand, the substances in mixtures such as apple juice, air, or salt water are evenly mixed. *A* **homogeneous mixture** *is a type of mixture in which the individual substances are evenly mixed.* In a homogeneous mixture, the particles of individual substances are so small and well mixed that you cannot see them, even under a microscope. The table below summarizes how a homogeneous mixture differs from a heterogeneous mixture.

Think it Over

8. Apply When you pour milk on your cereal in the morning, are you making a heterogeneous mixture or a homogeneous mixture? How do you know?

______

______

______

______

Types of Mixtures

Heterogeneous Mixture	Homogeneous Mixture
• The individual substances are not evenly mixed. • Different samples of a given heterogeneous mixture can have different combinations of the same substances.	• The individual substances are evenly mixed. • Different samples of a given homogeneous mixture will have the same combinations of the same substances.

A homogeneous mixture also is called a solution. The solvent in a solution is the substance present in the largest amount. Solutes are all other substances in a solution. The solutes dissolve in the solvent. *To* **dissolve** *means to form a solution by mixing evenly.*

Because the substances in a solution, or homogeneous mixture, are evenly mixed, two samples from a solution will have the same amounts of each substance. For example, imagine pouring two glasses of apple juice from the same container. Each glass will contain the same substances (water, sugar, and other substances) in the same amounts. However, because apple juice is a mixture, the amounts of the substances might vary from one container of apple juice to another.

Visual Check

9. Distinguish Highlight the key words that distinguish a heterogeneous mixture from a homogeneous mixture.

Key Concept Check

10. Contrast How do mixtures differ from substances?

______

______

______

______

Compounds v. Solutions

If you have a glass of pure water and a glass of salt water, can you tell which is which just by looking? You cannot. The compound (water) and the solution (salt water) appear identical. How do compounds and solutions differ?

Because water is a compound, its composition does not vary. Pure water is always made up of the same atoms in the same combinations. Therefore, a chemical formula can be used to describe the atoms that make up water (H_2O). Salt water is a homogeneous mixture, or solution. The solute (NaCl) and the solvent (H_2O) are evenly mixed but are not bonded together. Adding more salt or more water only changes the relative amounts of the substances. In other words, the composition varies. Because composition can vary in a mixture, a chemical formula cannot be used to describe mixtures.

Key Concept Check

11. Classify How can you classify matter?

____

____

____

Summarizing Matter

You have read in this lesson about classifying matter by the arrangement of its atoms. The figure below is a summary of this classification system. Where on this diagram would you classify the things you see each day?

Visual Check

12. Identify Circle the two classifications of substances.

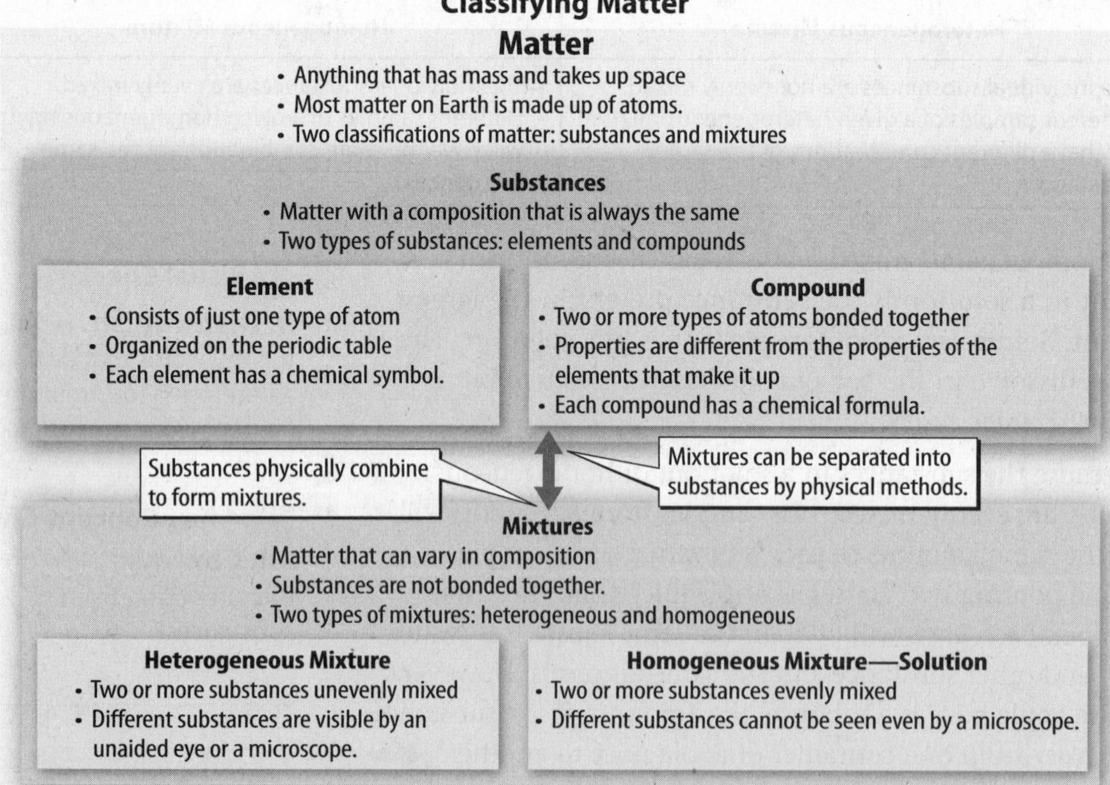

After You Read

Mini Glossary

atom: a small particle that is a building block of matter

compound: a type of substance containing atoms of two or more different elements chemically bonded together

dissolve: to form a solution by mixing evenly

element: a substance that consists of just one type of atom

heterogeneous mixture: a type of mixture in which the individual substances are not evenly mixed

homogeneous mixture: a type of mixture in which the individual substances are evenly mixed

matter: anything that has mass and takes up space

mixture: matter that can vary in composition

substance: matter with a composition that is always the same

1. Review the terms and their definitions in the Mini Glossary. Write a sentence that explains why light is not matter.

2. Complete the concept map by writing these terms in the correct boxes: *elements, solutions, substances, compounds.*

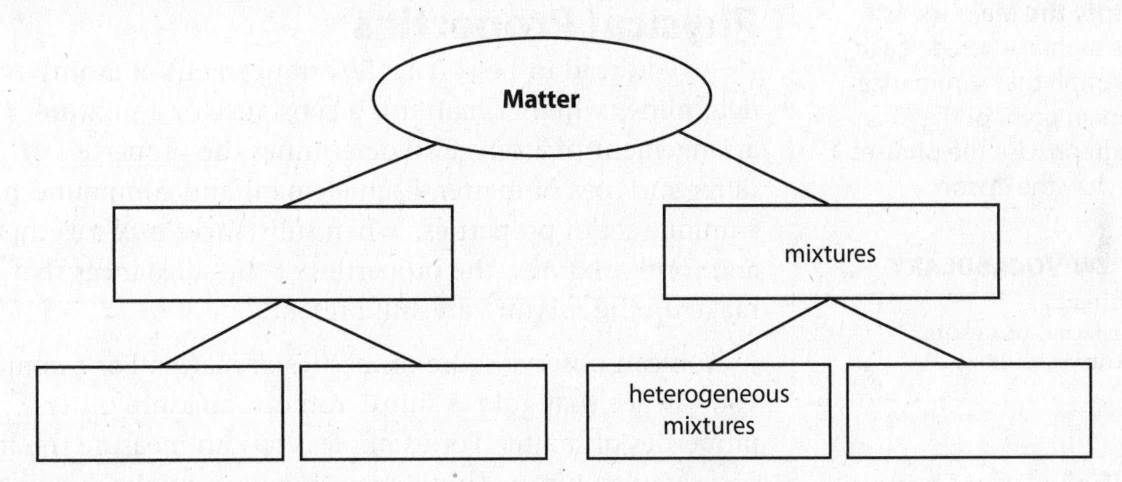

3. How many atoms of oxygen are in water, H_2O? Explain how you know.

What do you think NOW?

Reread the statements at the beginning of the lesson. Fill in the After column with an A if you agree with the statement or a D if you disagree. Did you change your mind?

Log on to ConnectED.mcgraw-hill.com and access your textbook to find this lesson's resources.

Foundations of Chemistry

Physical Properties

Key Concepts 

- What are some physical properties of matter?
- How are physical properties used to separate mixtures?

Before You Read

What do you think? Read the two statements below and decide whether you agree or disagree with them. Place an A in the Before column if you agree with the statement or a D if you disagree. After you've read this lesson, reread the statements to see if you have changed your mind.

Before	Statement	After
	3. The weight of a material never changes, regardless of where it is.	
	4. Boiling is one method used to separate parts of a mixture.	

Mark the Text

Identify the Main Ideas Write a phrase beside each paragraph that summarizes the main point of the paragraph. Use the phrases to review the lesson.

REVIEW VOCABULARY
property
a characteristic used to describe something

Think it Over

1. Observe Record two physical properties of the matter around you.

Read to Learn

Physical Properties

As you read in Lesson 1, the arrangement of atoms determines whether matter is a substance or a mixture. The arrangement of atoms also determines the properties of different types of matter. Each element and compound has a unique set of properties. When substances mix together and form mixtures, the properties of the substances that make up the mixture are still present.

You can observe some properties of matter. For example, you can see that gold is shiny. You can measure other properties of matter. For example, you can measure the mass of a sample of iron. Think about how you might describe the substances and mixtures in and around a flowing mountain stream. Could you describe some of the matter as a solid or a liquid? If you picked up a handful of pebbles from the bottom of the stream, why would the water leave your hand but not the pebbles? Could you describe the mass of the various rocks in the stream?

Each of these questions asks about the physical properties of matter. *A* **physical property** *is a characteristic of matter that you can observe or measure without changing the identity of the matter.* There are many types of physical properties. You will read about some types of physical properties in this lesson.

States of Matter

How do aluminum, water, and air differ? Recall that aluminum is an element, water is a compound, and air is a mixture. How else do these three types of matter differ?

At room temperature, aluminum is a solid, water is a liquid, and air is a gas. Solids, liquids, and gases are called states of matter.

The state of matter is a physical property of matter. Substances and mixtures can be solids, liquids, or gases. For example, water in the ocean is a liquid, but water in an iceberg is a solid. In addition, water vapor in the air above the ocean is a gas.

The particles (atoms or groups of atoms) that make up all matter move constantly and attract each other. Your pencil is made up of trillions of moving particles.

Every solid, liquid, and gas is made up of moving particles that attract one another. The state of matter depends on how close together the particles are and how fast they move.

The particles in a solid are very close together. They can move only by vibrating back and forth. This is why solids cannot easily change shape.

The particles in a liquid are slightly farther apart than in a solid. Therefore, the particles can move past one another. This is why you can pour a liquid. The particles in a gas are farther apart. They move quickly and spread out to fill their container.

Size-Dependent Properties

State is only one of many physical properties of matter. Some physical properties, such as mass and volume, depend on the size or the amount of matter. Measurements of these properties vary depending on how much matter is in a sample.

Mass Imagine holding a small dumbbell in one hand and a larger one in your other hand. What do you notice? The larger dumbbell seems heavier. The larger dumbbell has more mass than the smaller one.

Mass *is the amount of matter in an object*. Mass is a size-dependent property of a given substance because its value depends on the size of a sample.

FOLDABLES®

Make a three-tab book to record what you learn about different states of matter.

Think it Over

2. Consider How does water vapor in the air change state below the freezing point?

______

______

Reading Check

3. Contrast How do solids, liquids, and gases differ?

______

______

______

______

______

______

Think it Over

4. Analyze Does an astronaut have more mass on Earth than in space? Why or why not?

____

____

____

Math Skills

When you compare two numbers by division, you are using a ratio. Density can be written as a ratio of mass and volume. What is the density of a substance if a 5-mL sample has a mass of 25 g?

a. Set up a ratio.

$$\frac{25 \text{ g}}{5 \text{ mL}}$$

b. Divide the numerator by the denominator to get the mass (in g) of 1 mL.

$$\frac{25 \text{ g}}{5 \text{ mL}} = \frac{5 \text{ g}}{1 \text{ mL}}$$

c. The density is 5 g/mL.

5. Use Ratios A sample of wood has a mass of 12 g and a volume of 16 mL. What is the density of the wood?

____

Mass and Weight An object's mass and weight are not the same. Mass is an amount of matter in something. Weight is the pull of gravity on that matter. Weight changes with location, but mass does not. Suppose a dumbbell is on the Moon. The dumbbell would have the same mass on the Moon that it has on Earth. However, the Moon's gravity is much less than Earth's gravity. As a result, the dumbbell would weigh less on the Moon than on Earth.

Volume Another physical property that depends on the size or amount of a substance is volume. A unit often used to measure volume is the milliliter (mL). Volume is the amount of space something takes up. Suppose a full bottle of water contains 400 mL of water. If you pour exactly half of the water out, the bottle contains half of the original volume, or 200 mL, of water.

Size-Independent Properties

Some physical properties of a substance do not depend on the amount of matter present. These properties are the same for small samples and large samples. They are called size-independent properties. The table below and on the next page describes several physical properties of matter. The table provides examples of how physical properties can be used to separate mixtures. Notice that conductivity, boiling and melting points, state, density, solubility, and magnetism are size-independent properties.

Melting Point and Boiling Point The temperature at which a substance changes from a solid to a liquid is its melting point. The temperature at which a substance changes from a liquid to a gas is its boiling point. Different substances have different boiling points and melting points. For example, the boiling point for water is 100°C at sea level. The boiling point does not change for different volumes of water.

Physical Properties of Matter

Property	Mass	Conductivity	Volume
Size-dependent or size-independent	size-dependent	size-independent	size-dependent
Description of property	the amount of matter in an object	the ability of matter to conduct, or carry along, electricity or heat	the amount of space something occupies
How the property is used to separate a mixture (example)	Mass typically is not used to separate a mixture.	Conductivity typically is not used to separate a mixture.	Volume could be used to separate mixtures whose parts can be separated by filtration.

Density Imagine holding a bowling ball in one hand and a foam ball of the same size in the other. The bowling ball seems heavier because the density of the material that makes up the bowling ball is greater than the density of foam. **Density** *is the mass per unit volume of a substance*. Like melting point and boiling point, density is a size-independent property.

Conductivity Another property that is independent of the sample size is conductivity. Electrical conductivity is the ability of matter to conduct, or carry along, an electric current. Copper often is used for electrical wiring because it has high electrical conductivity.

Thermal conductivity is the ability of a material to conduct thermal energy. Metals tend to have high electrical and thermal conductivity. Stainless steel, for example, often is used to make cooking pots because of its high thermal conductivity. However, the handles on the pan probably are made out of wood, plastic, or some other substance that has low thermal conductivity.

Solubility Have you ever made lemonade by stirring a powdered drink mix into water? As you stir, the powder mixes evenly in the water. In other words, the powder dissolves in the water.

What would happen if you tried to dissolve sand in water? No matter how much you stir, the sand does not dissolve. **Solubility** *is the ability of one substance to dissolve in another*. The drink powder is soluble in water, but sand is not. The table below explains how physical properties such as conductivity and solubility can be used to identify objects and separate mixtures.

Key Concept Check

6. Name What are five different physical properties of matter?

____

____

____

____

____

Interpreting Tables

7. Consider How might you separate a mixture of iron filings and salt?

____

____

____

Physical Properties of Matter

Boiling/Melting Points	State of Matter	Density	Solubility	Magnetism
size-independent	size-dependent	size-independent	size-dependent	size-independent
the temperature at which a material changes state	whether something is a solid, a liquid, or a gas	the amount of mass per unit of volume	the ability of one substance to dissolve in another	attractive force for some metals, especially iron
Each part of a mixture will boil or melt at a different temperature.	A liquid can be poured off a solid.	Objects with greater density sink in objects with less density.	Dissolve a soluble material to separate it from a material with less solubility.	Use a magnet to attract iron shavings from a mixture of metals.

Separating Mixtures

In Lesson 1, you read about different types of mixtures. Recall that the substances that make up mixtures are not held together by chemical bonds. When substances form a mixture, the properties of the individual substances do not change.

You can separate the individual substances out of most mixtures by using differences in their physical properties. For example, when salt and water form a solution, the salt and the water do not lose any of their individual properties. Therefore, you can separate the salt from the water by using differences in their physical properties. Water has a lower boiling point than salt. When you boil salt water, the water evaporates, and the salt remains.

You cannot use physical properties to separate a compound into the elements it contains. The atoms that make up a compound are bonded together and cannot be separated by physical means. For example, you cannot separate the hydrogen atoms from the oxygen atoms in water by boiling water.

Key Concept Check
8. Explain How are physical properties used to separate mixtures?

____

____

____

____

____

____

After You Read

Mini Glossary

density: the mass per unit volume of a substance

mass: the amount of matter in an object

physical property: a characteristic of matter that you can observe or measure without changing the identity of the matter

solubility: the ability of one substance to dissolve in another

1. Review the terms and their definitions in the Mini Glossary. Write a sentence that explains how mass and weight are different.

____

____

2. Physical properties of matter can be classified as size-dependent or size-independent. Define each classification, and give at least two examples of each.

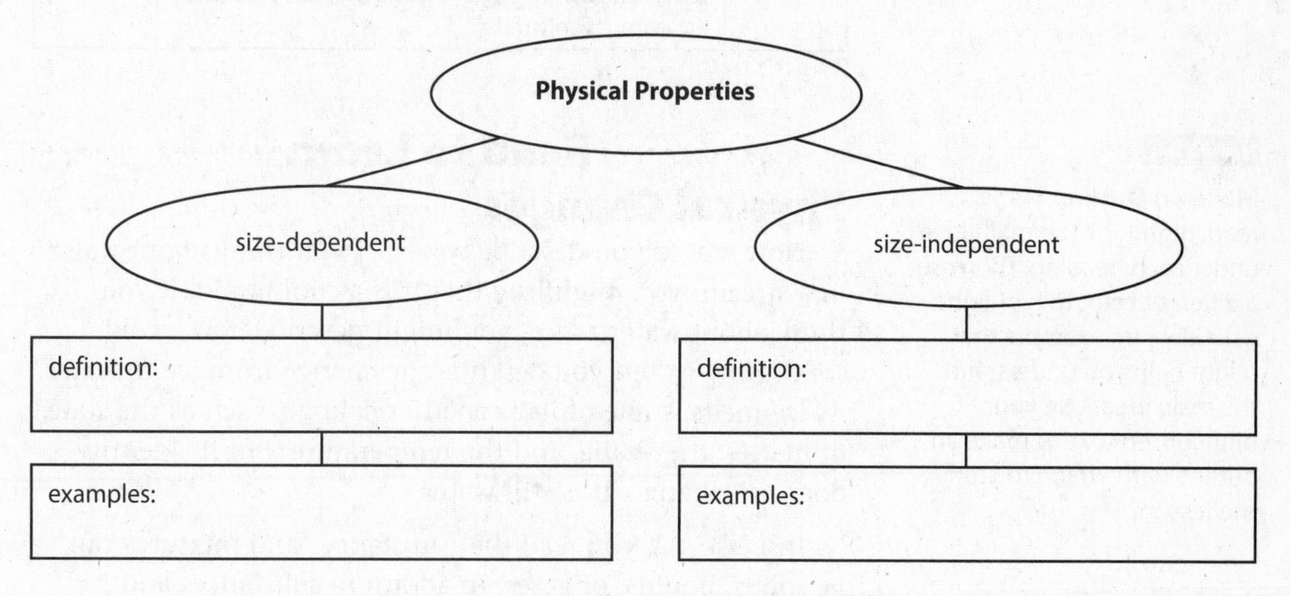

3. If you mix sand and water in a container, the sand will sink to the bottom. Explain why.

____

____

What do you think NOW?

Reread the statements at the beginning of the lesson. Fill in the After column with an A if you agree with the statement or a D if you disagree. Did you change your mind?

Log on to ConnectED.mcgraw-hill.com and access your textbook to find this lesson's resources.

Foundations of Chemistry

Physical Changes

Key Concepts
- How can a change in energy affect the state of matter?
- What happens when something dissolves?
- What is meant by conservation of mass?

Before You Read

What do you think? Read the two statements below and decide whether you agree or disagree with them. Place an A in the Before column if you agree with the statement or a D if you disagree. After you've read this lesson, reread the statements to see if you have changed your mind.

Before	Statement	After
	5. Heating a material decreases the energy of its particles.	
	6. When you stir sugar into water, the sugar and water evenly mix.	

Mark the Text

Make an Outline As you read, highlight the main idea under each heading. Then use a different color to highlight a detail or an example that might help you understand the main idea. Use your highlighted text to make an outline with which to study the lesson.

FOLDABLES®

Make a two-tab book to record specific examples of how adding or releasing thermal energy results in physical change.

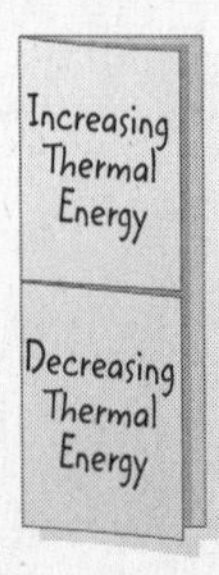

Read to Learn

Physical Changes

How would you describe water? If you think about water in a stream, you might say that it is a cool liquid. If you think about water as ice, you might describe it as a cold solid. How would you describe the change from ice to water? As ice melts, some of its properties change, such as the state of matter, the shape, and the temperature. But its identity does not change. It is still water.

In Lesson 2, you read that substances and mixtures can be solids, liquids, or gases. In addition, substances and mixtures can change from one state to another. *A* **physical change** *is a change in size, shape, form, or state of matter in which the matter's identity stays the same.* During a physical change, the matter does not become something different even though physical properties change.

Change in Shape and Size

Think about changes in the shapes and the sizes of materials you experience each day. When you chew food, you are breaking it into smaller pieces. This change in size helps make food easier to digest. When you pour juice from a bottle into a glass, you are changing the shape of the juice. Changes in shape and size are physical changes. The identity of the matter has not changed.

Change in State of Matter

Why does ice melt in your hand? Or, why does water turn to ice in the freezer? Matter, such as water, can change state. Recall how the particles in a solid, a liquid, and a gas behave. To change the state of matter, the movement of the particles has to change. To change the movement of particles, thermal energy must be added or removed.

Adding Thermal Energy When thermal energy is added to a solid, the particles in the solid move faster and faster, and the temperature increases. As the particles move faster, they are more likely to overcome the attractive forces that hold them tightly together. When the particles are moving too fast for attractive forces to hold them tightly together, the solid reaches its melting point. The melting point is the temperature at which a solid changes to a liquid.

After the entire solid has melted, the addition of more thermal energy causes the particles to move even faster. The temperature of the liquid increases. When the particles are moving so fast that attractive forces cannot hold them close together, the liquid reaches its boiling point. The boiling point is the temperature at which a liquid changes into a gas and the particles spread out. Some solids change directly to a gas without first becoming a liquid. This is called sublimation.

The figure below shows what happens as thermal energy is added to a material. Temperature increases when the state of matter is not changing. Temperature stays the same during a change of state.

Think it Over

1. Analyze Why does ice melt in your hand?

____

____

____

____

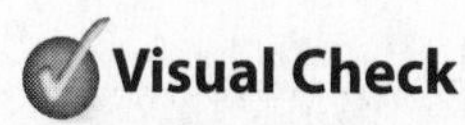

Visual Check

2. Point Out Circle the parts of the graph line that show a change of state is occurring.

Thermal Energy and the State of Matter

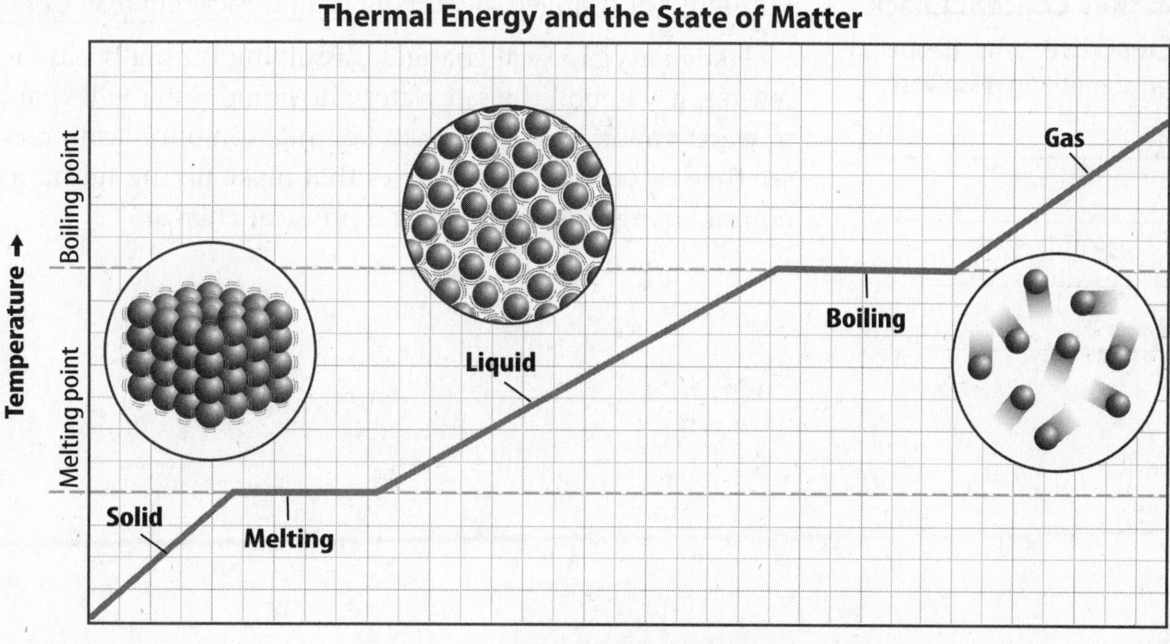

Removing Thermal Energy When thermal energy is removed from a gas such as water vapor, particles in the gas move more slowly and the temperature of the gas decreases. Condensation occurs when the particles are moving slowly enough for attractive forces to pull the particles close together. Recall that condensation is the process that occurs when a gas becomes a liquid.

After the gas has completely changed to a liquid, removing more thermal energy from the liquid causes the particles to move even more slowly. As the motion between the particles slows, the temperature decreases.

Freezing occurs when the particles are moving so slowly that attractive forces between the particles hold them tightly together. Now the particles only can vibrate in place. Recall that freezing is the process that occurs when a liquid becomes a solid.

Key Concept Check
3. Explain How can removing thermal energy affect the state of matter?

Freezing and melting are reverse processes, and they occur at the same temperature. The same is true of boiling and condensation. Another change of state is deposition. Deposition is the change from a gas directly to a solid. It is the opposite of sublimation. For example, deposition occurs when water vapor in the air forms frost.

Dissolving

Think about adding salt to water to create a saltwater aquarium. As you add the salt to the water, it gradually disappears. It is still there, but it dissolves, or mixes evenly, in the water. Because the identities of the substances—water and salt—are not changed, dissolving is a physical change.

Key Concept Check
4. Describe What happens when something dissolves?

Like many physical changes, dissolving is usually easy to reverse. If you boil the salt water, the liquid water will change to water vapor, leaving the salt behind. You once again can see the salt because the particles that make up the substances do not change identity during a physical change.

Conservation of Mass

During a physical change, the physical properties of matter change. The particles in matter that are present before a physical change are the same as those present after the physical change. Because the particles are the same before and after a physical change, the total mass before and after the change is also the same, as shown in the figure below. This is known as the conservation of mass. You will read in Lesson 4 that mass also is conserved during another type of change—a chemical change.

Key Concept Check

5. Define What is meant by conservation of mass?

____

____

____

____

Conservation of Mass

Visual Check

6. Calculate If a sample of water has a mass of 200 g and the final solution has a mass of 230 g, how much solute dissolved in the water?

____

After You Read

Mini Glossary

physical change: a change in size, shape, form, or state of matter in which the matter's identity stays the same

1. Review the terms and their definitions in the Mini Glossary. Write a sentence describing something you did today that resulted in a physical change of matter.

2. For each process listed in the diagram, identify the opposite process that occurs at the same temperature.

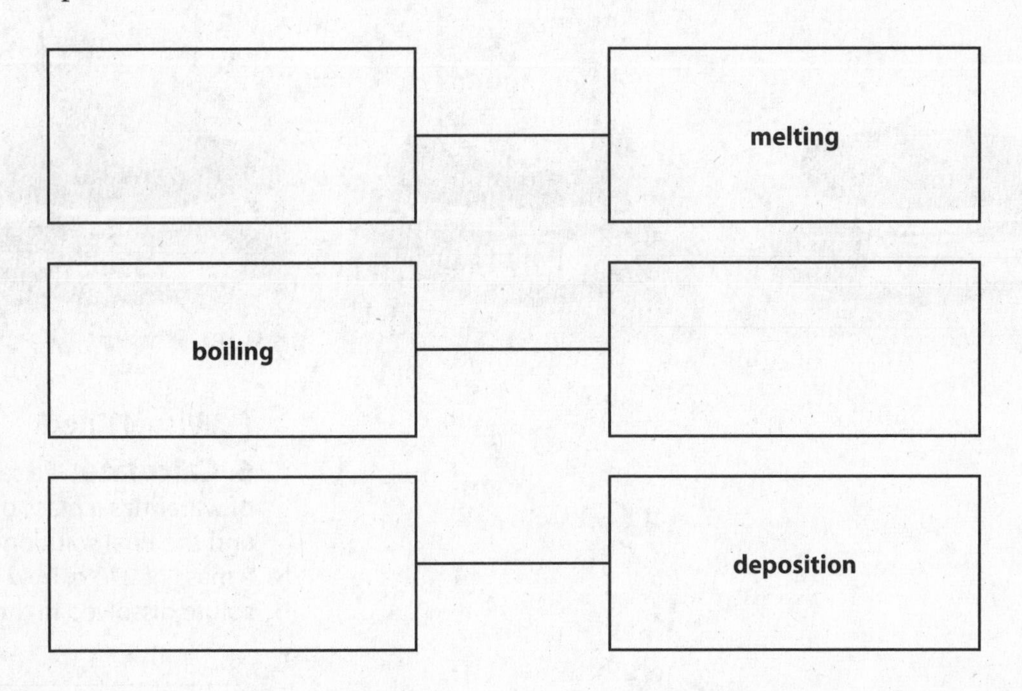

3. How did making an outline help you learn about the physical changes of matter? Write one main point that you highlighted and an example that helped you understand the main point.

What do you think NOW?

Reread the statements at the beginning of the lesson. Fill in the After column with an A if you agree with the statement or a D if you disagree. Did you change your mind?

Log on to ConnectED.mcgraw-hill.com and access your textbook to find this lesson's resources.

Foundations of Chemistry

Chemical Properties and Changes

Before You Read

What do you think? Read the two statements below and decide whether you agree or disagree with them. Place an A in the Before column if you agree with the statement or a D if you disagree. After you've read this lesson, reread the statements to see if you have changed your mind.

Before	Statement	After
	7. When wood burns, new materials form.	
	8. Temperature can affect the rate at which chemical changes occur.	

Key Concepts 

- What is a chemical property?
- What are some signs of chemical change?
- Why are chemical equations useful?
- What are some factors that affect the rate of chemical reactions?

Read to Learn

Chemical Properties

Recall that a physical property is a characteristic of matter that you can observe or measure without changing the identity of the matter. However, matter has other properties that you can observe only when the matter changes from one substance to another. *A* **chemical property** *is a characteristic of matter that can be observed as it changes to a different type of matter.* For example, can you tell by just looking at paper that it will burn easily? The only way to know that paper burns is to bring a flame near the paper and watch it burn. When paper burns, it changes into different types of matter. The ability of a substance to burn is a chemical property. The ability to rust is another chemical property.

Study Coach

Preview Headings Before you read the lesson, preview all the headings. Make a chart and write a question for each heading beginning with *What* or *How*. As you read, write the answers to your questions.

Comparing Properties

All matter can be described by its physical and chemical properties. For example, a wood log is solid, rounded, heavy, and rough. These are physical properties that you can observe with your senses. The log also has mass, volume, and density, which are physical properties that you can measure. The ability of wood to burn is a chemical property. This property is obvious only when you burn the wood. It also will rot, another chemical property you can observe when the log decomposes, becoming other substances. When you describe matter, you consider its physical and its chemical properties.

Key Concept Check

1. Identify What are some chemical properties of matter?

FOLDABLES®

Make a four-column chart to explain how the identity of matter changes during a chemical change.

Action/ Matter	Signs of Chemical Change	Explain the Chemical Reaction	What affects the reaction rate?

Key Concept Check
2. Recognize What are signs of a chemical change?

Key Concept Check
3. Explain Why are chemical equations useful?

Chemical Changes

Recall that during a physical change, the identity of matter does not change. However, *a* **chemical change** *is a change in matter in which the substances that make up the matter change into other substances with new physical and chemical properties.* When iron undergoes a chemical change with oxygen, rust forms. The substances that undergo a change no longer have the same properties because they no longer have the same identity.

Signs of Chemical Change

How do you know when a chemical change occurs? What signs show you that new types of matter have formed? Signs of chemical changes include the formation of bubbles or a change in odor, color, or energy. For example, the odor of fruit changes when it rots. Leaves change color in autumn. Energy changes when fireworks explode.

These signs do not always mean a chemical change has occurred. When you heat water on a stove, bubbles form as the water boils. In this case, bubbles show that the water is changing state, which is a physical change. Bubbles that form when you add an antacid tablet to water is evidence that a chemical change might have occurred. However, the only proof of chemical change is the formation of a new substance.

Explaining Chemical Reactions

Why do chemical changes create new substances? Recall that particles in matter move constantly. As particles move, they collide with each other. If the particles collide with enough force, the bonded atoms that make up the particles can break apart. These atoms then rearrange and bond with other atoms. When atoms bond together in new combinations, new substances form. This process is called a reaction. Chemical changes often are called chemical reactions.

Using Chemical Formulas

A chemical equation is a useful way to express what happens during a chemical reaction. A chemical equation shows the chemical formula of each substance in the reaction. Look at the chemical equation in the figure on the next page. The formulas to the left of the arrow represent the reactants. Reactants are the substances present before the reaction takes place. The formulas to the right of the arrow represent the products. Products are the new substances present after the reaction.

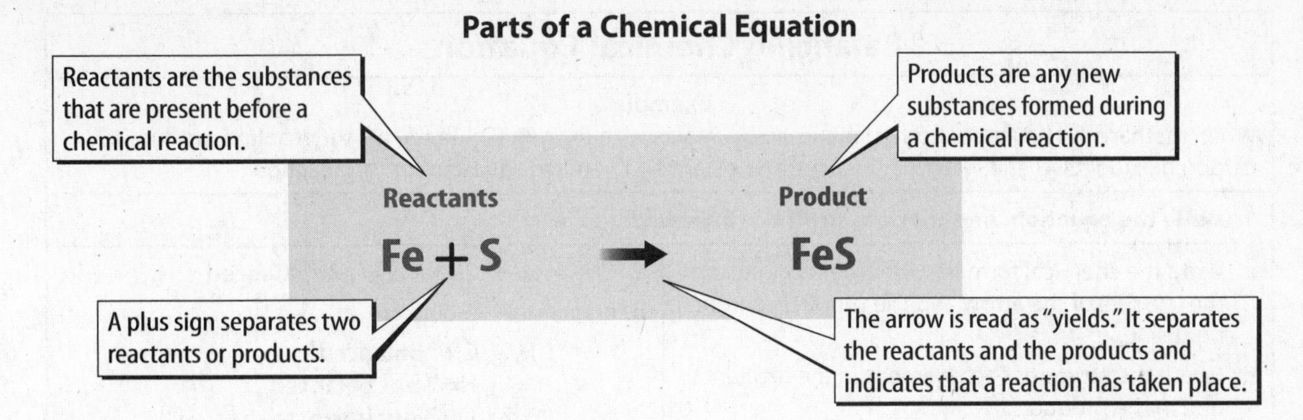

Balancing Chemical Equations

In the equation in the figure above, notice that one iron (Fe) atom is on the reactants side and one iron atom is on the product side. This is also true for the sulfur (S) atoms. One sulfur atom is on each side of the arrow. The arrow indicates that a reaction has taken place. In a chemical equation, the arrow is read as "yields." A reaction between the reactants to the left of the arrow yields, or produces, the new products on the right side of the arrow.

Recall that during physical and chemical changes, mass is conserved. This means that the total mass before and after a change must be equal. Therefore, in a chemical equation, the number of atoms of each element before a reaction must equal the number of atoms of each element after the reaction. This is called a balanced chemical equation, and it illustrates the conservation of mass.

When balancing an equation, you cannot change the chemical formula of any reactants or products. Changing a formula changes the identity of the substance. Instead, you can place coefficients, or multipliers, in front of formulas. Coefficients change the amount of the reactants and products present.

For example, an H_2O molecule has two H atoms and one O atom. Placing the coefficient *2* before H_2O (**2**H_2O) means that you double the number of H atoms and O atoms present:

$$\mathbf{2} \times 2 \text{ H atoms} = 4 \text{ H atoms}$$
$$\mathbf{2} \times 1 \text{ O atom} = 2 \text{ O atoms}$$

Note that $2H_2O$ is still water. However, it describes two water particles instead of one.

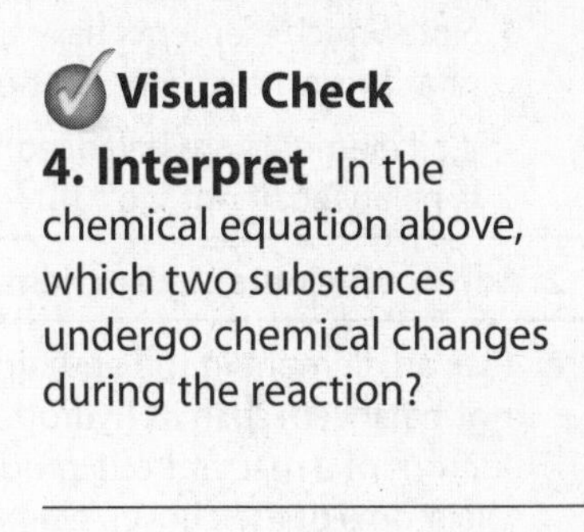

Visual Check

4. Interpret In the chemical equation above, which two substances undergo chemical changes during the reaction?

Think it Over

5. Apply Suppose you place the coefficient *4* before H_2O. How many atoms of hydrogen and how many atoms of oxygen will the formula have?

Balancing Chemical Equations	
Example When methane (CH_4)—a gas burned in furnaces—reacts with oxygen (O_2) in the air, the reaction produces carbon dioxide (CO_2) and water (H_2O). Write and balance a chemical equation for this reaction.	
1. Write the equation, and check to see if it is balanced.	
a. Write the chemical formulas with the reactants on the left side of the arrow and the products on the right side. **b.** Count the atoms of each element in the reactants and in the products. • Note which elements have a balanced number of atoms on each side of the equation. • If all elements are balanced, the overall equation is balanced. If not, go to step 2.	**a.** $CH_4 + O_2 \rightarrow CO_2 + H_2O$ **not balanced** **b.** reactants → products C=1 C=1 **balanced** H=4 H=2 **not balanced** O=2 O=3 **not balanced**
2. Add coefficients to the chemical formulas to balance the equation.	
a. Pick an element in the equation whose atoms are not balanced, such as hydrogen. Write a coefficient in front of a reactant or a product that will balance the atoms of the chosen element in the equation. **b.** Recount the atoms of each element in the reactants and the products, and note which are balanced on each side of the equation. **c.** Repeat steps 2a and 2b until all atoms of each element in the reactants equal those in the products.	**a.** $CH_4 + O_2 \rightarrow CO_2 + 2H_2O$ **not balanced** **b.** C=1 C=1 **balanced** H=4 H=4 **balanced** O=2 O=4 **not balanced** **c.** $CH_4 + 2O_2 \rightarrow CO_2 + 2H_2O$ **balanced** C=1 C=1 **balanced** H=4 H=4 **balanced** O=4 O=4 **balanced**
3. Write the balanced equation that includes the coefficients: $CH_4 + 2O_2 \rightarrow CO_2 + 2H_2O$	

Interpreting Tables

6. Recognize In the table, highlight the numbers of atoms that show the equation is not balanced.

The table above explains how to write and balance a chemical equation. Equations must balance because mass does not change during a chemical reaction.

The Rate of Chemical Reactions

Recall that the particles that make up matter are constantly moving and colliding with one another. Different factors can make these particles move faster and collide harder and more frequently. These factors increase the rate of a chemical reaction.

Reading Check

7. Explain Why does the rate of reaction increase when temperature increases?

Temperature A higher temperature usually increases the rate of reaction. For example, chemical reactions that occur during cooking happen at a faster rate when temperature increases. As temperature rises, the particles move faster. Therefore, the particles collide with greater force and more frequently.

Concentration *The amount of substance in a certain volume is the* **concentration** *of the substance.* A reaction occurs faster if the concentration of at least one reactant increases. When concentration increases, more particles are available to bump into each other and react.

For example, acid rain contains a higher concentration of acid than normal rain does. As a result, a statue that is exposed to acid rain is damaged more quickly than a statue that is exposed to normal rain.

Surface Area If at least one reactant is a solid, then surface area affects reaction rate. If you drop a whole antacid tablet into water, the tablet reacts with the water. However, if you break the tablet into several pieces and then add the pieces to the water, the reaction occurs more quickly. Smaller pieces have more total surface area, so more of the broken tablet is in contact with the water. An increase in total surface area makes more surface available for reactants to collide.

Chemistry

To understand chemistry, you need to understand matter. You need to know how the arrangement of atoms results in different types of matter. You also need to be able to distinguish physical properties from chemical properties and describe ways these properties can change. In later chemistry chapters and courses, you will examine each of these topics closely to gain a better understanding of matter.

After You Read

Mini Glossary

chemical change: a change in matter in which the substances that make up the matter change into other substances with new physical and chemical properties

chemical property: a characteristic of matter that can be observed as it changes to a different type of matter

concentration: the amount of substance in a certain volume

1. Review the terms and their definitions in the Mini Glossary. Write two sentences that explain the difference between a chemical change and a physical change.

____

____

____

2. Count the number of atoms of each element on both sides of the chemical equation below. Then determine whether the equation is balanced or not balanced.

$$C_2H_4O_2 + 2O_2 \longrightarrow 2CO_2 + 2H_2O$$

C = ______	C = ______
H = ______	H = ______
O = ______	O = ______

balanced or not balanced? ______________________

3. When a banana spoils, how can you tell that a chemical change has occurred?

____

____

____

What do you think NOW?

Reread the statements at the beginning of the lesson. Fill in the After column with an A if you agree with the statement or a D if you disagree. Did you change your mind?

Log on to ConnectED.mcgraw-hill.com and access your textbook to find this lesson's resources.

The Periodic Table

Using the Periodic Table

Before You Read

What do you think? Read the two statements below and decide whether you agree or disagree with them. Place an A in the Before column if you agree with the statement or a D if you disagree. After you've read this lesson, reread the statements to see if you have changed your mind.

Before	Statement	After
	1. The elements on the periodic table are arranged in rows in the order they were discovered.	
	2. The properties of an element are related to the element's location on the periodic table.	

Key Concepts 

- How are elements arranged on the periodic table?
- What can you learn about elements from the periodic table?

Read to Learn

What is the periodic table?

There are more than 100 elements. Each element has a unique set of properties. Scientists use a table, called the periodic (pihr ee AH dihk) table, to organize elements. The **periodic table** *is a chart of the elements arranged into rows and columns according to their physical and chemical properties.* The periodic table can be used to determine the relationships among the elements.

This chapter describes the development of the periodic table. It will show you how to use the periodic table to learn about the elements.

Study Coach

Create a Quiz As you study the information in this section, create questions about the information you read. Be sure to answer your questions. Refer to your questions and answers as you review the chapter.

Developing a Periodic Table

In 1869, a Russian chemist and teacher Dimitri Mendeleev (duh MEE tree · men duh LAY uf) put together an early periodic table. He studied the physical properties such as density, color, melting point, and atomic mass of each element. He also studied the chemical properties, such as how each element reacted with other elements. Mendeleev arranged the elements in rows of increasing atomic mass. He grouped elements with similar properties in the same column. 

Reading Check

1. Explain What physical property did Mendeleev use to place the elements in rows on the periodic table?

Make a top-tab book to organize your notes about the development of the periodic table.

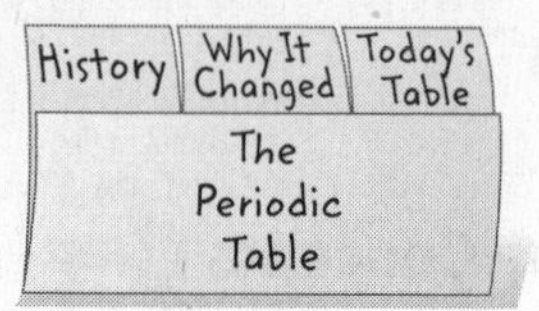

Reading Check

2. Describe What did Mendeleev predict about the properties of the elements missing from his periodic table?

Key Concept Check

3. Identify What determines where an element is located on the periodic table you use today?

Patterns in Properties

The word *periodic* means "repeating pattern." Seasons and months are periodic because they follow a repeating pattern every year. The days of the week are periodic because they repeat every seven days.

What were some of the repeating patterns Mendeleev noticed in his table? Melting point is one property that shows a repeating pattern. Melting point is the temperature at which a solid changes to a liquid. In the periodic table, melting points increase and then decrease across a row. Boiling points and reactivity also follow a periodic pattern.

Predicting Properties of Undiscovered Elements

When all of the elements known in Mendeleev's time were arranged in a periodic table, there were large gaps between some elements. Mendeleev predicted that scientists would discover elements that would fit into these spaces. He also predicted that the properties of those elements would be similar to the known elements in the same columns. Both of Mendeleev's predictions turned out to be true.

Changes to Mendeleev's Table

Mendeleev's periodic table made it possible for scientists to relate the properties of elements to their position on the table. However, the table had one big problem: some elements seemed to be out of place.

When elements were arranged in order of atomic mass, a few of the elements did not seem to belong in their columns. Their properties were similar to the properties of the elements in the next column on Mendeleev's table. What could be done to fix this problem on Mendeleev's table? The result is the periodic table we use today.

The Importance of Atomic Number

In the early 1900s, scientist Henry Moseley solved the problem with Mendeleev's table. Mendeleev had listed elements according to increasing atomic mass. Instead of listing elements according to increasing atomic mass, Moseley listed elements according to increasing atomic number.

The atomic number of an element is the number of protons in the nucleus of each of that element's atoms. When Mosely organized the table according to atomic number, he found that the columns contained elements with similar properties.

Today's Periodic Table

The periodic table is shown on the next two pages. You can identify many of the properties of an element from its placement on the periodic table. The table is organized into columns, rows, and blocks, which are based on certain patterns of properties. In the next two lessons, you will learn how an element's position on the periodic table can help you understand the element's physical and chemical properties.

What is on an element key?

Each element in the periodic table is represented by an element key. An element key shows important information about each element. The key shows the element's chemical symbol, atomic number, and atomic mass. The key also contains a symbol that shows the element's state of matter at room temperature. Look at the information given for helium in the figure on the right. It shows that helium is a gas at room temperature, it has the atomic number 2, its chemical symbol is He, and its atomic mass is 4.00.

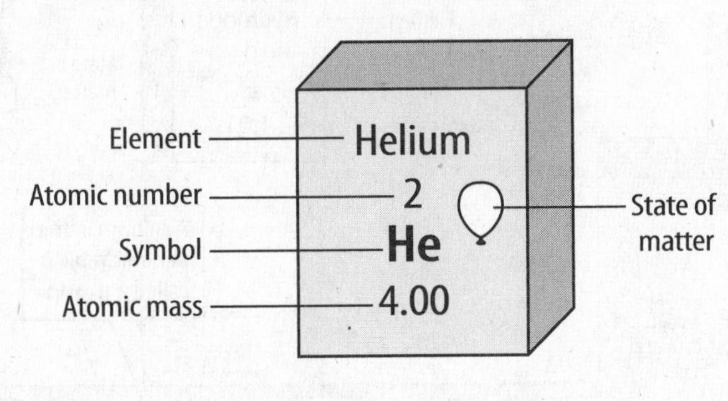

Groups

A **group** *is a column on the periodic table.* Elements in the same group have similar chemical properties. This means that the elements in a group react with other elements in similar ways. There are patterns in the physical properties of a group, such as density, melting point, and boiling point. The groups are numbered 1–18 at the top of each column on the periodic table.

Periods

The rows on the periodic table are called **periods**. The atomic number of each element increases by 1 as you read from left to right across each period. The physical and chemical properties of the elements also change as you move from left to right across a period.

Math Skills

The distance around a circle is the circumference (C). The distance across the circle, through its center, is the diameter (d). The radius (r) is half of the diameter. The circumference divided by the diameter for any circle is equal to π (pi), or 3.14. The formula for finding the circumference is:

$$C = \pi d \text{ or } C = 2\pi r$$

Example: The circumference of an iron (Fe) atom is:

$$C = 2 \times 3.14 \times 126 \text{ pm}$$

(picometers; 1 picometer = one-trillionth of a meter)

$$C = 791 \text{ pm}$$

4. Use Geometry The radius of a uranium (U) atom is 156 pm. What is its circumference?

____

Visual Check

5. Determine What does the key in the figure tell you about helium?

____

Key Concept Check

6. Describe What can you infer about the properties of two elements in the same group?

____

____

Metals, Nonmetals, and Metalloids

Almost three-fourths of the elements on the periodic table are metals. Metals are on the left side and in the middle of the table. Metals can have different properties, but all metals are shiny and conduct thermal energy and electricity.

Nonmetals, except for hydrogen, are located on the right side of the periodic table. The properties of nonmetals are different from those of metals. Nonmetals do not conduct thermal energy or electricity. Many nonmetals are gases.

Between the metals and the nonmetals on the periodic table are the metalloids. Metalloids have properties of both metals and nonmetals.

7. Identify How is the periodic table organized?

____

____

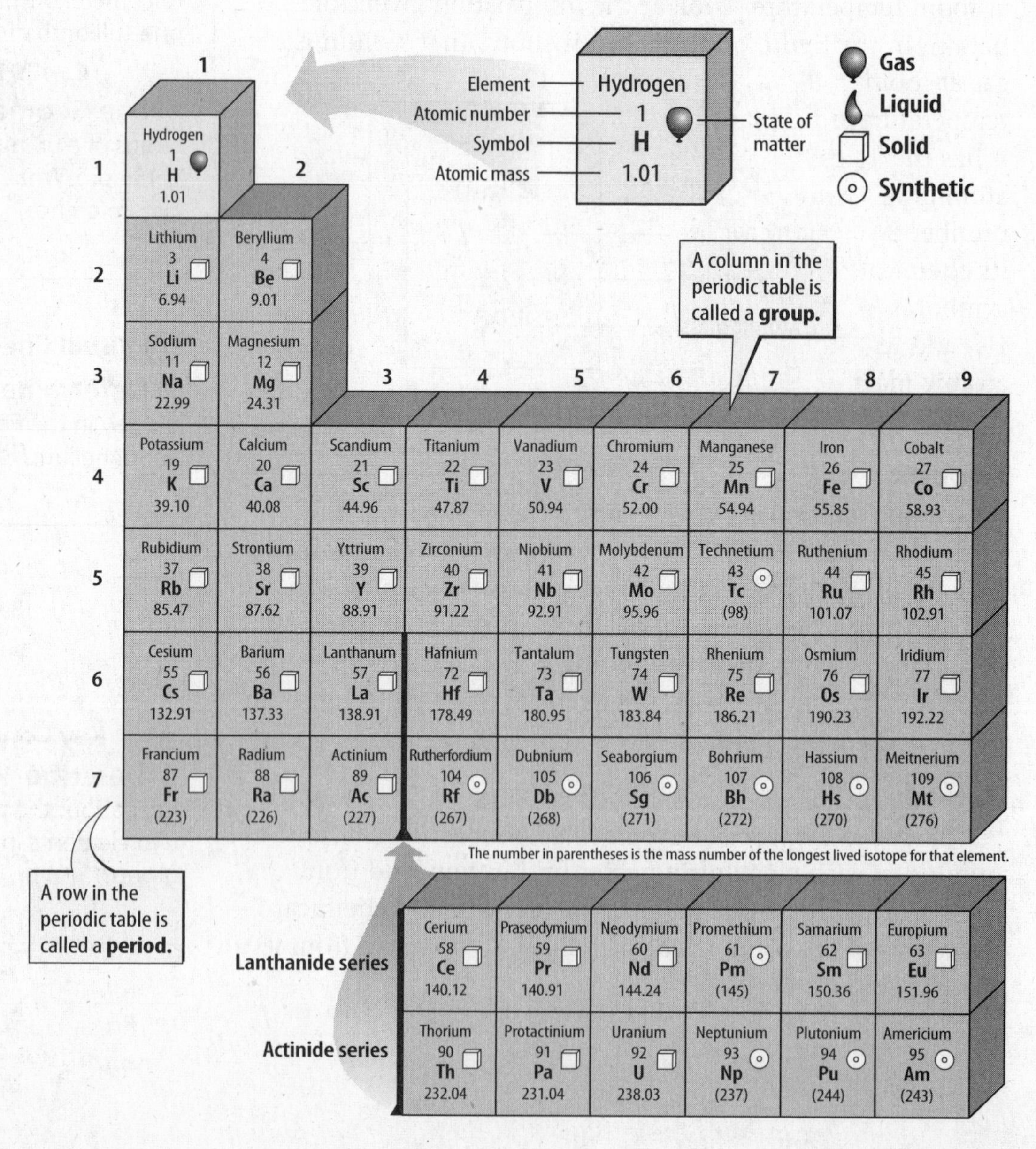

How Scientists Use the Periodic Table

More than 100 elements are known today. They are all listed on the periodic table. Each element has its own set of properties. It also has properties similar to the elements near it on the table. The periodic table shows how elements relate to each other and fit together into one organized chart. Scientists use the periodic table to understand and predict elements' properties.

The elements with the largest atomic masses are not found in nature. These are elements that can be made only by scientists in special laboratories. Elements that were created in laboratories are named to honor the scientists who created them or the laboratories in which they were created.

Reading Check

8. Explain How is the periodic table used to predict the properties of an element?

____

____

____

Metal

Metalloid

Nonmetal

Recently discovered

10	11	12	13	14	15	16	17	18
								Helium 2 He 4.00
			Boron 5 B 10.81	Carbon 6 C 12.01	Nitrogen 7 N 14.01	Oxygen 8 O 16.00	Fluorine 9 F 19.00	Neon 10 Ne 20.18
			Aluminum 13 Al 26.98	Silicon 14 Si 28.09	Phosphorus 15 P 30.97	Sulfur 16 S 32.07	Chlorine 17 Cl 35.45	Argon 18 Ar 39.95
Nickel 28 Ni 58.69	Copper 29 Cu 63.55	Zinc 30 Zn 65.38	Gallium 31 Ga 69.72	Germanium 32 Ge 72.64	Arsenic 33 As 74.92	Selenium 34 Se 78.96	Bromine 35 Br 79.90	Krypton 36 Kr 83.80
Palladium 46 Pd 106.42	Silver 47 Ag 107.87	Cadmium 48 Cd 112.41	Indium 49 In 114.82	Tin 50 Sn 118.71	Antimony 51 Sb 121.76	Tellurium 52 Te 127.60	Iodine 53 I 126.90	Xenon 54 Xe 131.29
Platinum 78 Pt 195.08	Gold 79 Au 196.97	Mercury 80 Hg 200.59	Thallium 81 Tl 204.38	Lead 82 Pb 207.20	Bismuth 83 Bi 208.98	Polonium 84 Po (209)	Astatine 85 At (210)	Radon 86 Rn (222)
Darmstadtium 110 Ds (281)	Roentgenium 111 Rg (280)	Ununbium * 112 Cn (285)	Ununtrium * 113 Uut (284)	Ununquadium * 114 Uuq (289)	Ununpentium * 115 Uup (288)	Ununhexium * 116 Uuh (293)		Ununoctium * 118 Uuo (294)

* The names and symbols for elements 112-116 and 118 are temporary. Final names will be selected when the elements' discoveries are verified.

Gadolinium 64 Gd 157.25	Terbium 65 Tb 158.93	Dysprosium 66 Dy 162.50	Holmium 67 Ho 164.93	Erbium 68 Er 167.26	Thulium 69 Tm 168.93	Ytterbium 70 Yb 173.05	Lutetium 71 Lu 174.97
Curium 96 Cm (247)	Berkelium 97 Bk (247)	Californium 98 Cf (251)	Einsteinium 99 Es (252)	Fermium 100 Fm (257)	Mendelevium 101 Md (258)	Nobelium 102 No (259)	Lawrencium 103 Lr (262)

After You Read

Mini Glossary

group: a column on the periodic table

period: a row on the periodic table

periodic (pihr ee AH dihk) table: a chart of the elements arranged into rows and columns according to their physical and chemical properties

1. Review the terms and their definitions in the Mini Glossary. Use all three words in the Mini Glossary to describe the periodic table and how it is arranged.

____

____

____

____

____

2. Examine the element key at right from the periodic table. From the element key, give all the information you can tell about the element shown.

____

____

____

3. How did preparing questions about the periodic table and the elements help you learn the information in the lesson?

____

____

____

____

What do you think NOW?

Reread the statements at the beginning of the lesson. Fill in the After column with an A if you agree with the statement or a D if you disagree. Did you change your mind?

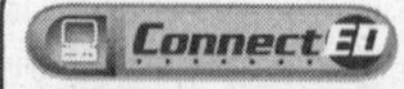

Log on to ConnectED.mcgraw-hill.com and access your textbook to find this lesson's resources.

The Periodic Table

Metals

Before You Read

What do you think? Read the two statements below and decide whether you agree or disagree with them. Place an A in the Before column if you agree with the statement or a D if you disagree. After you've read this lesson, reread the statements to see if you have changed your mind.

Before	Statement	After
	3. Fewer than half of the elements are metals.	
	4. Metals are usually good conductors of electricity.	

Key Concepts
- What elements are metals?
- What are the properties of metals?

Read to Learn

Mark the Text

Underline Main Ideas As you read, underline the main ideas under each heading. After you finish reading, review the main ideas that you have underlined.

What is a metal?

Metals are some of the most useful elements. Forks, knives, copper wire, aluminum foil, gold jewelry, and many other things are made of metal.

Most of the elements on the periodic table are metals. Except for hydrogen, all of the elements in groups 1–12 on the periodic table are metals. Some of the elements in groups 13–15 are metals also. To be a metal, an element must have certain properties.

Key Concept Check

1. Explain How does the position of an element on the periodic table allow you to determine if the element is a metal?

___

___

___

Physical Properties of Metals

Recall that physical properties are characteristics used to describe or identify something without changing its makeup. All metals share certain physical properties. *A* **metal** *is an element that is generally shiny. It is easily pulled into wires or hammered into thin sheets. A metal is a good conductor of electricity and thermal energy.* Gold exhibits the properties of metal.

Luster and Conductivity People use gold for jewelry because of its beautiful color and metallic luster. **Luster** *is the ability of a metal to reflect light.* Gold is also a good conductor of thermal energy and electricity. However, gold is too expensive to use in normal electrical wires or metal cookware. Copper is often used instead.

FOLDABLES

Make a two-tab book to record information about the physical and chemical properties of metals.

The Physical Properties of Metals | The Chemical Properties of Metals

Key Concept Check

2. Identify What are some physical properties of metals?

Visual Check

3. Identify What part of the periodic table is represented by the figure at right?

Ductility and Malleability Gold is the most ductile metal. **Ductility** (duk TIH luh tee) *is the ability of a substance to be pulled into thin wires.* A piece of gold with a mass the same as that of a paper clip can be pulled into a wire that is more than 3 km long.

Malleability (ma lee uh BIH luh tee) *is the ability of a substance to be hammered or rolled into sheets.* Gold is so malleable that it can be hammered into thin sheets. A pile of a million thin sheets of gold would be only as high as a coffee mug.

Other Physical Properties of Metals Metals have other physical properties. The density, strength, boiling point, and melting point of a metal are greater than those of other elements. Except for mercury, all metals are solid at room temperature. Many uses of a metal are determined by the metal's physical properties.

Chemical Properties of Metals

Recall that a chemical property is the ability or inability of a substance to change into one or more new substances. Most metals share similar physical properties. The chemical properties of metals, however, can vary greatly. Metals in the same group on the periodic table usually have similar chemical properties. The likelihood that one element will react with another is a chemical property.

Group 1: Alkali Metals

The elements in group 1 are called **alkali** (AL kuh li) **metals.** Group 1 elements are shown on the right. They include lithium, sodium, potassium, rubidium, cesium, and francium.

Because they are in the same group, alkali metals have similar chemical properties. Alkali metals are very reactive. Because they react quickly with other elements, alkali metals occur only in compounds in nature. Pure alkali metals must be stored so that they do not come into contact with oxygen and water vapor in the air. Alkali metals react violently with water. Alkali metals also have similar physical properties. Pure alkali metals have a silvery appearance and are soft enough to be cut with a knife. They also have the lowest densities of all metals. A block of pure sodium metal could float on water because of its very low density.

Element	Atomic number	Symbol	Atomic mass
Lithium	3	Li	6.94
Sodium	11	Na	22.99
Potassium	19	K	39.10
Rubidium	37	Rb	85.47
Cesium	55	Cs	132.91
Francium	87	Fr	(223)

Group 2: Alkaline Earth Metals

The elements in group 2 are called **alkaline** (AL kuh lun) **earth metals.** Group 2 elements are shown on the right. They include beryllium, magnesium, calcium, strontium, barium, and radium.

Beryllium
4
Be
9.01

Magnesium
12
Mg
24.31

Calcium
20
Ca
40.08

Strontium
38
Sr
87.62

Barium
56
Ba
137.33

Radium
88
Ra
(226)

Like alkali metals, alkaline earth metals react quickly with other elements. But they do not react as quickly as alkali metals do. Like alkali metals, pure alkaline earth metals do not occur naturally. They combine with other elements and form compounds.

The physical properties of the alkaline earth metals are also similar to those of the alkali metals. Alkaline earth metals are soft and silvery. They have low densities, but their densities are greater than those of alkali metals. ✓

Visual Check
4. Identify Circle the element in the figure with the highest atomic mass.

Reading Check
5. Identify Which element reacts faster with oxygen—barium or potassium?

____

____

____

Groups 3–12: Transition Elements

The elements in groups 3–12 are called **transition elements.** The transition elements are in two blocks on the periodic table. As shown below, one block is in the center of the periodic table. The other block is the two rows at the bottom of the periodic table.

Many colorful materials contain small amounts of transition elements. An emerald is green because it contains small amounts of chromium. A garnet is red because of the iron it contains.

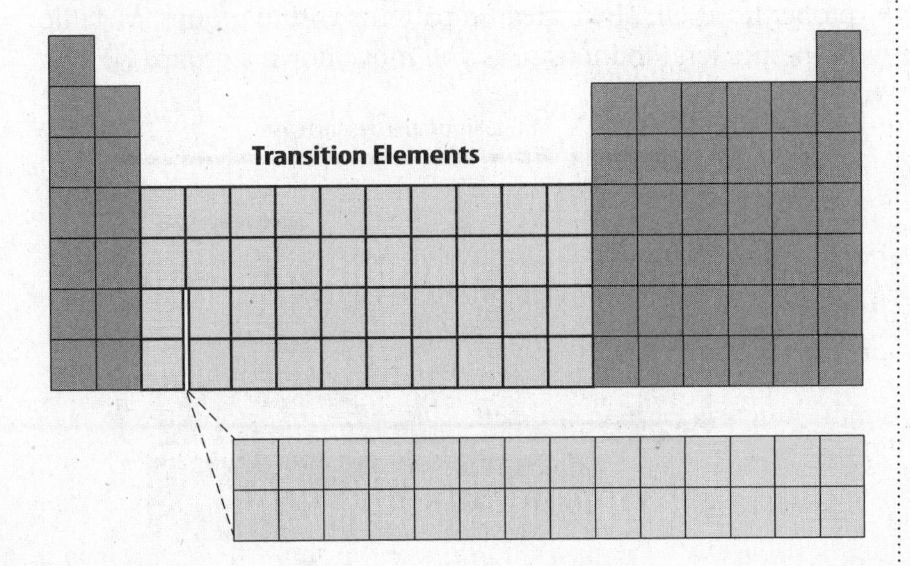

Visual Check
6. Identify How many periods of transition elements are there in the periodic table?

____

7. Contrast Describe two differences between transition elements and alkali metals.

Properties of Transition Elements

All transition elements are metals. They have higher melting points, greater strength, and higher densities than the alkali metals and the alkaline earth metals. Transition elements also react less quickly with oxygen. Some transition elements can exist in nature as free elements rather than in compounds. Free elements occur in pure form.

Uses of Transition Elements

Transition elements in the middle block of the periodic table have many important uses. Because they are dense, strong, and resist corrosion, transition elements such as iron make good building materials. Copper, silver, nickel, and gold are used to make coins. Many transition elements can react with other elements and form many compounds.

Lanthanide and Actinide Series

Two rows of transition elements are at the bottom of the periodic table. They are placed below the main table to keep the table from being too wide. Elements in the first row are called the lanthanide series, and elements in the second row are called the actinide series. Some elements from both series have valuable properties. Lanthanide series elements are used to make strong magnets. Plutonium, an actinide series element, is used as a fuel in some nuclear reactors.

Patterns in Properties of Metals

The properties of elements follow repeating patterns across the periods of the periodic table. The figure below shows these patterns. Potassium (K) has more luster, is the most malleable, and conducts electricity better than all the elements in period 4. All these properties decrease from left to right across the period. The elements on the far right have no metallic properties at all. There are also patterns within groups. Metallic properties tend to increase as you move down a group.

Reading Check

8. Locate Where on the periodic table would you expect to find elements with few or no metallic properties?

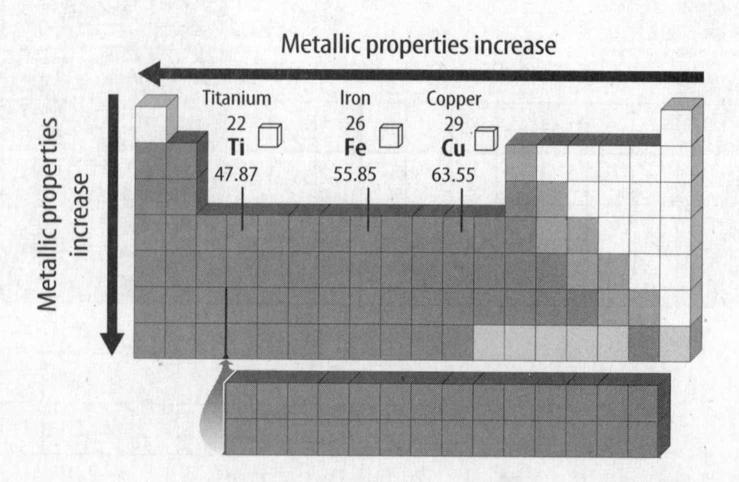

Visual Check

9. Identify Circle the most malleable metal: iron (FE), copper (Cu), or titanium (Ti).

After You Read

Mini Glossary

alkali (AL kuh li) metal: an element in group 1 on the periodic table

alkaline (AL kuh lun) earth metal: an element in group 2 on the periodic table

ductility (duk TIH luh tee): the ability of a substance to be pulled into thin wires

luster: the ability of a metal to reflect light

malleability (ma lee uh BIH luh tee): the ability of a substance to be hammered or rolled into sheets

metal: an element that is generally shiny, is easily pulled into wires or hammered into thin sheets, and is a good conductor of electricity and thermal energy

transition element: an element in groups 3–12 on the periodic table

1. Review the terms and their definitions in the Mini Glossary. Use three terms to tell what properties metals tend to have.

2. Examine the section of the periodic table at right. Which element has properties most similar to those of chromium (Cr)? Why?

Vanadium 23 **V**	Chromium 24 **Cr**	Maganese 25 **Mn**
Niobium 41 **Nb**	Molybdenum 42 **Mo**	Technetium 43 **Tc**

3. How did underlining the main ideas help you review the material?

What do you think NOW?

Reread the statements at the beginning of the lesson. Fill in the After column with an A if you agree with the statement or a D if you disagree. Did you change your mind?

Log on to ConnectED.mcgraw-hill.com and access your textbook to find this lesson's resources.

CHAPTER 11
LESSON 3

The Periodic Table

Nonmetals and Metalloids

Key Concepts

- Where are nonmetals and metalloids on the periodic table?
- What are the properties of nonmetals and metalloids?

Before You Read

What do you think? Read the two statements below and decide whether you agree or disagree with them. Place an A in the Before column if you agree with the statement or a D if you disagree. After you've read this lesson, reread the statements to see if you have changed your mind.

Before	Statement	After
	5. Most of the elements in living things are nonmetals.	
	6. Even though they look very different, oxygen and sulfur share some similar properties.	

Mark the Text

Underline Terms As you read this lesson, underline the names of important nonmetals. Highlight information about them. Use this information to review the lesson.

Read to Learn

The Elements of Life

More than 96 percent of the mass of your body comes from just four elements. They are shown in the figure. All four of these elements—oxygen, carbon, hydrogen, and nitrogen—are nonmetals. **Nonmetals** *are elements that have no metallic properties.*

Of the remaining elements in your body, the two most common are also nonmetals—phosphorus and sulfur. These six elements (oxygen, carbon, hydrogen, nitrogen, phosphorus, and sulfur) form the compounds in proteins, fats, nucleic acids, and other large molecules in your body. These elements also form the compounds in all other living things.

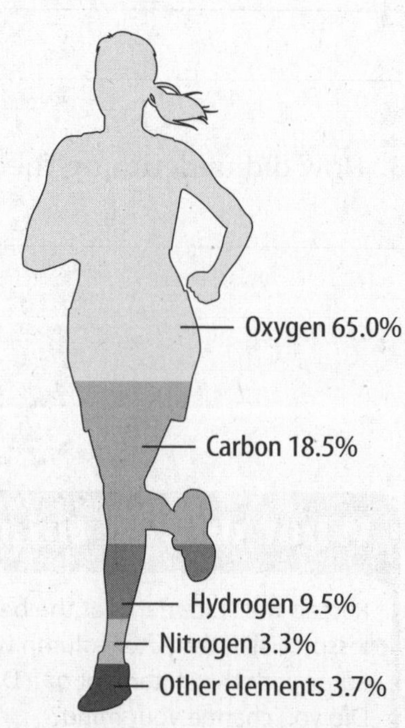

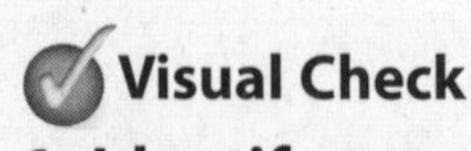

1. Identify Which three elements make up most of the mass of your body?

How are nonmetals different from metals?

Recall that metals have luster. They are ductile, malleable, and good conductors of electricity and thermal energy. All metals except mercury are solids at room temperature.

The properties of nonmetals are different from those of metals. Nonmetals do not conduct electricity or thermal energy well. Those that are solid at room temperature have no luster. Many of the nonmetals are gases at room temperature.

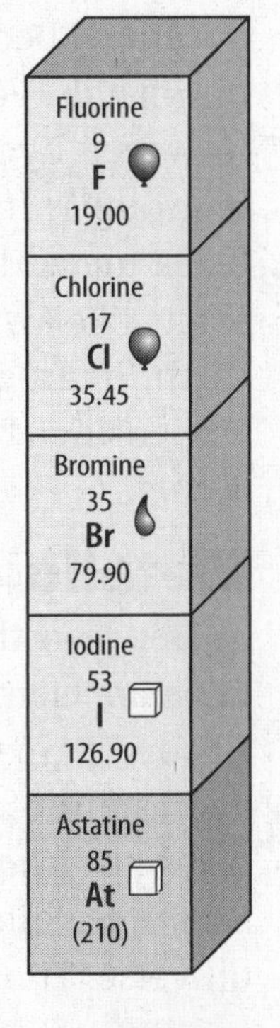

Nonmetals in Groups 14–16

Look at the periodic table in this chapter or in the back of this book. Notice that groups 14–16 contain metals, nonmetals, and metalloids. The chemical properties of the elements in each group are similar. However, the physical properties of the elements can be different. Nonmetals in the groups include carbon, nitrogen, phosphorus, oxygen, sulfur, and selenium.

Group 17: The Halogens

Group 17 of the periodic table is above on the right. *An element in group 17 of the periodic table is called a* **halogen** (HA luh jun). Halogens can react with a metal and form a salt. For example, chlorine gas reacts with solid sodium and forms sodium chloride, or table salt. All halogens react readily with other elements and form compounds. In fact, they can occur naturally only in compounds. They do not exist as free elements.

Group 18: The Noble Gases

Group 18 of the periodic table is shown at right. *The elements in group 18 are known as the* **noble gases.** The elements helium, neon, argon, krypton, xenon, and radon are the noble gases. Unlike the halogens, the only way noble gases react with other elements is under special conditions in a laboratory. They do not form compounds naturally.

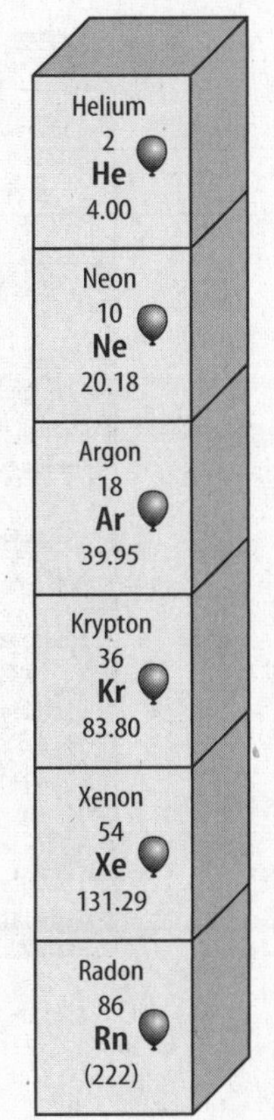

Key Concept Check

2. Identify What properties do nonmetals have?

FOLDABLES®

Make a chart with three columns and three rows to organize information about nonmetals and metalloids.

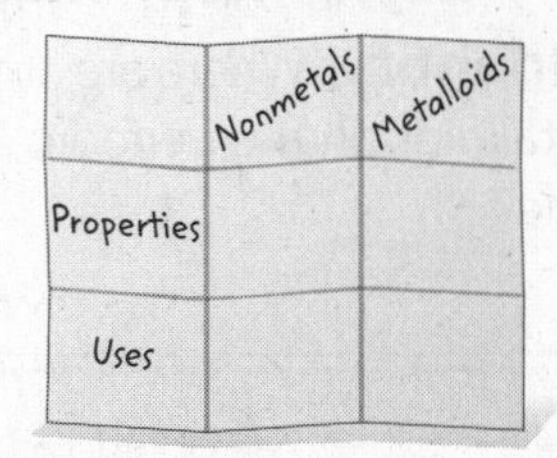

Reading Check

3. Predict Will bromine (Br) react with sodium (Na)? Explain your answer.

Visual Check

4. Identify Label each figure on the left with the correct group number from the periodic table.

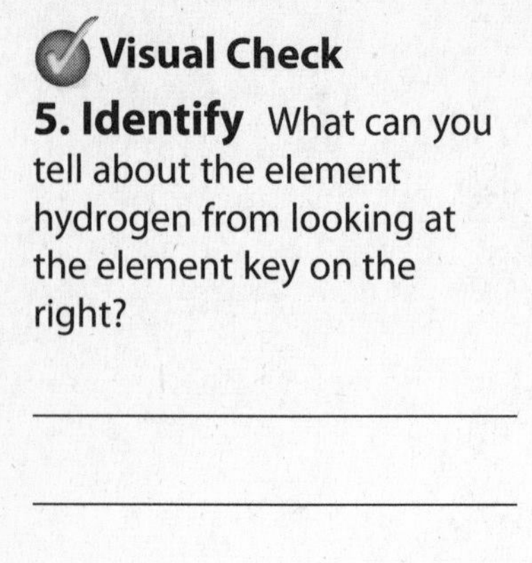

Visual Check

5. Identify What can you tell about the element hydrogen from looking at the element key on the right?

____

____

____

Hydrogen

The element key for hydrogen is shown at right. Hydrogen has the smallest atomic mass of any of the elements. Hydrogen is the most common element in the universe.

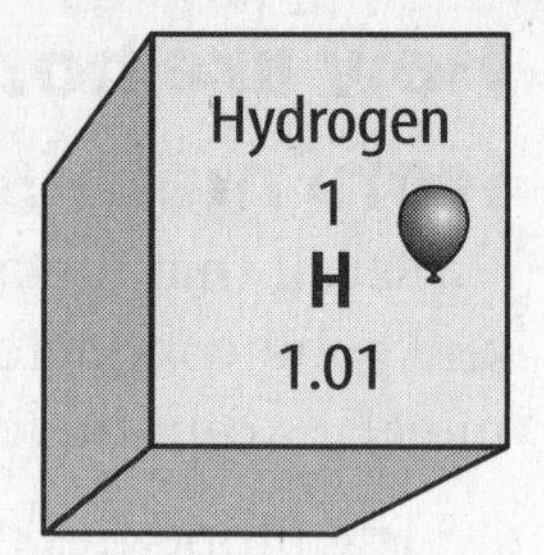

Hydrogen is most often classified as a nonmetal because it has many of the properties of nonmetals. For example, it is a gas at room temperature. However, hydrogen shares properties with the alkali metals in group 1. In liquid form, hydrogen conducts electricity just like a metal does. In some chemical reactions, hydrogen reacts like an alkali metal. However, under conditions on Earth, hydrogen usually behaves like a nonmetal. More than 90 percent of all the atoms in the universe are hydrogen atoms.

Metalloids

Between the metals and the nonmetals on the periodic table are elements known as metalloids. The elements boron, silicon, germanium, arsenic, antimony, tellurium, polonium, and astatine are metalloids. *A* **metalloid** (MEH tul oyd) *is an element that has physical and chemical properties of both metals and nonmetals.* Silicon is the most abundant metalloid in the universe. The metalloids are shown in the figure below.

Key Concept Check

6. Identify Where are metalloids on the periodic table?

____

____

____

Visual Check

7. Identify Circle the portion of the figure at right that contains the metalloid elements.

Boron 5 B 10.81	Carbon 6 C 12.01	Nitrogen 7 N 14.01	Oxygen 8 O 16.00	Fluorine 9 F 19.00
Aluminum 13 Al 26.98	Silicon 14 Si 28.09	Phosphorus 15 P 30.97	Sulfur 16 S 32.07	Chlorine 17 Cl 35.45
Gallium 31 Ga 69.72	Germanium 32 Ge 72.64	Arsenic 33 As 74.92	Selenium 34 Se 78.96	Bromine 35 Br 79.90
Indium 49 In 114.82	Tin 50 Sn 118.71	Antimony 51 Sb 121.76	Tellurium 52 Te 127.60	Iodine 53 I 126.90
Thallium 81 Tl 204.38	Lead 82 Pb 207.20	Bismuth 83 Bi 208.98	Polonium 84 Po (209)	Astatine 85 At (210)

Semiconductors

Recall that metals are good conductors of thermal energy and electricity. Nonmetals are poor conductors of thermal energy and electricity. But they are good insulators. A property of metalloids is the ability to act as a semiconductor. *A* **semiconductor** (seh mee kun DUK tur) *is an element that conducts electricity at high temperatures, but not at low temperatures.* At high temperatures, metalloids act like metals and conduct electricity. But at lower temperatures, they act like nonmetals and do not conduct electricity. This property is useful in electronic devices such as computers and televisions.

Properties and Uses of Metalloids

Silicon is one of the most abundant elements on Earth. Sand, clay, and many rocks and minerals are made of silicon compounds. Pure silicon is used in semiconductor devices for computers and other electronic products. Germanium is also used as a semiconductor. Semiconductors are an important use of metalloids. Metalloids also have other uses. Boron is used in water softeners and laundry products. Boron also glows bright green in fireworks.

Metals, Nonmetals, and Metalloids

You have read that all metallic elements have common characteristics, such as malleability, conductivity, and ductility. However, each metal has unique properties that make it different from other metals. The same is true for nonmetals and metalloids. How can knowing the properties of an element help you understand how to use it?

Look at the periodic table. An element's position on the periodic table tells you a lot about the element. By knowing that sulfur is a nonmetal, for example, you know that it breaks easily and does not conduct electricity. You would not choose sulfur to make a wire. You would not try to use oxygen as a semiconductor or sodium as a building material. You know that transition elements are strong, malleable, and do not react easily with oxygen or water. These elements make good building materials because they are strong and malleable. They are less reactive than other elements. Being familiar with the properties of metals and other elements can help you understand how they are used in different situations.

Think it Over

8. Explain What property makes semiconductors useful in electronic equipment?

Reading Check

9. Explain Why would you not use an element on the right side of the periodic table as a building material?

After You Read

Mini Glossary

halogen (HA luh jun): an element in group 17 of the periodic table

metalloid (MEH tul oyd): an element that has physical and chemical properties of both metals and nonmetals

noble gas: an element in group 18 of the periodic table

nonmetal: an element that has no metallic properties

semiconductor (seh mee kun DUK tur): an element that conducts electricity at high temperatures, but not at low temperatures

1. Review the terms and their definitions in the Mini Glossary. Write a sentence that compares nonmetals and metalloids.

---

2. In the graphic organizer below, write the names of five metalloids. In the largest box, write the name of the metalloid that is one of the most abundant elements on Earth.

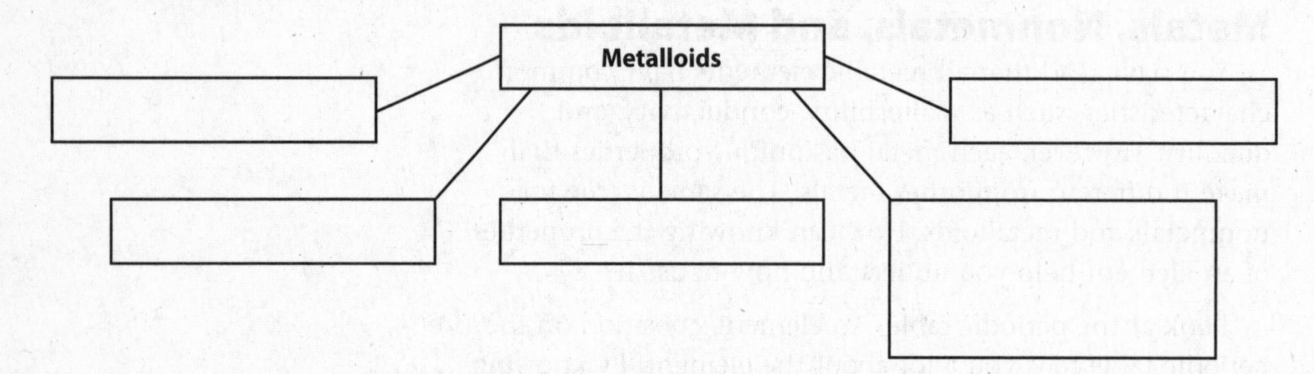

3. Review the names of the nonmetals that you underlined and the information that you highlighted. How did this strategy help you learn about nonmetals?

What do you think NOW?

Reread the statements at the beginning of the lesson. Fill in the After column with an A if you agree with the statement or a D if you disagree. Did you change your mind?

Log on to ConnectED.mcgraw-hill.com and access your textbook to find this lesson's resources.

Using Energy and Heat

Forms of Energy

Before You Read

What do you think? Read the two statements below and decide whether you agree or disagree with them. Place an A in the Before column if you agree with the statement or a D if you disagree. After you've read this lesson, reread the statements to see if you have changed your mind.

Before	Statement	After
	1. An object sitting on a high shelf has no energy.	
	2. There are many forms of energy.	

Key Concepts 

- How do potential energy and kinetic energy differ?
- How are mechanical energy and thermal energy similar?
- What two forms of energy are carried by waves?

Read to Learn

Energy

Some breakfast cereals promise to give you enough energy to get your day off to a great start. News reports often mention the price of oil, which is an energy source that provides fuel for cars and for transporting goods around the world. Weather reporters talk about the approach of a storm system that has a lot of energy. News anchors report on earthquakes and tsunamis, which carry so much energy they cause great damage. Politicians talk about the nation's energy policy and the need to conserve energy and to find new energy resources.

Energy influences everything in life, including the climate, the economy, and your body. Scientists define **energy** *as the ability to cause change.*

Mark the Text

Ask Questions As you read, write questions you have next to each paragraph. Read the lesson a second time and try to answer the questions. When you are done, ask your teacher any questions you still have.

Potential Energy

Think of a book balanced on the edge of a desk. The book's position could easily change, which means it has potential energy. **Potential energy** *is stored energy due to the interaction between objects or particles.* Particles include atoms, ions, and molecules. Objects have potential energy if they have the potential to cause change. Examples of potential energy include objects that could fall due to gravity and particles that could move because of electric or magnetic forces. ✓

Reading Check

1. Define What is potential energy?

______

______

______

______

2. Consider Which has the greater potential energy: a book on the top shelf of a bookcase or a book on the bottom shelf?

____

____

____

FOLDABLES®

Make a four-column chart book to organize your notes on the different forms of energy that fall into each of the categories.

Kinetic Energy	Potential Energy	Both Kinetic and Potential Energies	Energy from Waves

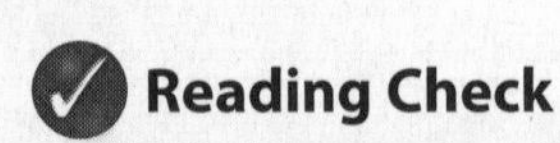

3. Recognize Energy due to motion is ___. (Circle the correct answer.)

a. potential energy

b. kinetic energy

c. chemical energy

Gravitational Potential Energy

Do the items stored on the shelves in your home have potential energy? Yes. Every item—including the shelves—has gravitational potential energy. Objects have gravitational potential energy if they have mass and height above Earth's surface. The gravitational potential energy of an object depends on two factors—the mass of an object and the distance of the object from Earth's surface.

Chemical Energy

Suppose you put on skates to play ice hockey. Where does your body get the energy it needs to play? Energy in your body comes from the foods you eat. All objects, including food, are made of atoms that are joined by chemical bonds. **Chemical energy** *is the energy stored in and released from the bonds between atoms.* Your body breaks chemical bonds in foods and converts the released energy into other forms of energy that your body can use.

Nuclear Energy

The energy stored in and released from the nucleus of an atom is called **nuclear energy.** When you watch the Sun set, you are experiencing nuclear energy. The Sun's energy is released through the process of nuclear fusion. During nuclear fusion, the nuclei of atoms join together and release large amounts of energy. Nuclear energy also is released when an atom breaks apart. This breaking apart of an atom is called nuclear fission. Nuclear fission is used in nuclear power plants to generate, or make, electricity. Nuclear fusion and nuclear fission are examples of nuclear energy.

Kinetic Energy

Are you moving your hand as you take notes? Are you squirming in your chair as you try to find a comfortable position? If so, you have **kinetic energy**—*energy due to motion.* All objects that have motion have kinetic energy. ✓

Kinetic Energy of Objects

An object's kinetic energy is related to the mass and the speed of the object. For example, suppose you hold a 3.6-kg bowling ball in your hands. Because the bowling ball is not moving, it has no speed and, therefore, no kinetic energy. Now suppose a friend rolls a 4.5-kg bowling ball at 8.0 m/s and another friend rolls a 5.5-kg ball at the same speed. The ball that has a greater mass has greater kinetic energy even though both balls are moving at the same speed.

Electric Energy

Even objects you cannot see have kinetic energy. Recall that all materials are made of atoms. In an atom, electrons move around a nucleus. Sometimes electrons move from one atom to another. Because electrons are moving, they have kinetic energy. When electrons move, they create an electric current. *The energy in an electric current is* **electric energy.** For example, in a simple circuit electrons move from one terminal of a battery through the copper wire and bulb to the other terminal of the battery. As the electrons move, their energy is transformed into light. Your brain and the nerves in your body that tell your arm and leg muscles to move also use electric energy.

Combined Kinetic Energy and Potential Energy

Your school is part of an education system. Earth is part of the solar system. A system is a collection of parts that interact and act together as a whole. In science, everything that is not in a given system is the environment. For example, a hockey player, the hockey stick, the hockey puck, and the ice under the player can be considered a system.

Mechanical Energy

Suppose the hockey player hits the hockey puck into the air. Does the puck have kinetic energy or potential energy? It has mass and motion, so it has kinetic energy. It also has height above Earth, so it has gravitational potential energy. Scientists often study the energy of systems, such as the one described above. *The sum of the potential energy and the kinetic energy in a system is* **mechanical energy.** You might think of mechanical energy as the ability to move another object. What happens when the hockey puck hits the net? The net moves. The hockey puck has mechanical energy that causes another object to move.

Thermal Energy

Even when the hockey puck is lying on the floor with no obvious motion, the particles that make up the solid puck are in motion. The particles vibrate back and forth in place. Therefore, the particles have kinetic energy. The particles also have potential energy because of attractive forces between the particles. An object's **thermal energy** *is the sum of the kinetic energy and the potential energy of the particles that make up the object.* Thermal energy of an object increases when the potential energy, the kinetic energy, or both increase.

Key Concept Check

4. Differentiate How do potential energy and kinetic energy differ?

Reading Check

5. Explain What is a system?

Key Concept Check

6. Compare mechanical energy and thermal energy.

Energy Carried by Waves

When a raindrop falls into a still pool of water, the raindrop disturbs the water's surface. It produces waves that move away from the place where the raindrop hit. *A* **wave** *is a disturbance that transfers energy from one place to another without transferring matter.* Energy, not matter, moves outward from the point where the raindrop hits the water.

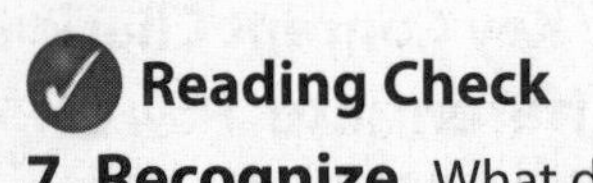

7. Recognize What do waves carry?

____

Sound Energy

When the raindrop hits the water, it disturbs the surface of the water, and it also disturbs the air. It creates a sound wave in the air similar to water waves. Sound waves move through matter. Each wave travels from particle to particle as the particles bump into each other, much like falling dominoes. **Sound energy** *is energy carried by sound waves.*

As a sound wave travels, it eventually reaches your ear. The sound energy moves tiny hairs inside your ear. This movement is transformed into an electric signal that travels to your brain. Your brain interprets the electric signal as the sound of a water splash.

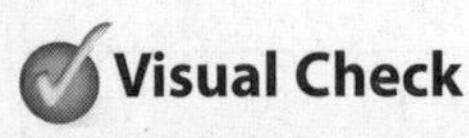

8. Examine Determine one type of radiant energy that has a shorter wavelength than visible light and another type that has a longer wavelength.

____

____

____

____

Radiant Energy

Have you ever wondered what light is? Light is a form of energy carried by electromagnetic waves—electric and magnetic waves moving perpendicularly to one another, as shown in the figure. *The energy carried by electromagnetic waves is* **radiant energy.**

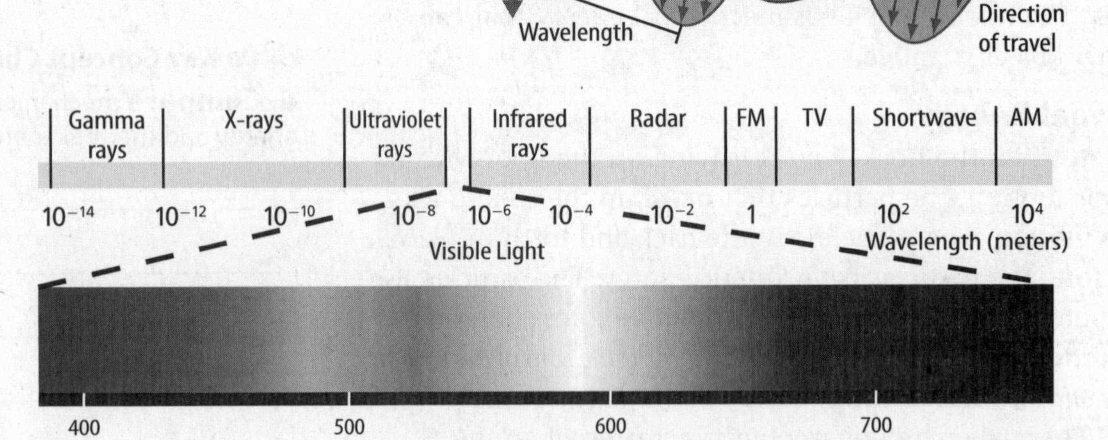

Electromagnetic Waves Waves that travel through matter and through spaces with little or no matter, such as outer space, are called electromagnetic waves. Electromagnetic waves often are described by their wavelengths. Wavelength is the distance from one point on a wave to the nearest point just like it.

Other Forms of Radiant Energy Visible light is only one form of radiant energy. Gamma rays and X-rays are electromagnetic waves with very short wavelengths. Gamma rays and X-rays often are used in medical procedures.

Ultraviolet rays have wavelengths that are a little shorter than those of light. This form of radiant energy is what causes your sunburn.

Infrared rays are the form of energy used by many television remote controls to change channels. They also provide the warmth you feel when the Sun shines on you. Radar, television, and radio waves have long wavelengths compared to the wavelength of visible light.

Key Concept Check

9. Identify What two forms of energy are carried by waves?

After You Read

Mini Glossary

chemical energy: energy that is stored in and released from the bonds between atoms

electric energy: energy in an electric current

energy: the ability to cause change

kinetic energy: energy due to motion

mechanical energy: the sum of the potential energy and the kinetic energy in a system

nuclear energy: energy stored in and released from the nucleus of an atom

potential energy: stored energy due to the interaction between objects or particles

radiant energy: energy carried by electromagnetic waves

sound energy: energy carried by sound waves

thermal energy: the sum of the kinetic energy and potential energy of the particles that make up an object

wave: a disturbance that transfers energy from one place to another without transferring matter

1. Review the terms and their definitions in the Mini Glossary. Write a sentence explaining how your body gets energy from food.

____

2. Identify each type of energy in the graphic organizer below.

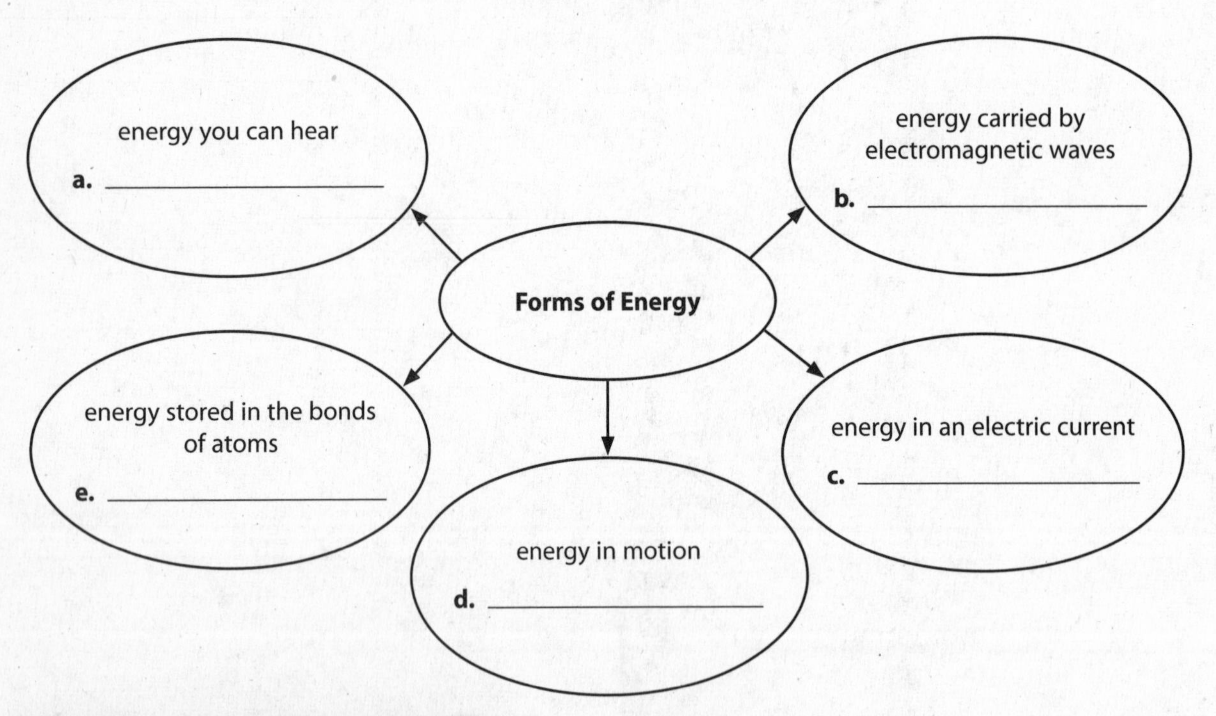

What do you think NOW?

Reread the statements at the beginning of the lesson. Fill in the After column with an A if you agree with the statement or a D if you disagree. Did you change your mind?

Log on to ConnectED.mcgraw-hill.com and access your textbook to find this lesson's resources.

Using Energy and Heat

Energy Transfers and Transformations

Before You Read

What do you think? Read the two statements below and decide whether you agree or disagree with them. Place an A in the Before column if you agree with the statement or a D if you disagree. After you've read this lesson, reread the statements to see if you have changed your mind.

Before	Statement	After
	3. In most systems, no energy is transferred to the environment.	
	4. Some forms of energy are replenished naturally.	

Key Concepts 

- What is the law of conservation of energy?
- How is energy transformed and transferred?
- What are renewable and nonrenewable energy resources?

Study Coach

Identify the Main Ideas Write one sentence to summarize the main idea in each paragraph. Write the main ideas on a sheet of paper or in your notebook to study later.

Visual Check

1. Trace the flow of energy from the batteries to the bulb.

Read to Learn

Law of Conservation of Energy

The **law of conservation of energy** *says that energy can be transformed from one form to another, but it cannot be created or destroyed.* In the flashlight below, chemical energy of the battery is transformed to electric energy that moves through the contact strip to the bulb. The electric energy is transformed into radiant energy and thermal energy in the lightbulb. The law of conservation of energy states that the amount of radiant energy shining out of the flashlight cannot be greater than the chemical energy in the battery.

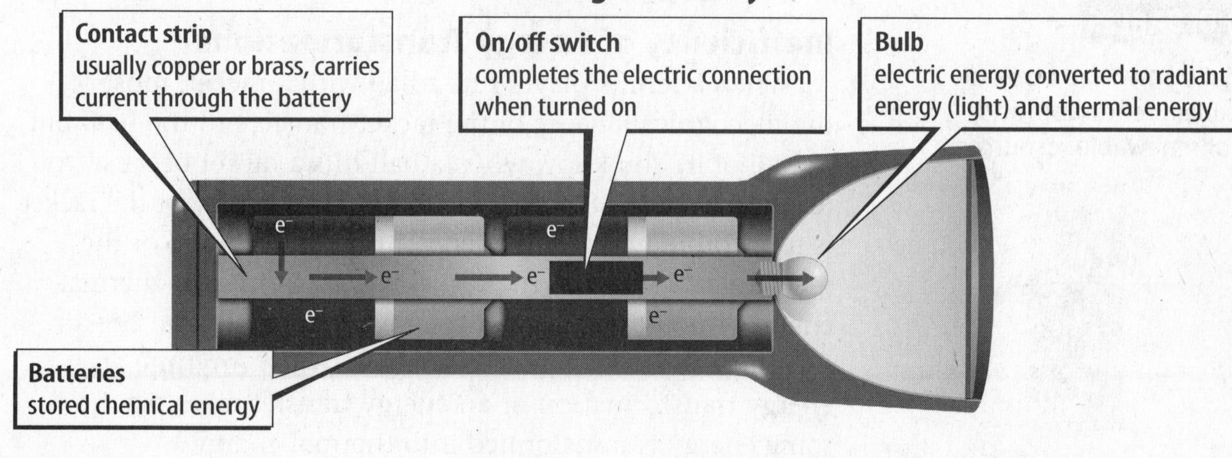

Key Concept Check
2. Explain What is the law of conservation of energy?

Key Concept Check
3. Identify an energy transfer and an energy transformation that occur when someone plays a guitar.

FOLDABLES

Make a two-tab book to explain renewable and nonrenewable resources.

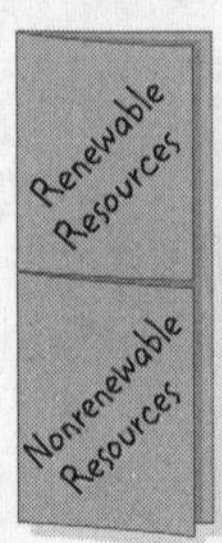

Efficiency The amount of radiant energy given off by a flashlight is less than the chemical energy in the battery. Where is the missing energy? In this lesson, you will learn that in every energy transformation, some energy transfers to the environment.

Energy Transfer

When a tennis player hits a ball with her racket, the mechanical energy of the racket changes the movement of the ball. The mechanical energy of the ball increases after it is hit. *When energy moves from one object to another without changing form, an* ***energy transfer*** *occurs.* The tennis racket transfers mechanical energy to the tennis ball.

Energy Transformation

Where does the mechanical energy in the tennis player's racket come from? Chemical energy stored in the player's muscles changes to mechanical energy when she swings her arm. *When one form of energy is converted to another form of energy, an* ***energy transformation*** *occurs.*

Energy and Work

Work *is the transfer of energy that occurs when a force makes an object move in the direction of the force. Work is only being done while the force is acting on the object.* As the tennis player swings the racket, the racket applies a force to the ball for about 1 m. Although the ball moves 10 m, work is done by the racket only as the racket applies a force to the ball. When the ball separates from the racket, the racket no longer does work.

Suppose the tennis player is standing still before she serves the ball. She is using her muscles to hold the ball. Is she doing work on the ball? No. She is not doing work because the ball is not moving. If a force does not make an object move in the direction of the force, it does no work on the object.

Inefficiency of Energy Transformations

When a tennis player hits a ball with a racket, most of the mechanical energy of the racket transfers to the ball, but not all of it. You know when a ball hits a racket because you can hear a sound. Some of the mechanical energy of the racket is transformed to sound energy. In addition, some of the mechanical energy of the racket is transformed to thermal energy. The temperature of the racket, the ball, and the air surrounding both objects increases slightly. Anytime an energy transformation or an energy transfer takes place, some energy is transformed into thermal energy.

Inefficiency in a Flashlight Recall the flashlight at the beginning of the lesson. The transformation of chemical energy of the battery to radiant energy from the lightbulb is inefficient, too. As the electric energy moves through the circuit, some electric energy transforms to thermal energy. When electric energy transforms to radiant energy in the lightbulb, more energy transforms to thermal energy. In some flashlights, the bulb is warm to the touch.

Inefficiency Defined The law of conservation of energy says that energy cannot be created or destroyed. When scientists say that energy transformations are inefficient, they do not mean that energy is destroyed. Energy transformations are inefficient because not all the energy that is transformed to another form of energy is usable.

Open Systems

In Lesson 1, you read that scientists often study the energy of systems. A car is a system. The chemical energy of the fuel is transformed to mechanical energy of the moving car. Because energy transformations are inefficient, some of the chemical energy transforms to thermal energy and sound energy, which are then released to the environment. An **open system,** such as a car engine, *is a system that exchanges matter or energy with the environment*.

Closed Systems

Can you think of a system that does not exchange energy with the environment? What about a flashlight? You have read that a flashlight releases radiant energy and thermal energy into the environment. What about your body? You eat food, which contains chemical energy and comes from the environment. Your body also releases several types of energy into the environment, including thermal energy, mechanical energy, and sound energy.

A **closed system** *is a system that does not exchange matter or energy with the environment*. In reality, there are no closed systems. Every physical system transfers some energy to or from its environment. Scientists use the idea of a closed system to study and model the movement of energy.

4. Identify What form of energy in a flashlight is considered to be useful energy?

____

Math Skills

The amount of work done on an object is calculated using the formula $W = F \times d$, where W = work, F = the force applied to the object, and d = the distance the object moves in the direction of the force and while the force is applied. For example, a student slides a library book across a table. The student pushed on the book with a force of 8.5 newtons (N), a distance of 0.3 m. The book slides a total distance of 1 m. How much work is done on the book?

$$W = F \times d$$

$$W = 8.5\text{ N} \times 0.30\text{ m} =$$

$$2.55\text{ N}\cdot\text{m} = 2.6\text{ J}$$

Recall that the distance used to calculate work is the distance the force was applied to the object.

Note: 1 N•m = 1 J, so 2.6 J of work was done on the book.

5. Calculate Work A student lifts a backpack straight up with a force of 53.5 N for a distance of 0.65 m. How much work is done on the backpack?

____

Energy Transformations and Electric Energy

You probably have heard someone say "turn off the lights, you're wasting energy." This form of energy is electric energy. Most appliances you use every day require electric energy. Where does this energy come from?

Renewable Energy Resources

If you think about all of the energy used in the United States, you realize that people need a lot of energy to continue living the way they do. This huge demand for energy and the desire to protect the environment has resulted in a search for renewable energy resources. *A* **renewable energy resource** *is an energy resource that is replaced as fast as, or faster than, it is used.* Several different kinds of renewable energy resources are available.

ACADEMIC VOCABULARY
resource
(noun) a stock or supply of materials, money, or other assets that can be used as needed

Solar Radiant energy from the Sun, or solar energy, is one energy resource that can be converted into electric energy. Some solar energy plants transform radiant energy into electric energy with using photovoltaic (foh toh vohl TAY ihk), or solar, cells. Photovoltaic cells are made from thin wafers of the element silicon. When radiant energy from the Sun hits the cells, it knocks electrons away from the silicon atoms. This movement of electrons is electric energy. Some homes, businesses, and small appliances such as calculators use photovoltaic cells to provide electricity.

Reading Check
6. Name the element from which photovoltaic cells are made.

In some solar energy plants, radiant energy from the Sun is transformed into thermal energy. The thermal energy is used to convert water to steam. The steam turns a generator, which transforms mechanical energy into electric energy.

Wind Have you ever driven along a highway and seen wind turbines? Wind turbines are built in places where winds blow almost continuously, such as the vast open spaces of the southwestern United States. Wind moves the blades of the turbine, turning a generator that transforms kinetic energy of the wind to electric energy. One of the drawbacks of wind energy is that wind does not blow steadily at all times. This source of electric energy is not very consistent or predictable.

Reading Check
7. State What energy transformations occur in a wind turbine?

Hydroelectric In hydroelectric plants, falling water from rivers and dams is channeled through a turbine. When the turbine spins, mechanical energy is transformed to electric energy. Most of the hydroelectric energy produced in the United States comes from the western part of the country. One drawback of hydroelectric energy is that dams and turbines can interrupt the natural movements of animals in rivers and lakes. The number of places where rivers are large enough for these energy plants to be built is also limited. ✓

Geothermal Earth's temperature increases with depth below Earth's surface. But in a few places, Earth is hot close to the surface. Geothermal plants are built where thermal energy from Earth is near Earth's surface. These plants transfer thermal energy to water, which creates steam. The steam turns turbines in electric generators. The states with the most geothermal reservoirs are Alaska, Hawaii, and California.

Biomass Biomass includes wood, plants, and even manure and garbage. All of these sources of stored chemical energy can be transformed to electric energy in energy plants. Burning biomass releases carbon dioxide into the atmosphere. Some scientists believe this contributes to climate change and global warming. However, when biomass crops are grown, the plants use carbon dioxide during photosynthesis. This reduces the overall amount of carbon dioxide from the process. ✓

Nonrenewable Energy Resources

Most of the energy used in homes, schools, stores, and businesses comes from fossil fuels and nuclear energy, as shown below. Fossil fuels and nuclear energy are **nonrenewable energy resources**—*energy resources that are available in limited amounts or that are used faster than they can be replaced in nature.*

Electric Energy Net Generation by Resources as of 2007

Nonrenewable Resources		Renewable Resources	
Resource	**Percentage**	**Resource**	**Percentage**
petroleum	1.6	biomass	about 1.0
natural gas	21.6	hydroelectric	5.8
coal	48.5	geothermal	<1.0
other gases	0.3	wind	<1.0
uranium (nuclear power)	19.4	solar and other	<1.0
Total	**91.4**	**Total**	**About 8.6**

✓ **Reading Check**

8. State What are some drawbacks of hydroelectric plants?

✓ **Reading Check**

9. Describe how biomass uses carbon dioxide.

Interpreting Tables

10. Examine Which resource is used to produce the most electric energy in the United States?

Fossil Fuels Did you know that when a car burns fuel, it is releasing chemical energy that has been stored for millions of years? Petroleum, natural gas, propane, and coal are fossil fuels. Ancient plants stored radiant energy from the Sun as chemical energy in their molecules. This chemical energy was passed on to the animals that ate the plants. Over millions of years, geological processes converted the remains of these ancient plants and animals into fossil fuels.

Fossil fuels are a concentrated form of chemical energy that easily transforms into other forms of energy. However, when fossil fuels burn, they release harmful wastes such as sulfur dioxide, nitrogen oxide, and carbon dioxide. Sulfur dioxide and nitrogen oxide contribute to acid rain. Carbon dioxide is suspected of contributing to global climate change. ✓

Nuclear Energy In nuclear energy plants, uranium atoms are split apart in a process called nuclear fission. Nuclear fission produces thermal energy, which heats water, producing steam. The steam turns turbines that produce electric energy. Nuclear energy plant emissions are not harmful, but the waste from these plants is radioactive. The safe disposal of radioactive waste is a major challenge associated with nuclear energy.

✓ **Reading Check**

11. Point Out What are two disadvantages of using fossil fuels?

Key Concept Check

12. Name What are renewable and nonrenewable energy resources?

After You Read

Mini Glossary

closed system: a system that does not exchange matter or energy with the environment

energy transfer: when energy moves from one object to another without changing form

energy transformation: when one form of energy is converted to another form of energy

law of conservation of energy: energy can be transformed from one form to another, but it cannot be created or destroyed

nonrenewable energy resource: an energy resource that is available in limited amounts or that is used faster than it can be replaced in nature

open system: a system that exchanges matter or energy with the environment

renewable energy resource: an energy resource that is replaced as fast as, or faster than, it is used

work: the transfer of energy that occurs when a force makes an object move in the direction of the force while the force is acting on the object

1. Review the terms and their definitions in the Mini Glossary. Write a sentence explaining the relationship between energy transfer and work.

2. In the table, list one advantage and one disadvantage of each energy resource.

Type of Resource	Advantage	Disadvantage
Wind		
Fossil fuels		
Biomass		

3. Select a sentence you wrote as you read the lesson. In the space below, write a question based on that sentence.

What do you think NOW?

Reread the statements at the beginning of the lesson. Fill in the After column with an A if you agree with the statement or a D if you disagree. Did you change your mind?

Log on to ConnectED.mcgraw-hill.com and access your textbook to find this lesson's resources.

Using Energy and Heat

Particles in Motion

Key Concepts
- What is the kinetic molecular theory?
- In what three ways is thermal energy transferred?
- How are thermal conductors and insulators different?

Before You Read

What do you think? Read the two statements below and decide whether you agree or disagree with them. Place an A in the Before column if you agree with the statement or a D if you disagree. After you've read this lesson, reread the statements to see if you have changed your mind.

Before	Statement	After
	5. Only particles that make up moving objects are in motion.	
	6. Thermal energy can be transferred in several ways.	

Study Coach

Sticky Notes As you read, use sticky notes to mark information that you do not understand. Read the text carefully a second time. If you still need help, write a list of questions to ask your teacher.

Read to Learn

Kinetic Molecular Theory

You read that in every energy transformation, some of the energy is transformed into thermal energy. Some of this thermal energy transfers to other materials. The transfer of thermal energy between materials depends on the movement of particles in the materials.

The kinetic molecular theory explains how particles move. It has three major points:

- All matter is made of particles.
- Particles are in constant, random motion.
- Particles constantly collide with each other and with the walls of their container.

The kinetic molecular theory explains that a carbonated beverage in a glass full of ice is made of particles. The particles move in different directions and at different speeds. They collide with each other, as well as with the particles that make up the ice and the glass. Particles that make up all matter, including carbonated beverages, are in constant motion. Solid particles move slowest, liquid particles move faster, and gas particles move the fastest.

Key Concept Check

1. Explain What are the three points of the kinetic molecular theory?

____

____

____

____

____

____

Temperature

When you pick up a glass of ice-cold soda, the glass feels cold. Could you estimate its temperature? The temperature of something depends on how much kinetic energy the particles that make up the material have.

The measure of the average kinetic energy of the particles in a material is **temperature.** If most of the drink particles have little kinetic energy, the drink has a low temperature and the glass feels cold.

The SI unit for temperature is the kelvin (K). However, scientists often use the Celsius temperature scale (°C) to measure temperature. When you do scientific investigations, you will most likely measure temperature in degrees Celsius too.

2. State What temperature scale do most scientists use?

____

Thermal Expansion

Suppose your teacher told everyone in your classroom to walk around slowly. How much space would you need? You probably would have enough space in your classroom. Now imagine your teacher told the students in your class to run around as fast as they could. There probably would not be enough space to run as fast as possible. But, if the class were in a large space, such as a gymnasium, everyone could run very quickly.

When the particles that make up a material move slowly, they occupy less volume than they do at a higher temperature. As the temperature of a material increases, particles begin to move faster. They collide with each other more often and push each other farther apart.

Reading Check

3. Identify As temperature rises, what happens to particles in a material?

____

____

____

Thermal expansion is the increase in volume of a material undergoes due to temperature increase. When the temperature of a material decreases, the volume of the material decreases. This process is known as thermal contraction.

Most materials contract as their temperature decreases, but water is an exception. Water does not contract when it is cooled. When water is cooled to near its freezing point, interactions between water molecules push the molecules apart. Empty spaces form among the molecules. This makes ice less dense than water.

Water expands as it freezes because of these molecular interactions. Because ice is less dense than water, ice floats on top of water instead of sinking.

FOLDABLES

Create a trifold book to explain the different ways in which thermal energy is transferred.

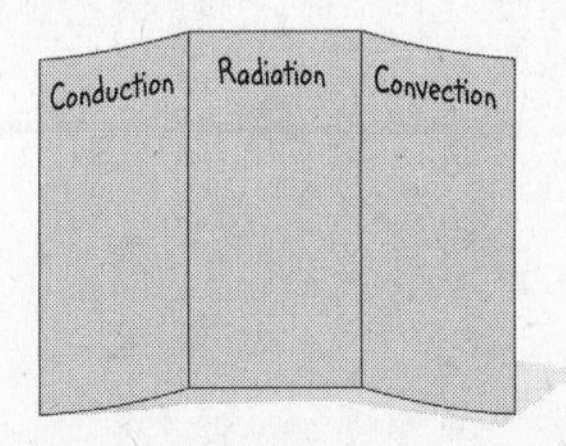

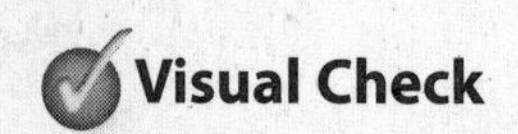

Visual Check

4. Trace the movement of thermal energy from the water to the bottle to the refrigerator air.

Thermal Energy in a Water Bottle

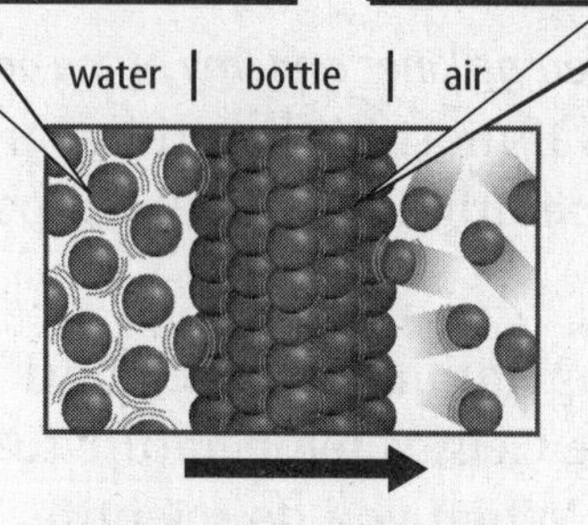

Thermal energy moves from the warm water to the cool air by the collision of particles.

Transferring Thermal Energy

Thermal energy is transferred through materials and from material to material by collisions of particles. Suppose you put a bottle of warm water in the refrigerator. As shown in the figure above, moving water molecules collide with the particles that make up the bottle. These collisions transfer kinetic energy to the particles that make up the bottle, and they vibrate faster. As the particles move faster, their average kinetic energy, or temperature increases. ✓

Reading Check

5. Recognize When average kinetic energy increases, temperature ___. (Circle the correct answer.)

a. increases
b. decreases
c. stabilizes

The particles that make up the bottle then collide with particles that make up the air in the refrigerator. The average kinetic energy of the particles that make up the air in the refrigerator increases. In other words, the temperature of the air in the refrigerator increases.

The average kinetic energy of the particles of water decreases as thermal energy moves from the water to the bottle. Therefore, the temperature of the water decreases.

As the kinetic energy of the particles that make up a material increases, the thermal energy of the particles increases. As the kinetic energy of the particles that make up a material decreases, the thermal energy of the particles decreases. So, when particles transfer kinetic energy, they transfer thermal energy.

Thermal Energy and Heat

Thermal energy moves from warmer materials, such as the warm water in the bottle, to cooler materials, such as the cool air in the refrigerator. *The movement of thermal energy from a region of higher temperature to a region of lower temperature is called* **heat.** Because your hand is warmer than the bottle, thermal energy moves from your hand to the bottle. When you place the bottle in the refrigerator, thermal energy moves from the warm bottle to the cool air in the refrigerator.

Thermal Equilibrium

What happens if the water remains in the refrigerator for several hours? The temperature of the water, the bottle, and the air in the refrigerator become the same. When the temperatures of materials that are in contact are the same, the materials are said to be in thermal equilibrium.

After the materials reach thermal equilibrium, the particles that make up the water, the bottle, and the air continue to collide with each other. The particles transfer kinetic energy back and forth, but the average kinetic energy of all the particles remains the same.

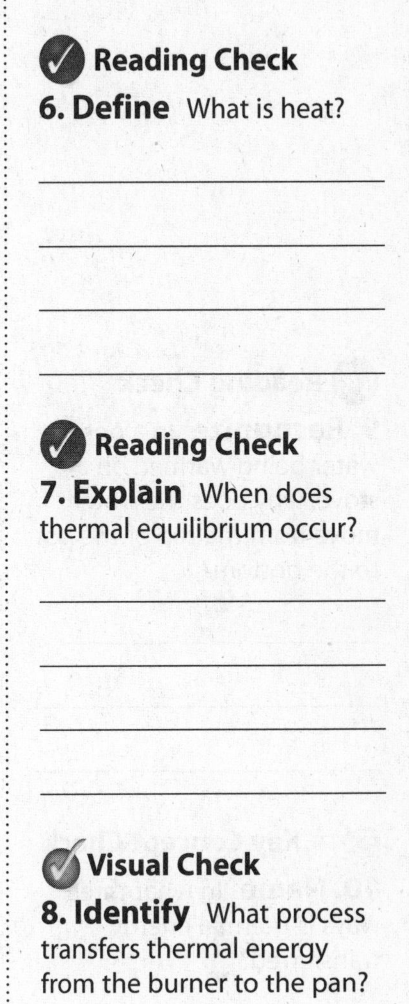

Reading Check

6. Define What is heat?

Reading Check

7. Explain When does thermal equilibrium occur?

Visual Check

8. Identify What process transfers thermal energy from the burner to the pan?

Heat Transfer

The figure below shows a pan of water heating on a burner. How is thermal energy transferred to the water?

1. Conduction Fast-moving particles of the gases in the flame collide with the particles that make up the pan. This transfers thermal energy to the pan. Then, the particles that make up the pan collide with particles of water, transferring thermal energy to the water. **Conduction** *is the transfer of thermal energy by collisions between particles in matter.*

Conduction, Radiation, and Convection

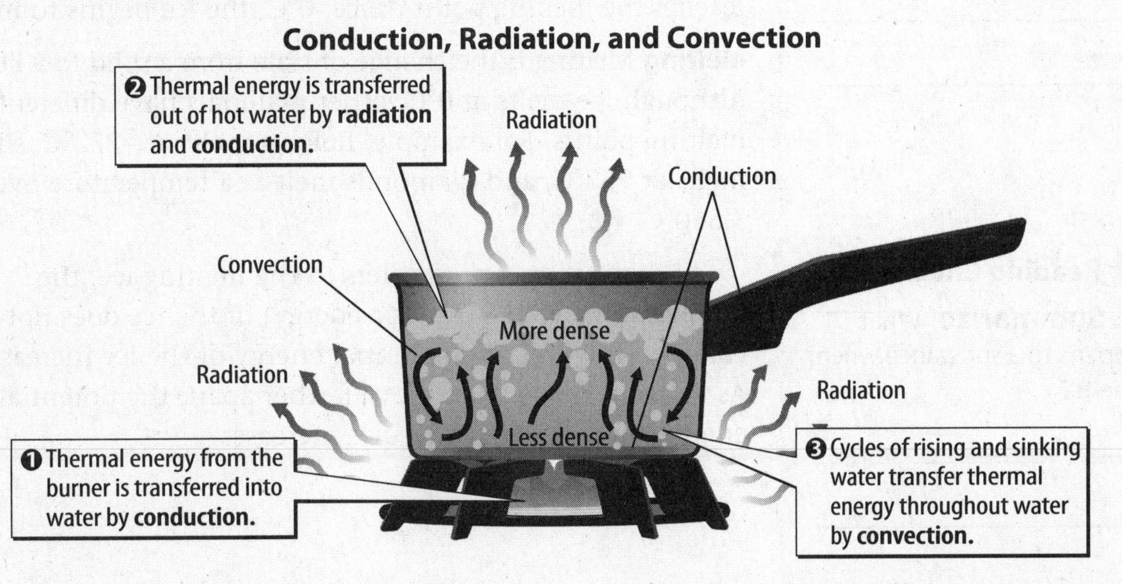

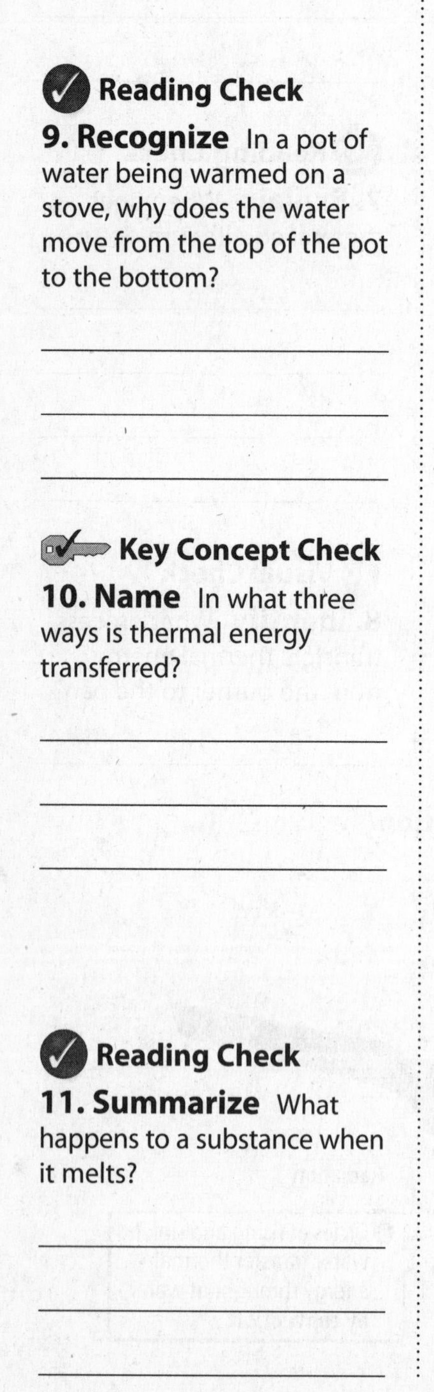

2. Radiation If you put your hands near the side of the pan on the burner, you feel warmth. The thermal energy you feel is from **radiation**—*the transfer of thermal energy by electromagnetic waves.* All objects emit radiation, but warmer materials, such as hot water, emit more radiation than cooler ones.

3. Convection The flame, or hot gases, heats water at the bottom of the pan. The water at the bottom of the pan undergoes thermal expansion and is now less dense than the water above it. The denser water sinks and forces the less dense, warmer water upward. The water continues this cycle of warming, rising, cooling, and sinking, as thermal energy moves throughout the water.

The transfer of thermal energy by the movement of the particles from one part of a material to another is **convection.** Convection also occurs in the atmosphere. Warm, less-dense air is forced upward by cooler, more-dense falling air. Thermal energy is transferred as the air rises and sinks.

Heat and Changes of State

When thermal energy is added or removed from a substance, sometimes only the temperature changes. At other times, a more dramatic change occurs—a change of state.

Changes Between Solids and Liquids

If enough thermal energy is added to a substance, the substance will change state. Ice (a solid) changes to water (a liquid) when thermal energy is added. Water changes to water vapor (a gas) if enough thermal energy is added. What happens if you place a flask of ice over a burner? Thermal energy moves from the burner to the flask and to the ice. The temperature of the ice increases. When the temperature reaches the melting point of ice, 0°C, the ice begins to melt.

Melting Melting is the change of state from a solid to a liquid. Although ice melts at 0°C, other materials have different melting points. For example, helium melts at –272°C, silver melts at 962°C, and diamonds melt at a temperature over 3,550°C.

As thermal energy transfers to the melting ice, the temperature (average kinetic energy) of the ice does not change. However, the potential energy of the ice increases. As the water molecules move farther apart, the potential energy between the molecules increases.

Freezing The reverse process occurs when thermal energy is removed from water. When water is placed in a freezer, thermal energy moves from the water to the colder air in the freezer. The average kinetic energy (temperature) of the water decreases.

When the temperature of the water reaches 0°C, the water begins to freeze. Freezing is the change of state from a liquid to a solid. Notice that the freezing point of water is the same as the melting point of ice. Freezing is the opposite of melting. ✓

While water is freezing, the temperature remains at 0°C until all the water is frozen. Once all the water freezes, the temperature of the ice begins to decrease. As the temperature decreases, the water molecules vibrate in place at a slower and slower rate. ✓

Changes Between Liquids and Gases

What happens when ice melts? As thermal energy transfers to the ice, the particles move faster and faster.

The average kinetic energy of the water particles that make up ice increases and the ice melts. The temperature continues of the ice continues to increase until it reaches 100°C. At 100°C, water begins to vaporize.

Vaporization *is the change of state from a liquid to a gas.* While the water is changing state—from a liquid to a gas—the kinetic energy of the particles remains constant. Liquids vaporize in two ways—boiling and evaporation. ✓

Boiling and Evaporation Vaporization that occurs within a liquid is called boiling. Vaporization that occurs at the surface of a liquid is called evaporation.

Have you heard the term *water vapor*? The gaseous state of a substance that is normally a liquid or a solid at room temperature is called vapor. Because water is liquid at room temperature, its gaseous state is referred to as water vapor.

✓ **Reading Check**

12. Explain Why does water become a solid when it is placed in a freezer?

✓ **Reading Check**

13. Recognize What is a change of state?

✓ **Reading Check**

14. Identify What are the two ways liquids vaporize?

Condensation The reverse process also can occur. Removing thermal energy from a gas changes it to a liquid. The change of state from a gas to a liquid is condensation. Water vapor often condenses on grass overnight. This condensation is called dew. ✓

Changes Between Solids and Gases

Usually, water transforms from a solid to a liquid and then to a gas as it absorbs thermal energy. However, this is not always the case.

Sublimation On cold winter days, ice often changes directly to water vapor without passing through the liquid state. The change of state that occurs when a solid changes to a gas without passing through the liquid state is called sublimation. Dry ice, or solid carbon dioxide, also sublimes. Dry ice is used to keep foods frozen when they are shipped.

Deposition When thermal energy is removed from some materials, they undergo deposition. Deposition is the change of state from a gas directly to a solid without passing through the liquid state. Water vapor undergoes deposition when it freezes and forms frost.

The table below summarizes six changes of state that matter can undergo.

Changes of State

Change of State	From	To	Thermal Energy
melting	solid	liquid	added
freezing	liquid	solid	removed
vaporization	liquid	gas	added
condensation	gas	liquid	removed
sublimation	solid	gas	added
deposition	gas	solid	removed

✓ **Reading Check**

15. Define What causes dew on the grass in the morning?

SCIENCE USE V. COMMON USE

sublime

Science Use to change from a solid state to a gas state without passing through the liquid state

Common Use inspiring awe; supreme, outstanding, or lofty in thought or language

Interpreting Tables

16. Name three changes of state that occur because thermal energy is removed from a substance.

Conductors and Insulators

When you put a metal pan on a burner, the pan gets hot. If the pan has a handle made of wood or plastic, the handle stays cool. Why doesn't the handle get hot like the pan as a result of thermal conduction? It would seem that conduction of thermal energy should transfer thermal energy to the handle as well as the pan.

The metal that makes up the pan is a **thermal conductor,** *a material in which thermal energy moves quickly.* The atoms that make up thermal conductors have electrons that are free to move, transferring thermal energy easily. The material that makes up the pan's handle is a **thermal insulator,** *a material in which thermal energy moves slowly.* The electrons in thermal insulators are held tightly in place and do not transfer thermal energy easily.

Think it Over

17. Name two commonly used items that are known as thermal insulators.

Key Concept Check

18. Contrast How do thermal conductors differ from thermal insulators?

Mini Glossary

conduction: the transfer of thermal energy due to collisions between particles in matter

convection: the transfer of thermal energy by the movement of the particles from one part of a material to another

heat: the movement of thermal energy from a region of higher temperature to a region of lower temperature

radiation: the transfer of thermal energy by electromagnetic waves

temperature: the measure of the average kinetic energy of the particles in a material

thermal conductor: a material in which thermal energy moves quickly

thermal insulator: a material in which thermal energy moves slowly

vaporization: the change of state from a liquid to a gas

1. Review the terms and their definitions in the Mini Glossary. Write a sentence comparing conduction and convection.

___

___

2. Indicate what could occur when thermal energy is added or removed from a substance by writing the letter of the action in the correct box of the graphic organizer. You may need to use the answer(s) more than once.

a. change of state
b. lower kinetic energy
c. thermal contraction
d. higher kinetic energy
e. thermal expansion
f. particles move faster

Thermal energy is added to a substance

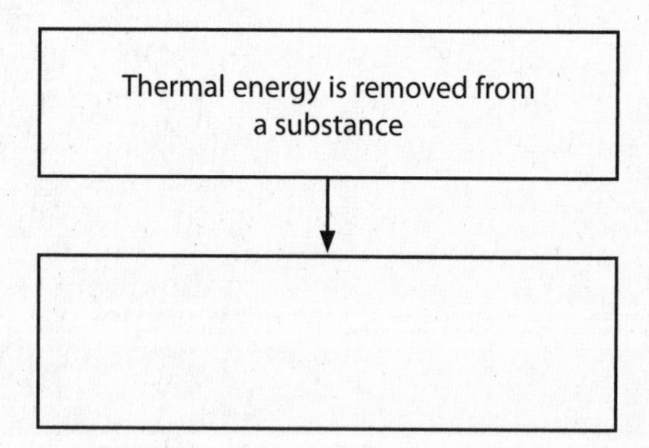

What do you think NOW?

Reread the statements at the beginning of the lesson. Fill in the After column with an A if you agree with the statement or a D if you disagree. Did you change your mind?

Log on to ConnectED.mcgraw-hill.com and access your textbook to find this lesson's resources.

The Earth System

Earth Systems and Interactions

Before You Read

What do you think? Read the three statements below and decide whether you agree or disagree with them. Place an A in the Before column if you agree with the statement or a D if you disagree. After you've read this lesson, reread the statements to see if you have changed your mind.

Before	Statement	After
	1. The amount of water on Earth remains constant over time.	
	2. Hydrogen makes up the hydrosphere.	
	3. Most carbon on Earth is in the atmosphere.	

Key Concepts 

- How do Earth systems interact in the carbon cycle?
- How do Earth systems interact in the phosphorus cycle?

Read to Learn

Earth Systems

Your body contains many systems. These systems work together and make one big system—your body. Earth is a system, too. Like you, Earth has smaller systems that work together, or interact, and make the larger Earth system. Four of these smaller systems are the atmosphere, the hydrosphere, the geosphere, and the biosphere. ✓

Study Coach

Make a Table Contrast the carbon cycle and the phosphorus cycle in a two-column table. Label one column Carbon Cycle and the other column Phosphorus Cycle. Complete the table as you read this lesson.

✓ **Reading Check**

1. Identify What systems make up the larger Earth system?

___

___

___

___

The Atmosphere

The outermost Earth system is a mixture of gases and particles of matter called the atmosphere. It forms a layer around the other Earth systems. The atmosphere is mainly nitrogen and oxygen. Gases in the atmosphere move freely, helping transport matter and energy among Earth systems.

The Hydrosphere

Below the atmosphere is the hydrosphere, the system that contains all of Earth's water. Most of the water is on Earth's surface—in oceans, glaciers, lakes, ice sheets, and rivers. Smaller amounts of water are deep beneath Earth's surface, in the atmosphere, and in living things. Like gases in the atmosphere, water in the hydrosphere continuously moves from place to place. Many substances dissolve easily in water. These dissolved substances move with the water.

The Geosphere

The largest Earth system is the geosphere, or the solid Earth. The geosphere includes the thin layer of soil and rocks on Earth's surface and all the underlying layers of Earth. Because the geosphere is mainly solid, materials in this system move more slowly than the gases in the atmosphere or the water in the hydrosphere. As the materials move, they slowly transport energy and matter. ✓

✓ Reading Check

2. Explain Why do materials in the geosphere move slowly?

____

____

____

The Biosphere

All living organisms on Earth make up the biosphere. Because organisms live in air, water, soil, and rocks, the biosphere is within all other Earth systems. Living organisms survive using gases from the atmosphere, water from the hydrosphere, and nutrients in soil and rocks.

Interactions Among Earth Systems

	Biosphere	Geosphere	Hydrosphere
Atmosphere	• The ozone layer helps protect organisms from harmful solar radiation. • Plants use oxygen and carbon dioxide during photosynthesis.	• Wind causes weathering and erosion. • Volcanic eruptions eject gas and debris into the air.	• The water cycle influences weather and climate. • Increasing global temperatures lead to melting polar ice caps.
Hydrosphere	• All organisms need water for life functions. • Rising sea levels change habitats.	• Water and ice cause weathering, erosion, and deposition. • Hurricanes and tsunamis change coastal landforms.	
Geosphere	• Materials in the geosphere provide nutrients for life functions. • Organisms contribute to weathering, erosion, and fossil fuel formation.		

Interpreting Tables

3. Name What other interactions can you name?

____

____

____

____

Interactions Among Earth Systems

Earth systems interact by exchanging matter and energy. The table above describes some of these interactions. Matter and energy often change in form as they flow between systems.

The Water Cycle

The water cycle is an example of interaction among Earth systems. It is the continuous movement of water on, above, and below Earth's surface, as shown in the figure at the top of the next page. Water moves within the hydrosphere and into other Earth systems.

The Water Cycle

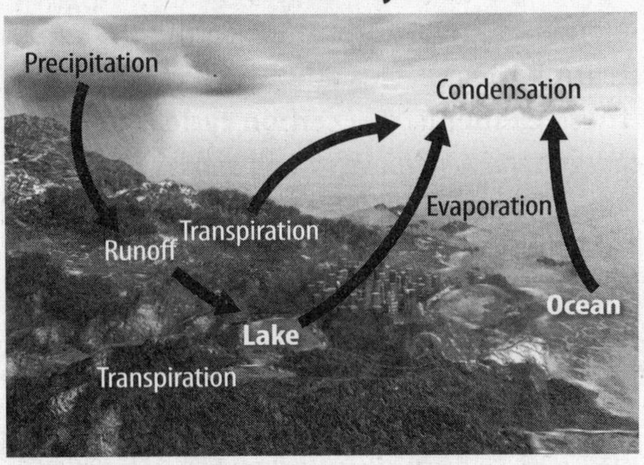

Water is not just a liquid. Sometimes it changes state and becomes solid ice or gaseous water vapor. As water flows or changes state, it moves thermal energy within the water cycle and among Earth systems.

The Rock Cycle

The rock cycle, shown below, is another type of interaction among Earth systems. It is the series of processes that change rocks from one form to another. Some processes happen deep within Earth. Others, like weathering and erosion, occur on or near Earth's surface. The atmosphere, hydrosphere, and biosphere interact with the geosphere through weathering and erosion. For example, rain and plants can weather rocks into sediments. Wind and flowing water can erode rocks and sediment and deposit them in new places. The amount of material cycling through each system usually stays the same, but it might change state or form—either quickly or over millions of years.

The Rock Cycle

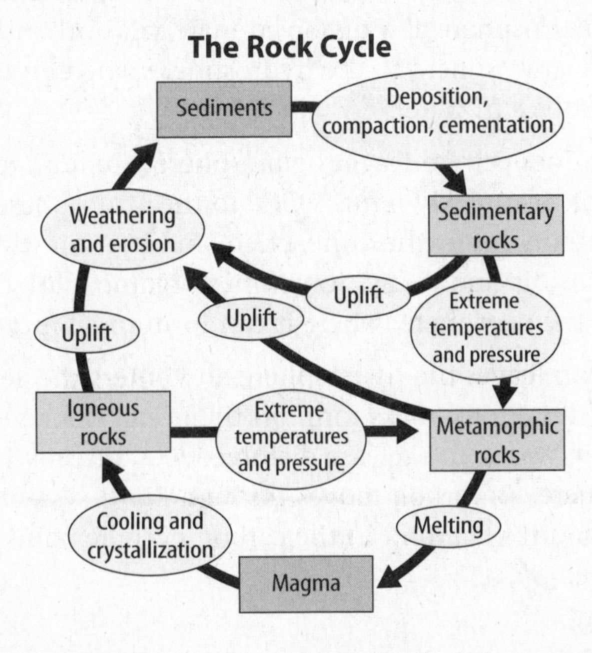

Visual Check

4. Identify What processes in the water cycle move liquid water from Earth's surface into the atmosphere as water vapor?

____

____

FOLDABLES®

Make a four-column chart book to organize your notes on Earth systems and interactions.

Water Cycle	Rock Cycle	Carbon Cycle	Phosphorus Cycle

Visual Check

5. Name What kind of rock is formed through extreme temperature and pressure?

____

The Carbon Cycle

Some elements are so important to life that scientists study their individual cycles among Earth systems. Carbon is one of these elements. *The* **carbon cycle** *is the series of processes that continuously move carbon among Earth systems.*

The Carbon Cycle

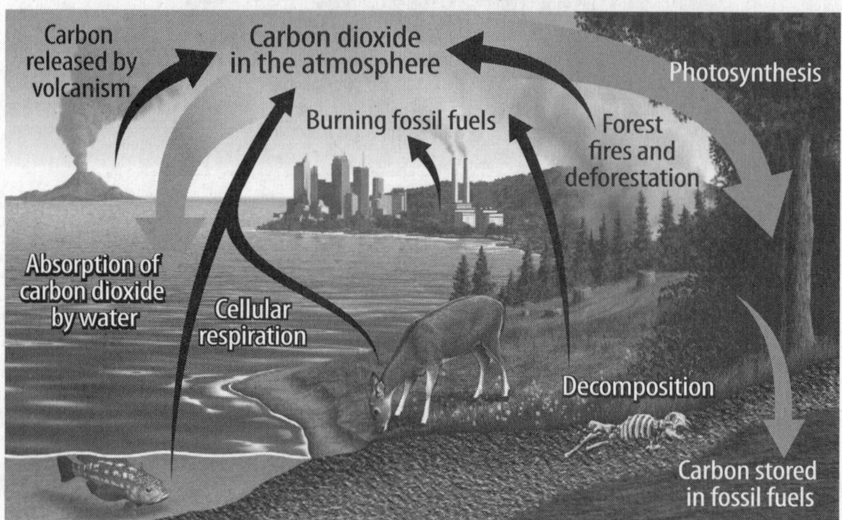

Processes of the Carbon Cycle Trace the path of carbon in the figure above as it is released from the geosphere during a volcanic eruption. Carbon from the geosphere enters the atmosphere as the trace gas carbon dioxide (CO_2). Several processes then remove CO_2 from the atmosphere.

During photosynthesis, plants use sunlight, CO_2, and water and make simple sugars. As a result, carbon leaves the atmosphere and enters the biosphere. Weathering of rocks also removes carbon from the atmosphere and transports it to the hydrosphere as a dissolved material. Carbon moves from the atmosphere to the hydrosphere when atmospheric CO_2 dissolves in water. ✓

How does carbon leave the biosphere? Cellular respiration in organisms quickly returns CO_2 to the atmosphere. Even more carbon enters the atmosphere and the soil when organisms die and decay. Sometimes organic matter is buried deep in the geosphere, where it can form fossil fuels.

Carbon leaves the hydrosphere and enters the geosphere when sedimentary rocks form on the ocean floor. Ocean water can warm and release dissolved CO_2 directly into the atmosphere. As carbon moves through Earth systems, the total amount of carbon in the carbon cycle remains about the same.

Visual Check

6. Examine Which processes add carbon dioxide to the atmosphere? Which processes remove it?

Reading Check

7. Describe What role does photosynthesis play in the carbon cycle?

Key Concept Check

8. Explain How do Earth systems interact in the carbon cycle?

Carbon Reservoirs		
Carbon Reservoirs	**Carbon (billions of tons)**	**Form**
Atmosphere	750	CO_2 gas
Biosphere	3,000	organic molecules
Hydrosphere	40,000	dissolved CO_2 gas
Geosphere (crust and upper mantle)	750,000	minerals and rocks
Geosphere (lower mantle)	750,000+	minerals and rocks

Carbon Reservoirs After water, carbon is the most abundant substance in living organisms. But as you just read, carbon is not limited to the biosphere. Carbon is in reservoirs, or storage places, within all Earth systems, as shown in the table above. On Earth, most carbon is combined with other elements in compounds. ✓

Carbon in the biosphere is stored in organisms. It does not exist as carbon atoms. It is combined with other elements in complex organic molecules, such as sugars and starches. Cells and tissues of all organisms are made of organic compounds.

In the atmosphere and the hydrosphere, carbon exists as carbon dioxide gas (CO_2). Though the atmosphere is the smallest carbon reservoir, atmospheric CO_2 is important. The amount of CO_2 in the atmosphere affects climate, as you will read later. CO_2 in the hydrosphere is dissolved in water.

Most of Earth's carbon is stored in the geosphere. Carbon is combined with other elements in minerals that form rocks. Limestone contains the mineral calcite, which contains carbon. Carbon also is stored as fossil fuels, such as coal, natural gas, and oil, which form underground.

Humans and the Carbon Cycle Some changes in the amount of CO_2 in the atmosphere occur naturally. For example, a volcanic eruption can release large amounts of CO_2 into the air. But not all changes in the amount of CO_2 in the atmosphere occur naturally.

Human activities cause some changes in levels of atmospheric CO_2. When people burn fossil fuels to generate electricity or to power vehicles, CO_2 is released directly into the atmosphere. Other activities can indirectly increase levels of atmospheric CO_2. For example, large tracts of forests might be cut down for agriculture or development. More CO_2 remains in the atmosphere because there are fewer trees to take in CO_2 during photosynthesis.

Interpreting Tables

9. Compare How does the amount of carbon in the atmosphere compare to the amount in the biosphere?

✓ Reading Check

10. Identify What are the two most abundant substances in living organisms?

REVIEW VOCABULARY

fossil fuels
fuels such as coal, oil, and natural gas that form in the Earth from plant or animal remains

Math Skills

You can use percentages to figure out many types of problems. For example, a lawn fertilizer labeled 24-2-8 contains 24 percent nitrogen, 2 percent phosphorus, and 8 percent potassium, in that order. How much phosphorus is in a 22-kg bag of fertilizer?

a. Change the percentage of phosphorus to a decimal by moving the decimal point two places to the left.

$2\% = 0.02$

b. Multiply the total mass of fertilizer by the decimal.

22 kg fertilizer $\times$ 0.02 = 0.44 kg phosphorus

11. Use Percentages How much phosphorus is in a 10-kg bag of 20-5-10 fertilizer?

____

Key Concept Check

12. Explain How do Earth systems interact in the phosphorus cycle?

____

____

____

____

____

Greenhouse Gases CO_2 is a greenhouse gas. **Greenhouse gases** *are gases in the atmosphere that absorb and reradiate thermal energy from the Sun*. These gases keep Earth from becoming too cold to support life.

When levels of CO_2 in the atmosphere increase, more thermal energy is absorbed and reradiated. Earth's average surface temperature increases. This phenomenon is called global warming.

Global warming can cause coastal flooding as ice caps melt and sea level rises. These changes might cause climates around the world to change, altering habitats and harming living organisms.

The Phosphorus Cycle

Some important elements do not cycle through all Earth systems. One example is phosphorus (FAHS fuh rus). Phosphorus does not exist in the atmosphere. *The* **phosphorus cycle** *is the series of processes that move phosphorus among Earth systems.*

Processes of the Phosphorus Cycle The processes in the phosphorus cycle help move phosphorus through the geosphere, the hydrosphere, and the biosphere. Phosphorus does not exist in nature as an element. It exists as phosphates (PO_4), compounds formed from phosphorus and oxygen.

The figure on the top of the next page illustrates the processes of the phosphorus cycle. Earth's phosphorus starts out in the geosphere. Rocks exposed at Earth's surface release phosphates when they weather. The phosphates either remain in the soil or dissolve and enter the hydrosphere.

Dissolved phosphate molecules move in liquid water through the water cycle. Eventually, the phosphates reach lake bottoms or the seafloor and are deposited along with sediment. The phosphorus becomes part of new sedimentary rocks that form from the deposited sediment.

Plants absorb phosphorus from soil or water. Animals take in phosphorus when they eat plants or when they eat other animals that have eaten plants. These phosphates return to the soil as part of animal waste or as part of decomposing organisms.

The Phosphorus Cycle

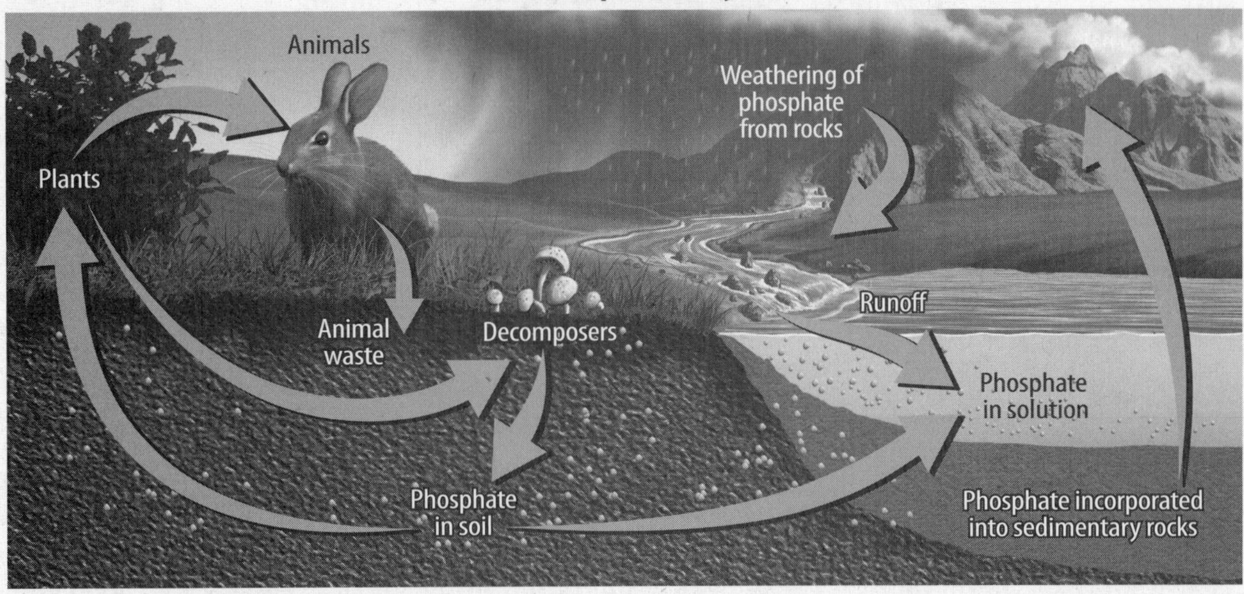

Phosphorus Reservoirs Rocks in the geosphere include minerals containing phosphates. The most common of these minerals is apatite (A puh tite). Turquoise is a mineral often used in jewelry. Turquoise is also a phosphate mineral.

The hydrosphere is another reservoir for phosphorus. As you just read, phosphates dissolve in water. Phosphorus moves through the water cycle as a dissolved substance in liquid water. However, it does not evaporate and enter the atmosphere.

Phosphorus cycles through the geosphere and the hydrosphere over long periods of time. Phosphates incorporated into sedimentary rocks on the ocean floor might not reenter the phosphorus cycle for millions of years.

In contrast, the phosphorus stored in organisms in the biosphere recycles fairly quickly. Like carbon, phosphorus is a necessary element for organisms. It is needed to make cell membranes and transfer energy. It also is an important component of teeth, bones, and shells. Animals store most of their phosphorus in these structures. Animal waste also is a major source of phosphorus. ✓

Humans and the Phosphorus Cycle Recall that humans can disturb the carbon cycle. Humans can also disturb the phosphorus cycle. For example, plants store most of the phosphorus in rain forests. As the plants drop their leaves or die, new plant growth quickly takes up the phosphorus. Clearing the trees in rain forests disturbs the phosphorus cycle.

Visual Check

13. Evaluate How do living things affect the phosphorus cycle?

____

____

____

____

____

Reading Check

14. Summarize How do living organisms use the element phosphorus?

____

____

____

____

Humans and the Phosphorus Cycle

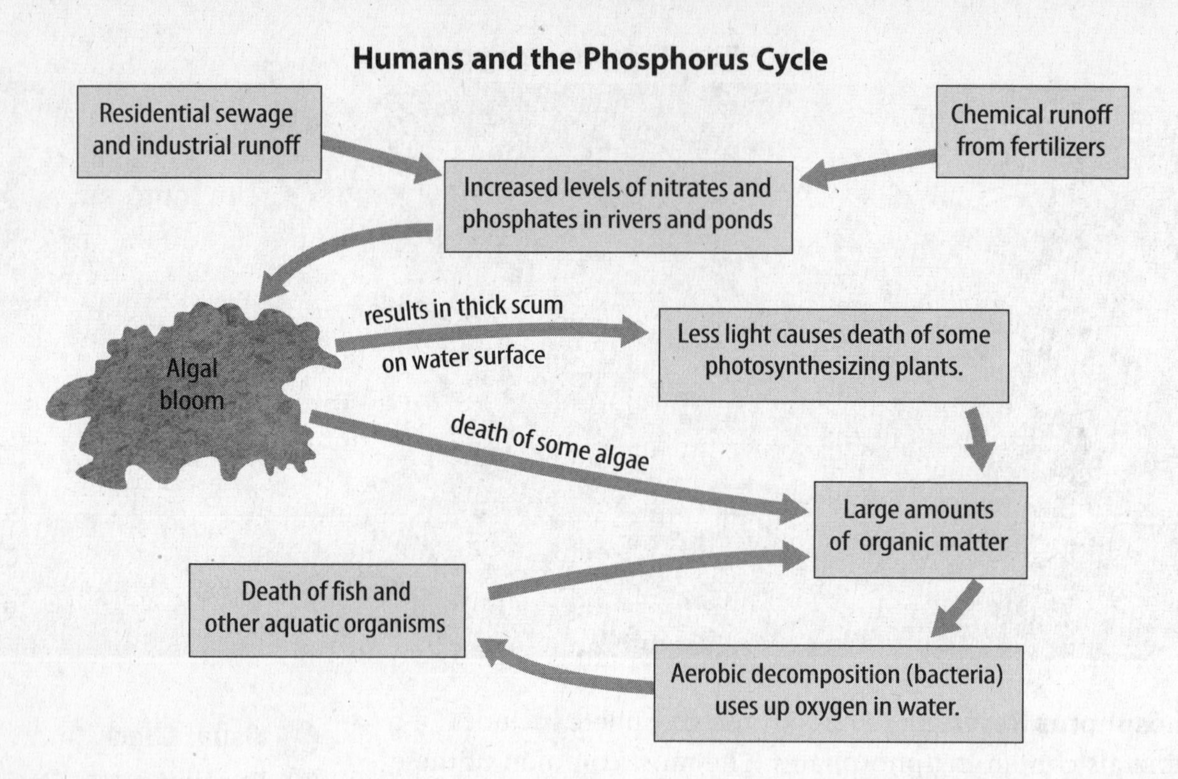

Visual Check
15. Trace the path of residential sewage.

Reading Check
16. Describe How can farming affect the phosphorus cycle?

Clearing forests also exposes soil. Heavy rains wash away the phosphorus released by decaying plants. The lack of this important element makes the soil unproductive for future crops.

Other practices also can impact the phosphorus cycle. As shown in the figure above, runoff from farms, homes, and factories can contain phosphorus. Rain can wash the phosphorus into rivers, streams, and lakes. Algae consume the phosphorus, and the algae population increases. As excess algae decompose, they use up oxygen in the water. This harms fish and other aquatic organisms.

After You Read

Mini Glossary

carbon cycle: the series of processes that continuously move carbon among Earth systems

greenhouse gas: a gas in the atmosphere that absorbs and reradiates thermal energy from the Sun

phosphorus cycle: the series of processes that move phosphorus among Earth systems

1. Review the terms and their definitions in the Mini Glossary. Write a sentence explaining the relationship between greenhouse gases and life on Earth.

2. The diagram on the left identifies one human activity that disturbs the carbon cycle and describes its impact. Fill in the graphic organizer on the right for the phosphorus cycle.

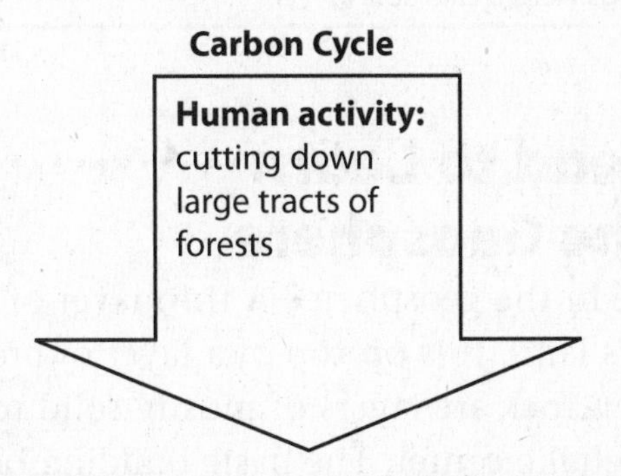

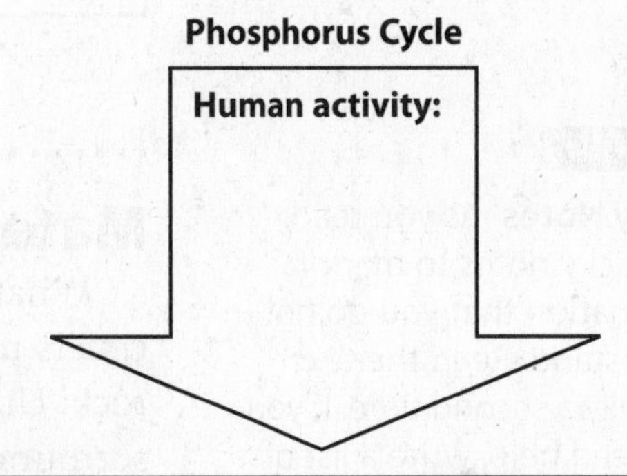

3. Review the information in the table you created as you read the lesson and write a question about the carbon or phosphorus cycle in the space below.

What do you think NOW?

Reread the statements at the beginning of the lesson. Fill in the After column with an A if you agree with the statement or a D if you disagree. Did you change your mind?

Log on to ConnectED.mcgraw-hill.com and access your textbook to find this lesson's resources.

The Earth System

The Geosphere

Key Concepts

- How do materials in the geosphere differ?
- Why does the geosphere have a layered structure?

Before You Read

What do you think? Read the three statements below and decide whether you agree or disagree with them. Place an A in the Before column if you agree with the statement or a D if you disagree. After you've read this lesson, reread the statements to see if you have changed your mind.

Before	Statement	After
	4. The inside of Earth is mostly solid rock.	
	5. Rocks make up minerals.	
	6. Living things help make soil.	

Mark the Text

Sticky Notes As you read, use sticky notes to mark information that you do not understand. Read the text carefully a second time. If you still need help, write a list of questions to ask your teacher.

Read to Learn

Materials in the Geosphere

What materials are in the geosphere? A thin layer of soil covers much of Earth's land. It is on top of a layer of broken rock. Under the broken rock are layers of mostly solid rock surrounding a hot, metallic center. The basic building blocks for soil, rocks, and metals are minerals. Minerals combine in different ways, forming the other materials in the geosphere.

Minerals

In science, the term *mineral* has a specific definition. A mineral is naturally occurring, inorganic, and solid, and it has a crystal structure and a definite chemical composition. Quartz is a mineral. Quartz formed naturally, was never living, is solid, has a crystal structure, and has a chemical composition of two oxygen atoms for each silicon atom.

Reading Check

1. Identify What are the characteristics of a mineral?

Mineral Properties Minerals have unique physical properties, such as color, that scientists use to identify them. Mineral properties include luster and streak. **Luster** *is the way a mineral's surface reflects light.* Some minerals reflect a lot of light and appear shiny. Other minerals do not reflect a lot of light and appear dull. **Streak** *is the color of a mineral's powder.* Scientists observe streak by scratching a mineral across a tile of unglazed porcelain. Although the color of a mineral often varies, its streak is always the same.

Hardness, Cleavage, and Fracture Hardness is another physical property of minerals. Certain minerals are harder than others. The Mohs scale ranks minerals on a scale of 1 to 10 based on their relative hardness. Talc, the softest mineral, has a hardness of 1. Diamond, the hardest mineral, has a hardness of 10. Scientists determine the hardness of a mineral by how easily another mineral or a common object scratches it.

Hardness affects how easily a mineral breaks. **Cleavage** *is the tendency of minerals to break along smooth, flat surfaces.* The mineral calcite exhibits cleavage. When calcite breaks, it has defined edges. **Fracture** *is the tendency of minerals to break along irregular surfaces.* Quartz exhibits fracture because it does not break along a flat plane.

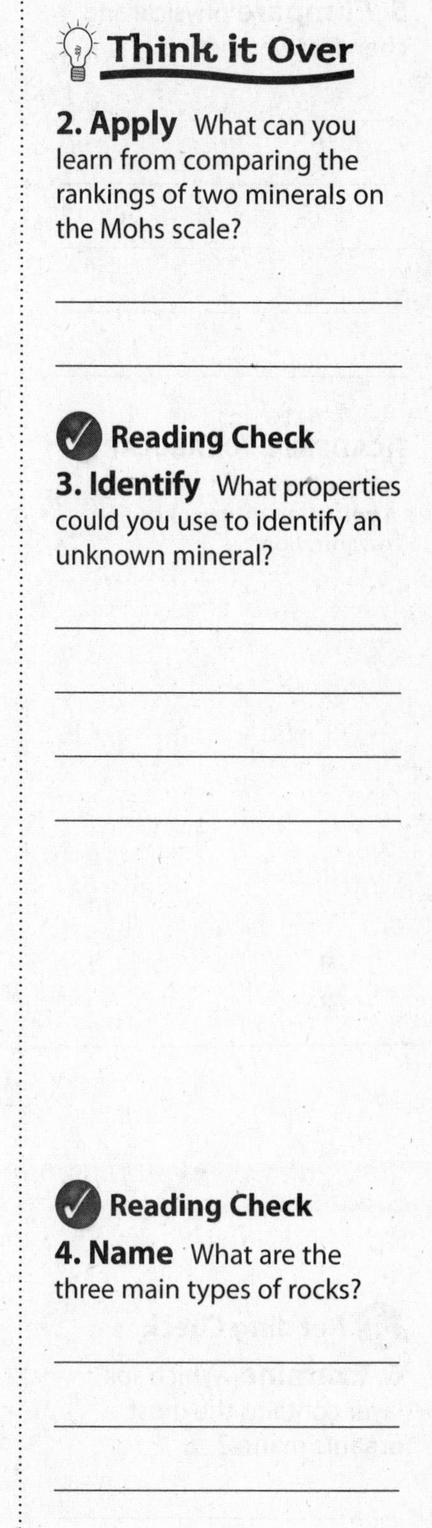

Think it Over

2. Apply What can you learn from comparing the rankings of two minerals on the Mohs scale?

Reading Check

3. Identify What properties could you use to identify an unknown mineral?

Reading Check

4. Name What are the three main types of rocks?

Mineral Interactions Some minerals tend to break apart and combine with other substances. Other minerals are more stable and durable. Calcite and quartz are two common minerals that display this difference. On the Mohs scale, calcite has a hardness of 3, and quartz has a hardness of 7. Calcite dissolves in water more easily than quartz. This increases its interactions with other materials. Quartz does not dissolve easily or break apart as easily as calcite.

Rocks

You might know that a rock is a naturally occurring solid comprised of minerals and other materials. Scientists classify rocks according to how they form.

Rock Types The three main types of rocks are igneous (IG nee us), metamorphic, and sedimentary. Igneous rock forms when molten rock material cools and hardens. This can happen deep inside Earth as magma cools, or when molten material called lava flows onto Earth's surface and cools.

Metamorphic rock forms when high temperatures and extreme pressure act on sedimentary, igneous, or other metamorphic rocks. High temperatures and pressure alter the texture or the chemical composition of the rocks and form new metamorphic rocks.

Rock fragments called sediment make up sedimentary rock. Sedimentary rocks form when water, wind, ice, or gravity erode sediment and deposit it in layers. Over time, the weight from upper layers of sediment compresses lower layers. Sediment compacts and cements together, forming sedimentary rock.

Reading Check

5. Compare physical and chemical weathering.

Interactions The formation of sedimentary rocks involves interactions among the geosphere, the atmosphere, the hydrosphere, and the biosphere. Recall that weathering breaks rock into small pieces. Physical weathering occurs when physical processes break down rock. For example, tree roots can grow in cracks in a rock, eventually breaking the rock. Chemical weathering results from chemical reactions on rock surfaces. Many of these chemical reactions include water.

Soil

ACADEMIC VOCABULARY

structure
(noun) arrangement or organization

Have you ever grown a garden? If so, you have used one of the most important materials in the geosphere—soil. Soil is the loose, weathered material in which plants grow. If you were to dig into the ground, you would see that soil has a layered structure. The layers form as Earth's processes slowly transform rock into soil.

How Soil Forms At Earth's surface, interactions among rocks, water, air, and organisms form soil. Soil formation begins when rocks weather into sediment. Water dissolves minerals and other materials from the sediment, and they become part of the developing soil. Animals and plants also affect soil formation. They weather sediment and create open spaces for air and water.

Wastes from organisms add nutrients to soil. Nutrients also enter soil when organisms die and their bodies decay. The organic matter makes soil more fertile and gives it a dark color. It takes hundreds to thousands of years to build thick layers of soil. Each layer has different properties.

A-Horizon The A-horizon is the part of the soil that you are most likely to see when you dig a shallow hole in the soil with your fingers. Organic matter from the decay of roots and the action of soil organisms often makes this horizon excellent for plant growth. Because the A-horizon contains most of the organic matter in the soil, it is usually darker than the other horizons.

Reading Check

6. Examine Which soil layer contains the most organic matter?

B-Horizon When water from rain or snow seeps through pores in the A-horizon, it carries clay particles. The clay is then deposited below the upper layer, forming a B-horizon. Other materials also accumulate in B-horizons.

C-Horizon The layer of weathered parent material below the B-horizon is called the C-horizon. Parent material can be rock or sediments.

Soil Interactions Soil contains minerals, water, air, and organisms, all in close contact. Therefore, interactions among all Earth systems take place in soil. Recall that plants need phosphorus and carbon to grow. Plants cannot get phosphorus from the air, but they can obtain it from soil or from water in the soil. A major part of the organic matter in soil is carbon that plants obtain from the atmosphere through photosynthesis. Soil plays a major role in the phosphorus and carbon cycles.

Structure of the Geosphere

What do you think you would see if you could look inside the solid Earth? You would see layers similar to those in the figure below. The geosphere has three main layers: the crust, the mantle, and the core. Each layer has a different density. Recall that density is a measure of the mass of a material divided by its volume. The densest layer of the geosphere is the center, or core. The least-dense layer is the outer crust. The density of the thick mantle varies.

Key Concept Check

7. Differentiate How do materials in the geosphere differ?

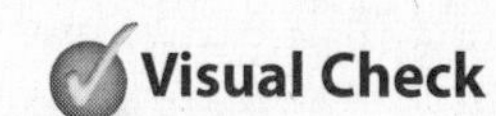

Visual Check

8. Identify Highlight the layers of the geosphere that are solid.

Layered Structure of the Geosphere

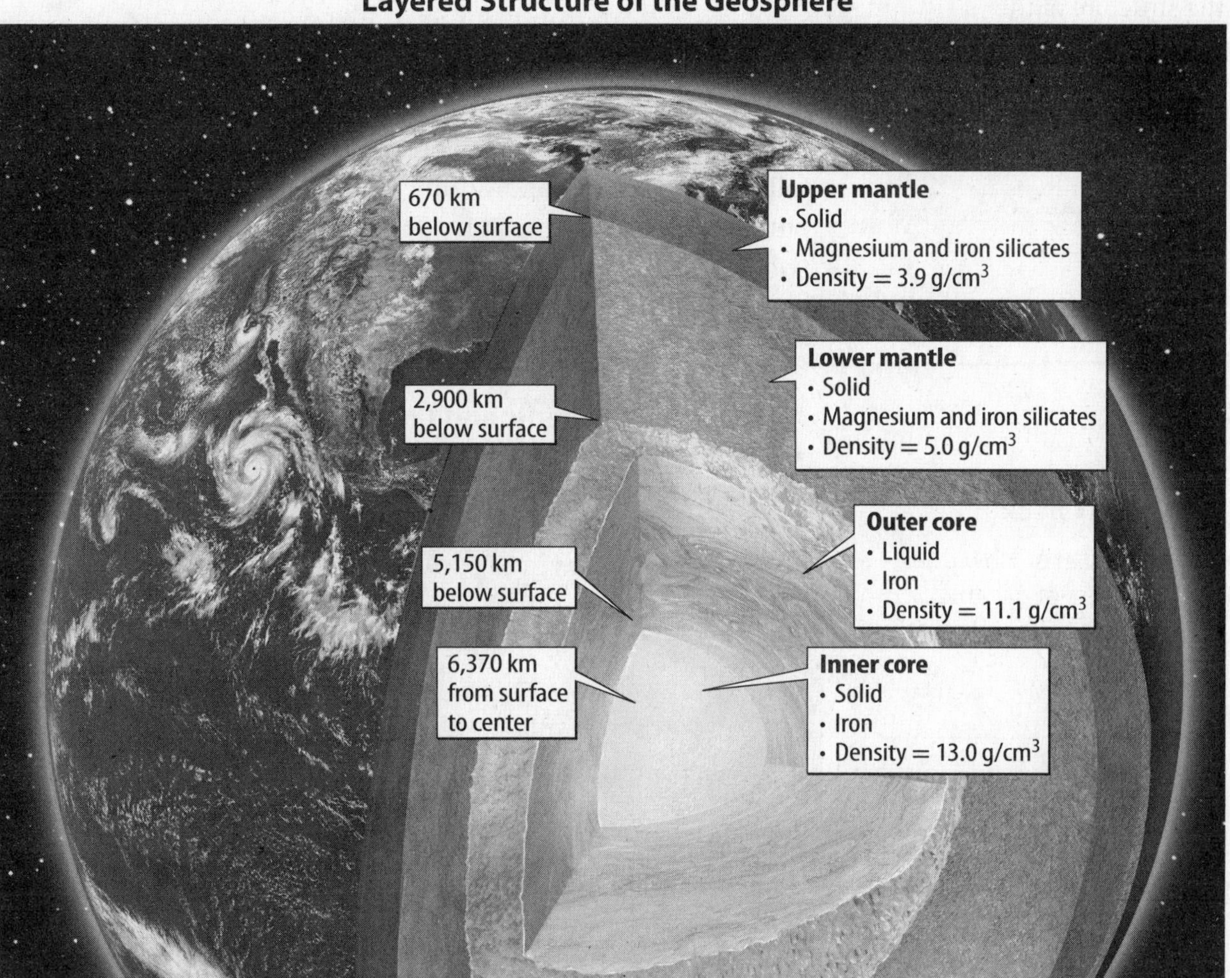

Key Concept Check
9. Explain Why is the geosphere layered?

____

____

____

____

____

FOLDABLES®

Make a three-tab and six-tab book to identify and describe the structure of the geosphere.

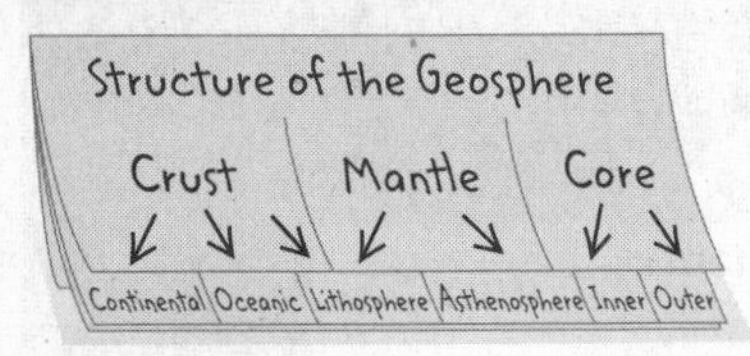

Visual Check
10. Compare How does the thickness of oceanic crust and continental crust compare?

____

____

____

Formation of Earth's Layers Scientists hypothesize that Earth's layers formed early in the planet's history. Ancient Earth was much hotter than it is today. Thermal energy melted some of the rock. Then, gravity pulled denser materials through the melted rock toward Earth's center, forming layers.

Makeup of Earth's Layers In addition to different densities, the layers of the geosphere have different compositions. Most of the geosphere is made of rock, but some of it is made of metal.

How do scientists know about the density and makeup of Earth's deep inner layers? Humans have never seen them. Scientists gather data by analyzing earthquake waves. As the waves travel through Earth, they change speed and direction as they pass through materials with different densities. Scientists use data about the waves to map Earth's interior.

Earth's Crust

The rocky cliffs you see exposed along the sides of a highway are part of Earth's crust. *The* ***crust*** *is the thin outer layer of the geosphere.* It is made of brittle rocks. These rocks are made of elements that combine and form minerals. Approximately 90 elements occur naturally in Earth's crust. Just eight of these elements make up about 98 percent of the crust. The most common element in Earth's crust is oxygen, followed by silicon, aluminum, iron, calcium, sodium, potassium, and magnesium. The following figure shows the two types of crust. One type is under the oceans. The other type makes up the continents. Oceanic crust is denser than continental crust.

Oceanic and Continental Crust

Oceanic Crust The crust under the oceans is about 7 km thick. Oceanic crust is made of the dense igneous rocks basalt and gabbro. These rocks are rich in the dense minerals iron and magnesium. This makes oceanic crust denser than continental crust.

Continental Crust The crust that makes up continents is thicker than oceanic crust. Continental crust has an average thickness of about 40 km. Under large mountains, continental crust is as much as 70 km thick.

Continental crust is not made of the same kinds of rocks as oceanic crust. Continental crust is made of igneous, metamorphic, and sedimentary rocks. Rocks in the continental crust are rich in silicon and oxygen. These elements are less dense than iron and magnesium. This makes continental crust less dense than oceanic crust. ✓

Earth's Mantle

Beneath Earth's crust is the mantle. *The* **mantle** *is the thick, rocky middle layer of the geosphere.* The mantle has the largest volume of any layer of Earth. Much of the mantle is made of the rock peridotite (puh RIH duh tite). Peridotite contains even more iron and magnesium than basalt and, therefore, is denser. The layers of Earth's mantle are shown in the figure below. ✓

Earth's Mantle

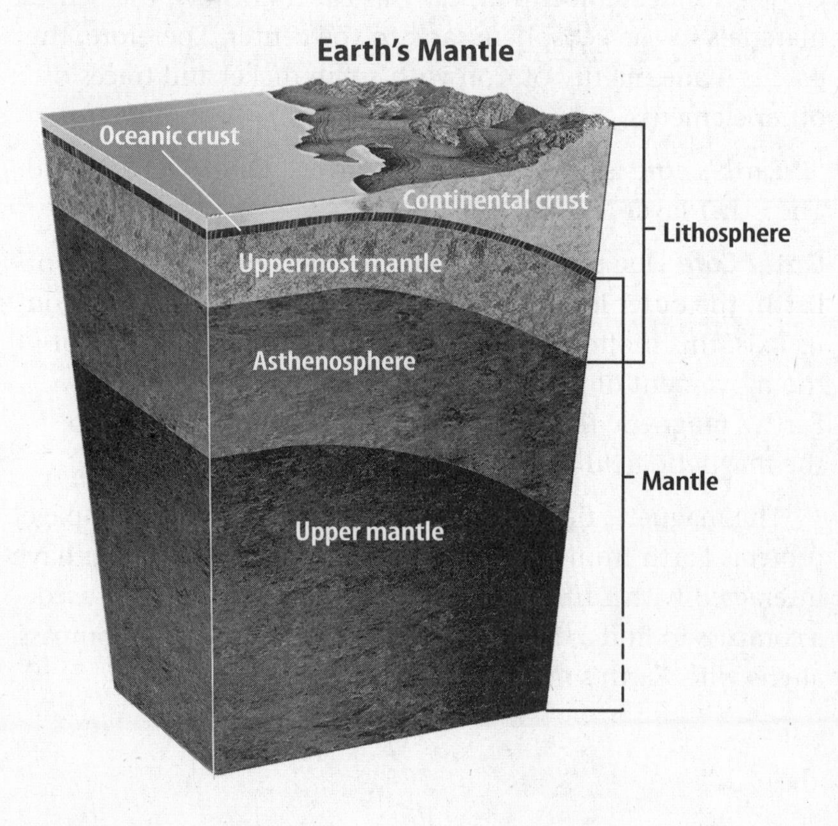

✓ **Reading Check**

11. Contrast How do oceanic crust and continental crust differ?

✓ **Reading Check**

12. Define What is the mantle?

✓ **Visual Check**

13. Name the layers of the lithosphere.

Reading Check

14. Explain Why are the crust and the uppermost mantle sometimes regarded as a single layer?

______

______

______

______

Reading Check

15. Contrast How is the lithosphere different from the asthenosphere?

______

______

______

Reading Check

16. Identify Which layer of Earth's core is liquid?

______

The Lithosphere Rocks in the uppermost mantle are more brittle than in the rest of the mantle and are similar to rocks in the crust. So the crust and the uppermost mantle are sometimes described as one layer, even though they have different compositions. *The crust and the uppermost mantle form a brittle outer layer called the* **lithosphere.**

Most of the rock below the lithosphere is solid. But high temperatures in the mantle make rock soft enough to flow. This is similar to the way a warm wax candle can bend instead of breaking. The mantle flows slowly, moving about as fast as your fingernails grow.

The Asthenosphere At a depth of about 100 km is an especially soft layer of the mantle. *This weak, partially melted layer of the mantle is called the* **asthenosphere** (as THEN uh sfihr). Less than 2 percent of the rock in the asthenosphere is melted. This small amount of molten rock makes the asthenosphere weaker than the rest of the mantle.

Earth's Core

The dense, metallic center of Earth is called the **core.** Note that the core is metallic and not rocky like the other layers of the geosphere.

Why is the core different than the other layers of the geosphere? Remember that, early in Earth's history, the densest materials in the geosphere sank to the center. Therefore, the core is made mainly of iron with some nickel and traces of other elements.

Earth's core is divided into two layers. One layer is liquid. The other layer is solid.

Outer Core Due to the high temperatures near the center of Earth, the outer layer of the core is liquid. As Earth spins on its axis, this molten iron flows. Scientists hypothesize that the movement of liquid iron in the outer core produces Earth's magnetic field. Earth's magnetic field is similar to the magnetic field of a huge bar magnet.

The magnetic field, shown in the figure on the next page, protects Earth from charged particles from the Sun. You have interacted with Earth's magnetic field if you have ever used a compass to find a direction. The metal needle in the compass aligns with Earth's magnetic field.

Inner Core Inside the outer core is a sphere of solid metal. Temperatures in this inner core are extremely hot, as high as 4,300°C. Despite the scorching heat, the metal in the inner core is not melted. The high pressure from the masses of all Earth's layers compresses the inner core, making it solid.

Earth's Core and Magnetic Field

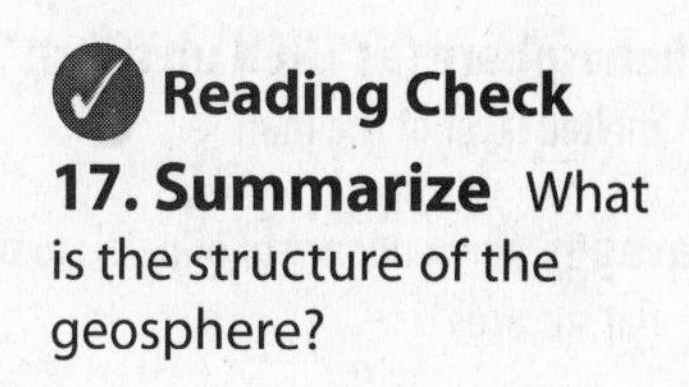

Reading Check

17. Summarize What is the structure of the geosphere?

____

____

____

____

Visual Check

18. Name Which layer, the inner or outer core, produces Earth's magnetic field?

____

After You Read

Mini Glossary

asthenosphere (as THEN uh sfihr): the weak, partially melted layer of the mantle

cleavage: the tendency of minerals to break along smooth, flat surfaces

core: the dense, metallic center of Earth

crust: the thin outer layer of the geosphere

fracture: the tendency of minerals to break along irregular surfaces

lithosphere: a brittle outer layer formed by the crust and the uppermost mantle

luster: the way a mineral's surface reflects light

mantle: the thick, rocky middle layer of the geosphere

streak: the color of a mineral's powder

1. Review the terms and their definitions in the Mini Glossary. Write a sentence explaining the difference between luster and streak.

2. Sequence the three major layers of the geosphere and state one fact about each.

upper → lower
Layer: **Fact:**
Layer: **Fact:**
Layer: **Fact:**

3. Which soil layer do you think is the best for growing plants? Why?

What do you think NOW?

Reread the statements at the beginning of the lesson. Fill in the After column with an A if you agree with the statement or a D if you disagree. Did you change your mind?

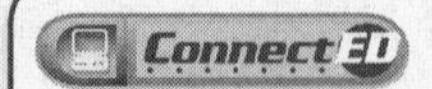

Log on to ConnectED.mcgraw-hill.com and access your textbook to find this lesson's resources.

Earth's Changing Surface

Plate Tectonics

Before You Read

What do you think? Read the two statements below and decide whether you agree or disagree with them. Place an A in the Before column if you agree with the statement or a D if you disagree. After you've read this lesson, reread the statements to see if you have changed your mind.

Before	Statement	After
	1. Continents do not move.	
	2. Earth's mantle is liquid.	

Key Concepts 
- What is the theory of plate tectonics?
- What evidence do scientists use to support the theory of plate tectonics?
- How do the forces created by plate motion change Earth's surface?

Read to Learn

Plate Motion

Even though you usually cannot feel it, Earth's surface is always moving. This motion can cause earthquakes and volcanic eruptions. It also can form mountains. The theory of **plate tectonics** *states that Earth's crust is broken into rigid plates that move slowly over Earth's surface*. The rigid plates are called tectonic plates. You live on the North American Plate, shown in the figure below. Tectonic plates slowly move over Earth's surface. The movement of one plate is described as either moving away from or toward another plate, or sliding past another plate. Plates move at speeds of only a few centimeters per year. At this rate, it takes moving plates millions of years to make new continents, new mountain ranges, or other landforms.

Study Coach

Make an Outline Summarize the information in the lesson by making an outline. Use the main headings in the lesson as the main headings in your outline. Use your outline to review the lesson.

Key Concept Check

1. State What is the theory of plate tectonics?

Evidence of Plate Motion

The theory of plate tectonics has helped geologists explain many observations about Earth and predict geologic events. Evidence gathered by scientists studying Earth for nearly 100 years supports the theory of plate tectonics. The theory replaced a hypothesis called continental drift.

Tectonic Plates

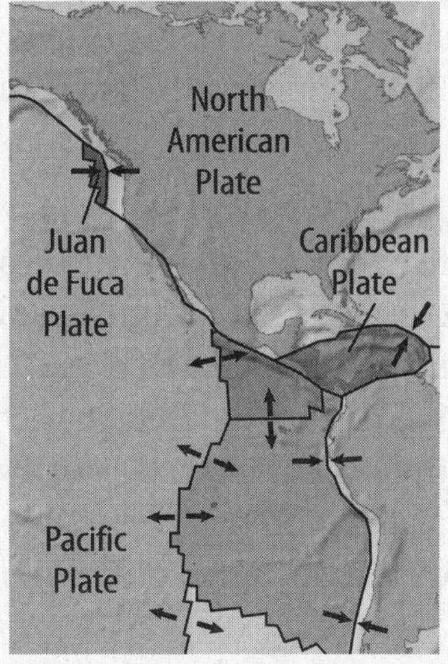

Visual Check

2. Identify What plates does the North American Plate interact with to the west?

FOLDABLES®

Make a horizontal six-door book to describe the changes in Earth's surface as the result of plate tectonics and forces.

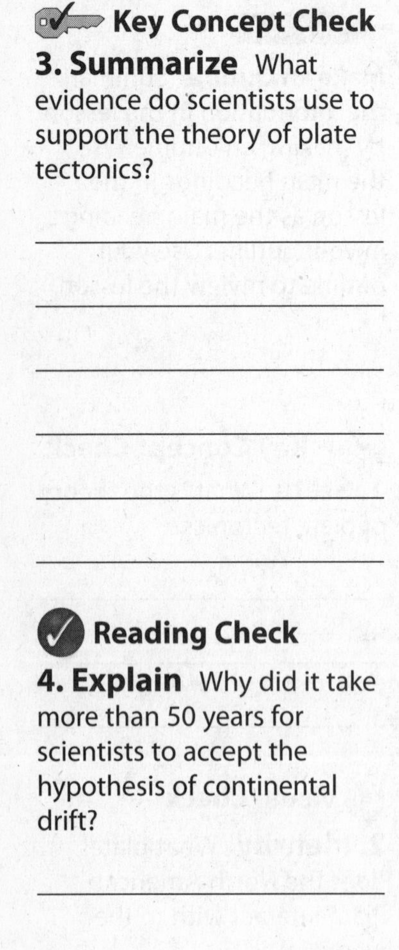

Key Concept Check

3. Summarize What evidence do scientists use to support the theory of plate tectonics?

Reading Check

4. Explain Why did it take more than 50 years for scientists to accept the hypothesis of continental drift?

Continental Drift Long before geologists proposed the theory of plate tectonics, they discovered evidence of continental movement. One piece of evidence is the shape of Earth's continents. Look at the outlines of South America and Africa on a map. If you could push these two continents together, they would fit together like two pieces of a puzzle. In 1912, Alfred Wegener developed *the hypothesis that continents move, called* **continental drift.**

Fossil Evidence Different plants and animals live on different continents. For example, lions live in Africa but not in South America. Many fossils of animals and plants show the same thing—some ancient organisms lived in certain areas but not in others. However, geologists have discovered the same types of fossils on continents that are now separated by vast oceans.

For example, fossils of the freshwater reptile *Mesosaurus* have been found in South America and in Africa. These two landmasses are separated by the Atlantic Ocean. So how did a freshwater reptile cross a saltwater ocean? Scientists hypothesize that when the two continents were together, *Mesosaurus* probably traveled in freshwater rivers from one area to the other.

Geological Evidence Rocks that are made of similar substances and mountains that formed at similar times are present on continents that are now far apart. Scientists can trace these rocks and mountains, as well as the locations of ancient glaciers, deserts, and coal swamps, from one continent to the next.

How Plates Move

The hypothesis of continental drift was not accepted for more than 50 years after it was proposed. The main reason was that the hypothesis did not explain how continents could move. Geologists knew that the mantle, the part of Earth underneath continents, was solid. How could a continent push its way through solid rock?

An Explanation New discoveries during the 1960s led scientists to propose the theory of plate tectonics. Recall that Earth's crust is broken into separate tectonic plates. These plates include the crust beneath the ocean and the continents. Scientists proposed that continents did not just float around the ocean. Instead, Earth's continents are part of tectonic plates. The plates move toward, away from, or past each other, carrying continents with them.

The Role of Convection The forces that move plates come from deep within Earth. Earth's mantle is so hot that rocks can deform and move without breaking, much like putty. Convection affects the mantle underneath tectonic plates. Hotter mantle rises toward Earth's surface and cooler mantle sinks deeper into the mantle, as shown in the figure at right. As the mantle moves, it pushes and pulls tectonic plates over Earth's surface. 

Earth's Mantle and Tectonic Plates

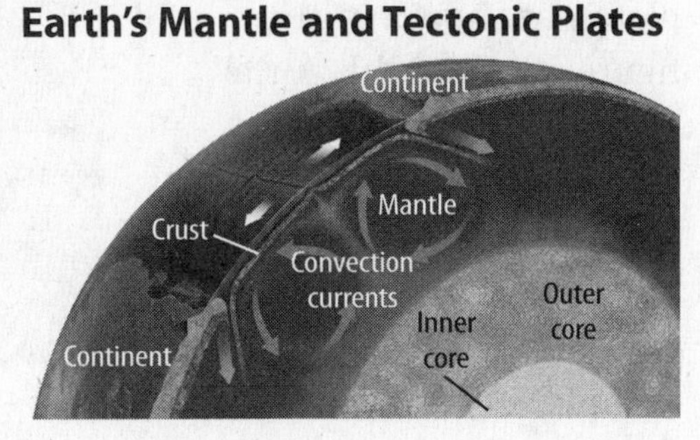

Visual Check

5. Identify Trace the movement of the mantle in the figure.

Reading Check

6. Describe How does Earth's mantle move tectonic plates?

____

____

____

____

Tectonic Plate Boundaries

The edges of tectonic plates are called plate boundaries. *A* **convergent boundary** *is where two plates move toward each other. A* **divergent boundary** *is where two plates move apart from each other. A* **transform boundary** *is where plates slide horizontally past each other.*

Convergent Boundaries

Recall that oceanic crust is denser than continental crust. This difference is important in areas where plates meet. When two plates come together, the denser oceanic plate usually is forced down into the mantle. The less dense continental plate remains on Earth's surface, as shown on the left in the figure below. *The area where one plate slides under another is called a* **subduction zone.** However, when two continents collide at a plate boundary, both continents remain on the surface. As two continents push together, the crust rises up and mountains form, as shown on the right in the figure below.

Plate Interactions

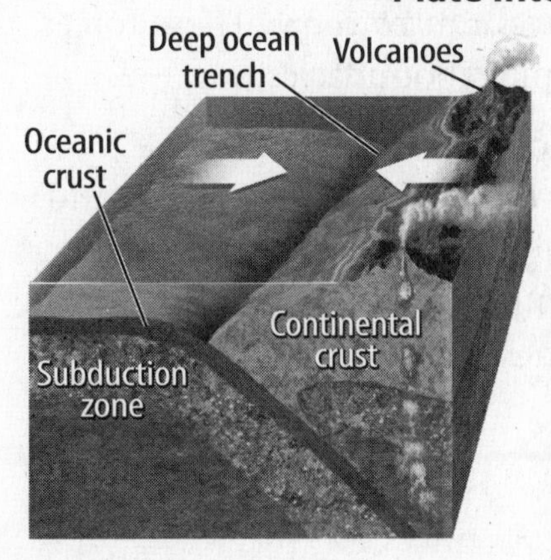

Convergent

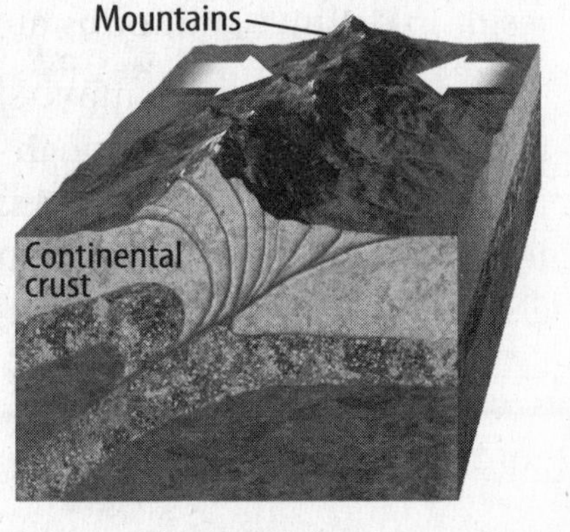

Convergent

Visual Check

7. Point Out Highlight the subduction zone in the left part of the figure.

Visual Check
8. Describe the movement of the plates at a divergent boundary.

Visual Check
9. Describe the movement of the plates at a transform boundary.

Key Concept Check
10. Summarize How do the forces created by plate motion change Earth's surface?

Divergent Boundaries

When plates move apart at divergent boundaries, a rift forms between the two plates, as shown in the figure on the right. A rift valley can form within continents when continental crust moves in opposite directions. A rift valley also can form at divergent boundaries on the ocean floor. As plates separate, molten rock can erupt from the rift. As the molten rock cools, it forms new crust.

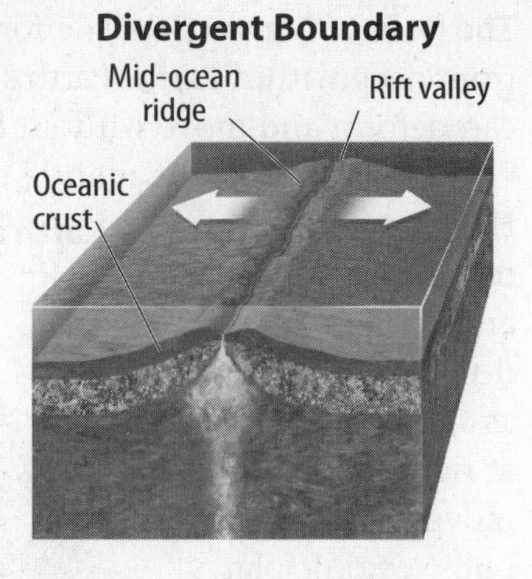

Transform Boundaries

Tectonic plates slide past each other at transform boundaries, as shown in the figure on the right. Each side of the boundary moves in an opposite direction. This movement can deform or break features such as fences, railways, or roads that cross the boundary.

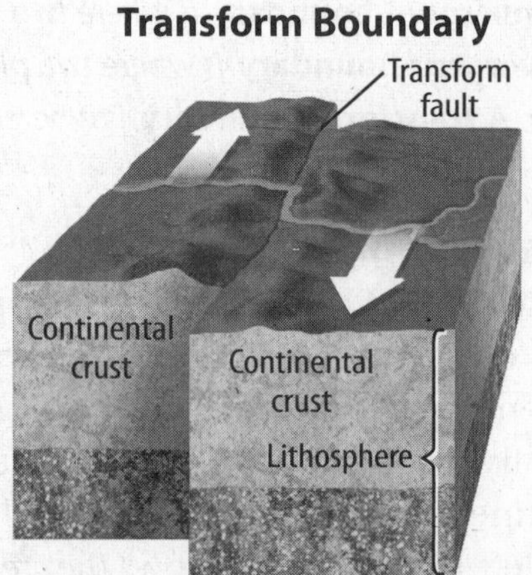

Forces Changing Earth's Surface

Forces within Earth cause plates to move. Different types of forces occur at the three types of plate boundaries. *The squeezing force at a convergent boundary is called* **compression.** *The pulling force at a divergent boundary is called* **tension.** *The side-by-side dragging force at transform boundaries is called* **shear.** These forces result in distinct landforms at plate boundaries.

Even though plates move slowly, the forces at plate boundaries are strong enough to form huge mountains and powerful earthquakes. Tensional forces pull the land apart and form rift valleys and mid-ocean ridges. Compressional forces also form mountains such as the Himalayas in India.

After You Read

Mini Glossary

compression: the squeezing force at a convergent boundary

continental drift: the hypothesis that continents move

convergent boundary: where two plates move toward each other

divergent boundary: where two plates move apart from each other

plate tectonics: theory stating that Earth's crust is broken into rigid plates that move slowly over Earth's surface

shear: the side-by-side dragging force at transform boundaries

subduction zone: the area where one plate slides under another

tension: the pulling force at a divergent boundary

transform boundary: where plates slide horizontally past each other

1. Review the terms and their definitions in the Mini Glossary. Write a sentence that shows you understand the theory of plate tectonics.

2. Write *shear, tension,* or *compression* in each box below to identify the type of force associated with each type of plate boundary.

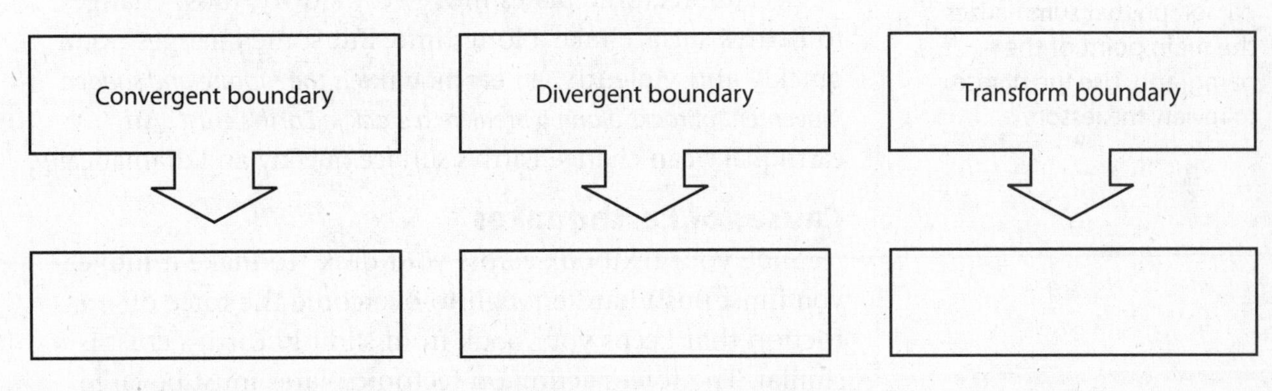

3. Select a term that appears in a main heading of the outline you created when you reviewed the lesson. In the space below, define that word.

What do you think NOW?

Reread the statements at the beginning of the lesson. Fill in the After column with an A if you agree with the statement or a D if you disagree. Did you change your mind?

Log on to ConnectED.mcgraw-hill.com and access your textbook to find this lesson's resources.

CHAPTER 14
LESSON 2

Earth's Changing Surface

Earthquakes and Volcanoes

Key Concepts

- What causes earthquakes?
- What causes volcanoes to form?
- How do earthquakes and volcanoes change Earth's surface?

Before You Read

What do you think? Read the two statements below and decide whether you agree or disagree with them. Place an A in the Before column if you agree with the statement or a D if you disagree. After you've read this lesson, reread the statements to see if you have changed your mind.

Before	Statement	After
	3. Earthquakes occur and volcanoes erupt only near plate boundaries.	
	4. Volcanoes erupt melted rock.	

Mark the Text

Identify the Main Ideas Write a phrase beside each paragraph that summarizes the main point of the paragraph. Use the phrases to review the lesson.

Read to Learn

Earthquakes

Because tectonic plates move very slowly, most changes to Earth's surface take a long time. But some changes occur quickly and violently. *An* **earthquake** *is the rupture and sudden movement of rocks along a break or a crack in Earth's crust.* An earthquake can change Earth's surface quickly and dramatically.

Causes of Earthquakes

Slide your textbook across your desk. To make it move, you must push hard enough to overcome the force of friction that keeps your book from sliding. Earth's crust is similar. The forces acting on tectonic plates must be large enough to make blocks of crust move. When these blocks move, earthquakes occur.

The surface along which the crust moves is called a **fault.** Movement along faults occurs when the forces pushing on the rock layers become large enough to cause movement along the fault.

Recall that compression and tension cause vertical motion on a fault, and shear forces cause horizontal motion. When pieces of crust slide past each other, energy is released, causing the ground to shake.

Key Concept Check
1. Explain What causes earthquakes?

Earthquake Locations

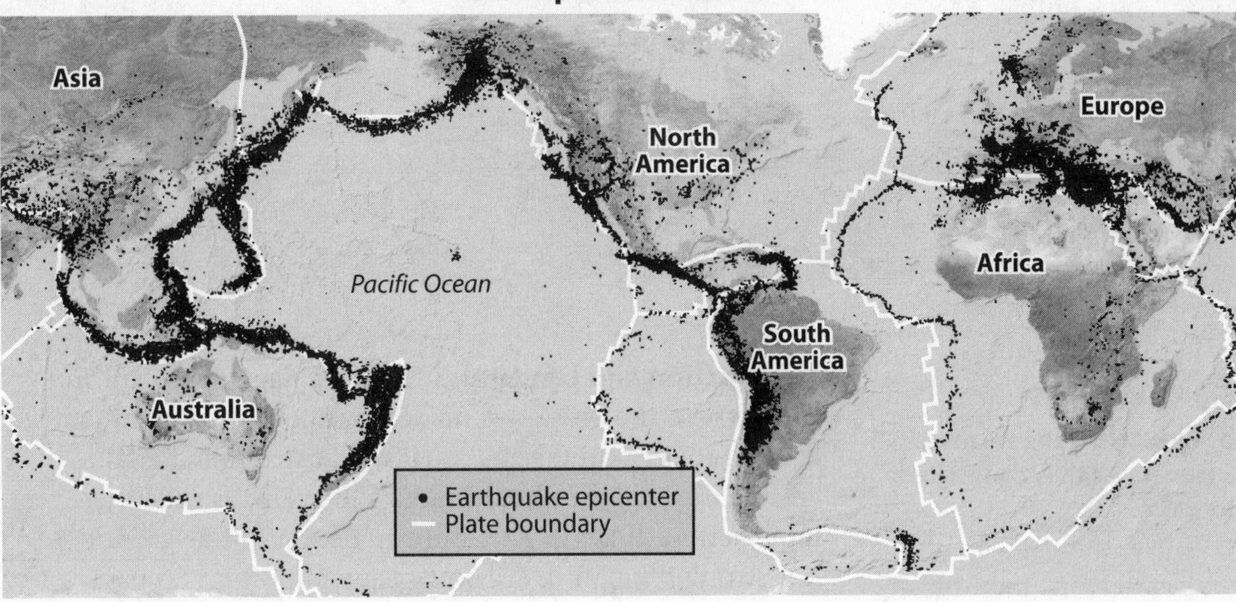

Where do earthquakes occur?

Most earthquakes occur near plate boundaries, as shown in the figure above. However, some of the largest earthquakes in the United States occurred far from plate boundaries. For example, in the winter of 1811–1812, three large earthquakes occurred in Missouri. An explanation of these earthquakes might be their nearness to an old fault system, the New Madrid fault system. This fault system is part of the beginning of a rift valley on the North American Plate.

Visual Check

2. Locate Most earthquakes occur _____. (Circle the correct answer.)

a. far from plate boundaries

b. near plate boundaries

c. only on plate boundaries

Changing Earth's Surface

You might be familiar with the damage earthquakes can cause. But earthquakes also can create landforms. Faults associated with earthquakes can be visible at Earth's surface. Some faults, such as the San Andreas Fault in California, are more than 1,000 km long. During the massive Sichuan (SI chwan), China, earthquake in 2008, blocks of crust moved as much as 9 m along a fault 240 km long and 20 km deep. Earthquakes also form mountains and change Earth's surface in other ways.

Mountains Every time blocks of crust slide past each other along a fault, an earthquake occurs. Blocks might move only 1–2 m. But after hundreds or thousands of earthquakes, blocks of crust will have moved a long distance. Compression and tension forces produce ridges and mountains as crust moves vertically.

FOLDABLES®

Make a three-tab Venn book to compare the causes and effects of volcanoes and earthquakes.

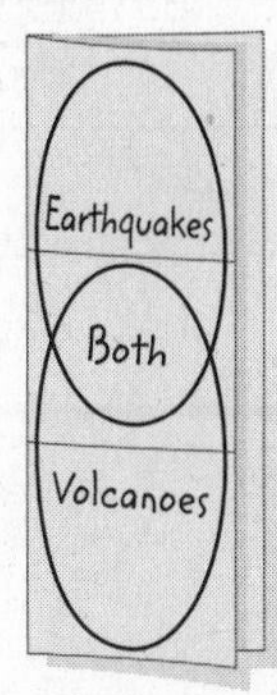

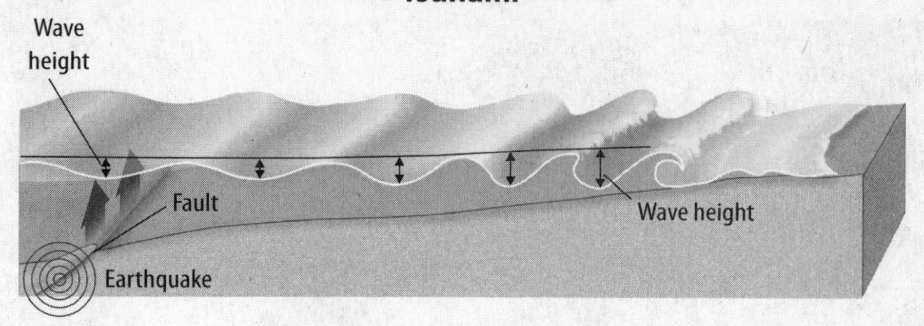

Visual Check
3. Describe the waves of a tsunami as they come closer to shore.

Reading Check
4. Define What is liquefaction?

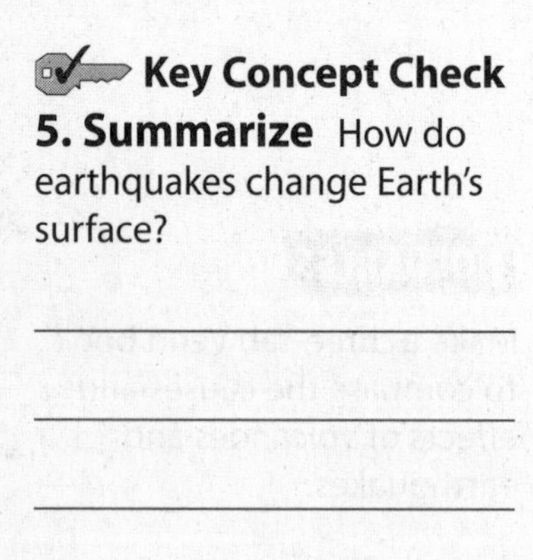

Key Concept Check
5. Summarize How do earthquakes change Earth's surface?

Liquefaction and Landslides Great damage can occur in areas where the ground is made up of loose sediment instead of solid rock. Extreme shaking can cause this material to behave more like a liquid than a solid. This is called liquefaction (li kwuh FAK shun). The liquid-like ground is not strong enough to support heavy buildings. So part of a building can sink into the ground, causing the building to collapse.

Liquefaction is responsible for most of the damage to buildings after an earthquake occurs. Shaking caused by earthquakes also can trigger landslides. Landslides bring rocks and soil from the tops of mountains into valleys.

Tsunamis Earthquakes that happen underwater can cause tsunamis (soo NAH meez), as shown in the figure above. Upward movement at a fault pushes the water up and creates huge ocean waves. These waves become taller as they reach shallower water near a shore. Tsunamis also can be caused by part of the ocean floor dropping down or by an underwater volcanic eruption.

Volcanoes

Recall that molten rock below Earth's surface is called magma. Because magma is hot, it is less dense than the surrounding rock and rises, or moves upward. Volcanoes are landforms that form when magma erupts onto Earth's surface as lava.

Volcanoes are common on Earth. Each year about 50–60 different volcanoes erupt somewhere on Earth. There are approximately 1,500 active volcanoes on Earth. Volcanoes can be destructive, but they also make new landforms.

Where Volcanoes Form

Volcanoes can occur at divergent plate boundaries, at convergent plate boundaries, and at hot spots. At a divergent boundary, lava flows into the rift formed by the separating plates, as shown in the figure below. New crust is made of the rocks that form as this lava cools. *The mountains that form as this lava builds up and cools are called* **mid-ocean ridges.**

Divergent Plate Boundaries

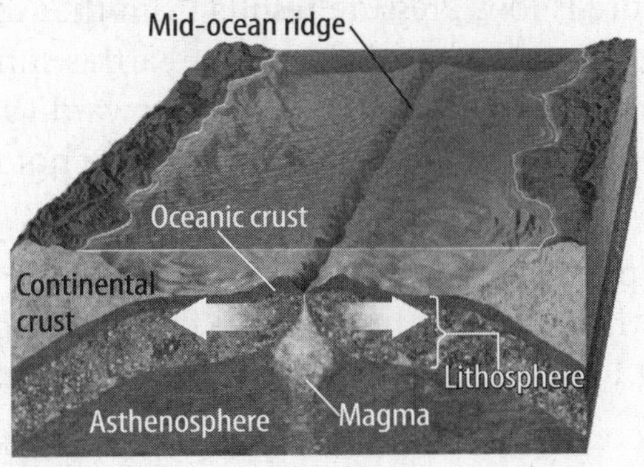

Visual Check

6. Draw a circle around the landforms that are created as the lava cools.

At some convergent boundaries, one tectonic plate sinks into the mantle. The sinking plate also carries water into the mantle. This causes the mantle to melt and form magma. The magma rises and erupts onto the plate that does not sink, as shown in the figure below. ✓

Reading Check

7. Summarize How do volcanoes form at convergent boundaries?

____

____

____

____

Convergent Plate Boundaries

Visual Check

8. Identify Highlight the path of the rising magma.

Hot Spots Not all volcanoes are near plate boundaries. In a few places, large volcanoes form near the center of a tectonic plate. These volcanoes form at **hot spots,** *locations where volcanoes form far from plate boundaries*. The Hawaiian Islands in the middle of the Pacific Ocean and Yellowstone National Park in Wyoming are hot spots.

Reading Check
9. State a hypothesis to explain why hot spots exist.

Reading Check
10. Summarize Why does rock melt more easily near Earth's surface?

Key Concept Check
11. Explain What causes volcanoes to form?

Why Hot Spots Exist Scientists do not fully understand the reason hot spots exist. One hypothesis is that hot spots occur above places where the mantle melts. The magma then rises toward the surface and eventually erupts through the crust.

Causes of Volcanic Eruptions

In order for magma to form, the crust and the mantle must become hot enough to melt. Rocks melt more easily when pressure is low. Pressure results from the weight of overlying rock, so pressure is lowest at Earth's surface. When hot rocks from deep inside Earth move toward Earth's surface, the decrease in pressure allows these hot rocks to melt.

The temperature at which rocks melt depends on the makeup of the rock and the presence or absence of water. If the melting temperature of rock is lowered, the rocks melt more easily.

Water enters the mantle at convergent boundaries. This allows the mantle to melt at a lower temperature. This is similar to adding salt to ice. If you put salt on ice, the ice melts at a lower, or colder, temperature.

Because magma is hot, it is less dense than the rock material around it. It moves upward and causes cracks to form in the solid rock.

Magma also contains dissolved gases. The rising magma plus the gases cause pressure to build up. Eventually, the magma erupts through cracks in Earth's surface and a volcano forms. Most of Earth's largest volcanoes are located at convergent plate boundaries.

Changing Earth's Surface

Volcanoes can be as small as a car. They also can be more than 10 km high. The shapes of volcanoes and the way lava erupts depend on where volcanoes form. What comes out of volcanoes, and how do volcanoes change Earth's surface?

Lava Flows Melted mantle material flows easily. When it erupts, it flows over Earth's surface. This creates *long streams of molten rock called* **lava flows.**

Lava eventually cools and solidifies, forming solid rock. Lava flows can be more than 10 km long. Over time and after repeated eruptions, lava flows build up as flat layers.

Explosive Eruptions At convergent plate boundaries, part of the continental crust can become mixed with magma from the mantle. When this mixture of molten materials erupts, it does not flow as easily as lava made only of melted mantle. Instead of forming lava flows, it often solidifies in the atmosphere, where it breaks into *small pieces of lava called* **volcanic ash.**

Ash can reach heights greater than 20 km. The ash eventually falls back to Earth's surface. Thick layers of these small pieces of lava can cover large areas that extend more than 100 km from the volcano. Eruptions that eject ash high into the atmosphere are called explosive eruptions. Lava is also produced during these eruptions.

Types of Volcanoes

Lava flows can build up and form large volcanoes. Shield volcanoes form after lava flows have occurred over time. Shield volcanoes tend to be large with gentle slopes, such as Mauna Loa in Hawaii.

Composite volcanoes also can form as lava flows and ash layers deposited by explosive eruptions build up. These types of volcanoes often have steep sides and are cone-shaped, such as Mount Adams in Washington. They are most common at convergent boundaries.

Before a volcano erupts, magma builds up in the crust in a reservoir called a magma chamber. What happens when large amounts of magma are removed from this chamber? Sometimes the surface above the chamber collapses. This creates *a large depression in the center of the volcano called a* **caldera** (kal DER uh). Some calderas can be more than 70 km wide.

Effects on the Atmosphere

Volcanoes also change Earth's atmosphere and climate. Volcanic ash and gases from explosive eruptions can blow high into the atmosphere. Some volcanic material remains in the atmosphere for years. This material can block sunlight. This can cause the temperature of the atmosphere near Earth's surface to decrease.

Math Skills

Geologists estimate the volume of a lava flow from a volcano by measuring the average depth and the radius of the hardened lava field. The volume of a cylinder is the area of the base times the height (h). The base of a cylinder is a circle with an area equal to the square of the radius (r^2) times π (3.14). Therefore, $V = \pi \times r^2 \times h$. For example, what is the volume of lava needed to produce a lava field with a radius of 100.0 m and an average depth of 20.0 m?

a. The formula for volume is

$$V = \pi \times r^2 \times h$$

b. Replace the values in the formula with the given values and calculate.

$V = 3.14 \times (100\text{ m})^2 \times 20.0\text{ m}$

$V = 3.14 \times 10{,}000\text{ m}^2 \times 20.0\text{ m}$

$V = 628{,}000\text{ m}^3$

12. Use Geometry What is the volume of lava in a field with a radius of 90 m and an average thickness of 10.0 m?

____

Key Concept Check

13. Describe How do volcanoes change Earth's surface?

____

____

____

____

____

After You Read

Mini Glossary

caldera (kal DER uh): a large depression in the center of a volcano

earthquake: the rupture and sudden movement of rocks along a break or a crack in Earth's crust

fault: the surface along which Earth's crust moves

hot spot: a location where a volcano forms far from plate boundaries

lava flow: a long stream of molten rock that erupts from a volcano

mid-ocean ridge: an undersea mountain that forms as lava builds up and cools on oceanic crust

volcanic ash: small pieces of lava

1. Review the terms and their definitions in the Mini Glossary. Write a sentence explaining the relationship between a fault and an earthquake.

2. Use the graphic organizer below to show three changes volcanoes make to Earth's surface, atmosphere, and/or climate.

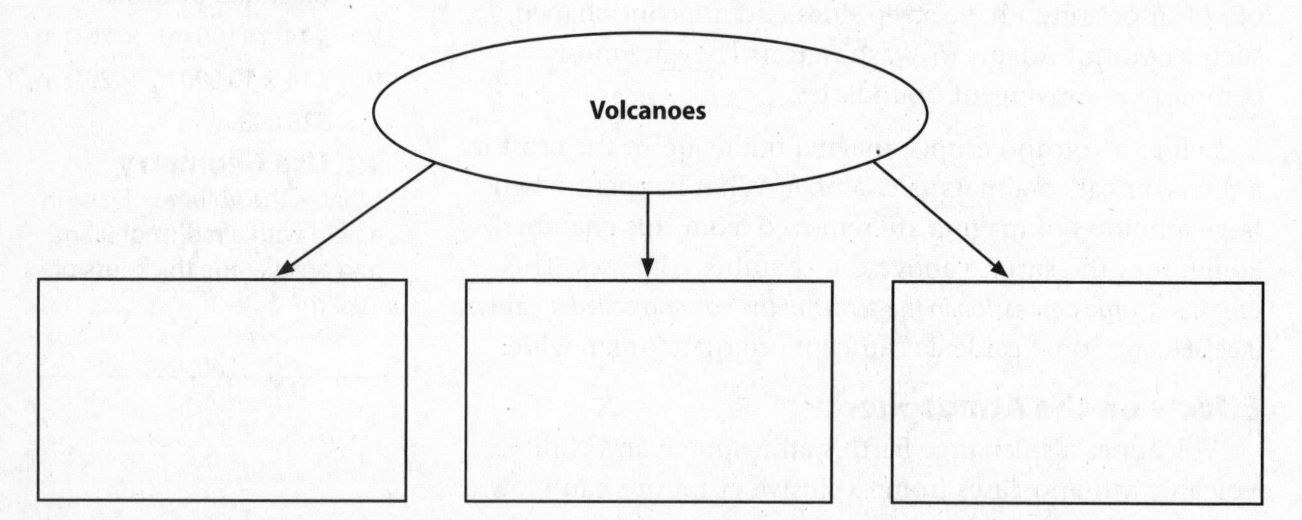

3. In the early 1800s, three large earthquakes occurred in Missouri. Do you think any of them caused a tsunami? Explain.

What do you think NOW?

Reread the statements at the beginning of the lesson. Fill in the After column with an A if you agree with the statement or a D if you disagree. Did you change your mind?

Log on to ConnectED.mcgraw-hill.com and access your textbook to find this lesson's resources.

Earth's Changing Surface

Weathering, Erosion, and Deposition

Before You Read

What do you think? Read the two statements below and decide whether you agree or disagree with them. Place an A in the Before column if you agree with the statement or a D if you disagree. After you've read this lesson, reread the statements to see if you have changed your mind.

Before	Statement	After
	5. Rocks cannot change.	
	6. Sediment can be transported by water, wind, and ice.	

Key Concepts 

- How are weathering and soil formation related?
- How do weathering, erosion, and deposition change Earth's surface?
- How are erosion and deposition related?

Read to Learn

Weathering

You have read that mountains can form as a result of plate motion and volcanoes. But why don't mountains last forever? **Weathering** *refers to the processes that break down rocks, changing Earth's surface over time.* **Erosion** *is the moving of weathered material, or sediment, from one location to another.* Slowly but surely, weathering and erosion wear down mountains. ✓

Mark the Text

Sticky Notes As you read, use sticky notes to mark information that you do not understand. Read the text carefully a second time. If you still need help, write a list of questions to ask your teacher.

Physical Weathering

The process of breaking rock into small pieces without changing the composition of the rock is **physical weathering.** Physical weathering can be caused by temperature and by plants.

Temperature and Physical Weathering Temperature is one factor that affects physical weathering. Most rocks contain water in cracks and spaces between the particles that make up the rock. During winter or at night, the water in rocks can freeze. When water freezes, it expands. If water in rocks freezes and melts repeatedly, it can break apart rocks. This is called frost wedging.

Plants and Physical Weathering Plants also can cause physical weathering. For example, the roots of plants can grow into cracks in rock. As the roots grow and take up more space in the cracks, the force they apply to the rock breaks the rock.

✓ **Reading Check**

1. Differentiate What is the difference between weathering and erosion?

____
____
____
____
____

FOLDABLES®

Make a horizontal two-tab book and use it to compare the different types of weathering.

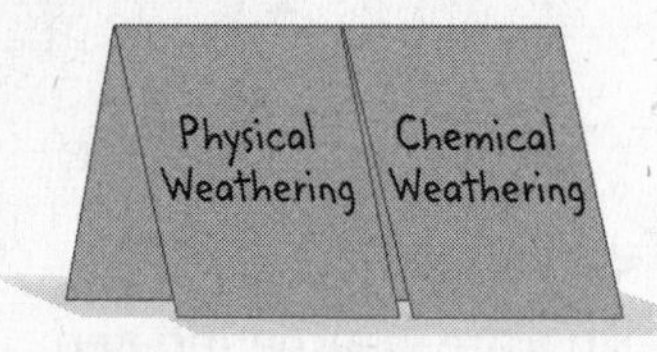

Key Concept Check

2. Relate How does weathering change Earth's surface?

Reading Check

3. Explain How do chemical and physical weathering work together?

Chemical Weathering

The process of changing the composition of rocks and minerals by exposure to water and the atmosphere is called **chemical weathering.** Some minerals chemically weather more easily than others. For example, calcite, the mineral that makes up limestone, dissolves readily in acidic rainwater. Feldspar, a common mineral in igneous rocks, easily weathers into the clay minerals, kaolinite. However, other minerals, such as quartz, are resistant to chemical weathering.

Gases and Chemical Weathering Gases in the atmosphere also can cause chemical weathering. Minerals containing iron react with oxygen in the atmosphere and form rust-colored minerals. Carbon dioxide in the atmosphere dissolves in water and makes acidic water. Limestone dissolves much faster in acidic water than in nonacidic water.

Temperature and Chemical Weathering Temperature also affects the rate of chemical weathering. You might know that chemical reactions happen faster at higher temperatures than at lower temperatures. That is why chemical weathering occurs fastest in hot, wet climates.

Weathering Interactions

Physical weathering exposes more surface area of rocks. This allows more water and atmospheric gases to enter rocks. Recall that water and gases help cause chemical weathering. Chemical weathering weakens rocks by changing the composition of some minerals and dissolving others. For example, clay formed by chemical weathering is weaker than the feldspar from which it formed. This weakening of rocks can increase the rate of physical weathering. In this way, chemical and physical weathering work together.

Soil Formation

Soil *consists of weathered rock, mineral material, water, air, and organic matter from the remains of organisms.* Soil forms directly on top of the rock layers from which it is made. The process of soil formation is illustrated in the figure at the top of the next page.

Soil formation takes a long time. It is the result of hundreds to thousands of years of weathering. The rock type that weathers, the biological activity, and the climate all affect soil formation.

Soil Formation

Weathering processes fracture and break down rock. Soil formation can take hundreds or thousands of years.

Plants, bacteria, and burrowing organisms help break down rock.

The upper part of the soil contains more organic material than the lower part. The lower part of the soil also can contain weathered rock.

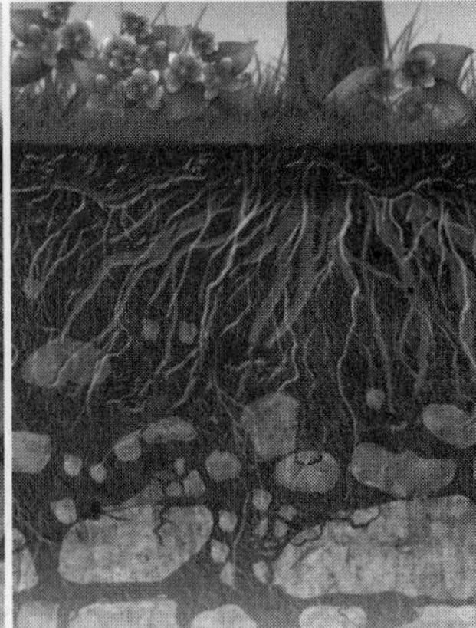
Over time, plants and other organisms in the soil die and decompose. The upper part of the soil contains nutrient-rich organic material.

Biological Activity and Soil Formation Biological activity plays an important role in making soil. Tunnels formed by worms and other organisms form pathways in soil for water and air. Decaying plants and animals also produce carbon dioxide and other acids that enhance chemical weathering. Eventually, the decayed plants and animals become part of the soil and make it better for plant growth.

Climate and Soil Formation Where do you think soil forms fastest? Soil forms fastest in warm, wet climates. Large amounts of rain can speed weathering of rocks. In addition, chemical reactions occur faster in warmer temperatures. Weathering also can happen quickly in areas where freezing and thawing break apart rocks.

Erosion

Weathering dissolves minerals and produces small particles of rock. *The minerals and small pieces of rock are called* **sediment.** What happens to sediment after it is made? The agents of erosion remove the sediment. Water, ice, and wind can transport sediment from one place to another.

Erosion by Water

Moving water causes erosion. The water picks up rock pieces and sediment. They then scrape along the ground, picking up more material. The faster the water flows, the larger the pieces of sediment the water can carry. Steep mountain streams carry away all sediment except large boulders. Water flowing in rivers as well as waves in lakes and oceans cause erosion.

Visual Check

4. Evaluate What happens to the solid rock layers during soil formation?

____

____

____

____

Key Concept Check

5. Compare How are weathering and soil formation related?

____

____

____

____

Erosion by Ice

Glaciers are large masses of ice. As a glacier flows down a mountain, it removes rocks and sediment beneath it and along its sides. This forms a smooth land surface underneath the ice.

Erosion by glaciers makes deep valleys and steep peaks. Some glaciers can be large enough to cover continents. The ice covering Antarctica is an example.

Erosion by Wind

Strong winds also can erode and move sediment. Soil and rock that are not protected by plants can be eroded by wind. In some places, wind has eroded the rocks and made them look so smooth that they seem to have been sculpted by an artist.

Key Concept Check
6. Describe How does erosion change Earth's surface?

Deposition

What happens to eroded sediment? Eventually, the moving water, ice, or wind slows down or stops. When this happens, the sediment is deposited. **Deposition** *is the process of laying down eroded material in a new location.*

Deposition by Water

Fast-flowing water carries sediment. If the speed of flowing water decreases, the water can no longer carry the sediment. The sediment will settle at the bottom of the water.

Floodplains form when sediment settles out of rivers that flood the areas next to them. The floodplain of the Hatchie River in Tennessee was formed in this way. Sediment also settles out of rivers where they enter lakes and oceans, forming deltas.

Think it Over
7. Infer Why might farming be an important activity in river delta regions?

Deposition by Ice

When glaciers melt, the water produced by the melting ice does not flow fast enough to carry sediment. The sediment is deposited where the ice melts. Glacial deposits of sediment are called moraines.

Some moraines form mounds at the front and sides of glaciers. Other moraines can cover the ground that was previously under the glacier. When the glaciers that once covered much of North America melted, they left moraines over most of the areas where they melted.

Reading Check
8. Define What is a moraine?

Deposition by Wind

Wind also can deposit sediment. Sand dunes are landforms made as wind continually moves and deposits sand grains. Wind moves the sand grains up one side of the sand dune and deposits them on the other side. Grain by grain, sand dunes migrate in the direction the wind blows.

The Erosion-Deposition Cycle

Weathering breaks rock into sediment that can be transported from high mountains to low areas. Sediment builds up on plains, at the bottom of lakes, and at the bottom of the ocean. Over time, thick layers of sediment form. The locations where sediment accumulates are called sedimentary basins. The Gulf of Mexico is a sedimentary basin into which the Mississippi River deposits sediment.

Recall that some minerals dissolve in water. If the water evaporates, the minerals form again. Over time, layers of salt can form in this way as water evaporates in sedimentary basins. The salt surrounding the Great Salt Lake in Utah is an example of minerals re-forming as water evaporates.

The cycle of weathering, erosion, and deposition has been repeated many times throughout Earth's history. The cycle continues today. The shapes of continents change. The locations of plate boundaries change. Sediment continues to be deposited in low areas and then forced upward as tectonic activity forms mountains. Earth's surface is continually changing.

Key Concept Check
9. Describe How does deposition change Earth's surface?

____
____
____
____
____

Key Concept Check
10. Compare How are erosion and deposition related?

____
____
____
____
____

After You Read

Mini Glossary

chemical weathering: the process of changing the composition of rocks and minerals by exposure to water and the atmosphere

deposition: the process of laying down eroded material in a new location

erosion: the moving of weathered material, or sediment, from one location to another

physical weathering: the process of breaking rock into small pieces without changing the composition of the rock

sediment: minerals and small pieces of rock produced by weathering

soil: consists of weathered rock, mineral material, water, air, and organic material from the remains of organisms

weathering: the processes that break down rocks, changing Earth's surface over time

1. Review the terms and their definitions in the Mini Glossary. Write a sentence explaining the relationship between deposition and sediment.

2. Write the letters in the diagram to compare physical and chemical weathering.

a. affected by temperature
b. brought about by acid rain
c. affected by atmospheric gases
d. does not change composition of rocks
e. might break down mountains
f. changes composition of rocks
g. breaks down rocks
h. caused by plant roots
i. caused by frost wedging

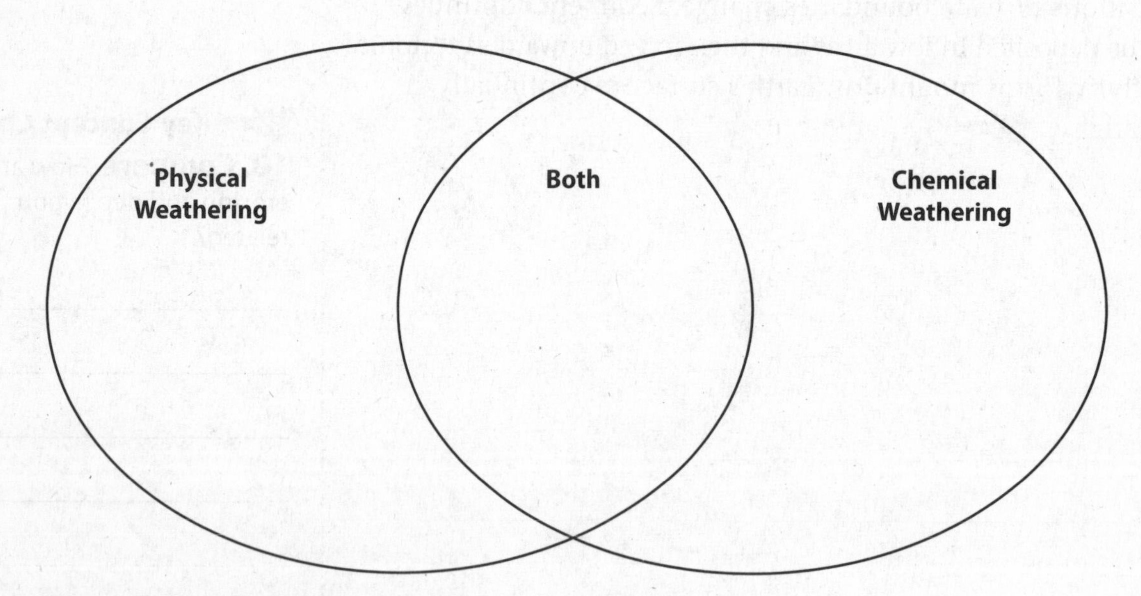

What do you think NOW?

Reread the statements at the beginning of the lesson. Fill in the After column with an A if you agree with the statement or a D if you disagree. Did you change your mind?

Log on to ConnectED.mcgraw-hill.com and access your textbook to find this lesson's resources.

Using Natural Resources

Earth's Resources

Before You Read

What do you think? Read the two statements below and decide whether you agree or disagree with them. Place an A in the Before column if you agree with the statement or a D if you disagree. After you've read this lesson, reread the statements to see if you have changed your mind.

Before	Statement	After
	1. The world's supply of oil will never run out.	
	2. You should include minerals in your diet.	

Key Concepts 

- What are natural resources?
- How do the three types of natural resources differ?

Read to Learn

Natural Resources

Where does the electricity for electric lights come from? It might come from a power plant that burns coal or natural gas. Or it might come from rooftop solar panels made with silicon, a mineral found in sand.

The smallest microbe and the largest whale rely on materials and energy from the environment. The same is true for humans. People depend on the environment for food, clothing, and fuels to heat and light their homes. *Parts of the environment that supply materials useful or necessary for the survival of living things are called* **natural resources.** Natural resources include land, air, minerals, and fuels. For example, trees and water are natural resources.

Study Coach

Building Vocabulary Work with another student to write a question about each vocabulary term in this lesson. Answer the questions and compare your answers. Reread the text to clarify the meaning of the terms.

Key Concept Check

1. Define What are natural resources?

____

____

____

____

____

Nonrenewable Resources

Do you travel in a vehicle that runs on gasoline? Do you drink soda from aluminum cans or water from plastic bottles? Gasoline, aluminum, and plastic are made from nonrenewable resources.

Nonrenewable resources *are natural resources that are being used up faster than they can be replaced by natural processes.* Nonrenewable resources form slowly, usually over thousands or millions of years. If they are used faster than they form, they will run out. Nonrenewable resources include fossil fuels and minerals.

FOLDABLES®

Make a horizontal tri-fold book and use it to identify similarities and differences among the types of resources.

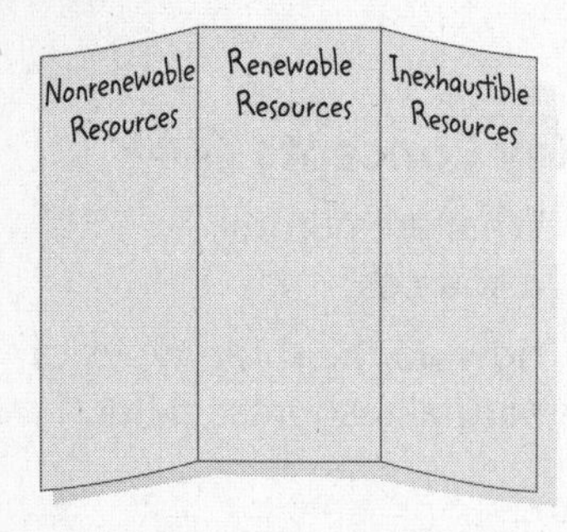

Fossil Fuels

Fossil fuels include coal, oil, and natural gas. The fossil fuels we use today formed from the decayed remains of organisms that died millions of years ago. Fossil fuels are forming all the time, but we use them much more quickly than nature replaces them.

Fossil fuels form underground. Coal is mined from the ground. Some coal mines are on the surface, and others are underground. The coal that is mined today formed from the decayed remains of trees, ferns, and other swamp plants that died 300–400 million years ago. The fossil fuels oil and natural gas are drilled from the ground.

Fossil fuels are mainly used as sources of energy. Many electric power plants burn coal or natural gas to heat water and make steam that powers generators. Natural gas also is used to heat homes and businesses. Gasoline, jet fuel, diesel fuel, kerosene, and other fuels are made from oil. Most plastics also are made from oil. ✓

✓ Reading Check

2. Identify three products that are made from oil.

____

____

____

Minerals

Have you ever added fertilizer to the soil around a plant? Fertilizers contain the minerals phosphorus and potassium. These minerals promote plant growth. Humans also need minerals for good health. Calcium and magnesium are two minerals the human body needs.

Minerals are nonliving substances found in Earth's crust. People use minerals for many purposes. The mineral gypsum is used in wallboard and cement. Silicon has many uses in industry. It is important for the manufacture of computers and other electronic devices. Copper is widely used in electrical wiring.

Uranium is a mineral that can be used as a source of energy. In a nuclear power plant, the nuclei of uranium atoms are split apart. This reaction is known as nuclear fission. Some of the energy that held the nuclei together is released as thermal energy. This thermal energy is then used to boil water and produce steam, which generates electricity.

Like fossil fuels, minerals are formed underground by geologic processes over millions of years. For that reason, most minerals are considered nonrenewable. Some minerals, such as calcium, are plentiful. Others, such as large rubies, are rare. ✓

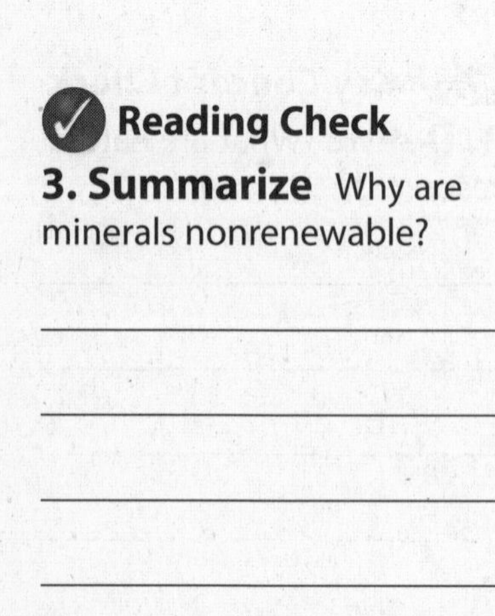

✓ Reading Check

3. Summarize Why are minerals nonrenewable?

____

____

____

____

____

____

Renewable Resources

Supplies of many natural resources are constantly renewed by natural cycles. The water cycle is an example. When liquid water evaporates, it rises into the atmosphere as water vapor. Water vapor condenses and falls back to the ground as rain or snow. Water is a renewable resource.

Renewable resources *are natural resources that can be replenished by natural processes at least as quickly as they are used.* These resources do not run out because they are replaced in a fairly short period of time. Renewable resources include water, air, land, and living things.

Renewable resources are replenished by natural processes. Still, they must be used wisely. If people use any resource faster than it is replaced, it becomes nonrenewable. A forest can be a nonrenewable resource if the trees are cut down faster than they can be replaced. ✓

Air

Green plants produce almost all of the oxygen in the air we breathe. Oxygen is a product of photosynthesis. Remember that photosynthesis is a series of chemical reactions in plants that use energy from light and produce sugars. Without plants, Earth's atmosphere would not contain enough oxygen to support most forms of life.

Air also contains carbon dioxide (CO_2), which plants need for photosynthesis. CO_2 is released into the air when dead plants and animals decay, when fossil fuels or wood are burned, and as a product of cellular respiration in plants and animals. Cellular respiration is a series of chemical reactions that convert energy from food into a form usable by cells. Without CO_2, photosynthesis would not be possible.

Land

Fertile soil is an important resource. Topsoil is the upper layer of soil that contains most of the nutrients plants need. Gardeners know that topsoil can be replenished by the decay of plant material. The carbon, nitrogen, and other elements in the decomposing plants become available for the growth of new plants.

Topsoil can be classified as a renewable resource. However, if it is carried away by water or wind, it can take hundreds of years to rebuild. Land resources also include wildlife and ecosystems such as forests, grasslands, deserts, and coral reefs. ✓

✓ **Reading Check**

4. Compare In what way are renewable and nonrenewable resources similar?

____

____

____

____

____

Think it Over

5. Generalize Do you think humans could survive if most of the plants on Earth died? Why or why not?

____

____

____

____

✓ **Reading Check**

6. Consider How is topsoil replenished by natural processes?

____

____

____

____

Math Skills

Converting a ratio to a percentage often makes it easier to visualize a set of numbers. For example, in 2007, 101.5 quadrillion units (quads) of energy were used in the United States. Of that, 6.813 quads were produced from renewable energy sources. What percentage of U.S. energy was produced from renewable energy sources?

Set up a ratio of the part over the whole.

$$\frac{6.813 \text{ quads}}{101.5 \text{ quads}}$$

Rewrite the fraction as a decimal.

$$\frac{6.813 \text{ quads}}{101.5 \text{ quads}} = 0.0671$$

Multiply by 100 and add %.

$$0.0671 \times 100 = 6.71\%$$

7. Use Percentages Of the 101.5 quads of energy used in 2007, 0.341 quads were from wind energy. What percentage of U.S. energy came from wind?

____

Key Concept Check

8. Differentiate How do inexhaustible resources differ from renewable and nonrenewable resources?

____

____

____

____

Water

Can you imagine a world without water? All organisms require water to live. People need a reliable supply of freshwater for drinking, washing, and irrigating crops. People also use water to run power plants and factories. Oceans, lakes, and rivers serve as major transportation routes and recreational areas. They are important habitats for many species, including some that people depend on for food.

Most of Earth's surface is covered by water. But only a small amount of that water is freshwater that people use. Freshwater is renewed through the water cycle. The total amount of water on Earth always remains the same.

Has your community ever been asked to conserve water because of a drought? A drought can cause a shortage in the supply of freshwater. In many large cities, water is transported from hundreds of miles away to meet the needs of residents. In some parts of the world, people must travel long distances every day to get water.

Inexhaustible Resources

An **inexhaustible resource** *is a natural resource that will not run out, no matter how much of it people use.* Energy from the Sun, solar energy, is inexhaustible. So is wind, which is generated by the Sun's uneven heating of Earth's lower atmosphere. Another inexhaustible resource is thermal energy from within Earth.

Solar Energy

Without the Sun's energy, life as it is on Earth would not be possible. If you've studied food chains, you know that energy from the Sun is used by plants and other producers during photosynthesis to make food. Consumers are organisms that get energy by eating producers or other consumers. The energy in food chains is always traced back to the Sun.

Solar energy can be used in many ways. Greenhouses trap thermal energy. They make it possible to grow warm-weather plants in cool climates. Solar cookers concentrate the Sun's thermal energy to cook food. Large solar-power plants provide electricity to many homes.

Solar energy also can be used to heat water in individual homes, as shown in the figure at the top of the next page. The hot water can be stored in a tank until it is needed.

Solar Energy from an Inexhaustible Resource

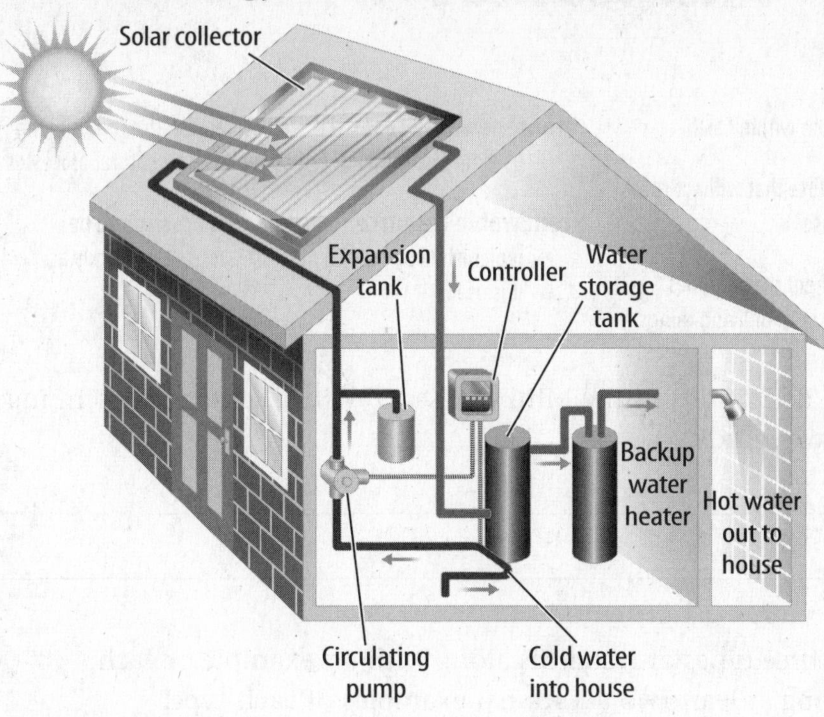

Wind Power

Sailboats, kites, and windmills are powered by wind. Wind is the movement of air over Earth's surface. Wind is an inexhaustible resource produced by the uneven heating of the atmosphere by the Sun.

If you live in an area that has strong winds, you might have seen giant wind turbines at work. In areas with frequent, strong winds, turbines can be used to produce electricity.

Geothermal Energy

Another type of inexhaustible resource is geothermal energy. **Geothermal energy** *is thermal energy from within Earth.* Molten rock that rises close to the surface of Earth's crust is called magma. Pockets of magma in some parts of Earth's crust heat underground water and rocks.

The heated water produces steam, which is used in geothermal power plants to generate electricity. People who live in California—as well as in other parts of the world—rely on geothermal energy to produce a large amount of their electricity.

Visual Check

9. Identify In which part of the system is water heated by the Sun?

Reading Check

10. Describe What causes wind?

Reading Check

11. Recognize What are three types of inexhaustible resources?

After You Read

Mini Glossary

geothermal energy: thermal energy from within Earth

inexhaustible resource: a natural resource that will not run out, no matter how much of it people use

natural resource: a part of the environment that supplies materials useful or necessary for the survival of living things

nonrenewable resource: a natural resource that is being used up faster than it can be replaced by natural processes

renewable resource: a natural resource that can be replenished by natural processes at least as quickly as it is used

1. Review the terms and their definitions in the Mini Glossary. Write a sentence describing natural resources in your own words.

____

____

2. The table below identifies three types of resources along with an example of each. Complete the table by writing at least two additional examples of each type.

Inexhaustible Resource	Nonrenewable Resource	Renewable Resource
geothermal energy	oil	water

3. Review the questions you wrote as you read the lesson. Select one, and write the answer below without referring to the text.

____

____

What do you think NOW?

Reread the statements at the beginning of the lesson. Fill in the After column with an A if you agree with the statement or a D if you disagree. Did you change your mind?

Log on to ConnectED.mcgraw-hill.com and access your textbook to find this lesson's resources.

Using Natural Resources

Pollution

Before You Read

What do you think? Read the two statements below and decide whether you agree or disagree with them. Place an A in the Before column if you agree with the statement or a D if you disagree. After you've read this lesson, reread the statements to see if you have changed your mind.

Before	Statement	After
	3. Global warming causes acid rain.	
	4. Smog can affect human health.	

Key Concepts 

- How does pollution affect air resources?
- How does pollution affect water resources?
- How does pollution affect land resources?

Read to Learn

What is pollution?

What happens when smoke gets in the air or toxic chemicals leak into soil? Smoke is a mixture of gases and tiny particles that make breathing difficult, especially for people who have health problems. Toxic chemicals that leak into soil can kill plants and soil organisms. These substances cause pollution. **Pollution** *is the contamination of the environment with substances that are harmful to life.*

Most pollution occurs because of human actions, such as burning fossil fuels or spilling toxic materials. However, pollution also can come from natural disasters. Wildfires create smoke. Volcanic eruptions send ash and toxic gases into the atmosphere. Regardless of its source, pollution affects air, water, and land resources.

Mark the Text

Main Ideas and Details
Highlight the main idea of each paragraph. Highlight two details that support each main idea with a different color. Use your highlighted copy to review what you studied in this lesson.

Air Pollution

Many large cities issue alerts about air quality when air pollution levels are high. On such days, people are asked to avoid activities that contribute to air pollution, such as driving, using gasoline-powered lawn mowers, or cooking on charcoal grills. To avoid breathing problems, people also are advised to exercise in the early morning when the air is cleaner. Air pollution that can affect human health and recreational activities can be caused by ozone loss, photochemical smog, global warming, and acid precipitation.

FOLDABLES®

Make a horizontal three-tab book and use it to explain the effects of pollution.

Effects of Pollution on ...
...Air Resources | ...Water Resources | ...Land Resources

Ozone Loss

Ozone is a molecule comprised of three oxygen atoms. In the upper atmosphere, it forms a protective layer around Earth. *The* **ozone layer** *prevents most harmful ultraviolet (UV) radiation from reaching Earth.* UV radiation from the Sun can cause cancer and cataracts. It can also damage crops. ✓

✓ Reading Check
1. Define What is ozone?

In the 1980s, scientists warned that Earth's protective ozone layer was getting thinner. The problem was caused mainly by chlorofluorocarbons (CFCs). CFCs are compounds used in refrigerators, air conditioners, and aerosol sprays.

Governments around the world have phased out the use of CFCs and other ozone-depleting gases. Because compounds such as CFCs are no longer widely used, the ozone layer is expected to recover within several decades. ✓

✓ Reading Check
2. Explain Why have many countries banned the use of CFCs?

Photochemical Smog

Sunlight reacts with waste gases from the burning of fossil fuels and forms a type of air pollution called **photochemical smog.** As shown in the figure below, smog makes the air dark. It also can smell bad.

Photochemical smog is formed of particles and gases that irritate the respiratory system, making it hard for people to breathe. Smog can worsen throughout the day as chemicals react with sunlight.

One of the gases in smog is ozone. In the upper atmosphere, ozone is helpful. But in the lower atmosphere, it is a pollutant that can harm plants and animals and cause lung damage.

✓ Visual Check
3. Identify What activities contribute to the formation of smog?

Smog

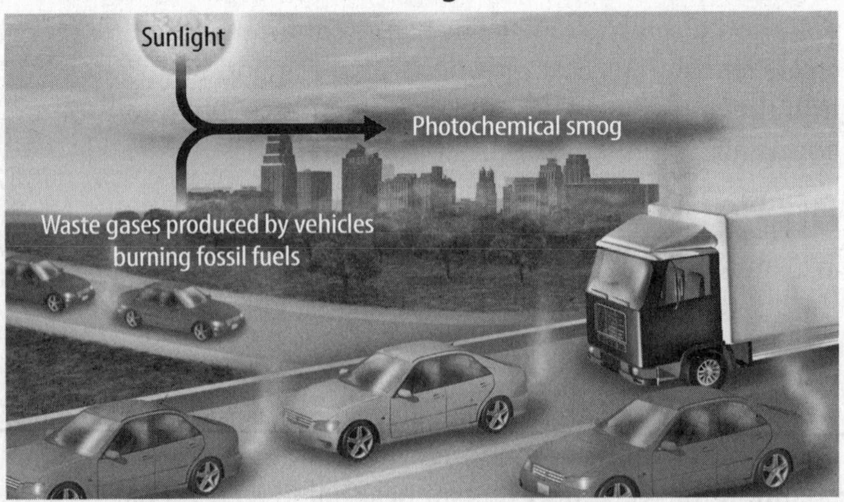

Global Warming

You might have heard news reports about the melting of glaciers and sea ice. Earth is getting warmer. **Global warming** *is the scientific observation that Earth's average surface temperature is increasing*. Global warming can change Earth's climate in many ways. It can

- change weather conditions;
- change ecosystems and food webs;
- increase the number and severity of floods and droughts; and
- increase coastal flooding as sea ice melts and sea levels rise.

Data indicate that Earth's average surface temperature and increases in atmospheric carbon dioxide (CO_2) follow the same general trend. CO_2 is a greenhouse gas. This means it traps thermal energy, helping to keep Earth warm. Greenhouse gases occur naturally. Without them, Earth would be too cold to support life. But human activities add greenhouse gases to the atmosphere, especially CO_2 from the burning of fossil fuels. Most scientists, including those on the United Nations Intergovernmental Panel on Climate Change, agree that increases in atmospheric CO_2 are contributing to global warming.

Acid Precipitation

Gases produced by the burning of fossil fuels also create other forms of air pollution, including acid precipitation. **Acid precipitation** *is acidic rain or snow that forms when waste gases from automobiles and power plants combine with moisture in the air.* Coal-burning power plants produce sulfur dioxide gas that combines with moisture to form sulfuric acid. Cars and trucks produce nitrous oxide gases that form nitric acid. Acid precipitation pollutes soil and can kill plants, including trees. It also contributes to water pollution and can damage buildings.

Water Pollution

Have you ever seen a stream covered with thick green algae? The stream might have been polluted with fertilizers from nearby lawns or farms. It might contain chemicals from nearby factories. Water pollution can come from chemical runoff and other agricultural, residential, and industrial sources.

Reading Check

4. Describe four possible effects of global warming.

ACADEMIC VOCABULARY

occur

(verb) to appear or happen

Key Concept Check

5. Summarize How does pollution affect air resources?

Wastewater

You probably already know that you should not pour paint or used motor oil into storm drains. In most cities, rainwater that flows into storm drains goes directly into nearby waterways. Materials that go in the storm drain, including grease and oil washed from the street, can contribute to water pollution.

The wastewater that drains from showers, sinks, and toilets contains harmful viruses and bacteria. To safeguard health, this wastewater usually is purified in a sewage-treatment plant before it is released into streams or used to irrigate crops. In some parts of the world, little or no sewage treatment occurs. People who live in these places might have to use polluted water. ✓

Wastewater that comes from industries and mining operations also contains pollutants. It requires treatment before it can be returned to the environment. Even after treatment, some harmful substances might remain and impact water quality. ✓

Reading Check
6. Consider Why is household wastewater treated in sewage-treatment plants?

Reading Check
7. Name three sources of wastewater.

Runoff and Sediments

When it rains, water can flow over the land. This water, called runoff, flows across lawns and farmland. Along the way, it picks up pesticides, herbicides, and fertilizers.

Runoff carries these pollutants into streams, where they can harm insects, fish, and other organisms. Runoff also carries sediment particles into streams. Too much sediment can damage stream habitats, clog waterways, and cause flooding.

Key Concept Check
8. Summarize How does pollution affect water resources?

Land Pollution

Have you ever helped clean up litter? Foam containers, plastic bags, bottles, cans, and even furniture and appliances get dumped along roadsides. Litter is more than an eyesore. It can pollute soil and water and disturb wildlife. Sources of land pollution include homes, farms, industry, and mines.

Agriculture

Farmers use pesticides and other agricultural chemicals to help plants grow. But these chemicals become pollutants if they are used in excess or disposed of improperly.

Herbicides kill weeds. But if they flow into streams, they can kill algae and plants and harm fish and amphibians. Some farming practices contaminate soil. Irrigation water contains salts that can build up in soil that is irrigated on a regular basis.

Industry and Mining

Many industrial facilities, including oil refineries and ore processors, produce toxic wastes. For example, power plants that burn coal produce coal ash sludge as waste. The sludge contains mercury, lead, arsenic, and other potentially harmful metals. If toxic wastes like these are incorrectly stored or disposed of, they can contaminate soil and water. The health of people, plants, and wildlife can be affected.

The mining of fossil fuels and minerals can disturb or destroy entire ecosystems. Some coal-mining techniques can release toxic substances that were buried in rock. After the coal has been removed, the area can be restored. But it is difficult or impossible to replace the original ecosystem.

Key Concept Check

9. Summarize How does pollution affect land resources?

After You Read

Mini Glossary

acid precipitation: acidic rain or snow that forms when waste gases from automobiles and power plants combine with moisture in the air

global warming: the scientific observation that Earth's average surface temperature is increasing

ozone layer: a protective layer around Earth that prevents most harmful ultraviolet (UV) radiation from reaching Earth

photochemical smog: a type of air pollution formed when sunlight reacts with waste gases from the burning of fossil fuels

pollution: the contamination of the environment with substances that are harmful to life

1. Review the terms and their definitions in the Mini Glossary. Write a sentence comparing and contrasting acid precipitation and photochemical smog.

____

____

2. Use the graphic organizer below to identify at least two sources of water pollution and two sources of land pollution.

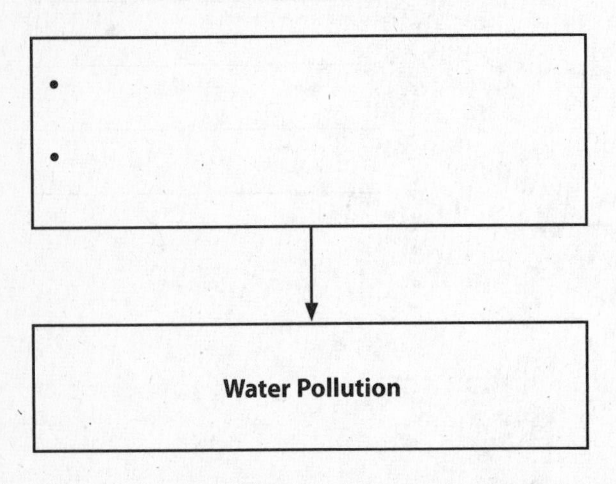

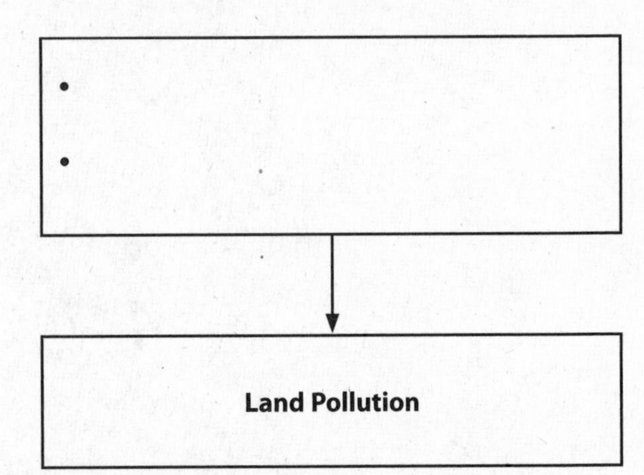

3. What do you think would happen to Earth's surface temperature if atmospheric CO_2 levels decreased? Why?

____

____

What do you think NOW?

Reread the statements at the beginning of the lesson. Fill in the After column with an A if you agree with the statement or a D if you disagree. Did you change your mind?

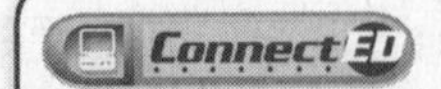

Log on to ConnectED.mcgraw-hill.com and access your textbook to find this lesson's resources.

Using Natural Resources

Protecting Earth

Before You Read

What do you think? Read the two statements below and decide whether you agree or disagree with them. Place an A in the Before column if you agree with the statement or a D if you disagree. After you've read this lesson, reread the statements to see if you have changed your mind.

Before	Statement	After
	5. Oil left over from frying potatoes can be used as automobile fuel.	
	6. Hybrid electric vehicles cannot travel far or go fast.	

Key Concepts
- How can people monitor resource use?
- How can people conserve resources?

Read to Learn

Monitoring Human Impact on Earth

As the human population increases, so does its impact on the planet. Scientists, governments, and concerned citizens around the world are working to identify environmental problems, educate people about them, and help find solutions.

Scientists collect data on environmental conditions by placing detectors on satellites, aircraft, high-altitude balloons, and ground-based monitoring stations. For example, the United States and the European Union have launched satellites into orbit around Earth. These satellites gather data on greenhouse gases, the ozone, ecosystem changes, melting glaciers and sea ice, climate patterns, and ocean health. The U.S. Environmental Protection Agency (EPA) is a government organization that watches the health of the environment and looks for ways to reduce the impact of humans. The EPA enforces environmental laws and supports research. It also identifies superfund sites—abandoned areas that have been contaminated by toxic wastes—and develops plans to clean them.

Mark the Text

Building Vocabulary As you read, underline the words and phrases that you do not understand. When you finish reading, discuss these words and phrases with another student or your teacher.

Key Concept Check

1. State How can people monitor resource use?

Developing Technologies

Many technologies have been developed to protect Earth's resources. These advances often focus on saving energy and reducing pollution.

FOLDABLES®

Make a small shutterfold book and use it to identify technology and methods that protect natural resources.

Monitoring Natural Resources

Conserving Natural Resources

Reading Check

2. Explain How does using CFLs help the environment?

______

______

______

Visual Check

3. Summarize How do CFCs affect ozone molecules?

______

______

Water-Saving Technologies

It takes energy to clean water and to transport it to homes and businesses. So technologies that conserve water also save energy. Low-flow showerheads and toilets as well as drip irrigation systems help reduce water use.

Energy-Saving Technologies

Saving energy can make Earth's supply of fossil fuels last longer. Using renewable energy sources reduces fossil fuel use. Some of these sources are expensive, but designs are improving and costs are falling. Solar electricity might soon cost the same as electricity produced by burning fossil fuels. Burning fewer fossil fuels also creates less pollution.

Other energy-saving advances include compact fluorescent lightbulbs (CFLs). They use about one-fourth the energy of incandescent bulbs and can last ten times longer. In 2007, Americans who switched to CFLs reduced greenhouse gas emissions by an amount equal to removing 2 million cars from the road.

CFC Replacements

CFCs thin the ozone layer because the chlorine atoms in CFC molecules react with sunlight to destroy ozone, as shown in the figure below. All CFCs soon will be phased out and replaced with chemicals that do not contain chlorine. Replacements include hydrofluorocarbons (HFCs) and perfluorocarbons (PFCs). Even after CFCs are no longer in use, it will take decades for the ozone layer to recover.

CFCs

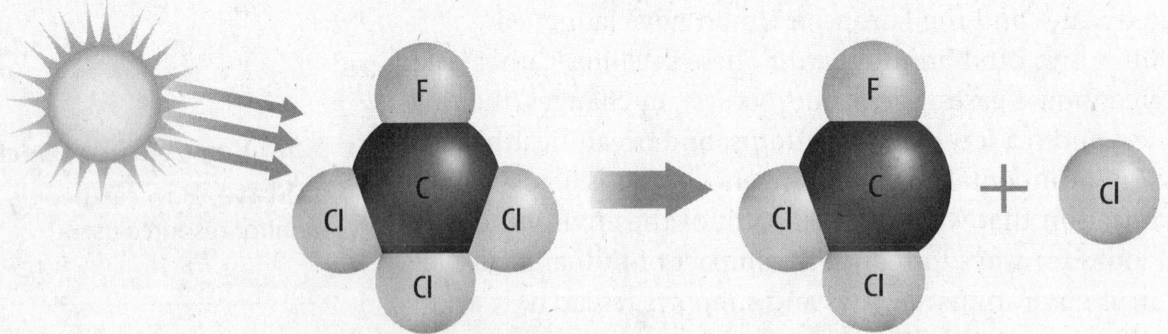

Sunlight reacts with a CFC molecule, causing a chlorine atom to break away.

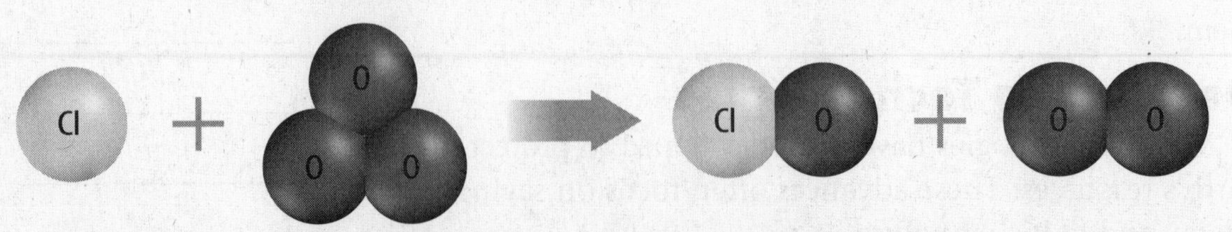

The chlorine atom reacts with and breaks apart an ozone molecule.

Alternative Fuels

Gasohol and biodiesel are alternative fuels that help reduce humans' use of fossil fuels. These alternative fuels also help reduce air pollution. 

Gasohol Gasohol is a mixture of 90 percent gasoline and 10 percent ethanol. Ethanol is alcohol made from corn, sugarcane, or other plants. Using gasohol in gasoline engines helps reduce emissions of carbon monoxide. Carbon monoxide is an air pollutant that contributes to smog.

The carbon in ethanol comes from plants instead of fossil fuels. So using gasohol can help reduce emissions that contribute to global warming.

Biodiesel Biodiesel is an alternative fuel that is made from renewable resources, primarily vegetable oils and animal fats. Biodiesel can even be made from oil that is left over from frying foods in restaurants!

Biodiesel can be burned in diesel engines in farm and industrial machinery, trucks, and cars. It produces fewer pollutants than regular diesel fuel, and it reduces CO_2 emissions by 78 percent.

Automobile Technologies

If you were buying a car, you would want to know how many miles it can travel per gallon of fuel. This measurement is called miles per gallon (mpg).

The higher a car's mpg rating, the less pollution it will add to the environment. A car with a high mpg rating also will use up fewer fossil-fuel resources.

HEVs Many people have decided to purchase a hybrid electric vehicle (HEV). HEVs combine a small gasoline engine with an electric motor powered by batteries.

HEVs run on battery power as much as possible. They get a boost from the gasoline engine for longer trips, higher speeds, and steep hills. The gasoline engine also charges the vehicle's batteries.

HEVs usually get excellent gas mileage. They can get up to twice the mileage of a regular car. Some recent models of HEVs get close to 50 mpg.

Reading Check

4. Name two alternative fuels that help reduce humans' use of fossil fuels.

____

____

SCIENCE USE V. COMMON USE

hybrid

Science Use an offspring of two animals or plants of different breeds or species

Common Use something that has two different components performing essentially the same task

Think it Over

5. Analyze Which do you think would probably be cheaper to own: a car that gets 30 mpg or one that gets 20 mpg? Why?

____

____

____

____

Hybrid Vehicle

Battery

Power split device

Electric motor

Generator

Internal combustion engine

Visual Check

6. Draw a circle around the power sources in the hybrid vehicle.

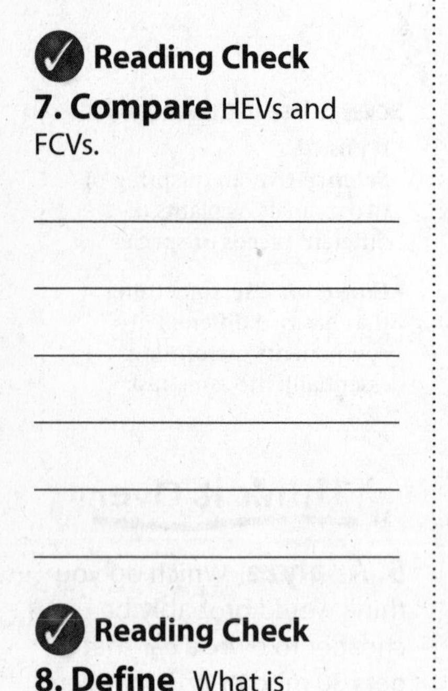

The figure above is a diagram of a typical hybrid electric vehicle. The battery powers the electric motor. The small gasoline engine can provide additional power.

FCVs In the future, another automobile alternative might be a fuel-cell vehicle (FCV). Inside a fuel cell, oxygen from the air chemically combines with hydrogen to produce electricity. The primary waste product is water. Tailpipe emissions from FCVs are nearly pollution-free. However, obtaining hydrogen fuel requires using methane or other fossil fuels. Researchers are looking for alternatives.

Reading Check

7. Compare HEVs and FCVs.

Making a Difference

Do you turn off the lights when you leave a room and recycle bottles and cans? If so, you are helping reduce your impact on the environment. You can help protect Earth's resources in other ways as well. Possibilities include cleaning up a stream, educating others about environmental issues, and analyzing the choices you make as a consumer.

Sustainability

When people talk about environmental issues, they often use the word *sustainability*. **Sustainability** *means meeting human needs in ways that ensure future generations also will be able to meet their needs*. When you turn off the lights as you leave a room, you are saving energy—and you are also helping to ensure a sustainable future. Other actions that promote a sustainable future include planting trees, composting, and picking up litter.

Reading Check

8. Define What is sustainability?

Restore and Rethink

Restoring damaged habitats and ecosystems to their original state is one way to make a difference. For example, picking up trash can restore water habitats.

You also can rethink the way you perform everyday activities. Instead of riding in a vehicle to nearby places, you could ride your bike or walk. ✓

Reduce and Reuse

You can reduce the amount of waste you create by reducing the amount of material you use. For example, you might avoid buying products with too much packaging.

Another way to reduce the amount of waste you create is to bring your own bags when you go shopping. Carrying your purchases in reusable bags, rather than using the plastic or paper bags the store offers, can help save energy and reduce waste.

Reusing items also helps reduce waste. Instead of buying new, reuse something that will work just as well. You also can donate used items to charities or sell them.

Recycle

If an item cannot be reused, you might be able to recycle it instead of throwing it away. **Recycling** *is manufacturing new products out of used products*. The recycling process reduces waste and extends our supply of natural resources. ✓

Computers and Electronics Computers and other electronics are examples of items that can be recycled. For example, these products contain valuable metals that can be used again. They also contain toxic materials that can contribute to pollution. So recycling also helps ensure that toxins are properly disposed of.

Compost Leaves, grass clippings, and vegetable scraps can be recycled by composting. In a compost pile, these materials decay into nutrient-rich soil that can be put back into the garden.

Buy Recycled Separating recyclables from the rest of the trash is just one step. To keep the cycle going, people need to buy and use recycled products. You can find shoes, clothing, paper, and carpets made from recycled materials.

✓ Reading Check

9. Consider How does using your bike for transportation benefit the environment?

____

____

____

____

✓ Reading Check

10. State How does recycling help the environment?

____

____

____

____

Key Concept Check

11. Explain How can you conserve resources?

____

____

____

____

____

____

After You Read

Mini Glossary

recycling: manufacturing new products out of used products

sustainability: meeting human needs in ways that ensure future generations also will be able to meet their needs

1. Review the terms and their definitions in the Mini Glossary. Write a sentence explaining how recycling contributes to sustainability.

___

___

___

2. Write the correct letter in each box within the graphic organizer to compare and contrast gasohol and biodiesel.

a. made from vegetable oils
b. burned in gasoline engines
c. made from gasoline and ethanol
d. burned in diesel engines
e. reduces harmful pollutants
f. alternative fuel

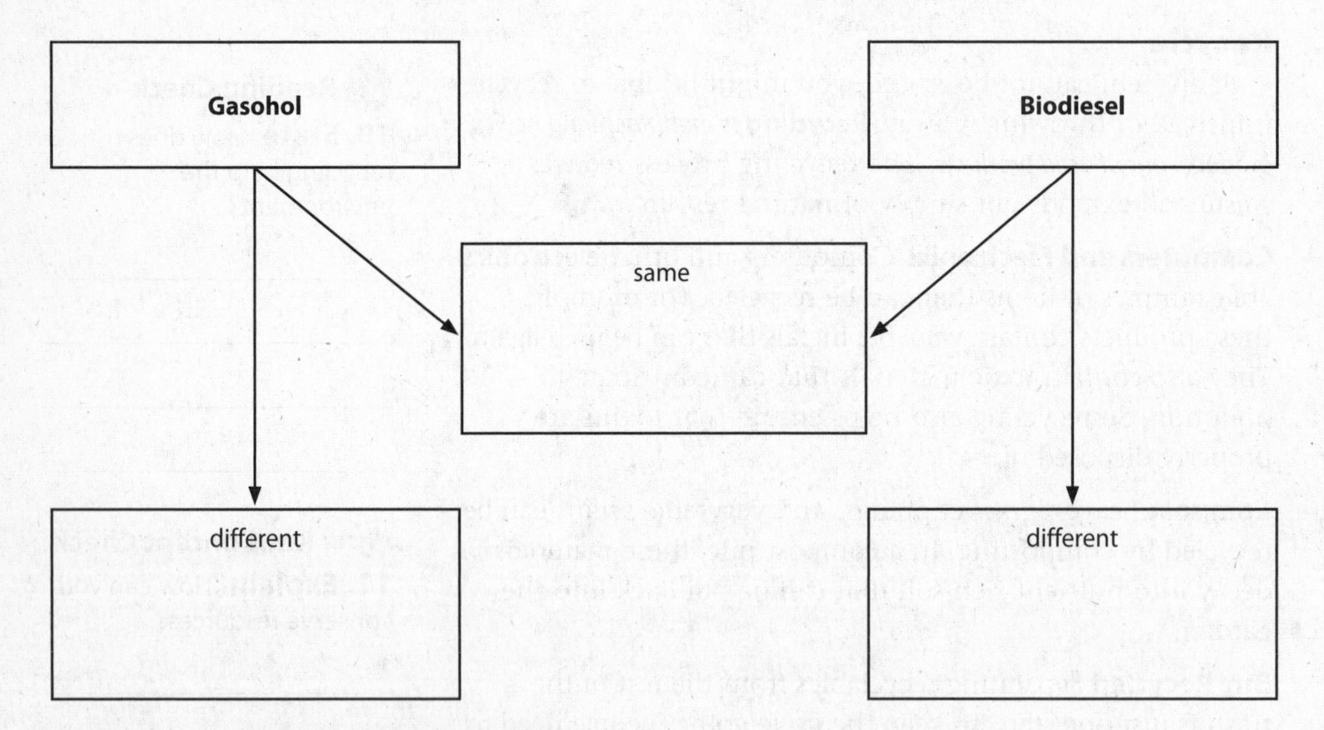

What do you think NOW?

Reread the statements at the beginning of the lesson. Fill in the After column with an A if you agree with the statement or a D if you disagree. Did you change your mind?

Log on to ConnectED.mcgraw-hill.com and access your textbook to find this lesson's resources.

Earth's Atmosphere

Describing Earth's Atmosphere

Before You Read

What do you think? Read the two statements below and decide whether you agree or disagree with them. Place an A in the Before column if you agree with the statement or a D if you disagree. After you've read this lesson, reread the statements to see if you have changed your mind.

Before	Statement	After
	1. Air is empty space.	
	2. Earth's atmosphere is important to living organisms.	

Key Concepts 
- How did Earth's atmosphere form?
- What is Earth's atmosphere made of?
- What are the layers of the atmosphere?
- How do air pressure and temperature change as altitude increases?

Read to Learn

Importance of Earth's Atmosphere

The **atmosphere** (AT muh sfihr) *is a thin layer of gases surrounding Earth*. Earth's atmosphere is hundreds of kilometers high. However, when compared to Earth's size, the atmosphere is about as thick as an apple's skin is to an apple.

The atmosphere contains the oxygen, carbon dioxide, and water that all life on Earth needs. Earth's atmosphere also acts like insulation in a house. The atmosphere helps keep temperatures within a range in which living organisms can survive.

Without an atmosphere, Earth's temperatures would vary greatly. Daytime temperatures would be very high, and nighttime temperatures would be very low.

Earth's atmosphere helps protect living organisms from some of the Sun's harmful rays. The atmosphere also helps protect Earth's surface from being struck by meteors. Most meteors that fall toward Earth enter the atmosphere and burn up before reaching Earth's surface. Friction with the atmosphere causes them to burn. Only the largest meteors strike Earth. ✓

Mark the Text

Identify Main Ideas To help you learn about Earth's atmosphere, highlight each heading. Then highlight the details that support and explain it. Use this highlighted text to review the lesson.

✓ **Reading Check**

1. Explain Why is Earth's atmosphere important to life on Earth?

Origins of Earth's Atmosphere

When Earth formed, it was a ball of molten rock. As Earth slowly cooled, its outer surface hardened. Erupting volcanoes released hot gases from inside Earth. These gases surrounded Earth, forming an atmosphere.

Reading Check
2. Explain How did Earth's ancient atmosphere form?

Ancient Earth's atmosphere was mainly water vapor with a little carbon dioxide (CO_2) and nitrogen. **Water vapor** *is water in its gaseous form.* Earth's ancient atmosphere did not have enough oxygen to support life as we know it. As Earth and its atmosphere continued to cool, the water vapor condensed into a liquid. Rain fell and then evaporated from Earth's surface over and over again for thousands of years. Oceans began to form as more and more water accumulated on Earth's surface. Most of the CO_2 from Earth's early atmosphere that dissolved in rain is in rocks on the ocean floor. Today, the atmosphere has more nitrogen than CO_2.

REVIEW VOCABULARY
liquid
matter with a definite volume but no definite shape that can flow from one place to another

Earth's first organisms could undergo photosynthesis. This changed the atmosphere. Recall that photosynthesis is a process that uses light energy to produce sugar and oxygen from CO_2 and water. Organisms that used photosynthesis removed CO_2 from the atmosphere and released oxygen. After a time, levels of CO_2 and oxygen in the atmosphere supported the development of other organisms.

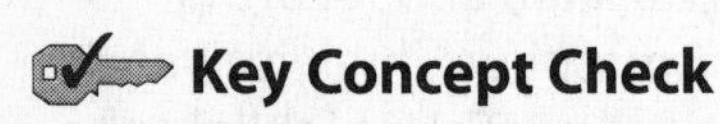
Key Concept Check
3. Explain How did Earth's present atmosphere form? (Circle the correct answer.)
a. from molten rock
b. from photosynthesis
c. from ancient volcanoes

Composition of the Atmosphere

Today's atmosphere is mostly made up of invisible gases with some solid and liquid particles. The gases include nitrogen, oxygen, and carbon dioxide. Some of the solid and liquid particles include ash from erupting volcanoes and water droplets.

Gases in the Atmosphere

The graph on the next page shows the gases in Earth's atmosphere. About 78 percent of Earth's atmosphere is made up of nitrogen. Oxygen makes up about 21 percent of Earth's atmosphere. Other gases, including argon, carbon dioxide, and water vapor, make up the remaining 1 percent of the atmosphere.

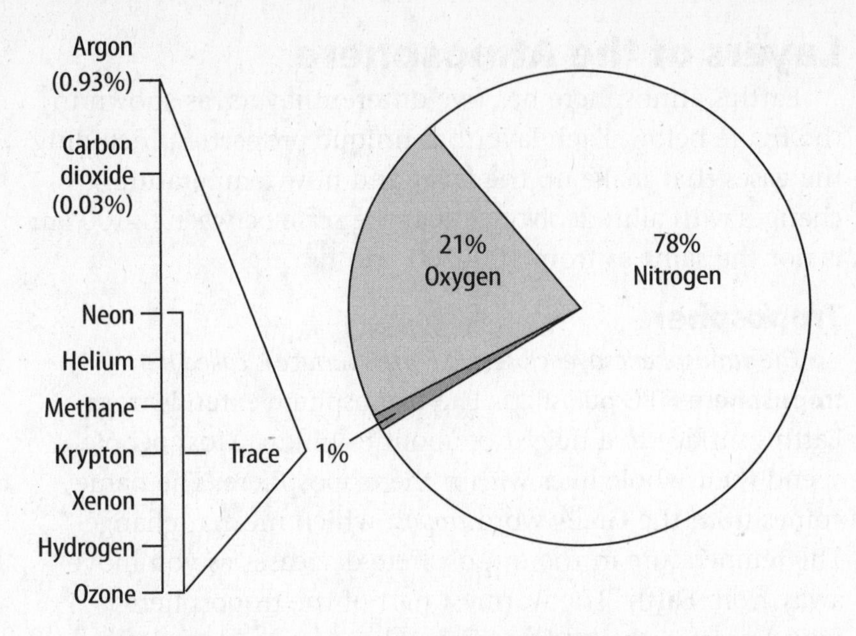

The concentrations, or amounts, of water vapor, carbon dioxide, and ozone vary. Concentrations can be different in different locations. The concentration of water vapor, for example, can be as little as 0 percent or as much as 4 percent. Carbon dioxide currently makes up 0.038 percent of the atmosphere. Ozone is a gas found in very small amounts at very high altitudes. Ozone also occurs near Earth's surface in urban areas.

Solids and Liquids in the Atmosphere

Earth's atmosphere also contains tiny solid particles. Many of these solids, such as pollen, dust, salt, and volcanic ash, enter the atmosphere through natural processes. Some solid particles enter Earth's atmosphere as the result of human activities, such as driving vehicles that exhaust soot.

Water droplets are the most common liquid particles in Earth's atmosphere. They are microscopic but visible when they form clouds. Other liquids include acids given off by erupting volcanoes and by the burning of fossil fuels. Sulfur dioxide and nitrous oxide combine with water vapor in the air and form the acids.

Visual Check

4. Read a Graph What percent of the atmosphere is made up of oxygen and nitrogen?

____

Think it Over

5. Calculate Use the graph to add up the percentages of different gases in Earth's atmosphere. What is the total? Why?

____

____

____

____

Key Concept Check

6. State What is Earth's atmosphere made of?

____

____

____

____

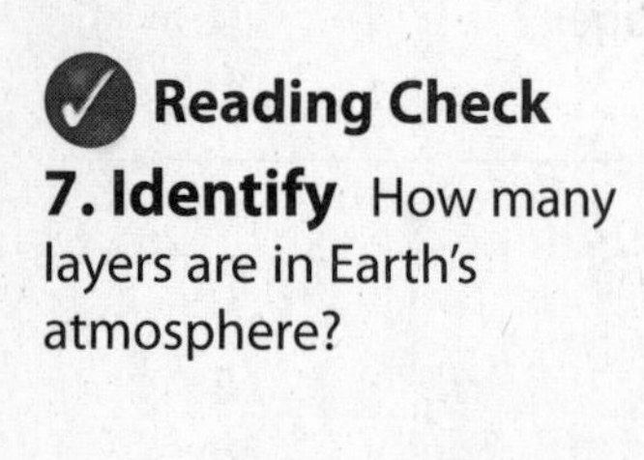

Reading Check

7. Identify How many layers are in Earth's atmosphere?

______

Reading Check

8. Describe the troposphere.

______

Visual Check

9. Identify In which layer of the atmosphere do planes fly? (Circle the correct answer.)

a. mesosphere

b. stratosphere

c. troposphere

Layers of the Atmosphere

Earth's atmosphere has five different layers, as shown in the figure below. Each layer has unique properties, including the gases that make up the layer and how temperature changes with altitude. Notice that the scale between 0–100 km is not the same as from 100–700 km. ✓

Troposphere

The atmospheric layer closest to Earth's surface is called the **troposphere** (TRO puh sfihr). The troposphere extends from Earth's surface to a height of about 8–15 km. Most people spend their whole lives within the troposphere. The name comes from the Greek word *tropos,* which means "change." The temperature in the troposphere decreases as you move away from Earth. The warmest part of the troposphere is found near Earth's surface. This is because most sunlight passes through the atmosphere and warms Earth's surface. The warmth radiates to the troposphere, causing weather. ✓

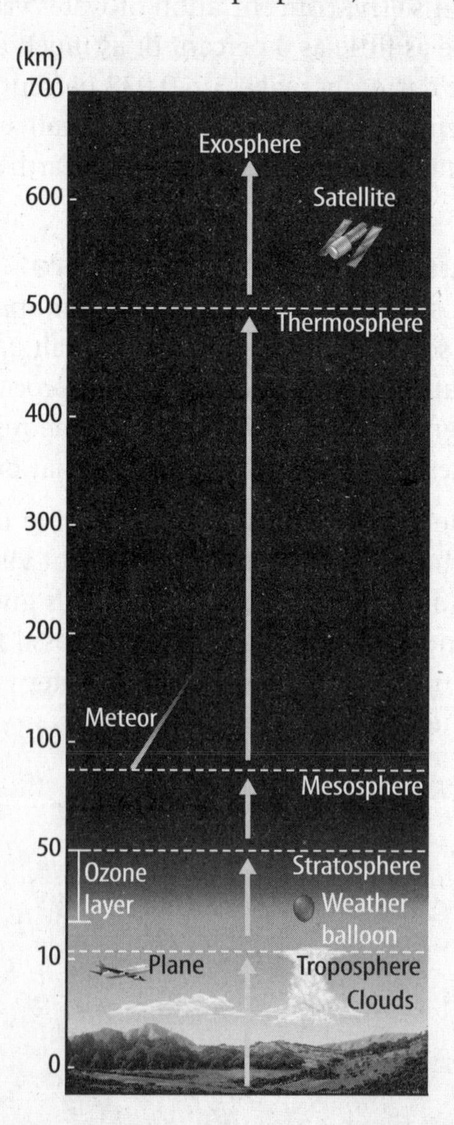

Stratosphere

The atmospheric layer directly above the troposphere is the **stratosphere** (STRA tuh sfihr). The stratosphere extends from about 15 km to about 50 km above Earth's surface. The bottom half of the stratosphere contains the highest concentration of ozone gas. *The area of the stratosphere with a high concentration of ozone is often referred to as the* **ozone layer.** The ozone layer causes temperatures in the stratosphere to increase as altitude increases.

An ozone (O_3) molecule is not the same as a molecule of the oxygen gas (O_2) that you breathe. Ozone has three oxygen atoms instead of two. This small difference is important. Ozone absorbs the Sun's ultraviolet rays more effectively than oxygen does. Ozone protects Earth from ultraviolet rays. Ultraviolet rays can kill plants, animals, and other organisms and they can cause skin cancers in humans.

Mesosphere and Thermosphere

The mesosphere extends from the stratosphere to about 85 km above Earth. Directly above the mesosphere is the thermosphere. The thermosphere can extend to more than 500 km above Earth. Combined, the mesosphere and the thermosphere are much broader than the troposphere and the stratosphere. However, only about 1 percent of the atmosphere's gas molecules are found in the mesosphere and the thermosphere. Most meteors burn up in these layers instead of striking Earth.

The Ionosphere *The* ***ionosphere*** *is a region within the mesosphere and thermosphere that contains ions.* Between 60 km and 500 km above Earth's surface, the ions in the ionosphere reflect AM radio waves transmitted from Earth. After sunset, when ions recombine, this reflection increases.

Auroras Displays of colored lights, called auroras, occur in the ionosphere. Auroras occur when ions from the Sun strike air molecules, causing them to give off bright colors of light. People who live in the higher latitudes, nearer to the North Pole and South Pole, are most likely to see auroras.

Exosphere

The exosphere is the atmospheric layer farthest from Earth's surface. Pressure and density are so low in the exosphere that individual gas molecules rarely strike one another. The molecules move at fast speeds after absorbing the Sun's radiation. These molecules can escape the pull of gravity and travel into space.

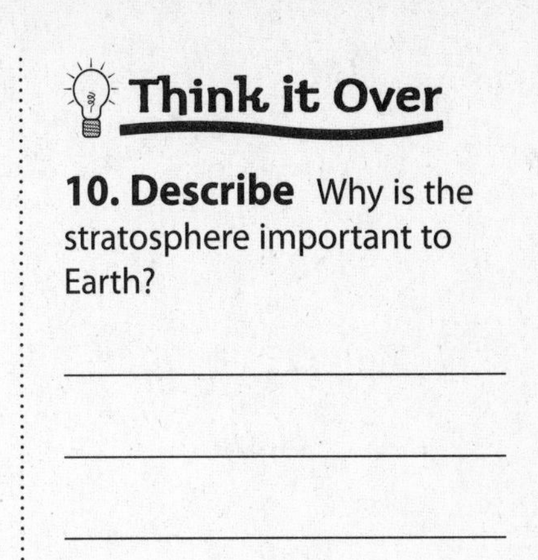

Think it Over

10. Describe Why is the stratosphere important to Earth?

FOLDABLES®

Make a vertical four-tab book to record similarities and differences among these four layers of the atmosphere.

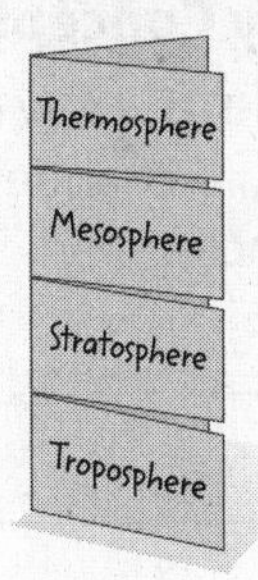

Key Concept Check

11. Name What are the layers of the atmosphere?

Air Pressure and Altitude

Gravity is the force that pulls all objects toward Earth. When you stand on a scale, you can read your weight. This is because gravity pulls you toward Earth. Gravity also pulls the atmosphere toward Earth. The pressure that a column of air exerts on anything below it is called air pressure. Gravity's pull on air increases its density. At higher altitudes, air is less dense. Air pressure is greatest near Earth's surface because the air molecules are closer together. This dense air exerts more force than the less-dense air near the top of the atmosphere. Mountain climbers sometimes carry oxygen tanks at high altitudes because fewer oxygen molecules are in the air at high altitudes.

Reading Check
12. Explain How does air pressure change as altitude increases?

Temperature and Altitude

The figure below shows how temperature changes with altitude in different layers of the atmosphere. If you have ever been hiking in the mountains, you know that the temperature cools as you reach higher elevations. In the troposphere, temperature decreases as altitude increases. Notice that the opposite is true in the stratosphere. As altitude increases in the stratosphere, the temperature increases. This happens because of high amounts of ozone in the stratosphere. Ozone absorbs energy from sunlight, which increases the temperature in the stratosphere.

In the mesosphere, as altitude increases, the temperature again decreases. In the thermosphere and exosphere, temperatures increase as altitude increases. The small number of particles in these layers absorbs large amounts of energy from the Sun. This creates high temperatures.

Key Concept Check
13. Explain How does temperature change as altitude increases?

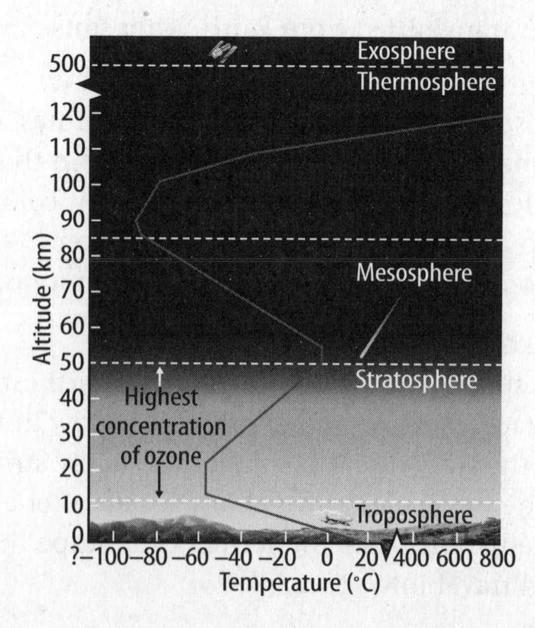

Visual Check
14. Identify Which layer has a temperature pattern most like the troposphere's?

After You Read

Mini Glossary

atmosphere (AT muh sfihr): a thin layer of gases surrounding Earth

ionosphere: the region within the mesosphere and thermosphere containing ions

ozone layer: the area of the stratosphere with a high concentration of ozone

stratosphere (STRA tuh sfihr): the atmospheric layer directly above the troposphere

troposphere (TRO puh sfihr): the atmospheric layer closest to Earth's surface

water vapor: water in a gaseous form

1. Review the terms and their definitions in the Mini Glossary. Write a sentence explaining why the ionosphere has that name.

____

____

2. Use what you have learned about the layers of Earth's atmosphere to complete the table.

Layer	Two Characteristics of the Layer
Troposphere	• •
Stratosphere	• •
Mesosphere	• •
Thermosphere	• •
Exosphere	• •

What do you think NOW?

Reread the statements at the beginning of the lesson. Fill in the After column with an A if you agree with the statement or a D if you disagree. Did you change your mind?

Log on to ConnectED.mcgraw-hill.com and access your textbook to find this lesson's resources.

CHAPTER 16
LESSON 2

Earth's Atmosphere

Energy Transfer in the Atmosphere

Key Concepts

- How does energy transfer from the Sun to Earth and to the atmosphere?
- How are air circulation patterns within the atmosphere created?

Before You Read

What do you think? Read the two statements below and decide whether you agree or disagree with them. Place an A in the Before column if you agree with the statement or a D if you disagree. After you've read this lesson, reread the statements to see if you have changed your mind.

Before	Statement	After
	3. All the energy from the Sun reaches Earth's surface.	
	4. Earth emits energy back into the atmosphere.	

K-W-L Fold a sheet of paper to form three columns. In the first column, write what you already know about energy transfer. In the second column, write what you want to know. In the third column, write what you have learned from reading this lesson.

Read to Learn

Energy from the Sun

The Sun's energy reaches Earth through the process of radiation. **Radiation** *is the transfer of energy by electromagnetic waves.* Ninety-nine percent of the radiant energy from the Sun consists of visible light, ultraviolet light, and infrared radiation.

Visible Light

Most sunlight is visible light. Visible light is light that you can see. Visible light passes through Earth's atmosphere. At Earth's surface, the Sun's energy is converted to thermal energy, commonly called heat.

Near-Visible Wavelengths

Ultraviolet (UV) light and infrared radiation (IR) are two other forms of radiant energy from the Sun. The wavelengths of UV and IR are just beyond the range of visibility to human eyes. UV light has short wavelengths and can break chemical bonds. A large dose of UV light will burn human skin and can cause skin cancer. Infrared radiation (IR) has longer wavelengths than visible light. You can feel IR as thermal energy or warmth. Earth absorbs energy from the Sun, and then it is radiated back as IR.

Reading Check

1. Contrast What is the difference between visible light and ultraviolet light?

____

____

____

____

Energy on Earth

As the Sun's energy passes through the atmosphere, some of it is absorbed, or taken in, by gases and particles. Some of it is reflected back into space. As a result, not all of the energy coming from the Sun reaches Earth's surface.

Absorption

Look at the figure below. Gases and particles in the atmosphere absorb about 20 percent of incoming solar radiation. Oxygen, ozone, and water vapor all absorb ultraviolet light. Water and carbon dioxide in the troposphere aborb some infrared radiation from the Sun. Earth's atmosphere does not absorb visible light. Visible light must be converted to infrared radiation before it can be absorbed.

Reflection

Look again at the figure below. Bright surfaces, especially clouds, <u>reflect</u> radiation as it enters the atmosphere. Clouds and other small particles in the air reflect about 25 percent of the Sun's radiation. Some of the radiation travels to Earth's surface. There, land and sea surfaces reflect it back. About 30 percent of all radiation that enters the atmosphere reflects back into space. If 30 percent of the incoming radiation reflects back into space and the atmosphere absorbs 20 percent, only about 50 percent of incoming solar radiation reaches Earth. Earth's surface then absorbs it.

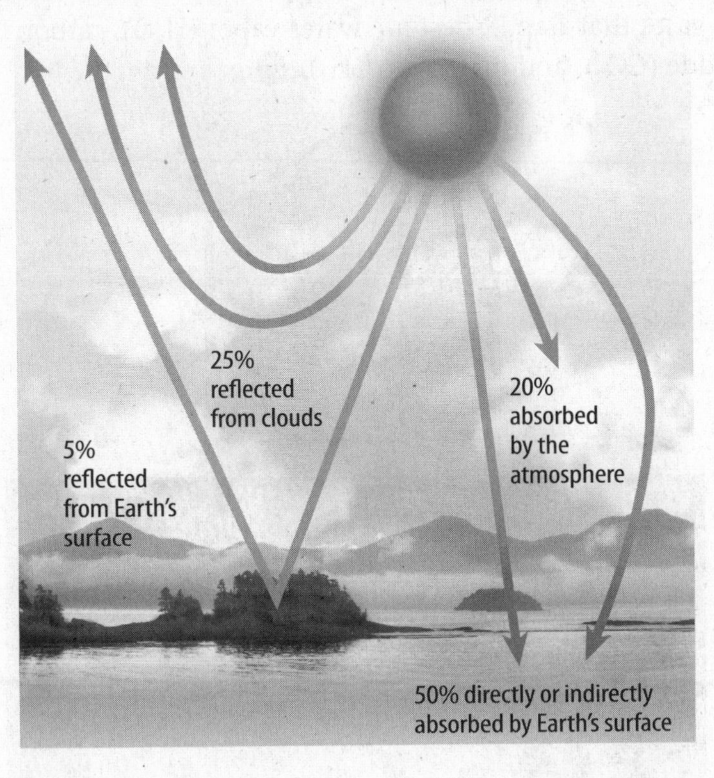

SCIENCE USE V. COMMON USE

reflect

Science Use to return light, heat, sound, and so on, after it strikes a surface

Common Use to think quietly and calmly

Think it Over

2. Express What is the total percentage of radiation that is reflected by Earth's atmosphere?

____

Visual Check

3. Identify What percent of incoming radiation is absorbed by gases and particles in the atmosphere? (Circle the correct answer.)

a. 20 percent

b. 25 percent

c. 50 percent

Radiation Balance

The Sun's radiation warms Earth. So why doesn't Earth get hotter and hotter? Over time, there is a balance between the amount of radiation that comes in from the Sun and the amount of radiation that leaves Earth.

The land, water, plants, and other organisms absorb solar radiation that reaches Earth's surface. After absorbing radiation, Earth reradiates, or bounces, it back into the outer atmosphere. The radiation from Earth is mostly in the form of infrared radiation (IR). The amount of radiation Earth receives from the Sun is the same amount Earth radiates back up into the atmosphere. Earth absorbs the Sun's energy and radiates it away until a balance is achieved. ✓

✓ Reading Check

4. Cause and Effect Why doesn't Earth get hotter as it continues to receive radiation from the Sun?

The Greenhouse Effect

The figure below shows how light passes through the glass of a greenhouse. The glass converts the light to infrared energy. The glass also stops the IR from escaping and it warms the greenhouse.

Some of the gases in Earth's atmosphere act like the glass of a greenhouse. These gases are called greenhouse gases. The gases let sunlight pass through, but they stop IR energy from escaping. When gases in Earth's atmosphere direct radiation back toward Earth's surface, this warms Earth's atmosphere more than normal. This is known as the greenhouse effect. The gases that trap IR best are water vapor (H_2O), carbon dioxide (CO_2), and methane (CH_4). ✓

✓ Reading Check

5. Describe What causes the greenhouse effect?

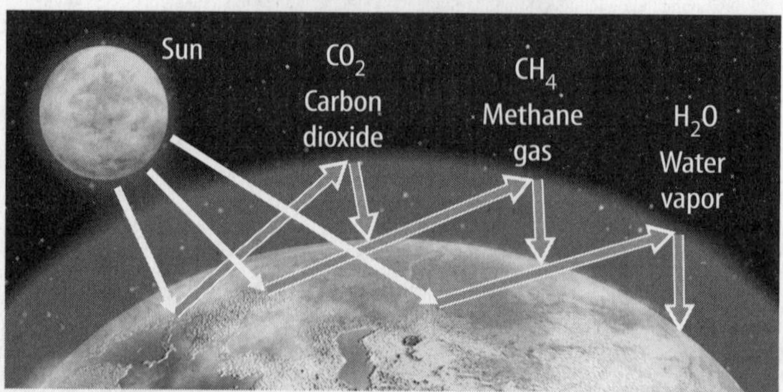

✓ Visual Check

6. Name three of Earth's greenhouse gases.

Thermal Energy Transfer

Recall that there are three types of thermal energy transfer—radiation, conduction, and convection. All three types of transfer occur in the atmosphere. Radiation is the process that transfers energy from the Sun to Earth.

Conduction

Thermal energy always moves from an object with a higher temperature to an object with a lower temperature. ***Conduction** is the transfer of thermal energy by collisions between particles of matter.* Particles must be close enough to touch to transfer energy by conduction. If you touched a pot of boiling water, thermal energy from the pot would move to your hand by conduction. Conduction occurs where the atmosphere touches Earth.

Convection

As molecules of air close to Earth's surface heat, they spread apart, and air becomes less dense. Less-dense air rises and transfers thermal energy to higher altitudes. *The transfer of thermal energy by the movement of matter from one place to another is called* **convection.** Convection occurs in the atmosphere when conduction heats air close to Earth's surface.

Latent Heat

More than 70 percent of Earth's surface is covered by water. Water is the only substance that can exist as a solid, a liquid, and a gas at the temperature ranges on Earth. Latent heat is the heat energy released or absorbed during the phase changes of water. When water changes from one phase to another, latent heat is exchanged also, as shown in the figure below. Like other forms of heat transfer, latent heat energy is transferred from Earth's surface to the atmosphere.

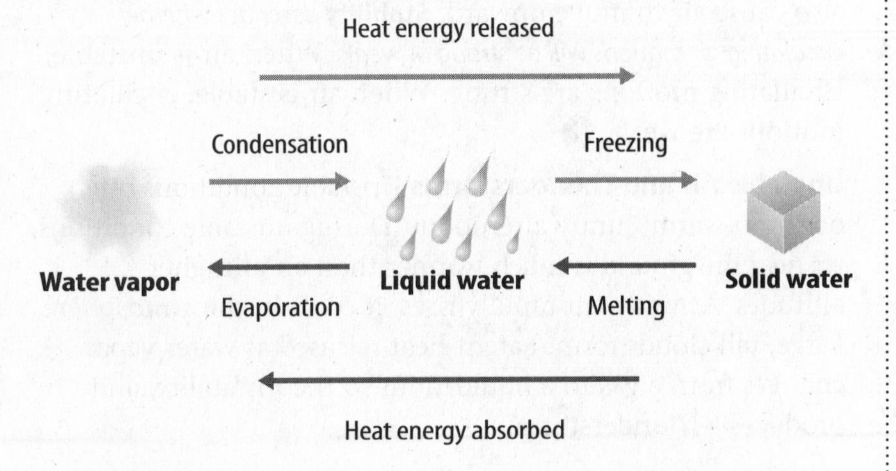

FOLDABLES

Make a four-column, four-row table. Use it to organize information about thermal energy transfer.

Energy Transfer by	Description	Everyday Example	Effect on the Atmosphere
Radiation			
Convection			
Conduction			

Key Concept Check

7. Identify How does energy transfer from the Sun to Earth and the atmosphere?

____

____

____

____

Visual Check

8. Explain What happens to water when heat energy is absorbed? (Circle the correct answer.)

a. Ice melts.

b. Liquid water freezes.

c. Water vapor condenses.

Circulating Air

You've already read that energy transfers through the atmosphere by convection. On a hot day, air that is heated becomes less dense. This creates a pressure difference. Cool, denser air pushes the warm air out of the way. The more-dense air replaces the warm air, as shown in the figure below. The warm air is often pushed upward. Warmer, rising air always comes with cooler, sinking air.

Visual Check
9. Describe what is happening in the figure.

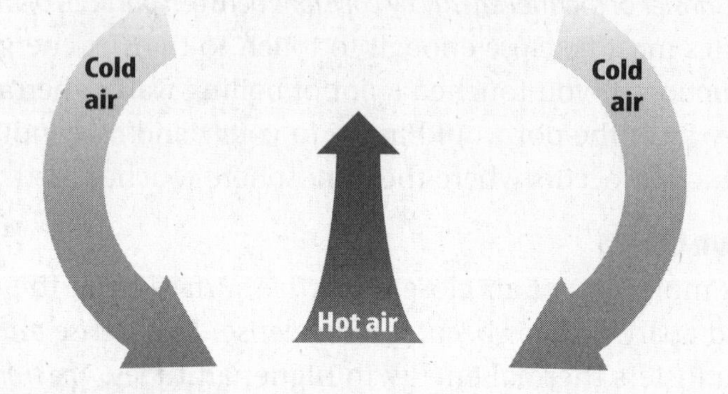

Air constantly moves in the atmosphere. For example, wind flowing into a mountain range rises and flows over the mountains. After the air reaches the top, it sinks on the other side. This up-and-down motion sets up an atmospheric condition called a mountain wave. The rising air within mountain waves can create saucer-shaped, or lenticular (len TIH kyuh lur), clouds. Circulating air affects weather and climates around the world.

Key Concept Check
10. Explain How are air circulation patterns within the atmosphere created?

Stability

When you stand outside in the wind, your body forces some of the air to move above you. The same is true for plants, hills, cars, and buildings. Conduction and convection also cause air to move upward. **Stability** *describes whether circulating air motions will be strong or weak.* When air is unstable, circulating motions are strong. When air is stable, circulating motions are weak.

Reading Check
11. Contrast How are circulating air motions different in stable and unstable air?

Unstable Air and Thunderstorms Unstable conditions often occur on warm, sunny afternoons. During unstable conditions, air near the ground is much warmer than air at higher altitudes. As warm air rapidly rises, it cools in the atmosphere. Large, tall clouds form. Latent heat released as water vapor changes from a gas to a liquid, adds to the instability, and produces a thunderstorm.

Stable Air and Temperature Inversions Sometimes, air near the ground is nearly the same temperature as air at higher altitudes. When this happens, the air is stable, and circulating motions are weak. A temperature inversion can take place under these conditions. *A* ***temperature inversion*** *occurs in the troposphere when temperature increases as altitude increases.*

The first figure below shows normal temperature conditions. The second figure shows a temperature inversion. During a temperature inversion, a layer of cooler air close to Earth is trapped by a layer of warm air above it. Temperature inversions prevent air from mixing. This can trap pollution in the air close to Earth's surface.

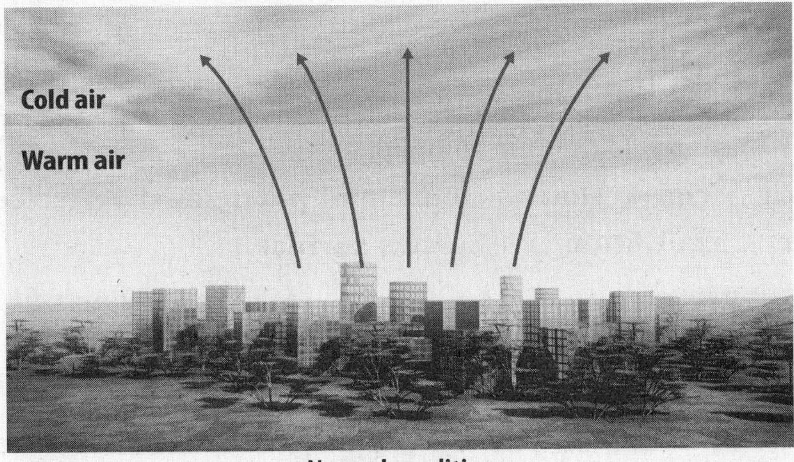

Normal conditions

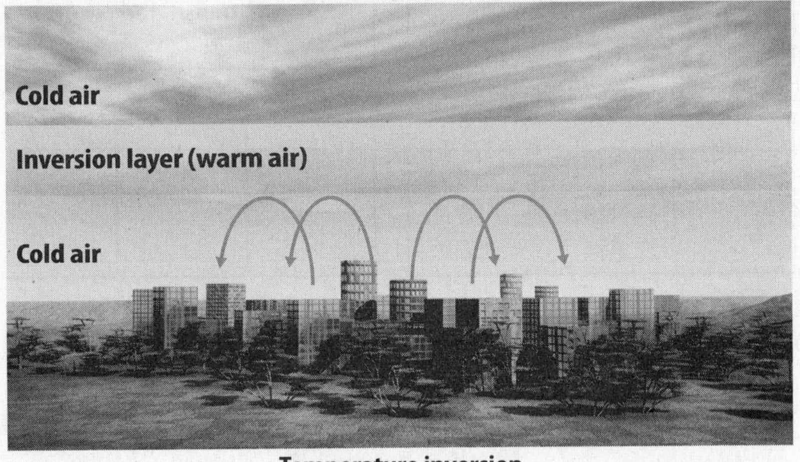

Temperature inversion

Visual Check

12. Contrast How do conditions during a temperature inversion differ from normal conditions?

____

____

____

____

After You Read

Mini Glossary

conduction: the transfer of thermal energy by collisions between particles of matter

convection: the transfer of thermal energy by the movement of matter from one place to another

radiation: the transfer of energy by electromagnetic waves

stability: describes whether circulating air motions will be strong or weak

temperature inversion: occurs in the troposphere when temperature increases as altitude increases

1. Review the terms and their definitions in the Mini Glossary. Use three words from the Mini Glossary to describe the transfer of thermal energy.

____

____

2. Use terms from the word bank to complete the concept map.

absorbed **land and sea** **convection** **clouds and particles** **reflected**
conduction **atmosphere** **radiation** **Earth's surface**

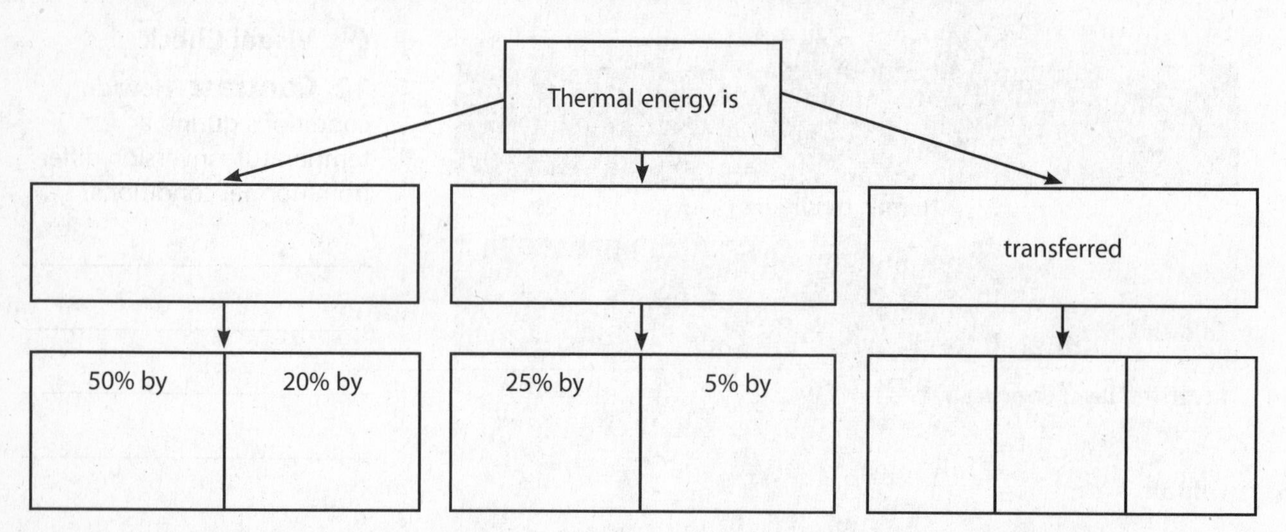

3. Give one example of each type of thermal energy transfer in your everyday life.

____

____

What do you think NOW?

Reread the statements at the beginning of the lesson. Fill in the After column with an A if you agree with the statement or a D if you disagree. Did you change your mind?

Log on to ConnectED.mcgraw-hill.com and access your textbook to find this lesson's resources.

Earth's Atmosphere

Air Currents

Before You Read

What do you think? Read the two statements below and decide whether you agree or disagree with them. Place an A in the Before column if you agree with the statement or a D if you disagree. After you've read this lesson, reread the statements to see if you have changed your mind.

Before	Statement	After
	5. Uneven heating in different parts of the atmosphere creates air circulation patterns.	
	6. Warm air sinks and cold air rises.	

Key Concepts 

- How does uneven heating of Earth's surface result in air movement?
- How are air currents on Earth affected by Earth's spin?
- What are the main wind belts on Earth?

Read to Learn

Global Winds

Wind patterns can be global or local. There are great wind belts that circle Earth. The energy that causes this large movement of air comes from the Sun. ✓

Unequal Heating of Earth's Surface

The Sun's energy warms Earth. Not all areas on Earth's surface receive the same amount of energy from the Sun. The amount of energy an area gets depends on the Sun's angle. Energy coming from the rising or setting Sun is not very strong. But Earth warms quickly when the Sun is high in the sky around noon.

Sunlight Areas in low latitudes near the equator are referred to as the tropics. Sunlight strikes Earth's surface there at a high angle. It is nearly a 90° angle all year-round. As a result, the tropics receive more sunlight per unit of surface area than other places on Earth. This causes the land, the water, and the air at the equator to always be warm.

At latitudes near the North Pole and the South Pole, sunlight strikes Earth's surface at a low angle. Sunlight is spread over a larger surface area than in the tropics. This means that the poles receive very little energy per unit of surface area and so they are cooler.

Study Coach

Finding Main Ideas Highlight the main idea of each paragraph in this lesson. Reread the sentences to review the lesson.

✓ **Reading Check**

1. Identify What is the source of energy of global wind belts?

____

____

Wind Recall that warm air rises and cold air sinks. Warm air is less dense than cold air. Rising warm air puts less pressure on Earth than cooler air. Air pressure is usually low over the tropics because it is usually warm there. Air pressure is usually high over colder areas such as the North and South Poles. This difference in pressure creates wind. ***Wind** is the movement of air from areas of high pressure to areas of low pressure.* Global wind belts influence both climate and weather on Earth.

Key Concept Check

2. Explain How does uneven heating of Earth's surface result in air movement?

Global Wind Belts

The figure below shows the three-cell model used to describe circulation in Earth's atmosphere. Three cells exist in both the northern hemisphere and the southern hemisphere.

In the first cell, hot air near the equator moves to the top of the troposphere. Then, the air moves toward the poles until it cools and moves back to Earth's surface near 30° latitude. Most of the air in this convection cell then returns to the equator near Earth's surface.

The third cell is at the highest latitudes and is also a convection cell. Air from the poles moves toward the equator along Earth's surface. Warmer air is pushed upward by cooler air near 60° latitude.

The second cell is between 30° and 60° latitude and is not a convection cell. Its motion is driven by the other two cells. Imagine rolling cookie dough between your hands. Your hand represent the first and third cells. The cookie dough is the second cell. This second cell moves in much the same way as the dough.

FOLDABLES®

Make a shutterfold book to describe Earth's global wind belts and to explain how they circulate.

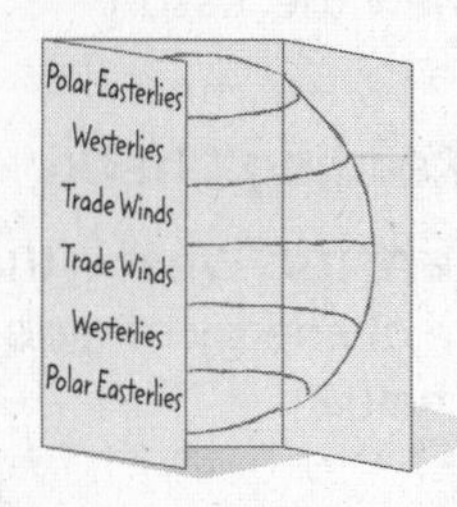

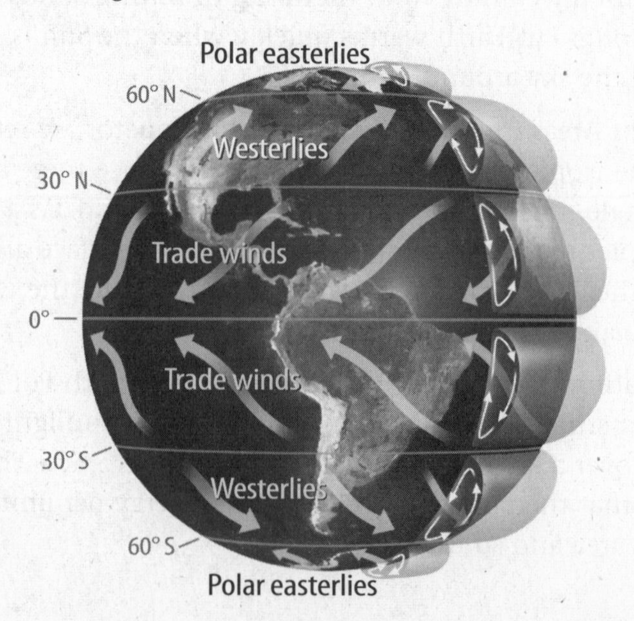

Visual Check

3. Identify Which wind belt do you live in?

The Coriolis Effect

What would happen if you threw a ball to someone sitting across from you on a moving merry-go-round? When the ball reached the opposite side, the person would have moved. The ball would have appeared to have curved. Like a merry-go-round, the rotation of Earth causes moving air and water to appear to move to the right in the northern hemisphere and to the left in the southern hemisphere. This is called the Coriolis effect. The difference between high and low pressure and the Coriolis effect create distinct wind patterns. These wind patterns are called prevailing winds.

Prevailing Winds

The three global wind cells in each hemisphere create northerly and southerly winds. When the Coriolis effect acts on the winds, the winds blow to the east or the west. These winds are relatively steady and predictable.

*The **trade winds** are steady winds that flow from east to west between 30°N latitude and 30°S latitude.* At about 30°N and 30°S latitude, air cools and sinks. This creates areas of high pressure and light, calm winds called the doldrums. Sailboats without engines can be stranded in the doldrums.

*The prevailing **westerlies** are steady winds that flow from west to east between latitudes 30°N and 60°N, and 30°S and 60°S. The **polar easterlies** are cold winds that blow from east to west near the North Pole and the South Pole.*

Jet Streams

*Near the top of the troposphere is a narrow band of high winds called the **jet stream.*** Jet streams flow around Earth from west to east, often making large loops to the north or the south. Jet streams also influence weather. They move cold air from the poles toward the tropics. Jet streams also move warm air from the tropics toward the poles. Jet streams can move as fast as 300 km/h.

Local Winds

Recall that global winds occur because of pressure differences around the globe. In the same way, local winds occur whenever air pressure is different from one location to another.

Key Concept Check

4. Describe How are air currents on Earth affected by Earth's spin?

Reading Check

5. Explain What might happen to a sailboat caught in the doldrums?

Key Concept Check

6. Summarize What are the main wind belts on Earth?

Sea and Land Breezes

If you have ever been to a lake or an ocean, you have probably experienced the connections among temperature, air pressure, and wind. *A* ***sea breeze*** *is wind that blows from the sea to the land due to local temperature and pressure differences.*

The left side of the figure below shows how a sea breeze forms. Land warms faster than water does. On sunny days, the air over the land warms by conduction and rises, creating an area of low pressure. The air over the water sinks. This creates an area of high pressure because it is cooler. The differences in pressure over the warm land and the cooler water results in a cool wind that blows from the sea onto the land.

A ***land breeze*** *is a wind that blows from the land to the sea due to local temperature and pressure differences.* The right side of the figure below shows how a land breeze forms. At night, the land cools more quickly than the water. The air above the land also cools more quickly than the air above the water. As a result, an area of lower pressure forms over the warmer water. In a land breeze, cool air over land moves toward lower pressure over the water. ✓

Reading Check

7. Compare and Contrast Explain how sea and land breezes are the same and different.

Visual Check

8. Sequence the steps involved in the formation of a land breeze.

Sea Breeze

Land Breeze

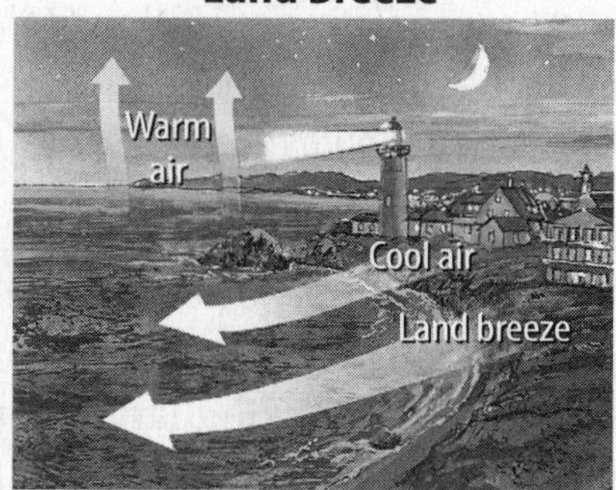

After You Read

Mini Glossary

jet stream: a narrow band of high winds located near the top of the troposphere

land breeze: wind that blows from the land to the sea due to local temperature and pressure differences

polar easterlies: cold winds that blow from the east to the west near the North and South Poles

sea breeze: wind that blows from the sea to the land due to local temperature and pressure differences

trade winds: steady winds that flow from east to west between 30°N latitude and 30°S latitude

westerlies: steady winds that flow from west to east between latitudes 30°N and 60°N, and 30°S and 60°S

wind: the movement of air from areas of high pressure to areas of low pressure

1. Review the terms and their definitions in the Mini Glossary. Choose two types of wind and explain how their names help you remember what they are.

____

____

____

2. Complete the table.

Winds	How They Blow and Where They Are
Trade winds	
Prevailing westerlies	
Polar easterlies	
Jet streams	
Sea breezes	
Land breezes	

What do you think NOW?

Reread the statements at the beginning of the lesson. Fill in the After column with an A if you agree with the statement or a D if you disagree. Did you change your mind?

Log on to ConnectED.mcgraw-hill.com and access your textbook to find this lesson's resources.

CHAPTER 16
LESSON 4

Earth's Atmosphere

Air Quality

Key Concepts
- How do humans impact air quality?
- Why do humans monitor air quality standards?

Before You Read

What do you think? Read the two statements below and decide whether you agree or disagree with them. Place an A in the Before column if you agree with the statement or a D if you disagree. After you've read this lesson, reread the statements to see if you have changed your mind.

Before	Statement	After
	7. If no humans lived on Earth, there would be no air pollution.	
	8. Pollution levels in the air are not measured or monitored.	

Study Coach

Asking Questions Read each head and write down one question you have about that topic. Try to find answers to your questions as you read.

Read to Learn

Sources of Air Pollution

The contamination of air by harmful substances, including gases and smoke, is called ***air pollution.*** Air pollution is harmful to humans and other living things. Years of exposure to polluted air can weaken a human's immune system. Air pollution can cause respiratory diseases such as asthma.

Air pollution comes from many sources. Point-source pollution is pollution that comes from a single, identifiable source. Examples of point sources include smokestacks of large factories and electric power plants that burn fossil fuels for energy. They release tons of polluting gases and particles into the air each day. An example of natural point-source pollution is an erupting volcano.

Nonpoint-source pollution is pollution that comes from a widespread area. An example of nonpoint-source pollution is air pollution in a large city. This is called nonpoint-source pollution because the pollution cannot be traced back to one source. Some bacteria that live in swamps and marshes are examples of natural sources of nonpoint-source pollution.

Key Concept Check
1. Compare point-source and nonpoint-source pollution.

Causes and Effects of Air Pollution

The harmful effects of air pollution are not limited to human health. Some pollutants, including ground-level ozone, can harm plants. Air pollution can also damage human-made structures. Sulfur dioxide pollution can change the color of stone, rust metal, and damage paint.

Acid Precipitation

When sulfur dioxide and nitrogen oxides combine with moisture in the atmosphere and form precipitation that has a pH lower than that of normal rainwater, it is called **acid precipitation.** Acid precipitation includes acid rain, snow, and fog. Acid precipitation affects the chemistry, or makeup, of water in lakes and rivers. This can harm the organisms living in the water. Acid precipitation also damages buildings and other structures made of stone.

Natural sources of sulfur dioxide include volcanoes and marshes. However, the most common sources of sulfur dioxide and nitrogen oxides are automobile exhausts and factory and power-plant smoke. ✓

Smog

Photochemical smog *is air pollution that forms from the interaction between chemicals in the air and sunlight.* Smog forms when nitrogen dioxide, released in gasoline-engine exhausts, reacts with sunlight. A series of chemical reactions produces ozone and other compounds that form smog. Recall that ozone in the stratosphere helps protect organisms from the Sun's harmful rays. Ground-level ozone can damage the tissues of plants and animals. Ground-level ozone is the main part of smog. Smog in urban areas reduces visibility and makes it difficult for some people to breathe the air.

Particulate Pollution

Although you can't see them, over 10,000 solid or liquid particles are in every cubic centimeter of air. A cubic centimeter is about the size of a sugar cube. This type of pollutant is called particulate matter. **Particulate** (par TIH kyuh lut) **matter** *is a mixture of dust, acids, and other chemicals that can be hazardous to human health.* The smallest particles are the most harmful. They can be inhaled and then enter your lungs. They can cause asthma and bronchitis and can lead to heart attacks. Children and older adults are most likely to experience health problems due to particulate matter.

Particulate matter in the atmosphere absorbs and scatters sunlight. This can create haze. Haze particles scatter light, make things blurry, and reduce visibility.

FOLDABLES®

Make a three-tab book to record information about the formation and effects of air pollution.

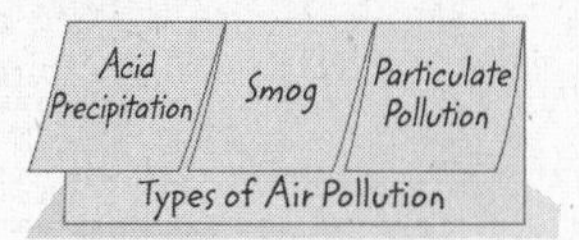

✓ Reading Check

2. Identify How does acid precipitation form? (Circle the correct answer.)

a. when gases mix with moisture in the atmosphere

b. when chemicals in the atmosphere react with sunlight

c. when ground-level ozone rusts metal

Key Concept Check

3. Explain How do humans impact air quality?

____

____

____

____

Movement of Air Pollution

Wind can influence the effects of air pollution. Because moving air carries pollution with it, some wind patterns cause more pollution problems than others. Weak winds or no wind prevents pollutants from mixing with the surrounding air. During weak wind conditions, pollution can grow to dangerous levels.

Temperature inversions can form during long winter nights when winds are weak and skies are clear. As land cools at night, the air above it also cools. Calm winds, however, prevent cool air from mixing with the warm air above it. Any pollution in the cool air stays close to Earth's surface and can cause problems.

Cool air, along with the pollution it contains, can be trapped in valleys. Cool air sinks down the sides of surrounding mountains into the valley preventing air layers from mixing. Pollution becomes trapped by the temperature inversion. ✓

✓ **Reading Check**

4. Cause and Effect How is pollution trapped by a temperature inversion?

____

____

____

____

Maintaining Healthful Air Quality

Preserving the quality of Earth's atmosphere requires the cooperation of government officials, scientists, and the public. The Clean Air Act is an example of how government can help fight air pollution. The act was passed in 1970. Since then, steps have been taken to reduce pollution from automobile exhaust. Pollutant levels have decreased greatly in the United States. Unfortunately, serious problems still remain. The amount of ground-level ozone is still too high in many large cities. Also, acid precipitation continues to form and harm organisms in lakes, streams, and forests.

Air Quality Standards

The Clean Air Act gives the United States government the power to set air-quality standards. The standards protect humans, animals, plants, and buildings from the effects of air pollution. All states must make sure pollutants do not exceed harmful levels. Some of the pollutants that are monitored include carbon monoxide, nitrogen oxides, particulate matter, ozone, and sulfur dioxide. ✓

✓ **Reading Check**

5. Describe What is the Clean Air Act?

____

____

____

____

Monitoring Air Pollution

Pollution levels are continuously monitored by hundreds of instruments in all major cities in the United States. If levels are too high, authorities might advise people to limit outdoor activities.

Air Quality Trends

Air quality in U.S. cities has improved over the last several decades, as shown in the graph below. Even though some pollution-producing processes have increased, such as burning fossil fuels and traveling in automobiles, levels of certain air pollutants have decreased. Levels of lead and carbon monoxide in the air have decreased the most. Levels of sulfur dioxide, nitrogen oxide, and particulate matter have also decreased.

Ground-level ozone has not decreased much. Recall that ozone can form from chemical reactions involving car exhaust. The large number of vehicles in use results in the level of ground-level ozone remaining high.

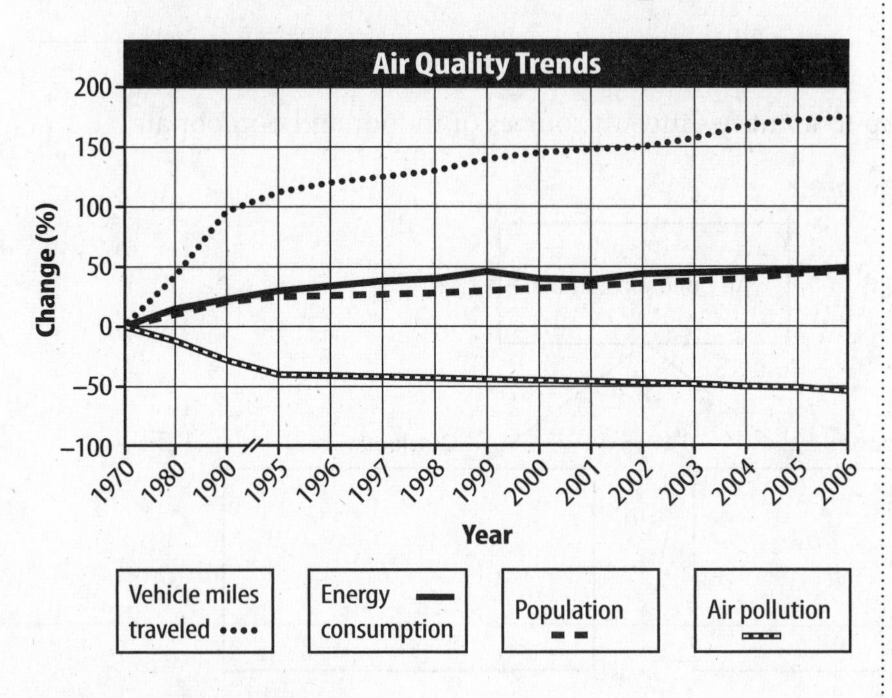

Indoor Air Pollution

Not all air pollution is outdoors. The air inside homes and other buildings can be as much as 50 times more polluted than outdoor air! Air quality indoors can affect human health more than air quality outdoors.

Indoor air pollution comes from many sources. Tobacco smoke, cleaning products, pesticides, and fireplaces are some common sources of indoor air pollutants. Furniture upholstery, carpets, and foam insulation also add pollutants to indoor air. Another indoor air pollutant is radon. Radon is an odorless gas given off by some soil and rocks. Radon leaks through cracks in the foundations of buildings and sometimes builds up to harmful levels inside homes. Harmful effects of radon come from breathing its particles.

Key Concept Check

6. Cause and Effect Why do humans monitor air quality standards?

____

____

____

____

Math Skills

The graph here shows the percent change in four different pollution factors from 1970 through 2006. All values are based on the 0% amount in 1970. For example, from 1970 to 1990, the number of vehicle miles driven increased by 100%. In other words, the vehicle miles doubled. During the same period, the amount of air pollution decreased by 30%.

7. Use Graphs What was the percent change in population between 1970 and 2006? (Circle the correct answer.)

a. 50 percent

b. 75 percent

c. 100 percent

After You Read

Mini Glossary

acid precipitation: precipitation that results when sulfur dioxide and nitrogen oxides combine with moisture in the atmosphere

air pollution: the contamination of air by harmful substances, including gases and smoke

particulate (par TIH kyuh lut) matter: a mixture of dust, acids, and other chemicals that can be hazardous to human health

photochemical smog: air pollution that forms from the interaction between chemicals in the atmosphere and sunlight

1. Review the terms and their definitions in the Mini Glossary. Explain in one sentence what photochemical smog is.

____

____

2. Complete the concept map to identify different sources of indoor and outdoor air pollution.

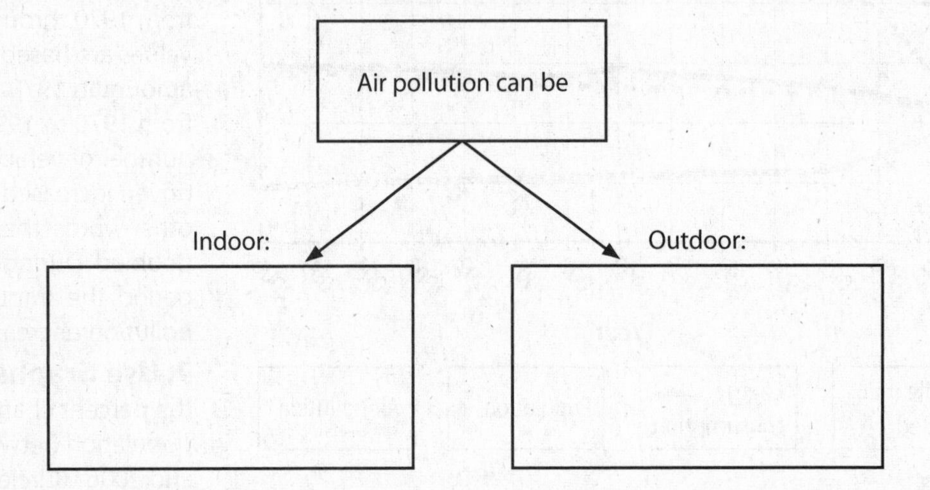

3. How did asking questions before you read the material help you better understand air quality?

____

____

____

What do you think NOW?

Reread the statements at the beginning of the lesson. Fill in the After column with an A if you agree with the statement or a D if you disagree. Did you change your mind?

Log on to ConnectED.mcgraw-hill.com and access your textbook to find this lesson's resources.

CHAPTER 17
LESSON 1

Weather

Describing Weather

Before You Read

What do you think? Read the two statements below and decide whether you agree or disagree with them. Place an A in the Before column if you agree with the statement or a D if you disagree. After you've read this lesson, reread the statements to see if you have changed your mind.

Before	Statement	After
	1. Weather is the long-term average of atmospheric patterns of an area.	
	2. All clouds are at the same altitude within the atmosphere.	

Key Concepts 

- What is weather?
- What variables are used to describe weather?
- How is weather related to the water cycle?

Read to Learn

What is weather?

Weather *is the atmospheric conditions, along with short-term changes, of a certain place at a certain time*. Have you ever been caught in a rainstorm on what began as a sunny day? If so, you know that weather can change quickly. It can also stay the same for days.

Study Coach

Summarize What You Read After you read each paragraph, write a sentence or two in your own words describing what you read. Use your sentences to review the lesson.

Weather Variables

Variables are things that can change. Temperature and rainfall are two of the variables used to describe weather. Meteorologists are scientists who study and predict weather. They use several variables that describe a variety of atmospheric conditions. These variables include air temperature, air pressure, wind speed and direction, humidity, cloud coverage, and precipitation.

Key Concept Check

1. Define What is weather?

Air Temperature

Air temperature is a measure of the average kinetic energy of molecules in the air. Kinetic energy is the energy an object has because it is moving. When the temperature is high, molecules have a high kinetic energy. Therefore, molecules in warm air move faster than molecules in cold air. Air temperatures vary with the time of day, season, location, and altitude.

Visual Check
2. Apply What happens to air pressure as altitude decreases?

____

____

____

Reading Check
3. Name What instrument measures air pressure?

____

Reading Check
4. Compare humidity and relative humidity.

____

____

____

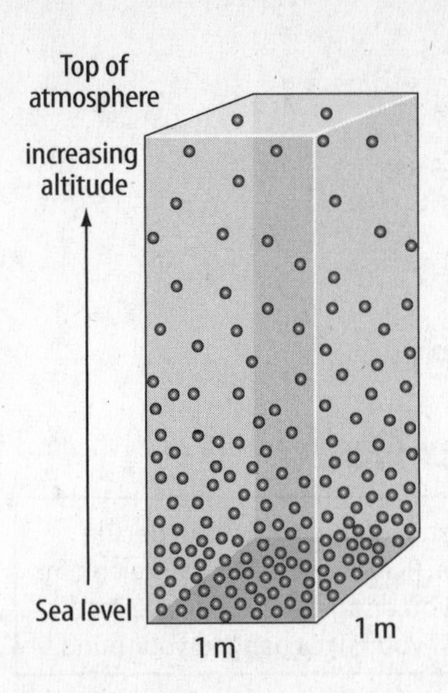

Air Pressure

Air pressure *is the pressure that a column of air exerts on the air, or surface, below it*. Look at the figure. Air pressure decreases as altitude increases. Therefore, air pressure is greater at low altitudes than it is at high altitudes.

Air pressure is measured with an instrument called a barometer. Air pressure is measured in millibars (mb). The term *barometric pressure* means "air pressure." Knowing the barometric pressure of different areas helps meteorologists predict the weather. ✓

Wind

Wind is created as air moves from areas of high pressure to areas of low pressure. Wind direction is given as the direction from which the wind is blowing. For example, winds that blow from west to east are called westerlies. Meteorologists measure wind speed using an instrument called an anemometer (a nuh MAH muh tur).

Humidity

The amount of water vapor in the air is called **humidity** (hyew MIH duh tee). Humidity can be measured in grams per cubic meter of air (g/m^3). When the humidity is high, there is more water vapor in the air. On a day with high humidity, your skin might feel sticky because sweat might not evaporate quickly from your skin.

Relative Humidity

A sponge can absorb water. When it becomes full, it cannot absorb any more water. In the same way, air can hold only a certain amount of gaseous water vapor. When air is saturated, it holds as much water vapor as possible. Temperature determines how much water vapor air can contain. Warm air can contain more water vapor than cold air can. **Relative humidity** *is the amount of water vapor present in the air compared to the maximum amount of water vapor the air could contain at that temperature*. Relative humidity is measured using a psychrometer. It is stated as a percent. A relative humidity of 50 percent means that the amount of water vapor in the air is one-half of the maximum the air can hold at that temperature. ✓

Dew Point

When a sponge becomes saturated with water, the water starts to drip from the sponge. Likewise, when air becomes saturated with water vapor, the water vapor condenses and forms water droplets. When air near the ground is saturated, water vapor condenses into a liquid. If the temperature is above 0°C, dew forms. If the temperature is below 0°C, ice crystals, or frost, form. Higher in the atmosphere, saturated air forms clouds.

When the temperature decreases, air can hold less moisture. Eventually, the air becomes saturated and dew forms. *The* **dew point** *is the temperature at which air becomes fully saturated because of decreasing temperatures while holding the amount of moisture constant.*

Clouds and Fog

Think about what happens when you exhale warm air on a cold day. The warm air you exhale cools. If it reaches its dew point, you can see the water vapor condense into a foggy cloud in front of your face. This also happens when warm air containing water vapor cools as it rises in the atmosphere. When the cooling air reaches its dew point, water vapor condenses on small particles in the air and forms droplets. The droplets block and reflect light. This makes them visible as clouds.

FOLDABLES®

Make a two-tab book and use it to collect information about the similarities and differences between clouds and fog.

Clouds are water droplets or ice crystals suspended in the atmosphere. Clouds can have different shapes. Clouds can form at different altitudes within the atmosphere. Read the table below that describes different types of clouds. As clouds move, water and thermal energy are transported from one location to another. Recall that clouds reflect some of the Sun's incoming radiation.

Types of Clouds

Stratus Clouds	Cumulus Clouds	Cirrus Clouds
• flat, white, and layered • altitude up to 2000 m	• fluffy, heaped, or piled up • 2,000 to 6,000 m altitude	• wispy • above 6,000 m

Visual Check

5. Illustrate Look out the window and find a cloud. Sketch the shape of the cloud in the space above. Then highlight in the table the type of cloud you drew.

A cloud that forms near Earth's surface is called fog. Fog is a suspension of water droplets or ice crystals close to or at Earth's surface. Fog reduces visibility. Visibility is the distance a person can see into the atmosphere. ✓

Reading Check

6. Describe What is fog?

____

____

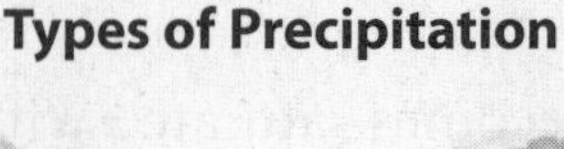

Rain

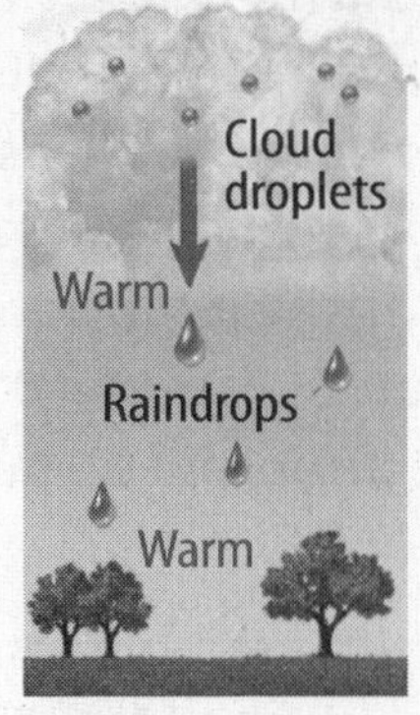

Snow

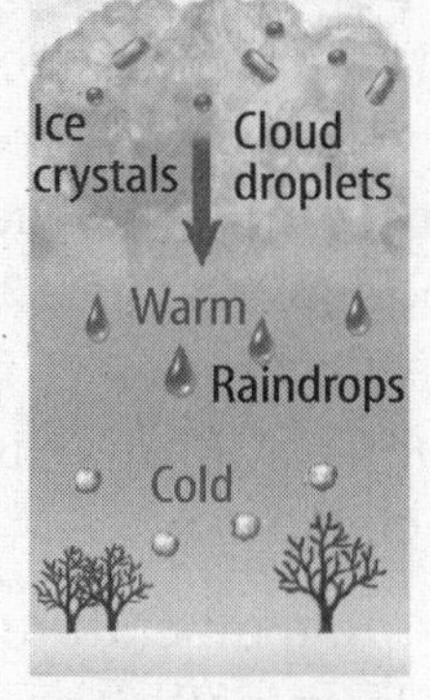

Sleet

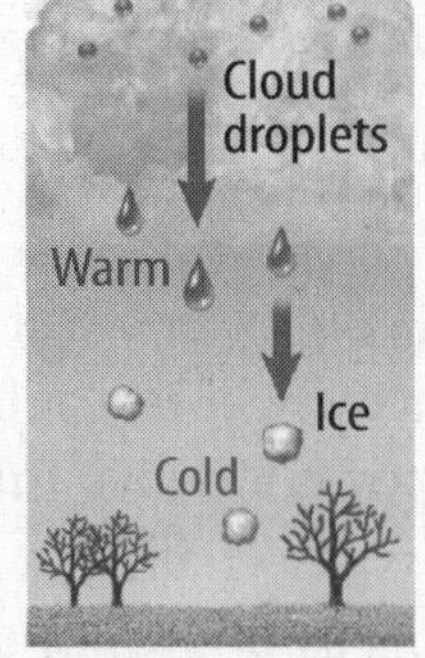

Hail

Visual Check

7. Compare What is the difference between snow and sleet?

____

____

____

Precipitation

Droplets in clouds form around small solid particles in the atmosphere. These particles might be dust, salt, or smoke. Precipitation occurs when cloud droplets combine and become large enough to fall to Earth's surface. **Precipitation** *is water in liquid or solid form that falls from the atmosphere*. Types of precipitation—rain, snow, sleet, and hail—are shown in the figure above. Rain is precipitation that reaches Earth's surface as droplets of water. Snow is precipitation that reaches Earth's surface as solid, frozen crystals of water. Sleet may start out as snow. The snow melts into rain as it passes through a layer of warm air and refreezes when it passes through a layer of below-freezing air. Other times, sleet is just freezing rain. Hail reaches Earth's surface as large ice pellets. Hail starts as a small piece of ice that is repeatedly caught in an updraft within a cloud. A layer of ice is added with each lifting. When it becomes too heavy, it falls to Earth.

Key Concept Check

8. Identify What variables are used to describe weather?

____

____

____

The Water Cycle

Precipitation is an important process in the water cycle, shown at the top of next page. Evaporation and condensation are also important processes in the water cycle. *The* **water cycle** *is the series of natural processes in which water continually moves among oceans, land, and the atmosphere.*

The Water Cycle

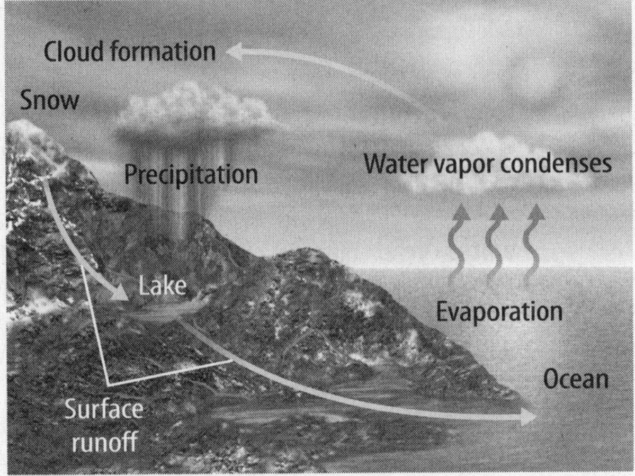

Most water vapor enters the atmosphere through evaporation. Water vapor forms as water is heated at the ocean's surface. Water vapor cools as it rises in the atmosphere. The cooled water vapor condenses back into liquid. Eventually droplets of liquid and solid water form clouds. Clouds produce precipitation. The precipitation falls to Earth's surface and later evaporates, continuing the cycle.

Visual Check

9. Identify Circle the name of the process in which liquid water changes into water vapor.

Key Concept Check

10. Describe How is weather related to the water cycle?

____

____

____

After You Read

Mini Glossary

air pressure: the pressure that a column of air exerts on the air, or surface, below it

dew point: the temperature at which air is fully saturated because of decreasing temperatures while holding the amount of moisture constant

humidity (hyew MIH duh tee): the amount of water vapor in the air

precipitation: water, in liquid or solid form, that falls from the atmosphere

relative humidity: the amount of water vapor present in the air compared to the maximum amount of water vapor the air could contain at that temperature

water cycle: the series of natural processes in which water continually moves among oceans, land, and the atmosphere

weather: the atmospheric conditions, along with short-term changes, of a certain place at a certain time

1. Review the terms and their definitions in the Mini Glossary. Write a sentence that explains how relative humidity and dew point are related.

____

____

2. Complete the graphic of the water cycle by writing the terms below in the correct order. You may begin with any process.

cloud formation **evaporation**
condensation **precipitation**

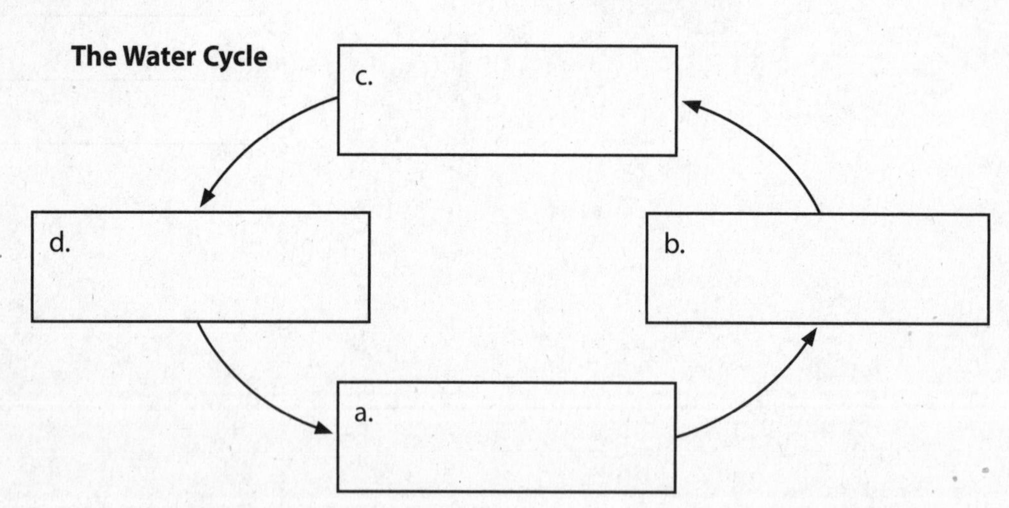

What do you think NOW?

Reread the statements at the beginning of the lesson. Fill in the After column with an A if you agree with the statement or a D if you disagree. Did you change your mind?

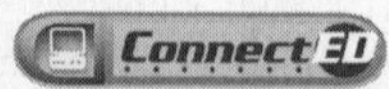

Log on to ConnectED.mcgraw-hill.com and access your textbook to find this lesson's resources.

Weather

Weather Patterns

Before You Read

What do you think? Read the two statements below and decide whether you agree or disagree with them. Place an A in the Before column if you agree with the statement or a D if you disagree. After you've read this lesson, reread the statements to see if you have changed your mind.

Before	Statement	After
	3. Precipitation often occurs at the boundaries of large air masses.	
	4. There are no safety precautions for severe weather, such as tornadoes and hurricanes.	

Key Concepts 
- What are two types of pressure systems?
- What drives weather patterns?
- Why is it useful to understand weather patterns?
- What are some examples of severe weather?

Read to Learn

Pressure Systems

Weather is often associated with pressure systems. Air pressure is the weight of the molecules in a large mass of air. Cool air molecules are closer together than warm air molecules. Cool air masses have high pressure, or more weight, than warm air masses do. Warm air masses have low pressure.

A **high-pressure system** *is a large body of circulating air with high pressure at its center and lower pressure outside of the system.* Air moves from high pressure to low pressure. Heavy, high-pressure air inside the system moves away from the center. Air moving from areas of high pressure to areas of low pressure is called wind. The dense air inside the high pressure system sinks and brings clear skies and fair weather.

A **low-pressure system** *is a large body of circulating air with low pressure at its center and higher pressure outside of the system.* Air on the outside of the system will spiral in toward the center. This causes air inside the low-pressure system to rise. The rising air cools and the water vapor condenses. Clouds form, and sometimes precipitation, such as rain or snow, also forms.

Study Coach

Learning with Graphics
Maps, diagrams, charts, and graphs can help you understand what you've read. Trace the details on each graphic with your finger after you read the description.

Key Concept Check
1. Compare and contrast two types of pressure systems.

Air Masses

Have you ever noticed that the weather sometimes stays the same for several days in a row? Air masses are responsible for this. **Air masses** *are large bodies of air with distinct temperature and moisture characteristics.* An air mass forms when a large, high-pressure system stays over an area for several days. The air circulating in the high-pressure system comes in contact with Earth. This air takes on the temperature and moisture characteristics of the surface below it.

Air masses, like high- and low-pressure systems, can extend for a thousand kilometers or more. Sometimes one air mass covers most of the United States. Air masses affect weather patterns.

FOLDABLES

Make a three-column book from a sheet of paper and record information about types of air masses.

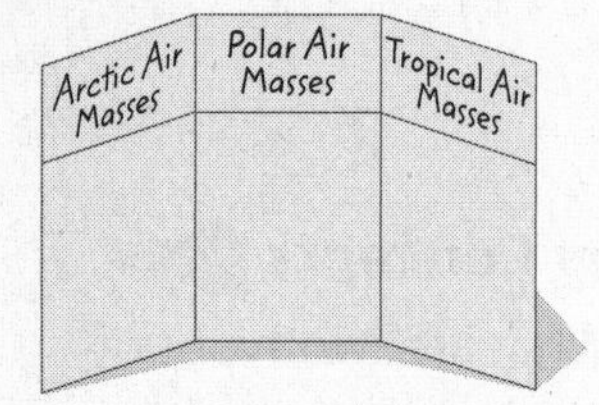

Air Mass Classification

The figure below identifies types of air masses and the regions where they form. The arrows on the map show the general paths that the air masses commonly follow. Air masses are classified by their temperature and moisture characteristics. Air masses that form over land are called continental air masses. Air masses that form over water are called maritime air masses. Air masses that form near the equator are called tropical air masses. Those air masses that form in cold regions are called polar air masses. Air masses that form near the poles are called arctic and antarctic air masses.

Air Mass Classifications

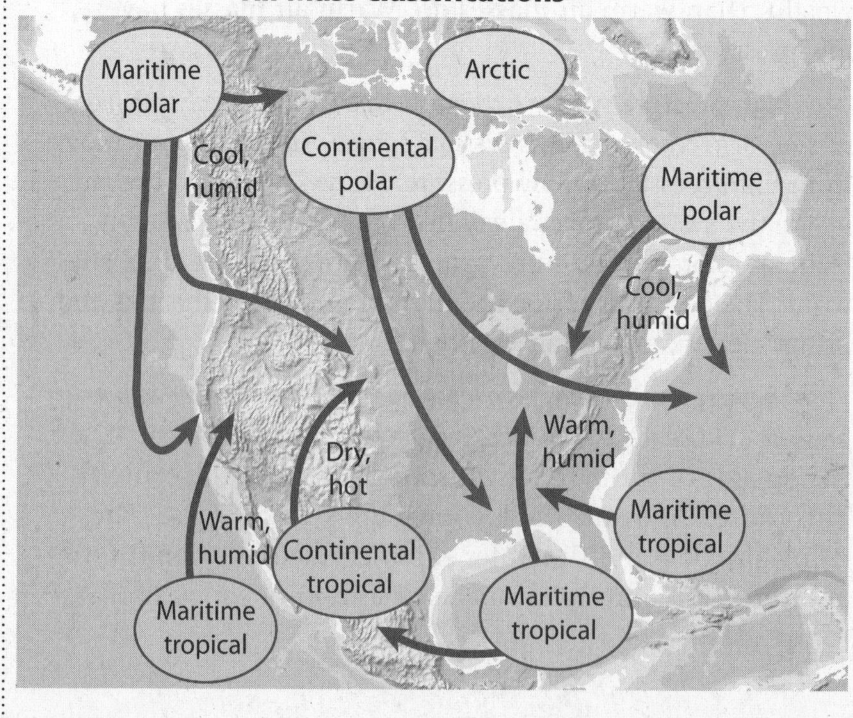

Visual Check

2. Classify Where does continental polar air come from?

____

____

Arctic Air Masses Arctic air masses form over Siberia and the Arctic. These air masses contain bitterly cold, dry air. During the winter, an arctic air mass can bring temperatures to −40°C.

Continental Polar Air Masses Land cannot transfer as much moisture to the air as oceans can. Thus, air masses that form over land are drier than air masses that form over oceans. Continental polar air masses are fast moving. They bring cold temperatures in winter and cool temperatures in summer. Polar air masses that affect North America often form over Alaska and Canada.

Maritime Polar Air Masses Air masses that form over the northern Atlantic and Pacific Oceans are maritime polar air masses. These air masses are cold and humid. Maritime polar air masses often bring cloudy, rainy weather.

Continental Tropical Air Masses Air masses forming in the tropics over dry, desert land are continental tropical air masses. These hot and dry air masses bring clear skies and high temperatures. Continental tropical air masses usually form only during summer.

Maritime Tropical Air Masses These air masses form over the Gulf of Mexico, the Caribbean Sea, and the eastern Pacific Ocean. Maritime tropical air masses are moist air masses. They bring hot, humid air to the southeastern United States in summer. In winter, they can bring heavy snowfall.

Air masses can change as they move over the land and ocean. Warm, moist air can lose its moisture and become cool. Cold, dry air can move over water and become moist and warm.

Fronts

In 1918, Norwegian Jacob Bjerknes (BYURK nuhs) and his coworkers developed a new method for forecasting the weather. Bjerknes noticed that specific types of weather occur at the boundaries between different air masses. He used the word *front,* a military term, to describe this boundary.

A military front is the boundary between opposing armies. *A weather* **front** *is the boundary between two air masses.* As wind carries an air mass away from the area where it formed, the air mass will eventually bump into another air mass. Major weather changes often occur at fronts. Changes in temperature, humidity, clouds, wind, and precipitation are common at fronts.

Math Skills

To convert Fahrenheit (°F) units to Celsius (°C) units, use this equation:

$$°C = \frac{(°F - 32)}{1.8}$$

Example: Covert 76°F to °C.

a. Always perform the operation in parentheses first.

$(76°F - 32 = 44°F)$

b. Divide the answer from Step a by 1.8.

$$\frac{(44°F)}{1.8} = 24°C$$

To convert Celsius (°C) units to Fahrenheit (°F) units, use this equation:

$$°F = (°C \times 1.8) + 32$$

3. Conversions

a. Convert 86°F to °C.

____

b. Convert 37°C to °F.

____

Key Concept Check

4. Describe What drives weather patterns?

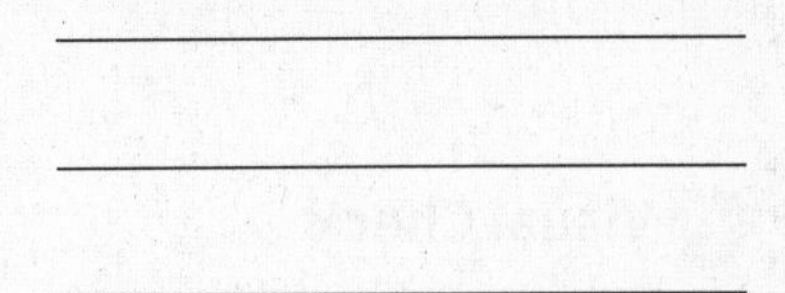

Cold Fronts

The figure below on the left shows a cold front. A cold front forms when a colder air mass moves toward a warmer air mass. Cold air is denser than warm air. As a result, the cold air pushes underneath the warm air mass. The warm air rises and begins to cool. Water vapor in the air condenses, and clouds form. Rain showers and thunderstorms often form along cold fronts. It is common for temperatures to decrease. The wind becomes gusty and changes direction. In many cases, cold fronts give rise to severe storms.

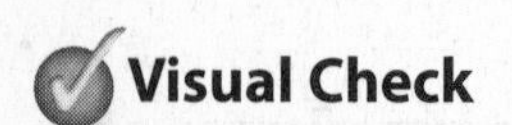

Reading Check

5. Specify What types of weather are associated with cold fronts?

______

______

______

Warm Fronts

The figure on the right shows a warm front. A warm front forms when less dense, warmer air moves toward colder, denser air. The warm air rises above the cold air mass. When the water vapor in the warm air condenses, a wide blanket of clouds forms. These clouds often bring steady rain or snow for several hours or days. A warm front brings warmer temperatures and causes the wind to shift directions.

Cold Front

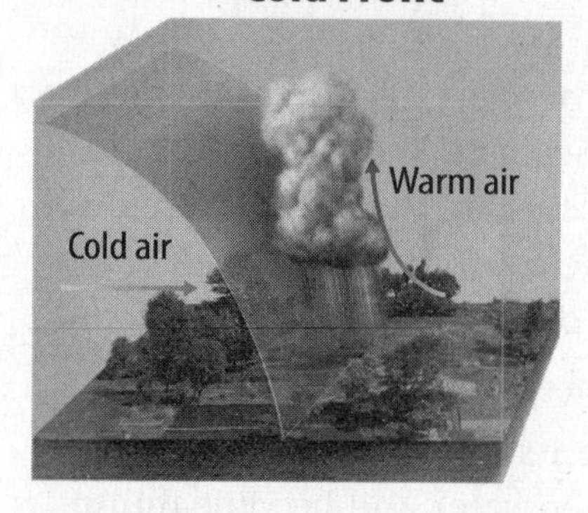

Warm Front

Visual Check

6. Describe the difference between a cold front and a warm front.

______

______

______

Stationary and Occluded Fronts

In addition to cold fronts and warm fronts, meteorologists have identified stationary fronts and occluded fronts. These two types of fronts are illustrated below and described on the next page.

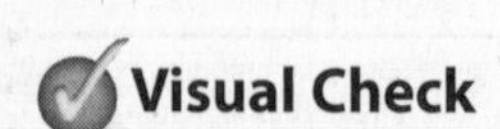

Visual Check

7. Contrast Highlight the label "Warm air" in both of the figures. How is the location of warm air different in the two types of fronts?

______

______

______

Stationary Front

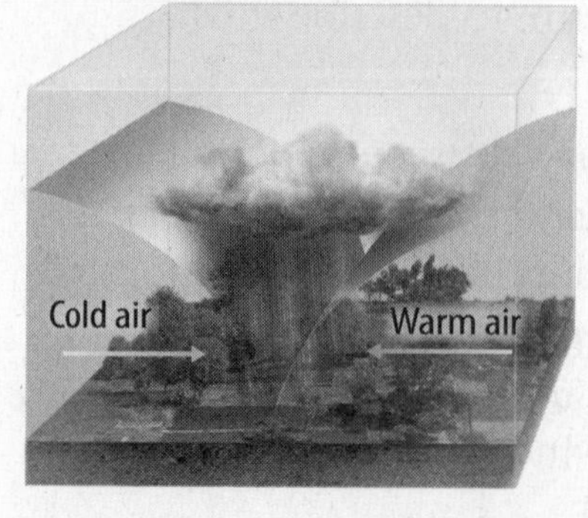

Occluded Front

Stationary Front Sometimes an approaching front stalls, or stops, for several days. Warm air is located on one side of the front and cold air on the other side. When the boundary between two air masses stalls, the front is called a stationary front. Cloudy skies and light rain are common along stationary fronts.

Occluded Front Cold fronts move faster than warm fronts. When a fast-moving cold front catches up with a slow-moving warm front, an occluded, or blocked, front forms. Occluded fronts usually bring precipitation.

Severe Weather

Severe weather can cause major damage, injuries, and death. Types of severe weather include thunderstorms, tornadoes, hurricanes, and blizzards.

Thunderstorms

Thunderstorms are also known as electrical storms because of their lightning. Thunderstorms have warm temperatures, moisture, and rising air. A low-pressure system brings these conditions. Thunderstorms can form quickly. For example, a cumulus cloud can grow into a 10-km-tall thundercloud in as little as 30 minutes.

A typical thunderstorm has three stages. During the cumulus stage, clouds form and updrafts occur. Updrafts are air currents that move vertically up from the ground. After the cumulus cloud has been created, downdrafts begin to appear. Downdrafts are air currents that move vertically down toward the ground. In the mature stage, heavy winds, rain, and lightning occur. Within 30 minutes of reaching the mature stage, the thunderstorm begins to fade, or dissipate. In the dissipation stage, updrafts stop, winds die down, lightning stops, and precipitation weakens.

Strong updrafts and downdrafts in a thunderstorm cause tiny ice crystals to crash into each other. This creates positively and negatively charged particles in the cloud. The difference between the charges of the particles in the cloud and the charges of the particles on the ground creates electricity. This electricity is seen as a bolt of lightning. Lightning can heat the nearby air to more than 27,000°C.

Lightning can move from cloud to cloud, cloud to ground, or ground to cloud. The extreme thermal energy from the lightning causes air molecules to rapidly expand and then contract. Thunder is the sound made by the rapid expansion and contractions of the air molecules.

Key Concept Check

8. Interpret Why is it useful to understand weather patterns associated with fronts?

Reading Check

9. Identify At which stage of the thunderstorm can you expect it to begin to die down? (Circle the correct answer.)

a. the cumulus stage
b. the mature stage
c. the dissipation stage

Think it Over

10. Apply What causes molecules in the air near lightning to make the sound known as thunder?

Tornadoes

A **tornado** *is a violent, whirling column of air in contact with the ground.* Most tornadoes have a diameter of several hundred meters. The largest tornadoes are more than 1,500 m in diameter.

Wind speeds within a tornado can reach more than 400 km/h. The strong, swirling wind in a tornado can send cars, trees, and houses flying through the air. Most tornadoes last only a few minutes. The more destructive ones, however, can last for several hours.

Formation of Tornadoes A tornado forms when thunderstorm updrafts begin to rotate. Swirling winds spiral downward from the base of the thunderstorm. This creates a funnel cloud. When the funnel reaches the ground, it becomes a tornado. Swirling air is invisible. The funnel cloud you see is the dirt and debris lifted by the tornado. ✓

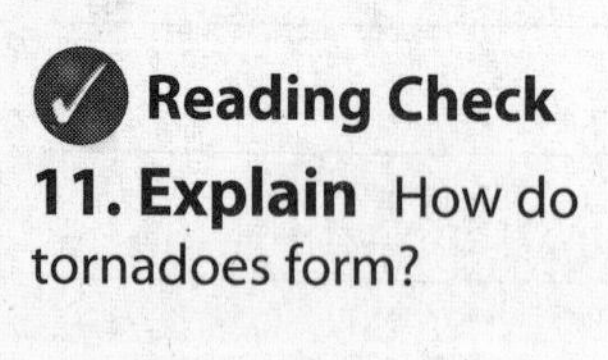

Reading Check

11. Explain How do tornadoes form?

____

____

____

Tornado Alley More tornadoes occur in the United States than anywhere else on Earth. The most tornadoes occur in an area in the central United States. This area has been named Tornado Alley. It extends from Nebraska to Texas. In Tornado Alley, cold air blowing southward from Canada often bumps into warm, moist air moving northward from the Gulf of Mexico. These conditions are ideal for severe thunderstorms and tornadoes.

Classifying Tornadoes Dr. Ted Fujita developed a system for classifying the strength of tornadoes. Tornadoes are classified on the Fujita intensity scale based on the damage they cause. F0 tornadoes cause little damage. Damage might include broken tree branches and damaged billboards. F1 through F4 tornadoes cause moderate to devastating damage. F5 tornadoes cause incredible damage. Concrete and steel buildings can be destroyed. F5 tornados can pull bark from trees.

Think it Over

12. Apply Would you expect an F4 tornado to cause more damage or less damage than an F3 tornado?

____

Hurricanes

Hurricanes are the most destructive storms on Earth. *A* **hurricane** *is an intense tropical storm with winds exceeding 119 km/h.* Hurricanes typically form in late summer over warm, tropical ocean water.

Hurricanes, like tornadoes, have strong, swirling winds. A hurricane is much larger than a tornado. A typical hurricane is 480 km across, more than 150 thousand times larger than a tornado. At the center of a hurricane is the eye. The eye is an area of clear skies and light winds.

WORD ORIGIN

hurricane
from Spanish *huracan,* means "tempest"

Hurricane Formation

1. Low-Pressure Area	2. Tropical Depression	3. Tropical Storm	4. Hurricane
Warm, moist air rises. As air rises, it cools. Water vapor condenses and clouds form. More rising air creates an area of low pressure over the ocean.	Air moves toward the low pressure in the center. The center begins to rotate. The storm becomes a tropical depression with winds of 37–62 km/h.	Air continues to rise and rotate. The storm builds to a tropical storm with winds of more than 63 km/h. The storm produces strong thunderstorms.	When winds exceed 119 km/h, the storm becomes a hurricane. Only one percent of tropical storms become hurricanes.

The figure above shows how a hurricane forms. Damage from hurricanes occurs as the result of strong winds and flooding. Hurricanes create high waves that can flood coastal areas. As a hurricane crosses the coastline, strong rains contribute to flooding that can damage or destroy entire areas. Once a hurricane moves over land or colder water, it loses energy and dies out. In Asia, this type of storm is called a typhoon. In Australia, it is called a tropical cyclone.

Visual Check

13. Identify How do hurricanes form?

____

____

____

Winter Storms

Winter weather can be severe. When temperatures are close to freezing (0°C), rain can freeze when it hits the ground. Ice storms coat the ground, trees, and buildings with a layer of ice. The weight of the ice can break trees and power lines.

A ***blizzard*** *is a violent winter storm characterized by freezing temperatures, strong winds, and blowing snow.* The blowing snow can reduce visibility to a few meters or less. Strong winds and cold temperatures can rapidly cool exposed skin. The loss of body heat can result in frostbite and hypothermia (hi poh THER mee uh), a dangerous condition in which a person's body temperature is lowered.

Key Concept Check

14. Summarize What are examples of severe weather?

____

____

____

Severe Weather Safety

The U.S. National Weather Service issues watches and warnings for severe weather. A watch means that severe weather is possible. A warning means that severe weather is already occurring. Paying attention to watches and warnings is important and could save your life.

During thunderstorms, stay inside and away from metal objects and electrical cords. If you are outside, stay away from water, high places, and trees that stand alone. When wind-chill temperatures are below −20°C, dress in layers, keep your head and fingers covered, and limit your time outdoors.

After You Read

Mini Glossary

air mass: a large body of air with distinct temperature and moisture characteristics

blizzard: a violent winter storm characterized by freezing temperatures, strong winds, and blowing snow

front: a boundary between two air masses

high-pressure system: a large body of circulating air with high pressure at its center and lower pressure outside of the system

hurricane: an intense tropical storm with winds exceeding 119 km/h

low-pressure system: a large body of circulating air with low pressure at the center and higher pressure outside of the system

tornado: a violent, whirling column of air in contact with the ground

1. Review the terms and their definitions in the Mini Glossary. Write a sentence that explains how an air mass and a front are related.

____

____

2. Use the terms below to complete the chart about air masses.

arctic
continental polar
continental tropical
maritime polar
maritime tropical

Air masses that form over . . .	are . . .
warm ocean water	(a) air masses.
warm land	(b) air masses.
cold land	(c) air masses.
the coldest areas	(d) air masses.
cold ocean water	(e) air masses.

3. Describe how tracing the details on a graphic helped you better understand one of the processes described in this lesson.

____

____

What do you think NOW?

Reread the statements at the beginning of the lesson. Fill in the After column with an A if you agree with the statement or a D if you disagree. Did you change your mind?

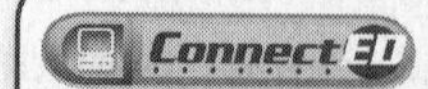

Log on to ConnectED.mcgraw-hill.com and access your textbook to find this lesson's resources.

Weather

Weather Forecasts

Before You Read

What do you think? Read the two statements below and decide whether you agree or disagree with them. Place an A in the Before column if you agree with the statement or a D if you disagree. After you've read this lesson, reread the statements to see if you have changed your mind.

Before	Statement	After
	5. Weather variables are measured every day at locations around the world.	
	6. Modern weather forecasts are done using computers.	

Key Concepts 

- What instruments are used to measure weather variables?
- How are computer models used to predict the weather?

Read to Learn

Measuring the Weather

The first step in making a weather forecast is to measure the conditions of the atmosphere. You learned in Lesson 1 that a variety of instruments is used to measure weather variables. A thermometer measures temperature. A barometer measures air pressure. A psychrometer measures relative humidity, and an anemometer measures wind speed. Meteorologists use the data from these instruments to make weather forecasts.

Study Coach

K-W-L Divide a sheet of paper into three columns. In the first column, write what you know about weather forecasts. In the second column, write what you want to know about weather forecasts. Fill in the third column with facts you learned about weather forecasts after you have read this lesson.

Surface and Upper-Air Reports

A **surface report** *describes a set of weather measurements made on Earth's surface.* Weather variables are measured by a weather station. A weather station is a collection of instruments that report temperature, air pressure, humidity, precipitation, and wind speed and direction. Cloud amounts and visibility are often measured by human observers.

An **upper-air report** *describes wind, temperature, and humidity conditions above Earth's surface.* These atmospheric conditions are measured by a radiosonde (RAY dee oh sahnd). A radiosonde is a package of weather instruments carried high above ground by a weather balloon. Radiosonde reports are made twice daily, all at the same time, at hundreds of locations around the world.

Think it Over

1. Relate How does a weather balloon assist with radiosonde reports?

______

______

______

Make a two-tab book, and then use it to collect information on satellite and radar images.

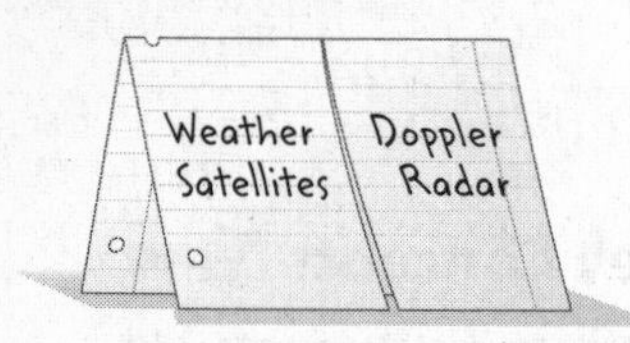

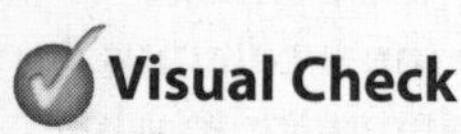

2. Identify the weather variables that radiosondes, infrared satellites, and Doppler radar measure.

Visual Check

3. Name two weather variable measurements included on the station model.

Satellite and Radar Images

Images taken from satellites orbiting at 35,000 km above Earth provide information about weather conditions on Earth. A visible light image shows white clouds over Earth. An infrared satellite image shows infrared energy in a false color. The infrared energy comes from Earth. It is stored in the atmosphere as latent heat. Monitoring infrared energy provides information about cloud height and atmospheric temperature. Meteorologists use these satellite images to identify fronts and air masses.

Radar measures precipitation when radio waves bounce off raindrops and snowflakes. **Doppler radar** *is a specialized type of radar that can detect precipitation as well as the movement of small particles, which can be used to approximate wind speed.* Wind causes the movement of precipitation. Thus, Doppler radar can be used to estimate wind speed. Measuring wind speed is especially important during severe weather, such as tornadoes or thunderstorms.

Weather Maps

Every day, thousands of surface reports, upper-air reports, and satellite and radar observations are made around the world. Meteorologists have tools that help them understand the large amount of weather data collected.

The Station Model

The figure below shows the station model. The station model diagram displays data from many different weather measurements for one location. It uses numbers and symbols, and it displays all observations from surface reports and upper-air reports.

The Station Model

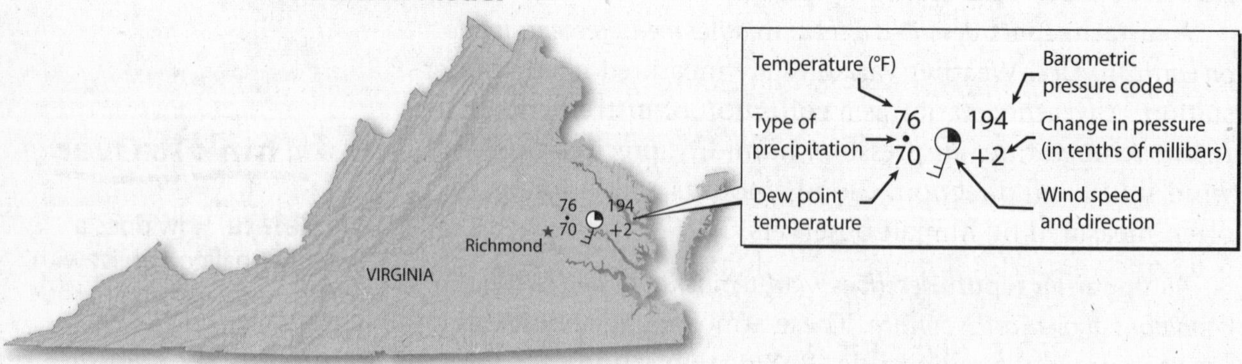

Mapping Temperature and Pressure

Weather maps use more symbols than the station model. For example, weather maps show isobars. **Isobars** *are lines that connect all places on a map where pressure has the same value.* Find an isobar on the map below. Isobars show the location of high-pressure and low-pressure systems. Isobars also give information about wind speed. Winds are strong when isobars are close together. Winds are weaker when isobars are farther apart.

In a similar way, isotherms (not shown on the map below) are lines that connect places with the same temperature. Isotherms show which areas are warm and which are cold. Lines with symbols on them show fronts. Look at the map below to see the symbols used for different types of fronts.

Reading Check

4. Compare isobars and isotherms.

Weather Map

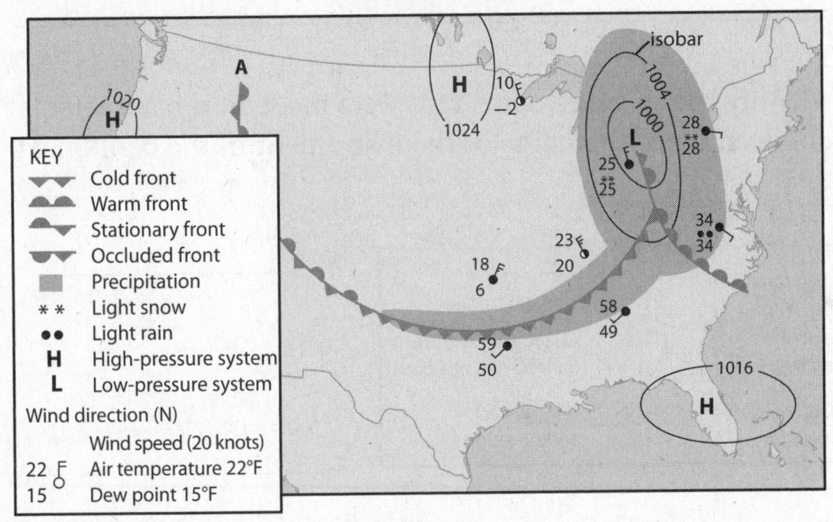

Visual Check

5. Identify Which symbols represent high-pressure and low-pressure systems?

Predicting the Weather

Meteorologists use computer models to help them forecast the weather. **Computer models** *are detailed computer programs that solve a set of complex mathematical formulas.* The formulas predict what temperatures and winds might occur, when and where it will rain or snow, and what types of clouds will form.

Government meteorological offices use computers and the Internet to share weather measurements throughout the day. Weather maps are drawn and forecasts are made using computer models. The maps and forecasts are made available to the public through television, radio, newspapers, and the Internet.

Key Concept Check

6. Describe How are computers used to predict the weather?

After You Read

Mini Glossary

computer model: a detailed computer program that solves a set of complex mathematical formulas

Doppler radar: a specialized type of radar that can detect precipitation as well as the movement of small particles, which can be used to approximate wind speed

isobar: a line that connects all places on a map where pressure has the same value

surface report: a set of weather measurements made on Earth's surface

upper-air report: a description of wind, temperature, and humidity conditions above Earth's surface

1. Review the terms and their definitions in the Mini Glossary. Write a sentence that compares and contrasts a surface report and an upper-air report.

____

____

____

2. In the station model below, identify the measurements that were made by a barometer, a thermometer, and an anemometer. Write the name of the instrument in the blank next to its measurement.

a. _______________

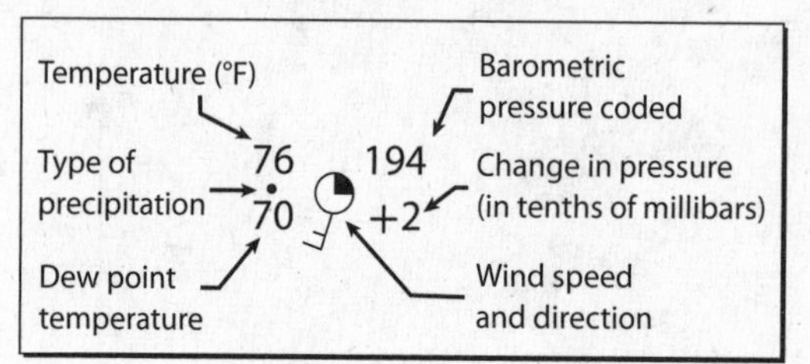

b. _______________

c. _______________

3. What is infrared energy and how does an infrared satellite image display it?

____

____

____

What do you think NOW?

Reread the statements at the beginning of the lesson. Fill in the After column with an A if you agree with the statement or a D if you disagree. Did you change your mind?

Log on to ConnectED.mcgraw-hill.com and access your textbook to find this lesson's resources.

Climate

Climates of Earth

Before You Read

What do you think? Read the two statements below and decide whether you agree or disagree with them. Place an A in the Before column if you agree with the statement or a D if you disagree. After you've read this lesson, reread the statements and see if you have changed your mind.

Before	Statement	After
	1. Locations at the center of large continents usually have the same climate as locations along the coast.	
	2. Latitude does not affect climate.	

Key Concepts 
- What is climate?
- Why is one climate different from another?
- How are climates classified?

Read to Learn

What is climate?

Weather is the atmospheric conditions and short term changes of a certain place at a certain time. The weather changes from day to day in many places on Earth. However, the weather is more constant in other places. In Antarctica, temperatures are rarely above 0°C. Areas in Africa's Sahara have temperatures above 20°C every day of the year.

Climate *is the long-term average weather conditions that occur in a particular region*. A region's climate depends on average temperature and precipitation. It also depends on how these variables change throughout the year.

Mark the Text

Summarize Write a short phrase beside each heading that summarizes the main point of the section. Use the summaries as you review the lesson.

Key Concept Check
1. Define What is climate?

What affects climate?

Several factors determine a region's climate. The latitude of a location affects climate. For example, areas close to the equator have the warmest climates. Large bodies of water, such as lakes and oceans, also influence the climate of a region. Along coastlines, weather is more constant throughout the year. Hot summers and cold winters often occur in the center of continents. The altitude of an area affects climate. Mountainous areas are often rainy and snowy. Temperatures are often higher in urban areas because the buildings and concrete retain solar energy.

FOLDABLES®

Make a layered book to organize your notes about the factors that determine a region's climate.

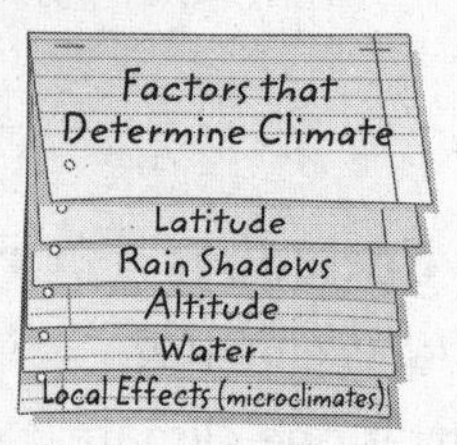

Latitude

The equator is 0° latitude. Latitude increases from 0° to 90° as you move toward the North Pole or the South Pole. The amount of solar energy per unit of Earth's surface area depends on latitude. The figure below shows that locations close to the equator receive more solar energy per unit of surface area each year than locations farther north or south. This is mainly because Earth's curved surface causes the angle of the Sun's rays to spread out over a larger area. Locations near the equator also tend to have warmer climates than locations at higher latitudes. Polar regions are colder because they receive less solar energy per unit of surface area annually. In the middle latitudes, between 30° and 60°, summers are generally hot and winters are usually cold.

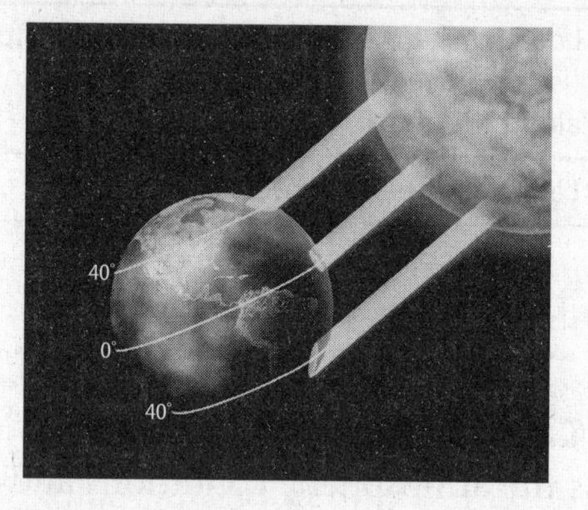

Visual Check

2. Explain why the Sun's rays are more spread out at 40° than they are at the equator.

____

____

____

Altitude

Altitude can also influence climate. Temperature decreases as altitude increases in the troposphere. So, as you climb a tall mountain, you might experience the same cold, snowy climate that is near the poles.

Rain Shadows

Mountains influence climate because they are barriers to prevailing winds. This leads to unique precipitation patterns called rain shadows. *An area of low rainfall on the downwind slope of a mountain is called a* **rain shadow.** Rain shadows form as prevailing winds carry moist, warm air over Earth's surface. As the air nears mountains, it rises and cools. Water vapor in the air condenses and rain or snow falls on the upwind slope. The dry air then passes over the mountains. The air warms as it sinks. This causes dry weather on the downwind slope.

Because there are different amounts of precipitation on either side of a mountain, there are different types and amounts of vegetation. Large quantities of vegetation grow on the side of the mountain that receives precipitation. Very little vegetation grows on the dry downwind slope.

Think it Over

3. Analyze Why don't rain shadows form on the upwind slope of a mountain?

____

____

____

Large Bodies of Water

On a sunny day at the beach, you might notice that the sand feels warmer than the water. The water is cooler than the sand on a sunny day because water has a high specific heat. **Specific heat** *is the amount (measured in joules) of thermal energy needed to raise the temperature of 1 kg of a material by 1°C.*

The specific heat of water is about six times higher than the specific heat of sand. This means that the water would have to absorb six times as much thermal energy to be the same temperature as the sand.

The high specific heat of water causes the climates along coastlines to remain more constant than climates in the middle of a continent. For example, the West Coast of the United States has moderate temperatures year-round.

Ocean currents can also affect climate. The Gulf Stream is a warm current that flows northward along the coast of eastern North America. This warm current causes temperatures to be warmer along parts of the East Coast of the United States and parts of Europe. ✓

Classifying Climates

Many factors affect climate. In 1918, Wladimir Köppen (vlah DEE mihr • KAWP pehn) developed a system for classifying the world's many climates. Köppen, a German scientist, classified a region's climate by its temperature, precipitation, and native vegetation.

Native vegetation is often limited to particular climate conditions. For example, you would not expect to find a warm-desert cactus growing in the cold, snowy Arctic. Wladimir Köppen identified five climate types. The characteristics of these five types are given below.

Climate	Temperature	Precipitation
Polar	cold year-round	little precipitation
Continental	warm summers, cold winters	moderate precipitation
Dry	hot summers, cooler winters	very little precipitation
Tropical	warm year-round	high precipitation
Mild	warm summers, mild winters	humid, with high precipitation

Reading Check

4. Summarize How do large bodies of water influence climate?

Key Concept Check

5. Explain How are climates classified?

Interpreting Tables

6. Identify the climate that has a large variation between summer and winter temperatures, and moderate precipitation. (Circle the correct answer.)

a. continental

b. tropical

c. polar

Key Concept Check
7. Discuss Why is one climate different from another?

Visual Check
8. Interpret What is the temperature difference between downtown and rural farmland?

Reading Check
9. Name one example of how an organism is adapted to a particular climate.

Microclimates

Cities have more roads and buildings made of concrete than the rural areas around them. Concrete absorbs solar radiation. This causes warmer temperatures in urban areas than in the surrounding countryside. The result is a common microclimate called an urban heat island, shown by the graph below. *A* **microclimate** *is a localized climate that is different from the climate of the larger area surrounding it.* A forest is a microclimate. Forests are often cooler and less windy than the surrounding countryside. Hilltops are microclimates as well. Hilltops are windier than nearby lower land.

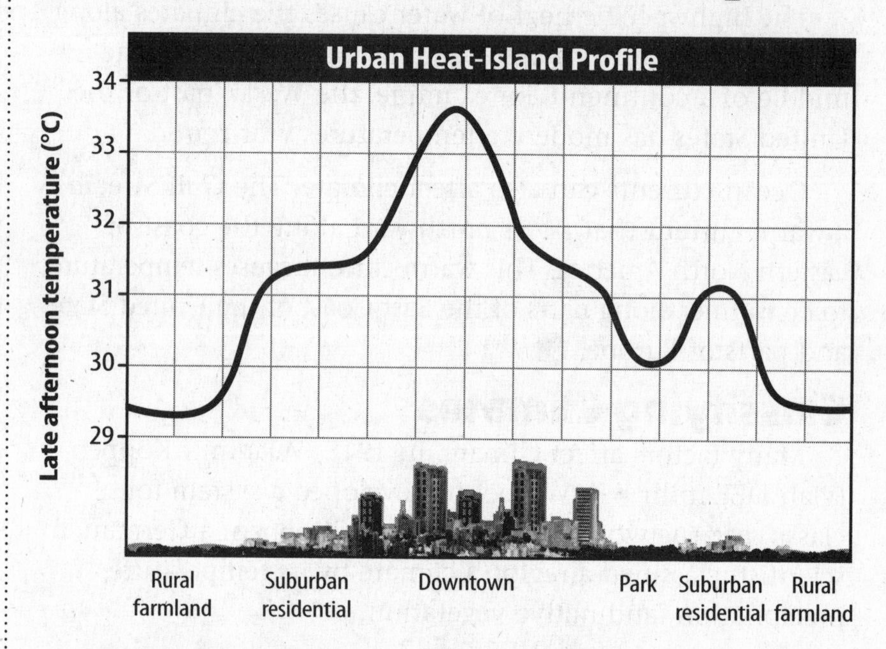

How Climate Affects Living Organisms

Organisms have adaptations for the climates where they live. Polar bears have thick fur and a layer of fat that helps keep them warm. Many animals that live in deserts, such as camels, have adaptations for hot, dry conditions. Some desert plants have large, shallow root systems that collect rainwater. Deciduous trees in continental climates lose their leaves during the winter. This reduces water loss.

Climate also influences humans in many ways. Average temperature and rainfall help determine what types of crops humans will grow in an area. Orange trees grow in Florida, where the climate is mild. Wisconsin's continental climate is good for growing cranberries.

Humans often design buildings with climate in mind. In polar climates, the soil is frozen year-round. This is a condition called permafrost. Humans build houses and other buildings on stilts so that thermal energy from the buildings does not melt the permafrost.

After You Read

Mini Glossary

climate: the long-term average weather conditions that occur in a particular region

microclimate: a localized climate that is different from the climate of the larger area surrounding it

rain shadow: an area of low rainfall on the downwind slope of a mountain

specific heat: the amount (joules) of thermal energy needed to raise the temperature of 1 kg of a material by 1°C

1. Review the terms and their definitions in the Mini Glossary. Write a sentence describing why a rain shadow forms on the downwind slope of a mountain.

____

____

2. Fill in the chart below to identify factors that can influence climate.

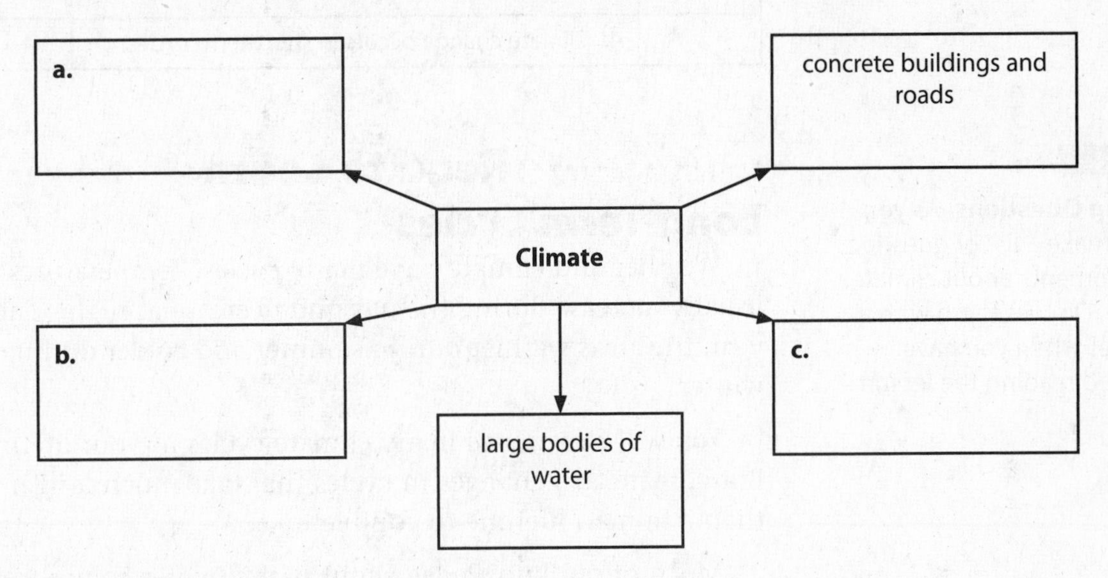

3. Explain the difference between vegetation growth on the upwind side and the rain shadow side of a mountain.

____

____

What do you think NOW?

Reread the statements at the beginning of the lesson. Fill in the After column with an A if you agree with the statement or a D if you disagree. Did you change your mind?

Log on to ConnectED.mcgraw-hill.com and access your textbook to find this lesson's resources.

Climate

Climate Cycles

Key Concepts
- How has climate varied over time?
- What causes seasons?
- How does the ocean affect climate?

Before You Read

What do you think? Read the two statements below and decide whether you agree or disagree with them. Place an A in the Before column if you agree with the statement or a D if you disagree. After you've read this lesson, reread the statements and see if you have changed your mind.

Before	Statement	After
	3. Climate on Earth today is the same as it has been in the past.	
	4. Climate change occurs in short-term cycles.	

Study Coach

Asking Questions As you read, make a list of questions or comments about climate cycles. Discuss them with a partner when you have finished reading the lesson.

Read to Learn

Long-Term Cycles

Weather and climate have many cycles. Temperatures usually increase during the day and decrease at night. Each year, the air is warmer during summer and colder during winter.

You will experience many climate cycles in your lifetime. But climate also changes in cycles that take much longer than a human lifetime to complete.

Much of our knowledge about past climates comes from natural records of climate. Scientists study ice cores drilled from ice layers in glaciers and ice sheets to gain information about past climate changes. They also study fossilized pollen, ocean sediments, and the growth rings of trees to learn about climate changes. Scientists use the information to compare present-day climates to those that occurred many thousands of years ago.

Reading Check

1. Discuss how scientists find information about past climates on Earth.

Ice Ages and Interglacials

Many major atmospheric and climate changes have occurred during Earth's history. **Ice ages** *are cold periods lasting from hundreds to millions of years when glaciers cover much of Earth.* Glaciers and ice sheets advance during cold periods and retreat during interglacials. **Interglacials** *are warm periods that occur during ice ages.*

Major Ice Ages and Warm Periods

The most recent ice age began about 2 million years ago. Much of Earth was covered with ice about 20,000 years ago. Then the ice sheets started to shrink. About 10,000 years ago, the current interglacial period, called the Holocene Epoch, began.

Temperatures on Earth have varied during the Holocene. For example, the period between 950 and 1100 was one of the warmest in Europe. The Little Ice Age, which lasted from 1250 to about 1850, was bitterly cold.

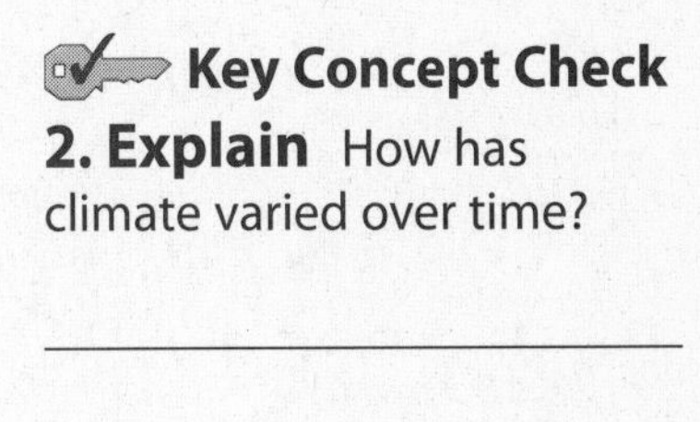

Key Concept Check

2. Explain How has climate varied over time?

Causes of Long-Term Climate Cycles

The amount of solar energy reaching Earth changes over time. Earth's climate changes with changes in the amount of solar energy that Earth receives.

The shape of Earth's orbit affects the amount of solar energy that Earth receives. Earth's orbit varies between an elliptical and circular shape over the course of about 100,000 years. The figure below shows the shapes of these two orbits. When Earth's orbit is more circular, Earth averages a greater distance from the Sun. This results in below-average temperatures on Earth.

Visual Check

3. Circle the location in the diagram where Earth would be farthest from the Sun.

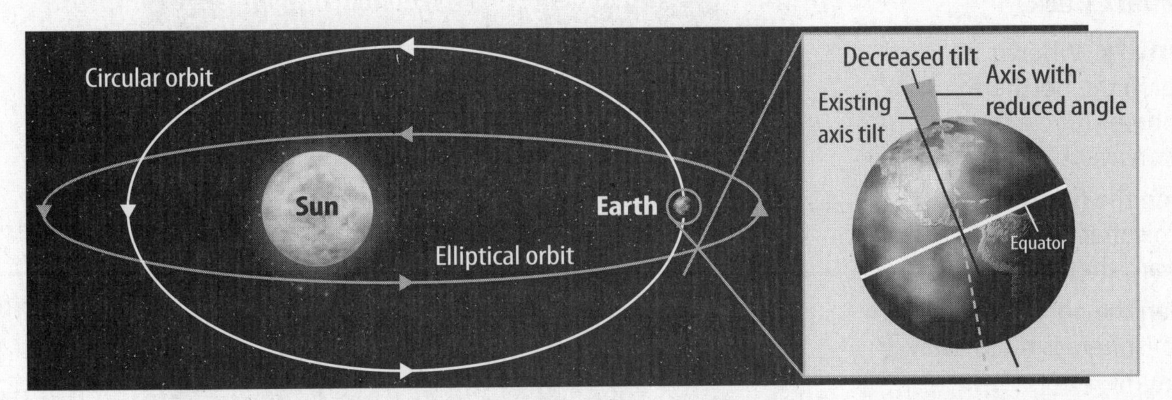

Another factor that scientists suspect influences climate change is changes in the tilt of Earth's axis. The tilt of Earth's axis changes in 41,000-year cycles. Currently, Earth is tilted at an angle of 23.5°. The tilt angle varies from 22° to 24.5°.

The angle of Earth's tilt affects the range of temperatures throughout the year. For example, a decrease in the tilt angle, as shown in the figure above, could result in a decrease in temperature differences between summer and winter. The slow movement of Earth's continents and changes in ocean circulation also affect Earth's long-term climate cycles.

Reading Check

4. State two factors that affect long-term climate change.

FOLDABLES®

Make a three-tab book to organize information about short-term climate cycles.

Short-term Climate cycles
Seasons | ENSO | Monsoons

Short-Term Cycles

Climate also changes in short-term cycles. Seasonal changes are the most common short-term cycle. Changes that result from the interaction between the ocean and the atmosphere are also short-term climate changes.

Seasons

Seasons occur because the amount of solar energy that Earth receives at different latitudes changes during the year. Seasonal changes include regular changes in temperature and the number of hours of daylight.

The amount of solar energy per unit of Earth's surface is related to latitude. Earth's tilt on its axis also affects the amount of solar energy an area receives, as shown below.

When the northern hemisphere is tilted toward the Sun, it receives more direct solar energy. There are more daylight hours than dark hours, and temperatures are warmer. It is summer in the northern hemisphere. During this time, the southern hemisphere receives less overall solar energy. It is winter there.

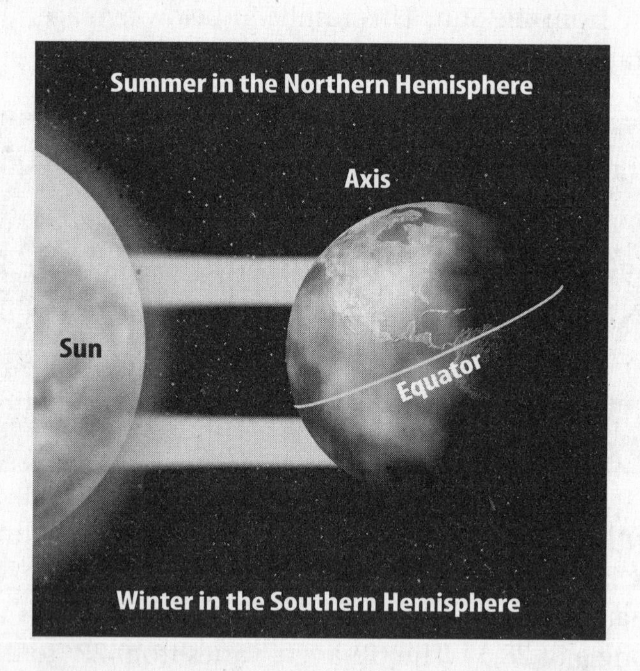

Visual Check

5. Identify When is it summer in the northern hemisphere? (Circle the correct answer.)

a. when the northern hemisphere is tilted toward the Sun

b. when the northern hemisphere is tilted away from the Sun

c. when Earth is not tilted toward or away from the Sun

The opposite occurs in six months when the northern hemisphere is tilted away from the Sun. This is shown in the figure on the top of the next page. There are fewer daylight hours than nighttime hours. Temperatures are colder. The solar energy that reaches the northern hemisphere is indirect and less intense. It is winter in the northern hemisphere. The southern hemisphere receives more direct solar energy, and it is summer there.

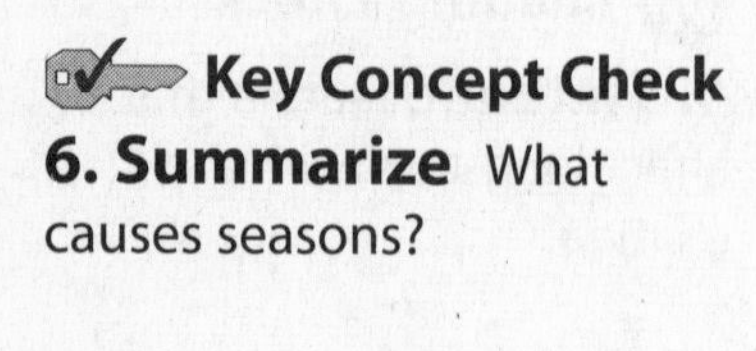

Key Concept Check

6. Summarize What causes seasons?

____

____

____

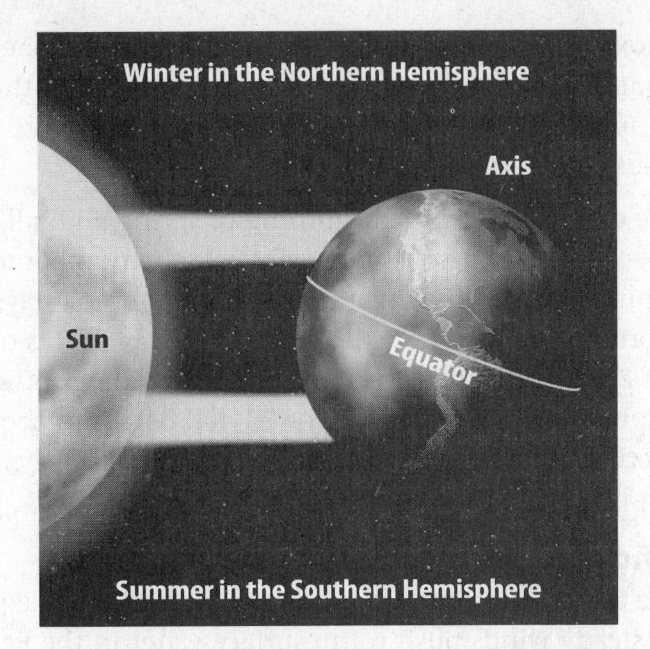

Visual Check

7. Describe the tilt of Earth on its axis when it is winter in the northern hemisphere.

Solstices and Equinoxes

Earth revolves around the Sun once about every 365 days. During Earth's revolution, there are four days that mark the beginning of each of the seasons. These days are a summer solstice, a fall equinox, a winter solstice, and a spring equinox. These days are shown in the figure below.

SCIENCE USE V. COMMON USE
revolution
Science Use the action by a celestial body of going around in an orbit or elliptical path

Common Use a sudden, radical, or complete change

Solstices The solstices mark the beginnings of summer and winter. In the northern hemisphere, the summer solstice occurs on June 21 or 22. On this day, the northern hemisphere is tilted toward the Sun. In the southern hemisphere, this day marks the beginning of winter.

The winter solstice begins on December 21 or 22 in the northern hemisphere. On this day, the northern hemisphere is titled away from the Sun. Summer begins in the southern hemisphere.

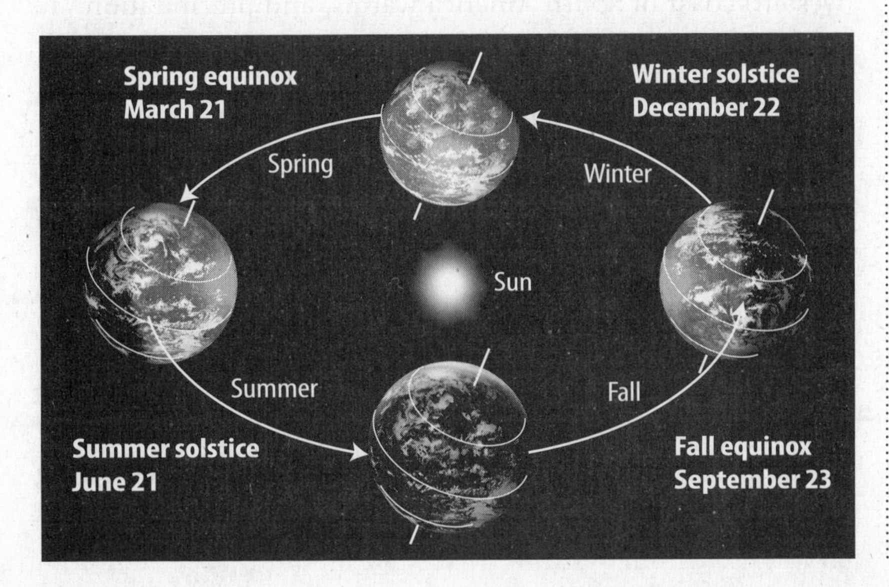

Visual Check

8. Discover How does the amount of sunlight striking the North Pole change from summer to winter?

Equinoxes Days when neither the northern hemisphere nor the southern hemisphere tilts toward or away from the Sun are equinoxes. The equinoxes are shown in the figure on the previous page.

The equinoxes are the beginning of spring and fall. The number of daylight hours almost equals the number of nighttime hours everywhere on Earth on equinox days. In the northern hemisphere, the spring equinox occurs on March 21 or 22. Fall begins the same day in the southern hemisphere. On September 22 or 23, fall begins in the northern hemisphere and spring begins in the southern hemisphere. ✓

✓ Reading Check

9. Compare and contrast solstices and equinoxes.

____

____

El Niño and the Southern Oscillation

The trade winds blow from east to west near the equator. These steady winds push warm surface water in the Pacific Ocean away from the western coast of South America. This allows cold water to rush upward from below in a process called upwelling. The air above the cold, upwelling water cools and sinks, creating a high-pressure area. On the other side of the Pacific Ocean, air rises over the warm waters around the equator creating a low-pressure area. This difference in air pressures across the Pacific Ocean helps keep the trade winds blowing.

Sometimes the trade winds weaken. This reverses the normal pattern of high and low pressures across the Pacific Ocean. Warm water surges back toward South America, preventing cold water from upwelling. This phenomenon is called El Niño.

ACADEMIC VOCABULARY

phenomenon
(noun) an observable fact or event

El Niño shows the connection between the atmosphere and the ocean. During El Niño, the normally dry, cool western coast of South America warms, and precipitation increases. Climate changes can be seen around the world. Droughts occur in areas that are normally wet. The number of violent storms in California and the southern United States increases. ✓

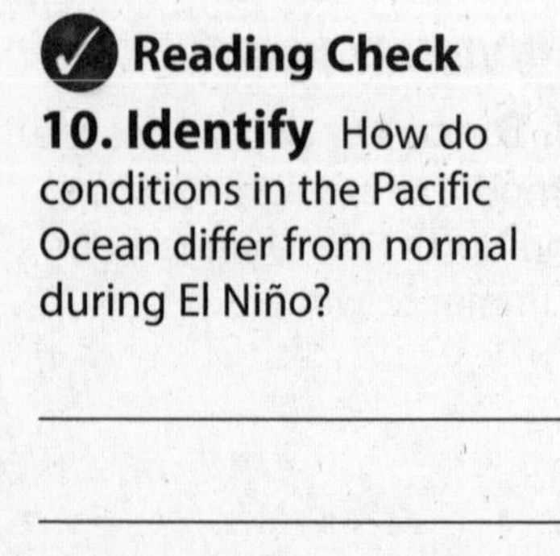

✓ Reading Check

10. Identify How do conditions in the Pacific Ocean differ from normal during El Niño?

____

____

____

The combined ocean and atmospheric cycle that results in weakened trade winds across the Pacific Ocean is called **El Niño/Southern Oscillation**, or ENSO. A complete ENSO cycle occurs every 3–8 years.

The North Atlantic Oscillation (NAO) is another cycle that can change the climate for long periods. The NAO affects the strength of storms in North America and Europe by changing the position of the jet stream.

Monsoons

Another climate cycle involving the atmosphere and the ocean is a monsoon. *A* **monsoon** *is a wind circulation pattern that changes direction with the seasons.* Temperature differences between the ocean and the land cause winds. During summer, warm air over land rises and creates low pressure. Cooler, heavier air sinks over the water, creating high pressure. The winds blow from the water toward the land, bringing heavy rains. You can see this in the figure to the left below. During the winter, the pattern reverses. The winds blow from the land toward the water, as shown in the figure on the right.

The world's largest monsoon is in Asia. Cherrapunji, India, is one of the world's wettest locations. It receives an average of 10 m of monsoon rainfall each year. Even more rain falls during El Niño events.

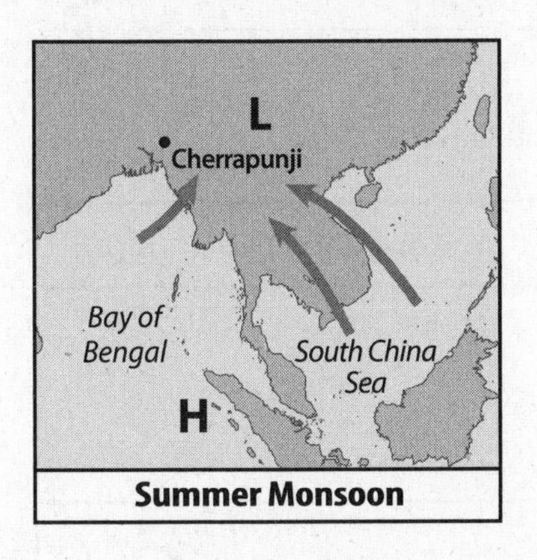

Summer Monsoon

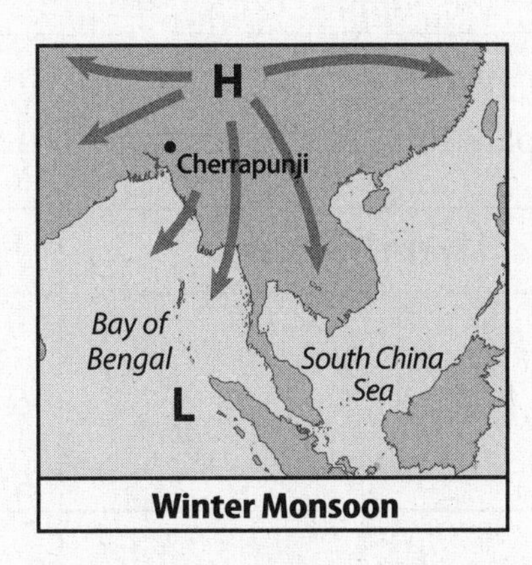

Winter Monsoon

A smaller monsoon occurs in southern Arizona. As a result, weather is dry during spring and early summer. Thunderstorms occur more often from July to September.

Droughts, Heat Waves, and Cold Waves

A **drought** *is a period with below-average precipitation.* Droughts can cause crop damage and water shortages.

Droughts often occur at the same time as heat waves, which are periods of unusually high temperatures. Droughts and heat waves occur when large hot-air masses remain in one place for weeks or months.

Cold waves are long periods of unusually cold temperatures. A cold wave occurs when a large continental polar air mass stays over a region for days or weeks. Severe weather of the types discussed can be the result of climatic changes on Earth or extremes in the average weather of a climate.

Key Concept Check

11. Analyze How does the ocean affect climate?

Visual Check

12. Locate Where is the area of high pressure during a winter monsoon?

Reading Check

13. State two things that can result from a period of drought.

After You Read

Mini Glossary

drought: a period with below-average precipitation

El Niño/Southern Oscillation: the combined ocean and atmospheric cycle that results in weakened trade winds across the Pacific Ocean

ice age: a cold period lasting from hundreds to millions of years when glaciers cover much of Earth

interglacial: a warm period that occurs during ice ages

monsoon: a wind circulation pattern that changes direction with the seasons

1. Review the terms and their definitions in the Mini Glossary. Write a sentence comparing an ice age to an interglacial.

____

____

2. Fill in the table below to compare long-term and short-term climate cycles.

	Short-Term Cycle	Long-Term Cycle
Examples		ice ages
Causes		
Results	seasonal variations, abnormal weather	

3. What phenomenon occurs when trade winds in the Pacific Ocean grow weak?

____

What do you think NOW?

Reread the statements at the beginning of the lesson. Fill in the After column with an A if you agree with the statement or a D if you disagree. Did you change your mind?

Log on to ConnectED.mcgraw-hill.com and access your textbook to find this lesson's resources.

Climate

Recent Climate Change

Before You Read

What do you think? Read the two statements below and decide whether you agree or disagree with them. Place an A in the Before column if you agree with the statement or a D if you disagree. After you've read this lesson, reread the statements and see if you have changed your mind.

Before	Statement	After
	5. Human activities can impact climate.	
	6. You can help reduce the amount of greenhouse gases released into the atmosphere.	

Key Concepts 

- How can human activities affect climate?
- How are predictions for future climate change made?

Read to Learn

Regional and Global Climate Change

Average temperatures on Earth have been increasing for the past 100 years. The graph below shows that the warming has not been steady.

Since about 1975, global average temperatures have steadily increased. The greatest warming has been in the northern hemisphere. However, temperatures have been steady in some areas of the southern hemisphere. Parts of Antarctica have cooled.

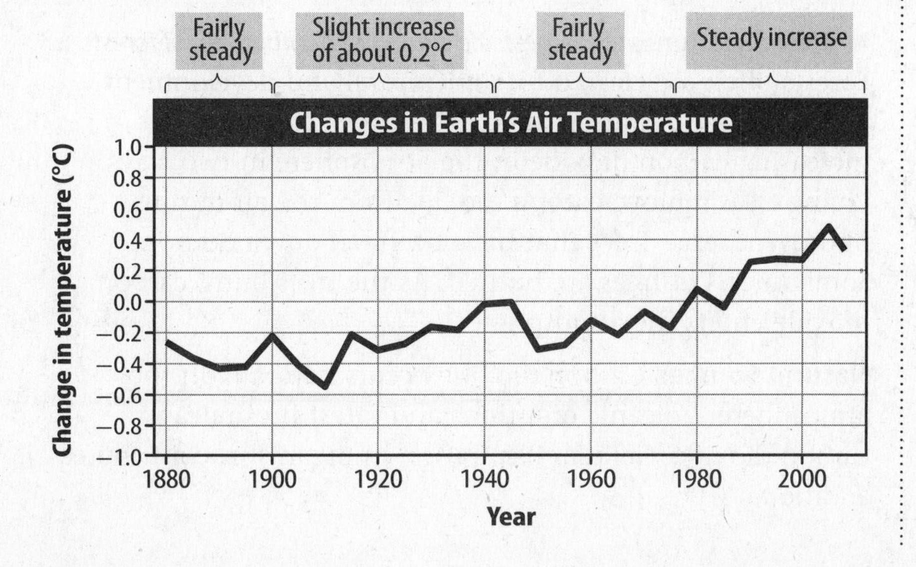

Study Coach

Organize Your Notes Make a table with two columns. As you read, write the main idea about each heading in the left column. List the details that support the main idea in the right column.

Visual Check

1. Identify What 20-year period shows the most change?

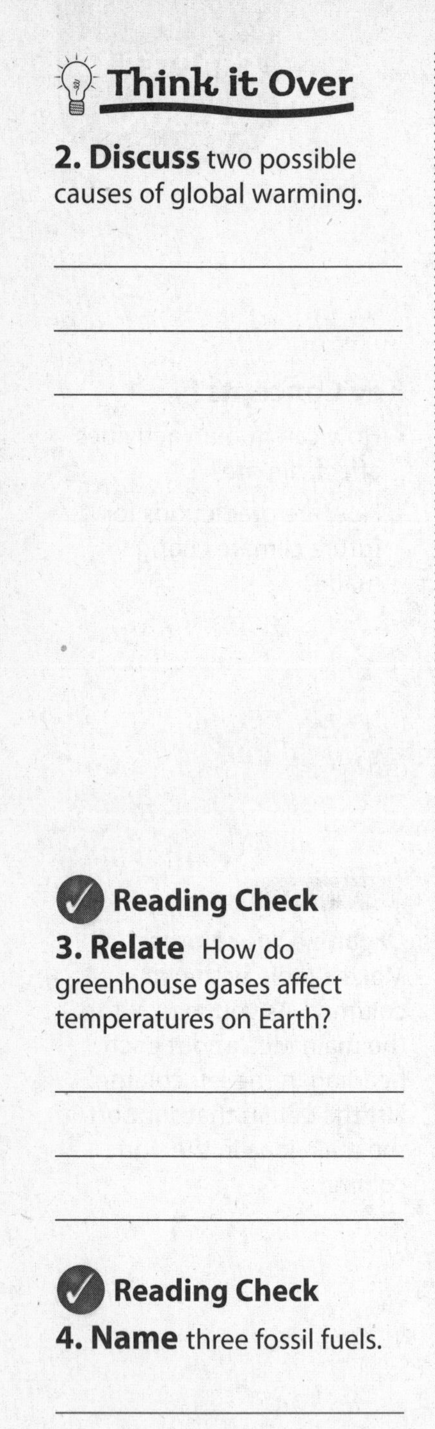

Think it Over

2. Discuss two possible causes of global warming.

Reading Check

3. Relate How do greenhouse gases affect temperatures on Earth?

Reading Check

4. Name three fossil fuels.

Human Impact on Climate Change

The rise in Earth's average surface temperature during the past 100 years is often referred to as **global warming.** Scientists have studied the possible causes of global warming. The Intergovernmental Panel on Climate Change (IPCC) is an international organization created to study global warming. In 2007, the IPCC concluded that most of this temperature increase is a result of human activities. These acitivities include the release of increasing amounts of greenhouse gases into the atmosphere through the burning of fossil fuels and the cutting and burning of forests. Although many scientists agree with the IPCC, some scientists propose that global warming is a result of natural climate cycles.

Greenhouse Gases

Gases in the atmosphere that absorb Earth's outgoing infrared radiation are **greenhouse gases.** Greenhouse gases help keep temperatures on Earth warm enough for living things to survive. This phenomenon is called the greenhouse effect. Without greenhouse gases, the average temperature on Earth would be much colder, about −18°C. Carbon dioxide (CO_2), methane, and water vapor are all greenhouse gases.

Levels of CO_2 in the atmosphere have been increasing for the past 120 years. Higher levels of greenhouse gases create a greater greenhouse effect. Most scientists suggest that global warming is a result of the greater greenhouse effect. What are the sources of excess CO_2?

Human-Caused Sources Coal, oil, and natural gases are fossil fuels. Carbon dioxide enters the atmosphere when fossil fuels burn. Burning fossil fuels releases energy that provides electricity, heats homes and buildings, and powers automobiles.

Deforestation *is the large-scale cutting and/or burning of forests.* Forests often are cleared for agricultural and development purposes. Deforestation affects the global climate by increasing carbon dioxide in the atmosphere in two ways. Living trees remove carbon dioxide from the air during photosynthesis. Trees that have been cut down do not. Sometimes, cut trees are burned. As the trees burn, carbon dioxide enters the atmosphere.

Natural Sources Carbon dioxide occurs naturally in the atmosphere. Volcanic eruptions and forest fires release carbon dioxide. Cellular respiration in organisms contributes additional CO_2.

Aerosols

The burning of fossil fuels also releases aerosols into the atmosphere. Aerosols are tiny liquid or solid particles. They reflect sunlight back into space. This prevents some of the Sun's energy from reaching Earth. Aerosols can cool the climate over time.

Clouds that form in areas with large amounts of aerosols have smaller cloud droplets. Clouds with small droplets reflect more sunlight than clouds with larger droplets. Clouds with small droplets prevent sunlight from reaching Earth's surface, helping to cool the climate.

Climate and Society

A changing climate can present serious problems for society. Heat waves and droughts can cause food and water shortages. Too much rainfall can cause flooding and mudslides.

Climate change can also benefit society. Warmer temperatures can mean longer growing seasons. Food crops can grow in areas that were previously too cold. Governments around the world are responding to the problems and opportunities created by climate change.

Environmental Impacts of Climate Change

Warmer temperatures can cause more water to evaporate from the ocean surface. The increased water vapor in the atmosphere has resulted in heavy rainfall and frequent storms in parts of North America. Precipitation has decreased over parts of southern Africa, the Mediterranean, and southern Asia.

Increasing temperatures can affect the environment in many ways. As glaciers and polar ice sheets melt, the sea level rises.

Ecosystems can be disturbed as coastal areas flood. Coastal flooding is a serious concern for the people living in low-lying areas on Earth.

Extreme weather events are becoming more common. What effects will heat waves, droughts, and heavy rainfall have on infectious disease, existing plants and animals, and other systems of nature? Will increased CO_2 levels have the same effects?

Permanently higher temperatures can have worldwide effects. Previously frozen soils are beginning to thaw and then refreeze. Other ecosystem changes can affect the migration patterns of insects, birds, fish, and mammals.

Reading Check
5. Define aerosols.

Key Concept Check
6. Evaluate How can human activities affect climate?

FOLDABLES®

Make a tri-fold book and use it to organize your notes about climate change and its possible causes.

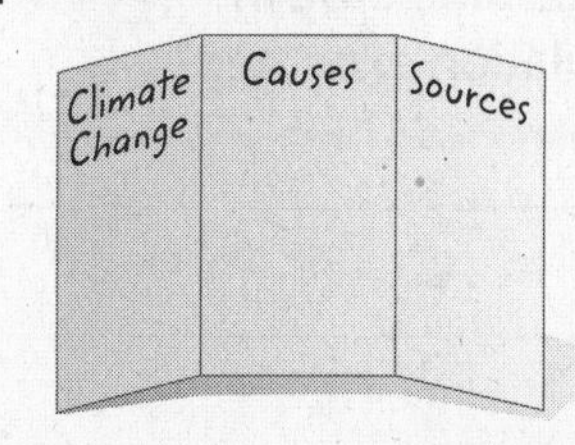

Predicting Climate Change

Weather forecasts help people make daily choices about their clothing and activities. In a similar way, climate forecasts help governments plan how to respond to future climate changes.

A **global climate model,** *or GCM, is a set of complex equations used to predict future climates.* While weather forecasts are short-term and regional, GCMs are long-term and global. GCMs use mathematics and physics to predict temperatures, amounts of precipitation, wind speeds, and other characteristics of climates. Supercomputers solve the mathematical equations and display the results as maps. ✓

Reading Check
7. Explain what a GCM is.

GCMs Predictions The forecasts and predictions given by GCMs cannot be immediately compared to real data. This is a drawback of GCMs. Weather forecasters can compare their model's predictions with the actual weather results. Then they can adjust their models to be more accurate. GCMs predict climate conditions for several decades into the future. For this reason, the accuracy of climate models is not yet known.

GCMs and Global Warming Most GCMs predict further global warming as a result of greenhouse gas emissions. By the year 2100, temperatures are expected to rise between 1°C and 4°C. Polar regions are expected to warm more than the tropics. Summer arctic sea ice is predicted to disappear by the end of the twenty-first century. Global warming and the rise in sea level are expected to continue for several centuries.

Key Concept Check
8. Describe How are predictions for future climate change made?

Human Population

In 2000, more than 6 billion people lived on Earth. The graph shows that Earth's population is expected to increase to 9 billion by the year 2050.

By the year 2030, it is likely that two of every three people on Earth will live in urban areas. Many of these areas will be in developing countries in Africa and Asia. Large areas of forests are already being cleared for expanding cities. More greenhouse gases and other air pollutants will be added to the atmosphere.

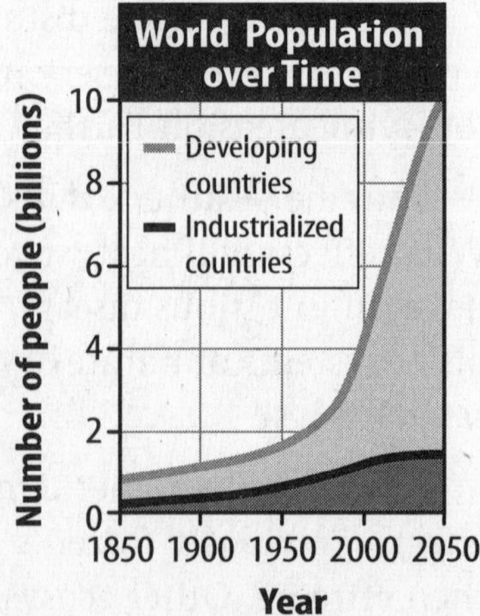

Visual Check
9. Analyze Where is the greatest increase in population expected?

Ways to Reduce Greenhouse Gases

There are many ways to reduce levels of pollution and greenhouse gases. People are developing alternative sources of energy that do not release carbon dioxide into the atmosphere. Solar energy and wind energy are two alternatives. Hybrid vehicles reduce greenhouse gas emissions because they use electric motors part of the time. This reduces the amount of fuel that is used.

Green Building Emissions of greenhouse gases can be reduced by green building. Green building is the practice of creating energy-efficient buildings. Planting trees in deforested areas is another example of a green activity that helps remove carbon dioxide from the atmosphere.

Conservation and Recycling You can help control greenhouse gases and pollution by conserving fuel and recycling. The table below shows the average number of watts of electricity used by common household appliances. Turning off lights and electronic equipment that you are not using reduces the amount of electricity you use. Recycling metal, paper, plastic, and glass reduces the amount of fuel required to manufacture these materials.

Average Amount of Electricity Used by Common Household Appliances

Appliance	Average Electricity Used per Hour	Appliance	Average Electricity Used per Hour
Standard lightbulb	100	Oven	1,300
Stereo	100	Air conditioner	1,500
Television	230	Hair dryer	1,500
Washing machine	250	Microwave	1,500
Vacuum cleaner	750	Clothes dryer	4,000
Dishwasher	1,000	Freezer	5,100
Toaster	1,200	Refrigerator/freezer	6,000

Math Skills

If Earth's population increases from 6 billion to 9 billion, what percent is the increase?

a. Subtract the initial value from the final value:

9 billion − 6 billion = 3 billion

b. Divide the difference by the starting value:

$\frac{3}{6} = 0.50$

c. Multiply by 100 and add a percent sign:

$0.50 \times 100 = 50\%$

10. Use Percents If a climate's mean temperature changes from 18.2°C to 18.6°C, what is the percentage of increase?

____

Think it Over

11. Explain what changes you could make to reduce greenhouse gases.

____

____

____

Visual Check

12. Estimate, using the chart, the average number of watts per hour you use when you study.

____

After You Read

Mini Glossary

deforestation: the large-scale cutting and/or burning of forests

global climate model (GCM): a set of complex equations used to predict future climates

global warming: the rise in Earth's average surface temperature during the past 100 years

greenhouse gas: a gas in Earth's atmosphere that absorbs Earth's outgoing infrared radiation

1. Review the terms and their definitions in the Mini Glossary. Describe the connection between greenhouse gases and global warming.

____

____

2. Fill in the chart below to identify two ways that humans bring about changes in climate. Then describe actions that humans can take to limit the changes.

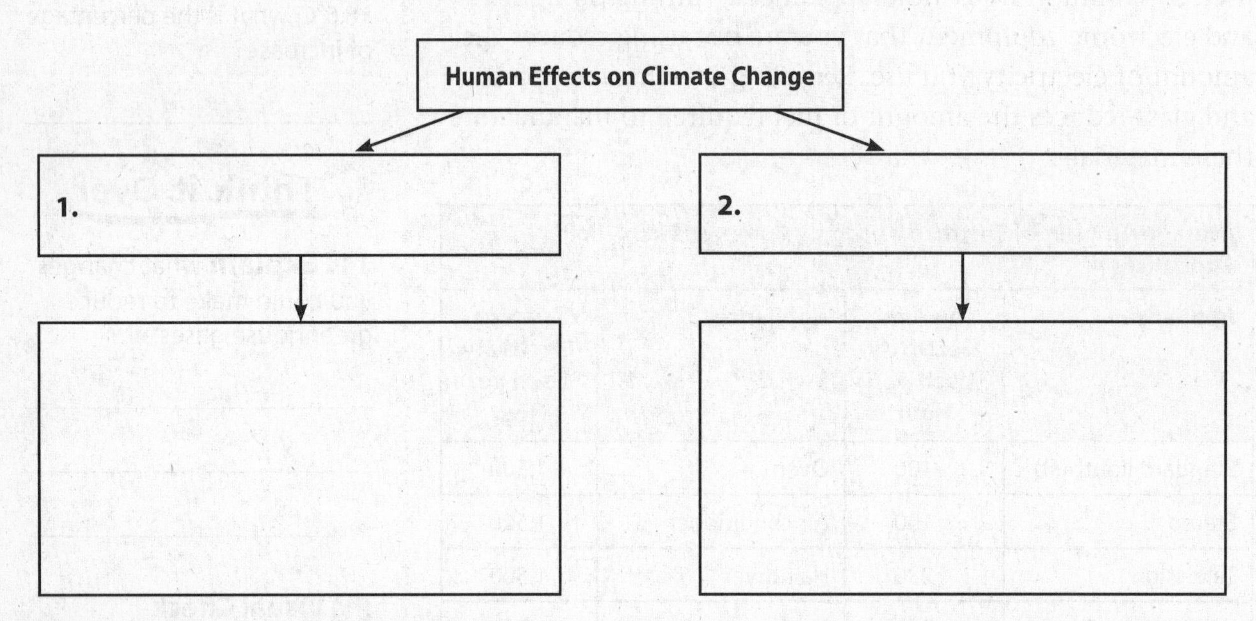

3. How did organizing your notes into two columns help you understand recent changes in climate?

____

____

What do you think NOW?

Reread the statements at the beginning of the lesson. Fill in the After column with an A if you agree with the statement or a D if you disagree. Did you change your mind?

Log on to ConnectED.mcgraw-hill.com and access your textbook to find this lesson's resources.

Motion, Forces, and Newton's Laws

Describing Motion

Before You Read

What do you think? Read the two statements below and decide whether you agree or disagree with them. Place an A in the Before column if you agree with the statement or a D if you disagree. After you've read this lesson, reread the statements to see if you have changed your mind.

Before	Statement	After
	1. You must use a reference point to describe an object's motion.	
	2. An object that is accelerating must be speeding up.	

Key Concepts 

- What information do you need to describe the motion of an object?
- How are speed, velocity, and acceleration related?
- How can a graph help you understand the motion of an object?

Read to Learn

Motion

Suppose you have been playing a shuffleboard game in an arcade. You decide you want to try something new, so you walk to a racing game. As you walk to the new game, your position in the room changes. **Motion** *is the process of changing position*. If the games are 5 m apart, your motion changed your position by a distance of 5 m.

Study Coach

Building Vocabulary Make a vocabulary card for each bold term in this lesson. Write each term on one side of the card. On the other side, write the definition. Use these terms to review the vocabulary for the lesson.

Motion and Reference Points

How would you describe your motion to someone else? You could say you walked 5 m away from the shuffleboard game. Or, you could say that you moved 5 m toward the racing game. *The starting point you use to describe the motion or position of an object is called the* **reference point.**

You describe motion differently depending on the reference point you choose. You can choose any point as a reference point.

In addition to using a reference point to describe motion, you also need a direction. Identifying whether you move toward or away from a reference point describes a direction. Other descriptions of direction might include east or west, or up or down. When raising your hand, you could use a part of your body as a reference point. ✓

Reading Check

1. Describe your motion as you walk from your desk to the door. Use a reference point and a direction.

Distance and Displacement

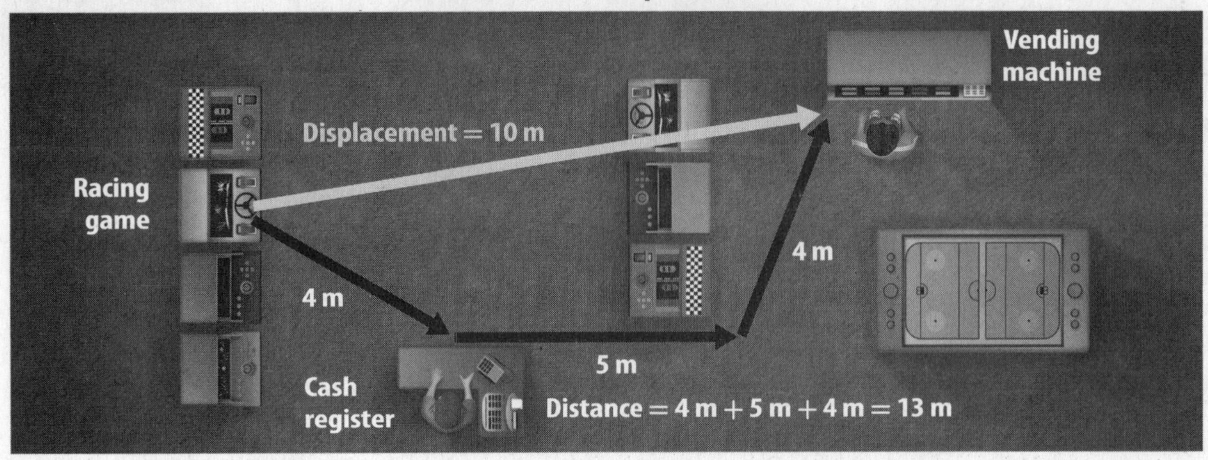

Visual Check
2. Contrast Why is the total displacement in the figure different from the total distance?

Key Concept Check
3. Identify What information do you need to describe an object's motion?

Reading Check
4. Generalize What does it mean to say that an object is moving at a "constant speed"?

Distance and Displacement

Suppose you go from the racing game to the cash register to get more tokens. Then you go to the vending machine for a snack. The three dark arrows in the figure above show your path. How far did you travel? **Distance** *is the total length of your path*. Your total distance is 4 m + 5 m + 4 m = 13 m.

Your **displacement** *is the distance between your initial, or starting, position and your final position*. A straight arrow from the starting point to the ending point represents displacement. The light upper arrow in the figure shows the displacement between the racing game where you started and the vending machine where you stopped. Your displacement is 10 m. A complete description of your motion includes a reference point, your displacement, and your direction.

Speed

Suppose you leave the arcade. You walk the first block slowly. Then you realize that you promised to meet a friend at the library in 15 min. You begin to run. You travel the distance of the next block in a much shorter time. What was different between your motion in the first block and your motion in the second block? Your speed was different. **Speed** *is the distance an object moves divided by the time it took to move that distance.*

Constant and Changing Speed

Speed is either constant or changing. Look at the figure at the top of the next page. The stopwatches show the girl's motion every second for 6 seconds. In the first 4 seconds, she travels the same distance during each second. This means that she was moving with constant, or unchanging, speed. When the girl starts running, the distance she travels each second gets larger and larger. The girl's speed changes.

Constant and Changing Speed

Average Speed

Suppose you want to know how fast you traveled from the arcade to the library. As you moved, your speed changed from second to second. Therefore, in order to describe your speed, you describe the average speed of the entire trip. Average speed is a ratio. It is the distance an object moves divided by the time it takes for the object to move that distance. If you traveled the 1-km distance to the library in 15 min, or 0.25 h, your average speed was 1 km/0.25 h, or 4 km/h.

Visual Check

5. Interpret When does the girl's speed begin to change?

____

____

FOLDABLES®

Make a three-tab concept map book to organize your notes on motion.

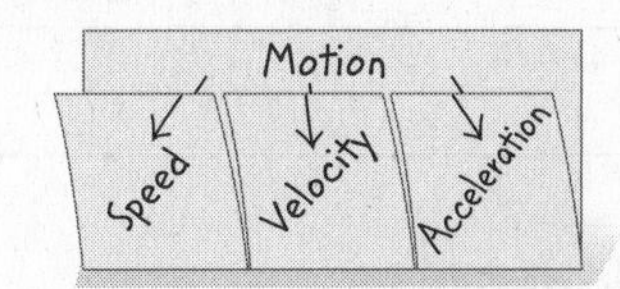

Velocity

When you describe your motion to a friend, you might say how fast you are traveling. You are describing your speed. You could give your friend a better description of your motion if you also state the direction in which you are moving. **Velocity** *is the speed and direction of an object's motion.*

Velocity often is shown by using an arrow, as illustrated in the figure below. The length of the arrow represents the speed of an object. The direction in which the arrow points represents the direction in which the object is moving.

Velocity

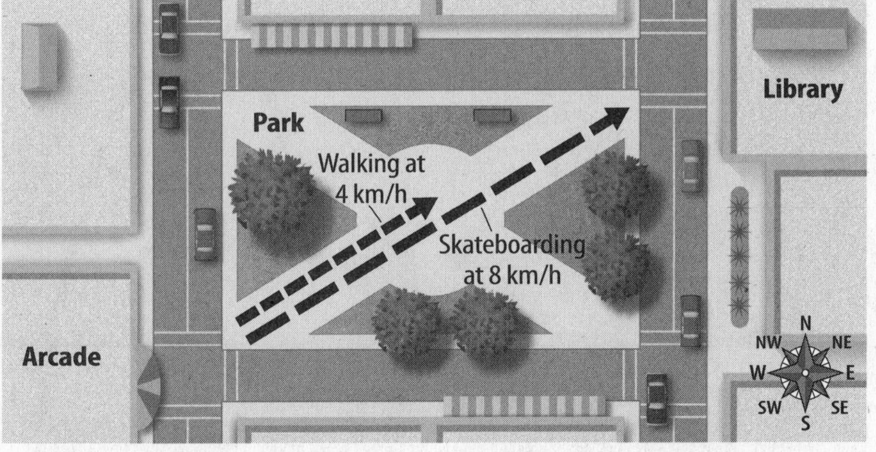

Visual Check

6. Interpret Which velocity shown in the figure is greater? (Circle the correct answer.)

a. skateboarding

b. walking

c. equal

Reading Check

7. Point Out What two factors must be constant in order for velocity to be constant?

Key Concept Check

8. Explain Can an object traveling at a constant speed have a changing velocity? Why or why not?

Visual Check

9. Identify In the caption of each panel, highlight the factor that changes, resulting in a changing velocity.

Constant Velocity

Velocity is constant, or does not change, when an object's speed and direction of movement do not change. If you use an arrow to describe velocity, you can divide the arrow into segments to show whether velocity is constant.

Look at the skateboarding arrow in the figure at the bottom of the previous page. Each segment of the arrow shows the distance and the direction the skateboarder moves in a given unit of time. Notice that each segment is the same length. This means the skateboarder is moving the same distance and in the same direction during each unit of time. Both speed and direction of movement are constant, so the skateboarder is moving at a constant velocity.

Look at the walking arrow in the figure at the bottom of the previous page. The skateboarder's velocity is greater than the walker's velocity. Both velocities are constant because the arrows show a constant speed and direction.

Changing Velocity

Velocity can change even if the speed of an object remains constant. Recall that velocity includes an object's speed and the direction it is traveling. The velocity changes when speed changes, direction changes, or both change. The figure below shows examples of changing velocity.

Change in Speed In the first panel, the ball drops toward the ground in a straight line, or constant direction. The increasing length of each arrow shows that the speed of the ball increases as it falls. As speed changes, velocity changes, even when direction remains constant.

Changes in Velocity

Speed changes, direction remains constant

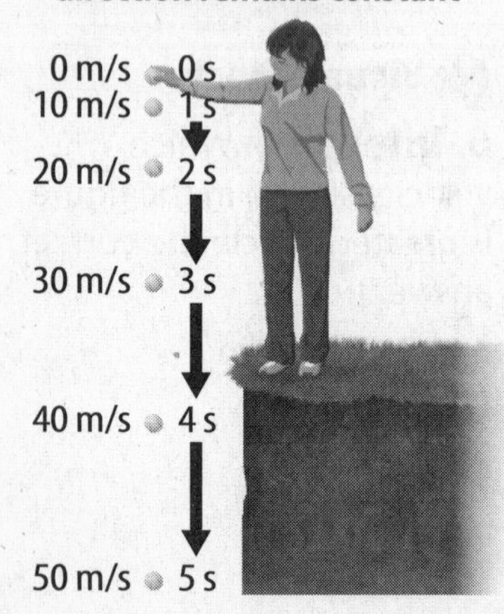

Speed remains constant, direction changes

Speed changes, direction changes

Change in Direction See the second panel of the figure on the previous page. Each arrow is the same length. This tells you that the Ferris-wheel cars travel around the circle at a constant speed. However, the arrows show that the cars are changing direction. As direction changes, velocity changes, even when speed remains constant.

Change in Speed and Direction The third panel of the figure on the previous page shows the path of a ball thrown into the air. The arrows show that the ball's speed and its direction change, so its velocity changes.

Acceleration When an object's speed or its velocity changes, the object is accelerating. ***Acceleration** is a measure of how quickly the velocity of an object changes.*

Calculating Acceleration

As shown in the first panel of the figure on the previous page, a ball speeds up as it falls toward the ground. The velocity of the ball is changing. Objects accelerate any time their velocity changes. You can calculate average acceleration using the following equation:

$$\bar{a} = \frac{v_f - v_i}{t}$$

Notice that this equation refers only to a change in speed, not direction. The symbol for average acceleration is $\bar{a}$. The symbol v_f means the final velocity. The symbol v_i means the initial, or starting, velocity. The symbol t stands for the period of time it takes to make that change in velocity.

Positive Acceleration

When an object speeds up, as when a ball rolls down a hill, its final velocity (v_f) is greater than its initial velocity (v_i). If you calculate the ball's average acceleration, the numerator (final velocity minus initial velocity) is positive. When you divide the positive numerator by time (t) to find the average acceleration, the result will also be positive. Therefore, when an object speeds up, it has positive acceleration.

Negative Acceleration

When the ball from the above example starts to roll up the next hill, it slows down. The initial velocity of the ball is greater than its final velocity. The numerator in the equation is negative. Thus, the average acceleration is negative. As an object slows down, it has negative acceleration. Some people refer to negative acceleration as deceleration.

Key Concept Check

10. Contrast How does acceleration differ from velocity?

____

____

____

____

____

Math Skills

A skateboarder moves at 2 m/s as he or she starts moving down a ramp. As the skateboarder heads down the ramp, he or she accelerates to a speed of 6 m/s in 4 seconds. What is the skateboarder's acceleration?

a. This is what you know:

final velocity: $v_f = 6$ m/s

initial velocity: $v_i = 2$ m/s

time: $t = 4$ s

b. You need to find out:

average acceleration: $\bar{a}$

c. Use this formula:

$$\bar{a} = \frac{v_f - v_i}{t}$$

d. Substitute: the values for v_f, v_i, and t: $\frac{6 \text{ m/s} - 2 \text{ m/s}}{4 \text{ s}}$

subtract and divide:

$$\frac{4 \text{ m/s}}{4 \text{ s}} = 1 \text{ m/s}^2$$

Answer: The average acceleration is 1 m/s^2.

11. Solve for Average Acceleration As the skateboarder starts moving up the other side of the ramp, his or her velocity drops from 6 m/s to 0 m/s in 3 seconds. What was his or her acceleration?

____

Using Graphs to Represent Motion

How can you track the motion of an animal that can move hundreds of miles without being seen by humans? To understand the movements of animals, such as the polar bear shown in the figure below, biologists put tracking devices on them. These devices constantly send information about the position of the animal to satellites. Biologists download the data from the satellites. They use the data to create graphs of motion, such as those shown below and on the next page.

ACADEMIC VOCABULARY
satellite
(noun) an object in orbit around another object

Displacement-Time Graphs

The graph below is a displacement-time graph of a polar bear's motion. The *x*-axis shows the time. The *y*-axis shows the displacement of the polar bear from a reference point. The displacement-time graph shows the bear's speed and displacement from the reference point at any point in time.

The line on the displacement-time graph represents the average speed of the bear at that moment in time. The line does not show the actual path of motion. As the average speed of the bear changes, the slope of the line on the graph changes. Because of this, you can use a displacement-time graph to describe the motion of an object.

Visual Check
12. Interpreting Graphs What was the average speed of the bear between hours 7 and 11?

____

Displacement-Time Graph

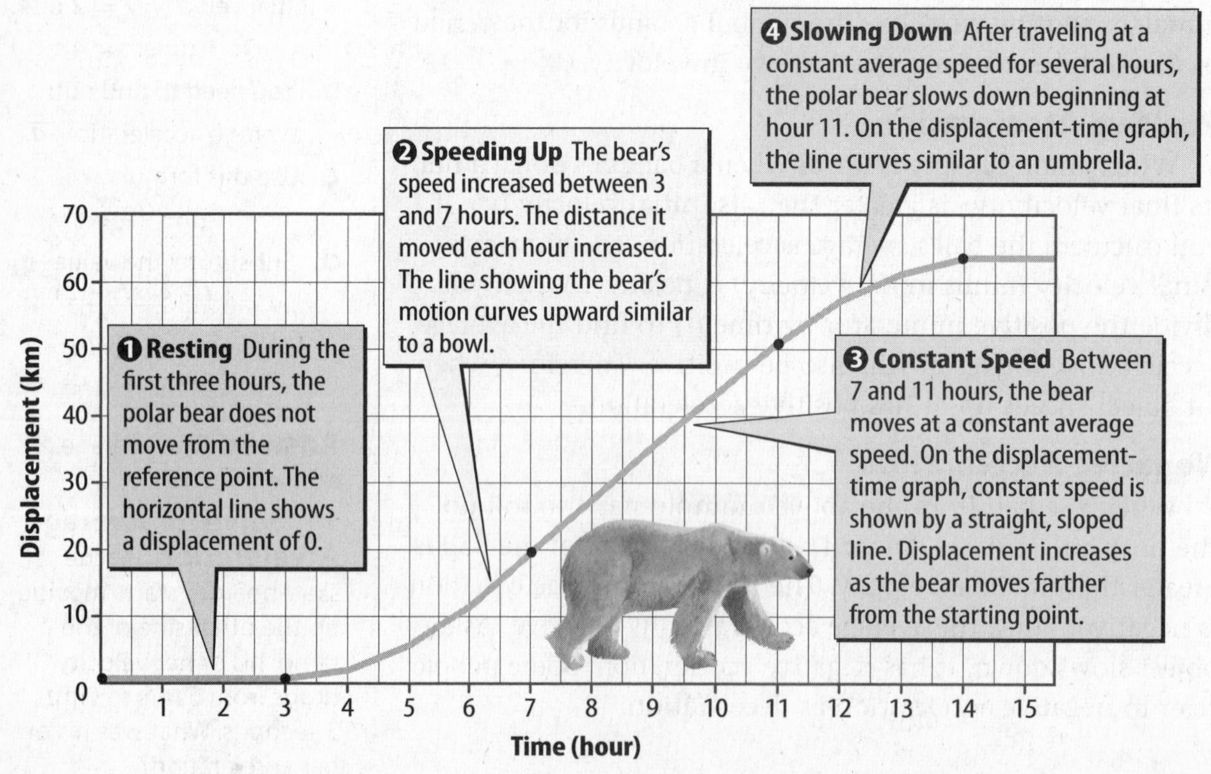

Speed-Time Graphs

The graph below is a speed-time graph of the polar bear's motion. The x-axis shows the time, and the y-axis shows the speed of the bear. In this case, the line shows how the speed changes as the bear moves. It does not show how displacement changes. The speed-time graph shows the speed of the bear at any point during its journey.

A horizontal line on a speed-time graph shows an object in constant motion—either at rest or moving at a constant speed. On the graph below, the horizontal line at $y = 0$ means the bear is at rest, because its speed is 0 km/h. Notice that a horizontal line at $y = 0$ on a displacement-time graph or a speed-time graph represents an object at rest.

Keep in mind that "constant speed" describes average speed. The bear might have sped up or slowed down slightly each second. But, during hours 7–11, you could describe that the bear's average remained constant since it covered the same distance each hour.

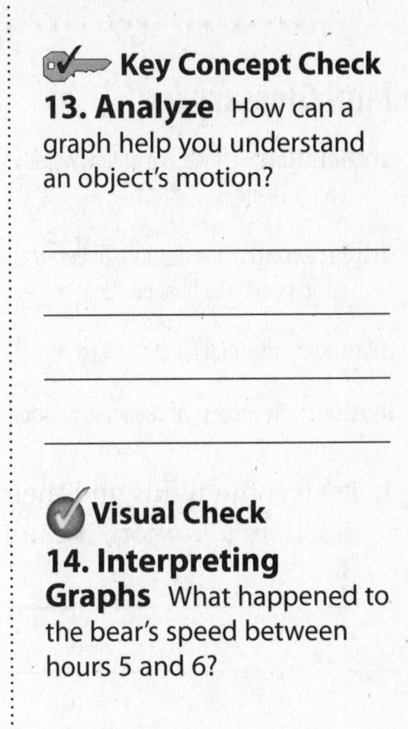

Key Concept Check

13. Analyze How can a graph help you understand an object's motion?

Visual Check

14. Interpreting Graphs What happened to the bear's speed between hours 5 and 6?

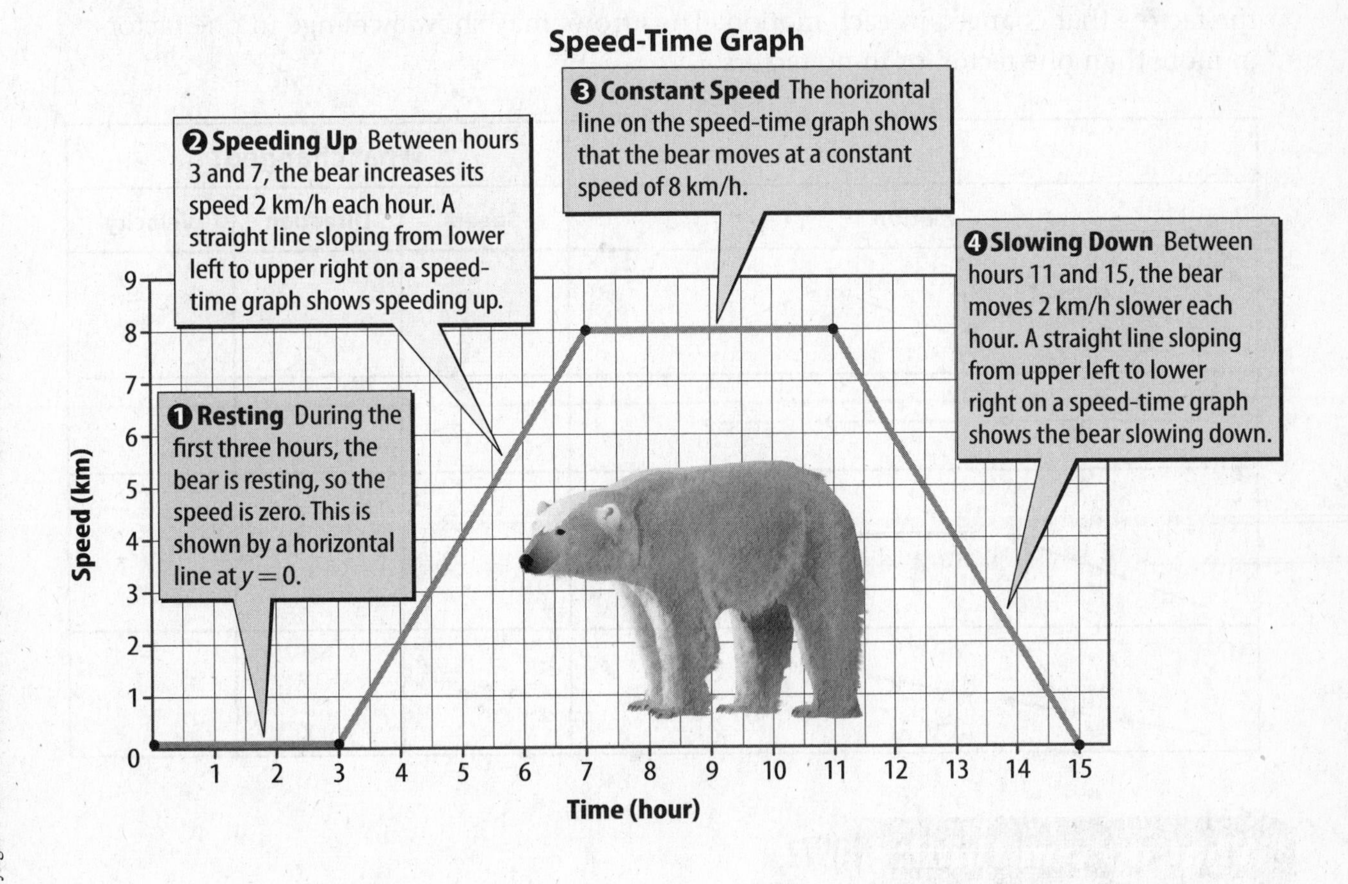

After You Read

Mini Glossary

acceleration: a measure of how quickly the velocity of an object changes

displacement: the distance between an initial, or starting, position and the final position

distance: the total length of a path

motion: the process of changing position

reference point: the starting point you use to describe the motion or the position of an object

speed: the distance an object moves divided by the time it took to move that distance

velocity: the speed and direction of an object's motion

1. Review the terms and their definitions in the Mini Glossary. Write a sentence that describes a motion, including a reference point, displacement, and direction.

____

____

2. The arrows in the table below describe motion. Put a check in the columns to identify the factors that changed in each motion. The arrows may show a change in one factor, in more than one factor, or in no factors.

	What changed?		
Motion	Speed	Direction	Velocity
a.			
b.			
c.			
d.			

What do you think NOW?

Reread the statements at the beginning of the lesson. Fill in the After column with an A if you agree with the statement or a D if you disagree. Did you change your mind?

Log on to ConnectED.mcgraw-hill.com and access your textbook to find this lesson's resources.

Motion, Forces, and Newton's Laws

Forces

Before You Read

What do you think? Read the two statements below and decide whether you agree or disagree with them. Place an A in the Before column if you agree with the statement or a D if you disagree. After you've read this lesson, reread the statements to see if you have changed your mind.

Before	Statement	After
	3. Objects must be in contact with one another to exert a force.	
	4. Gravity is a force that depends on the masses of two objects and the distance between them.	

Key Concepts 

- How do different types of forces affect objects?
- What factors affect the way gravity acts on objects?
- How do balanced and unbalanced forces differ?

Read to Learn

What are forces?

What do typing on a computer, lifting a bike, and putting on a sweater have in common? They involve an interaction between you and another object. You push on the keys. You push or pull on the bike. You pull on the sweater. *A push or pull on an object is a* **force.**

A force has a size and a direction. An arrow can represent force in a diagram. The length of the arrow represents the size of the force. A longer arrow means greater force than a shorter arrow. The arrow points in the direction of the force. The unit of force is the newton (N). It takes about 4 N of force to lift a can of soda.

A force can affect an object in several ways. A force can change an object's speed. It also can change the direction in which the object is moving. In other words, a force might cause acceleration. Recall that acceleration is a change in an object's velocity—its speed, its direction, or both. When you apply a force to a tennis ball with a racket, the force first stops the motion of the ball. The force then causes the ball to accelerate in the opposite direction. This changes the speed and the direction of the ball.

Mark the Text

Make an Outline As you read, highlight the main idea under each heading. Then use a different color to highlight a detail or an example that might help you understand the main idea. Use your highlighted text to make an outline to organize the main ideas.

Reading Check

1. Summarize What are some ways in which forces can affect objects?

Types of Forces

Some forces are easy to recognize. You can see a hammer apply a force as it hits a nail. Other forces seem to act on objects without touching them. For example, what force causes a dropped object to fall toward the ground?

Contact Forces

Suppose you observe a baker pressing bread dough into a pan. He pushes his fingers into the dough, causing the top of the dough to accelerate downward. The baker's hand and the dough come into contact with each other. *A* **contact force** *is a push or a pull applied by one object to another object that is touching it.* Contact forces also are called mechanical forces. The table below describes other contact forces.

Interpreting Tables

2. Apply When you open a door, what type of contact force are you exerting?

____

Types of Contact (Mechanical) Forces

Force	Description	Example
applied force	a force in which one object directly pushes or pulls on another object	A baker pushes down on the top of dough.
elastic force or spring force	the force exerted by a compressed or stretched object	As you stretch an elastic band, the band exerts a force on you.
normal force	the support force exerted on an object that touches another stable object	A gymnast pushes down on a pommel horse to support her body above the horse. At the same time, the horse exerts upward force on her arms.

Noncontact Forces

Does your hair ever stick out after you brush it? Electric forces can pull your hair toward the brush, even though your hair isn't touching the brush. *A force that pushes or pulls an object without touching it is a* **noncontact force.**

Electric forces are one type of noncontact force. Gravity and magnetism are other types. Magnetic forces hold the like ends of two magnets apart. Gravity pulls a dropped object toward the ground.

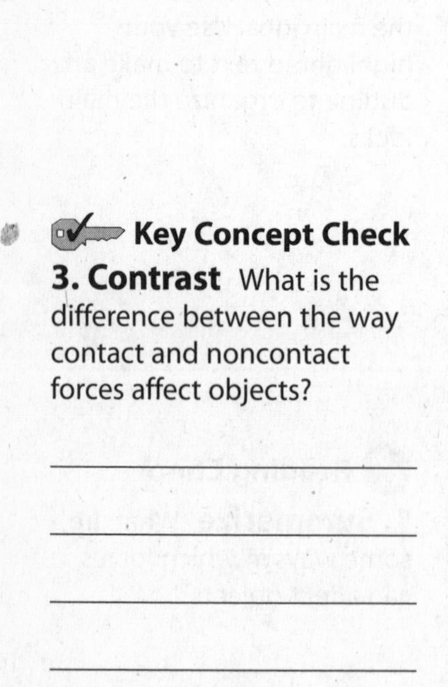

Key Concept Check

3. Contrast What is the difference between the way contact and noncontact forces affect objects?

____

____

____

____

____

____

Friction

Why does a baseball player slow down as he slides into a base? **Friction** *is a contact force that resists the sliding motion between two objects that are touching.* The force of friction acts in the opposite direction of the motion. As he slides to the base, the player must overcome friction from the ground, which is exerting a force away from the base.

Rougher surfaces produce greater friction than smooth surfaces. Other factors, such as the surface area and the weight of an object, also affect the force of friction.

Gravity

If you drop a pencil anywhere on Earth, it will fall. **Gravity** *is a noncontact attractive force that exists between all objects that have mass.*

Mass is the amount of matter in an object. Your pencil and Earth have mass. They exert the same gravitational force on each other. Your pencil actually pulls Earth toward it. The pencil has very little mass, so the force of gravity causes it to rapidly accelerate downward toward Earth's surface. Earth "falls" upward toward the pencil at the same time. But because of its mass, Earth's motion is too small to see.

Distance and Gravity

Are astronauts truly "weightless" in space? No, but they do weigh less in space than on Earth. Weight is a measure of the force of gravity acting on an object. As two objects get farther apart, the gravitational force between the objects decreases. The figure below shows how the weight of an astronaut changes as he or she moves farther from Earth.

If an astronaut drops a hammer on the Moon, will it fall toward Earth? The hammer is much closer to the Moon than to Earth. The attraction between the hammer and the Moon is stronger than the attraction between the hammer and Earth. The hammer will fall toward the Moon.

Force of Gravity

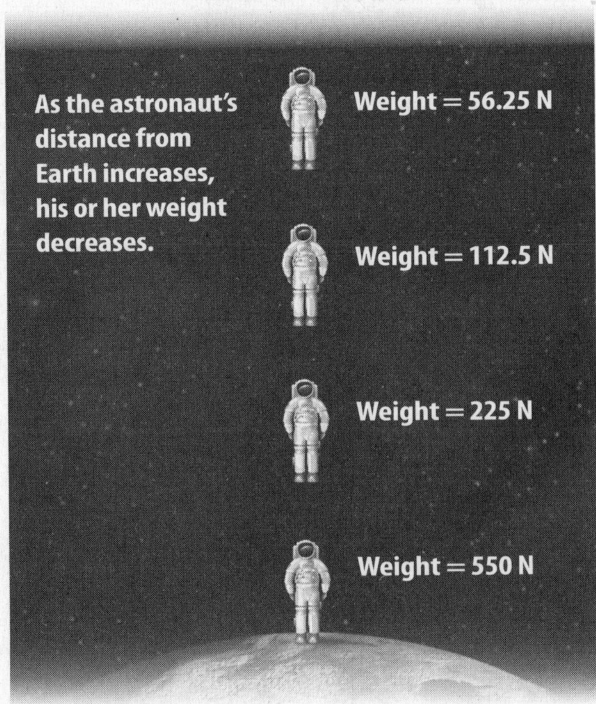

FOLDABLES®

Make a two-tab book to organize your notes on forces.

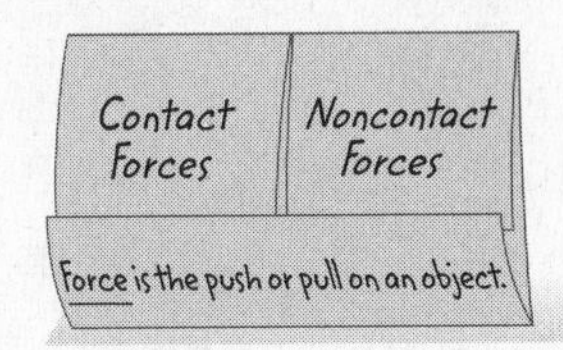

Reading Check

4. Explain When you drop a pencil, why does it move toward the floor?

Visual Check

5. Interpret As an astronaut descends toward Earth, how does his or her weight change?

Mass and Gravity

The mass of the objects also affects the force of gravity between two objects. As the mass of one or both objects increases, the gravitational force between them increases. For example, doubling the mass of one of the objects doubles the force of attraction, as shown in the figure below. The force of attraction between the bottom two objects is twice as much as between the top two objects.

The effect of mass on gravity is easiest to see when one object is much more massive than the other. For example, Earth is massive. A person has much less mass. Even though the force of gravity acts equally on both objects, the less-massive object accelerates more quickly due to its smaller mass. Therefore, when you jump off a step, you seem to "fall" toward the object with greater mass, Earth.

Key Concept Check
6. Name What factors affect the way gravity acts on objects?

____

____

Mass and Gravity

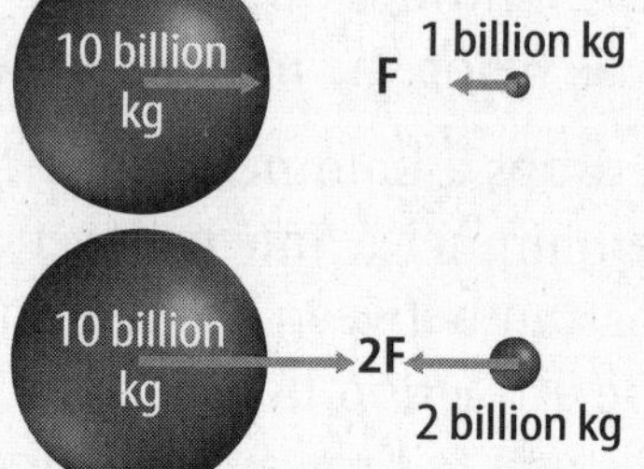

Visual Check
7. Describe the acceleration of the bottom spheres due to the gravitational force between them.

____

____

____

____

Combining Forces

Have you ever played tug-of-war? If one person pulls against a team, the team will probably pull that person over the line. However, if you are on a team, your team might pull hard enough to cause the other team to move in your direction. When several forces act on an object, the forces combine to act as a single force, as shown below. The sum of the forces acting on an object is called the net force.

Forces in the Same Direction

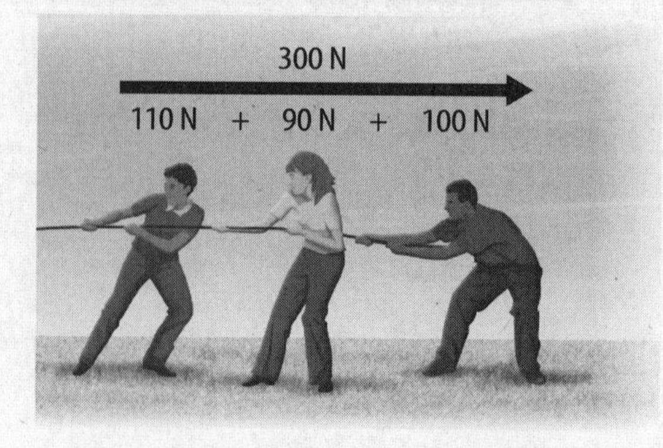

Visual Check
8. Calculate What would the total force be if the person on the right stopped pulling?

____

Forces in the Same Direction

When different forces act on an object in the same direction, the net force is the forces added together. In the figure at the bottom of the previous page, each team member is pulling in the same direction. The net force on the rope is 110 N + 90 N + 100 N = 300 N.

Forces in Opposite Directions

When forces act in opposite directions, you must include the directions when you add the forces. Like numbers on a number line, forces to the right are usually considered to be positive values. Forces to the left are negative values. In the first panel below, the team on the right pulls with a force of 300 N. The team on the left pulls with a force of –300 N. The net force is 300 N + (–300 N) = 0.

Forces in the Opposite Direction

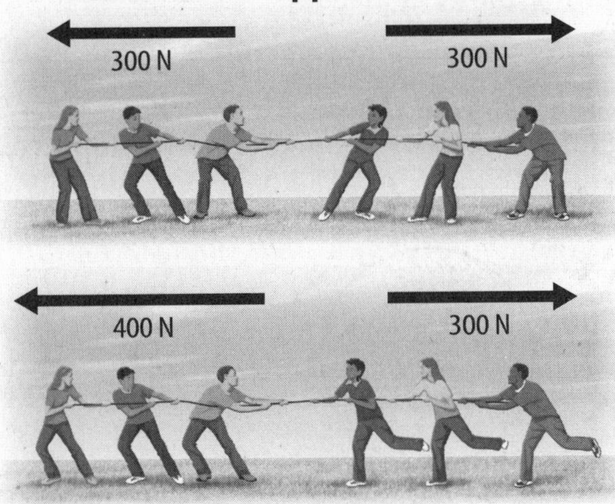

Balanced and Unbalanced Forces

The net force on the rope in the top panel of the figure above is 0. *When the net force on an object is 0, the forces acting on it are* **balanced forces.** If the forces acting on an object are balanced, the will not affect the motion of the object.

When the net force acting on an object is not 0, the forces acting on the object are **unbalanced forces.** The forces acting on the rope in the bottom panel of the figure above are unbalanced. Unbalanced forces cause objects to change their motion. As you can see in the second panel of the figure above, unbalanced forces cause the team on the right to accelerate to the left.

Think it Over

9. Analyze How does a net force of 0 N affect the motion of the objects?

____

____

Visual Check

10. Calculate In the second panel, suppose that "to the left" is the positive direction. What is the net force?

____

Key Concept Check

11. Contrast How do balanced and unbalanced forces differ?

____

____

____

____

____

____

After You Read

Mini Glossary

balanced forces: forces acting on an object with the net force of 0

contact force: a push or a pull applied by one object to another object that is touching it

force: a push or a pull on an object

friction: a contact force that resists the sliding motion between two objects that are touching

gravity: a noncontact attractive force that exists between all objects that have mass

noncontact force: a force that pushes or pulls an object without touching it

unbalanced forces: forces acting on an object with a net force that is not 0

1. Review the terms and their definitions in the Mini Glossary. Write a sentence that gives an example of friction that you have experienced.

2. Below each diagram, describe how the force of gravity changes from the top two sets of objects to the bottom two sets of objects.

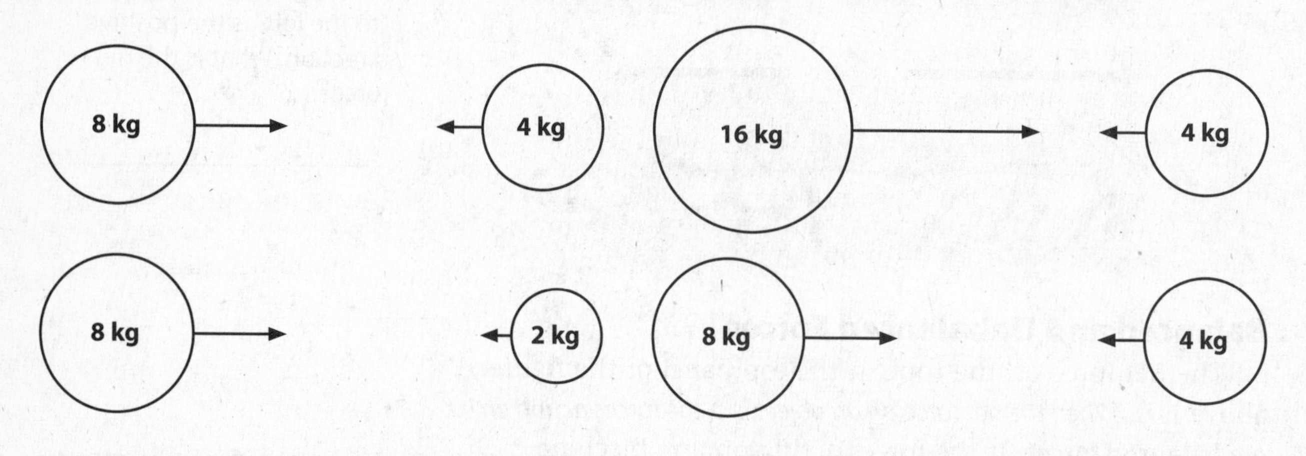

3. How does an increase in distance between objects affect the gravitational force?

What do you think NOW?

Reread the statements at the beginning of the lesson. Fill in the After column with an A if you agree with the statement or a D if you disagree. Did you change your mind?

Log on to ConnectED.mcgraw-hill.com and access your textbook to find this lesson's resources.

Motion, Forces, and Newton's Laws

Newton's Laws of Motion

Before You Read

What do you think? Read the two statements below and decide whether you agree or disagree with them. Place an A in the Before column if you agree with the statement or a D if you disagree. After you've read this lesson, reread the statements to see if you have changed your mind.

Before	Statement	After
	5. All forces change the motion of objects.	
	6. The net force on an object is equal to the mass of the object times the acceleration of the object.	

Key Concepts 

- How do unbalanced forces affect an object's motion?
- How are the acceleration, the net force, and the mass of an object related?
- What happens to an object when another object exerts a force on it?

Read to Learn

Newton's Laws

Recall that forces are measured in a unit called a newton (N). This unit is named after English scientist Isaac Newton, who studied the motion of objects. Newton summarized his findings in three laws of motion. You demonstrate Newton's laws when you run to catch a baseball or ride your bike. You can use Newton's laws to explain how the rides and games at an amusement park work.

Mark the Text

Sticky Notes As you read, use sticky notes to mark information that you do not understand. Read that text carefully a second time. If you still need help, write a list of questions to ask your teacher.

Newton's First Law

What causes the motion of amusement park rides to give riders a thrill? Without protective devices to hold you in your seat, you could fly off the ride! *The tendency of an object to resist a change in motion is called* **inertia.** Inertia acts to keep you at rest when the ride starts moving. It also keeps you moving in a straight line when the ride stops or changes direction. Your safety belt keeps you in the seat and moving with the ride.

Newton's first law of motion *states that if the net force acting on an object is zero, the motion of the object does not change.* In other words, an object remains at rest or in constant motion unless an outside, unbalanced force acts on it. Newton's first law of motion is sometimes called the law of inertia.

SCIENCE USE V. COMMON USE
inertia
Science Use *the tendency to resist a change in motion*
Common Use lack of action

Effects of Balanced Forces

Suppose you are at an amusement park and you want to ride a free-fall car. How does a free-fall ride illustrate Newton's first law of motion? Recall that when the forces acting on an object are balanced, the object is either at rest or moving with a constant velocity.

Reading Check

1. Predict Suppose you know that the forces acting on an object are balanced. What motion would you expect the object to display?

Objects at Rest At the top of the ride, the force of the cable pulling upward on the car is equal to the force of gravity pulling downward on the car. Gravity and the cables pull on the car equally, but in opposite directions. As a result, the forces are balanced. The car is at rest. As long as the forces remain balanced, the car remains at rest.

Objects in Motion To lift the car to the top of the ride, the cable pulls upward. After a short acceleration, the car moves upward at a constant speed. The force of the cable pulling upward is the same size as the force of gravity pulling downward. With the forces once again balanced, the car rises to the top of the ride at a constant velocity. Newton's first law describes the car's motion when the forces applied to it are balanced.

Reading Check

2. Describe What happens to the velocity of the car when the upward pull of the cable is greater than the downward pull of gravity as the car rises toward the top?

Balanced forces act on the car only when it is at rest or moving with a constant velocity. When the car reaches the top of the ride, it doesn't remain at rest for long. When the operator releases the upward pull on the cable, the forces become unbalanced. Gravity causes the car to accelerate toward the ground. Because inertia tends to keep you at rest, the car feels as if it falls out from under you. Your safety belt acts as an outside force to keep you attached to the car.

Effects of Unbalanced Forces

You continue your visit to the amusement park with a ride on the reverse bungee jump. This ride propels you upward like a slingshot on elastic bungee cords.

According to Newton's first law of motion, the motion of an object changes only when a net force acts on it. This ride gives you two chances to experience what a net force can do. Unbalanced forces cause a bungee jumper to speed up during part of the ride and slow down during another part of the ride.

Speeding Up After the ride attendant releases you, the upward force of the bungee cord is greater than the downward force of gravity. The forces are unbalanced. The net force acting on you is upward, and you accelerate upward. This is positive acceleration because you are speeding up.

FOLDABLES®

Make a layered book to organize your notes on Newton's laws.

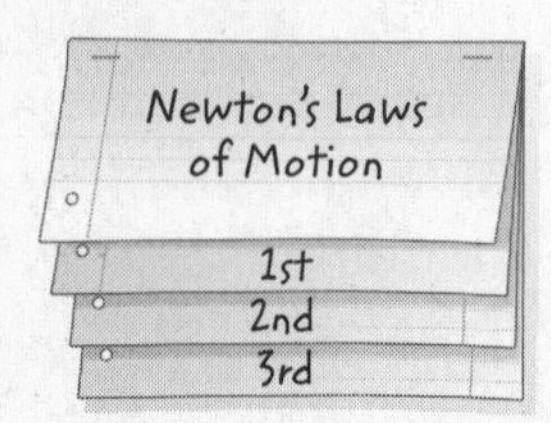

Slowing Down As you approach the top of your bungee ride, the cords become slack. The upward force of the cords becomes less than the downward force of gravity. Even though you still are moving upward because of inertia, the net force is now due to the downward force of gravity. You slow down, or decelerate. 

Changing Direction Your next stop is a swing ride. You sit in a chair attached by cables to a canopy on a center post. The post and canopy turn in a circle, propelling your chair outward on its cables.

When the ride starts to turn, the force of the cables pulls your chair toward the center of the ride. The force of gravity acts downward. Because these forces don't act in opposite directions, the unbalanced force constantly changes your direction. The unbalanced force of the cable pulling toward the center makes you accelerate in a circular direction. ✓

The designers of amusement-park rides use inertia to create excitement. Much of what makes a swing ride fun is the feeling that you might fly off the ride with constant velocity if your safety belt didn't hold you in place.

Newton's Second Law of Motion

Suppose you play a game in which you throw a baseball to knock over wooden milk bottles. You have seen that unbalanced forces cause objects to accelerate. To knock over the bottles, the ball has to accelerate fast enough to overcome the forces keeping the bottles upright.

On your first try, you don't throw the ball with enough force. On your second try, you throw the ball as hard as you can. As the ball flies out of your hand, it accelerates rapidly and knocks over the bottles, winning you a prize. You won the prize by using another of Newton's laws. ✓

Newton described the relationship between an object's acceleration (change in velocity) and the net force exerted on the object. **Newton's second law of motion** *states that the acceleration of an object is equal to the net force exerted on the object divided by the object's mass.* The direction of acceleration is the same as the direction of the net force. The following formula expresses Newton's second law of motion.

$$\text{acceleration} = \frac{\text{force}}{\text{mass}} \qquad a = \frac{F}{m}$$

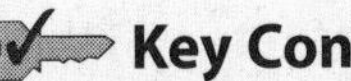 **Key Concept Check**

3. Calculate If one force on an object is 5 N upward and the other is 10 N downward, what is the object's motion?

____

✓ **Reading Check**

4. Explain On the swing ride, your chair constantly changes direction. Why?

____

____

____

✓ **Reading Check**

5. State Why did the first ball fail to knock over the milk bottles?

____

____

____

____

____

6. Calculate Suppose you throw a ball with a force of 2 N. The ball has a mass of 0.4 kg. At what rate will the ball accelerate?

Key Concept Check
7. Explain How are the acceleration, the net force, and the mass of an object related?

Key Concept Check
8. Describe What happens when one object applies a force on a second object?

Calculating Acceleration

You can use the equation to calculate the acceleration of the ball. If you apply a force of 1.5 N to a ball with a mass of 0.3 kg, what is the ball's acceleration?

$$\text{acceleration} = \frac{\text{force}}{\text{mass}} \qquad \text{acceleration} = \frac{1.5\ \text{N}}{0.3\ \text{kg}} = \frac{5\ \text{m}}{\text{s}^2}$$

What would happen to the acceleration if you doubled the force on the ball? The acceleration would also double.

$$\text{acceleration} = \frac{\text{force}}{\text{mass}} \qquad \text{acceleration} = \frac{3.0\ \text{N}}{0.3\ \text{kg}} = \frac{10\ \text{m}}{\text{s}^2}$$

Changing the Mass

What would happen to the acceleration if the force you apply stays the same, but the mass of the ball changes? Instead of 0.3 kg, the ball has a mass of 0.6 kg.

$$\text{acceleration} = \frac{\text{force}}{\text{mass}} \qquad \text{acceleration} = \frac{1.5\ \text{N}}{0.6\ \text{kg}} = \frac{2.5\ \text{m}}{\text{s}^2}$$

A ball with twice the mass has half the acceleration. Newton's second law lets you predict what combination of force and mass will achieve the acceleration you need.

Newton's Third Law

Suppose you are driving a bumper car. What happens when you crash into another car? **Newton's third law of motion** *says that when one object applies a force on a second object, the second object applies a force of the same size, but in the opposite direction, on the first object*. According to Newton's third law, the bumper cars apply forces to each other that are equal but are in opposite directions.

Action and Reaction Forces

When two objects apply forces on each other, one force is the action force, and the other is the reaction force. The bumper car that accelerates into the other car applies the action force. The car that takes the hit applies the reaction force.

Force Pairs

As you walk around the amusement park, your shoes push against the ground. If the ground did not push back with equal force, gravity would pull you down into the ground! *When two objects apply forces on each other, the two forces are a* **force pair.**

Opposite Forces The opposite forces of the bumper cars hitting each other are a force pair. Force pairs are not the same as balanced forces. Balanced forces combine or cancel each other out because they act on the same object. Each force in a force pair acts on a different object.

When a soccer player hits the ball with her head, her head applies a force on the ball. The ball applies an equal but opposite force on her head. Newton's laws work together. Newton's first law explains that a force is needed to change an object's motion. Newton's third law describes the action-reaction forces. Newton's second law explains why the effect of the force is greater on the ball. The mass of the ball is much less than the player's mass. Therefore, a force of the same size produces a much greater acceleration in the ball.

Newton's Laws in Action

Newton's laws are useful because they apply to the moving objects that you observe each day. Using Newton's laws, humans have traveled to other planets and invented many useful tools and machines. You can often see the effects of all three laws at the same time. The table below describes everyday examples of Newton's laws in action.

9. Apply If the force of the soccer player's head on the ball is 1.5 N upward, what is the force of the ball on the player's head? (Circle the correct answer.)

a. 0 N
b. 1.5 N
c. 3.0 N

Visual Check

10. Explain In the first example, how do you know that the table is applying a force on the bowl of fruit?

____

____

____

Newton's Laws in Action

Example	Newton's First Law	Newton's Second Law	Newton's Third Law
A bowl of fruit with a mass of 2 kg sits on a table.	The upward and downward forces on the bowl are balanced. The motion of the bowl is not changing. It is at rest.	Because the bowl is at rest, its acceleration is 0 m/s^2. You can use Newton's second law to calculate the net force on the bowl: $F = m \times a$ $F = 2\text{ kg} \times 0\text{ m/s}^2$ $F = 0\text{ N}$	The force of gravity pulls the bowl down so it applies a force on the table. The table pushes up on the bowl with a force that is the same size but in the opposite direction.
A woman is walking on a sidewalk.	The forces acting on the woman are balanced. Her inertia keeps her moving at a constant speed in a straight line.	When an object moves at a constant velocity, there is no acceleration. A net force would have to act on the woman before she would speed up or slow down.	The woman's feet push against the sidewalk as she walks. The sidewalk pushes on the woman's feet with equal force, moving her forward.
A dog on a skateboard is pushing off with its paw.	Inertia keeps the dog and the skateboard at rest until the dog produces a net force by pushing its paw on the road.	When net forces act on the dog and the road, which is part of Earth, the dog will accelerate at a much greater rate because its mass is much less than that of Earth.	The dog's paw applies a backward force on the road. The road applies an equal but opposite force on the dog's paw, pushing it forward.

After You Read

Mini Glossary

force pair: the two forces when two objects apply forces on each other

inertia: the tendency of an object to resist a change in motion

Newton's first law of motion: states that if the net force acting on an object is zero, the motion of the object does not change

Newton's second law of motion: states that the acceleration of an object is equal to the net force applied on the object divided by the object's mass

Newton's third law of motion: states that when one object applies a force on a second object, the second object applies a force of the same size, but in the opposite direction, on the first object

1. Review the terms and their definitions in the Mini Glossary. Write a sentence to explain how force pairs work as you walk across the ground.

2. Use the formula to determine whether the change in force or mass will cause acceleration to increase or decrease. Write *increase* or *decrease* in the Acceleration column.

Newton's Second Law: $\text{acceleration} = \frac{\text{force}}{\text{mass}}$

	Acceleration	Force	Mass
a.		increase	no change
b.		no change	decrease
c.		decrease	no change

3. Record a question from your sticky notes that was especially difficult for you. Then write the answer that you learned from rereading the material or asking your teacher.

What do you think NOW?

Reread the statements at the beginning of the lesson. Fill in the After column with an A if you agree with the statement or a D if you disagree. Did you change your mind?

Log on to ConnectED.mcgraw-hill.com and access your textbook to find this lesson's resources.

The Sun-Earth-Moon System

Earth's Motion

Before You Read

What do you think? Read the two statements below and decide whether you agree or disagree with them. Place an A in the Before column if you agree with the statement or a D if you disagree. After you've read this lesson, reread the statements to see if you have changed your mind.

Before	Statement	After
	1. Earth's movement around the Sun causes sunrises and sunsets.	
	2. Earth has seasons because its distance from the Sun changes throughout the year.	

Key Concepts 

- How does Earth move?
- Why is Earth warmer at the equator and colder at the poles?
- Why do the seasons change as Earth moves around the Sun?

Read to Learn

Earth and the Sun

If you look around you, it does not seem as if Earth is moving. The ground, trees, and buildings do not seem to be moving. But Earth is always in motion. It spins and moves around the Sun. Earth's motion causes changes on Earth. As Earth spins, day changes to night and back to day again. The seasons change as Earth moves around the Sun. Summer turns to winter because Earth's motion changes how energy from the Sun spreads out over Earth's surface.

Mark the Text

Identify Main Ideas Highlight each head in one color. Use another color to highlight key words in the paragraphs under the head that explain or support the head. Use your highlighting to review the lesson.

The Sun

The nearest star to Earth is the Sun. The Sun is about 150 million km from Earth. The Sun is much larger than Earth. The Sun's diameter is more than 100 times greater than Earth's diameter. The Sun's mass is more than 300,000 times greater than Earth's mass.

The Sun is a giant ball of hot gases. It emits light and energy. Inside the Sun, the nuclei of atoms combine to produce huge amounts of energy. This process is called nuclear fusion. Nuclear fusion produces so much energy that the Sun's core temperature is more than 15,000,000°C. Even at the Sun's surface, the temperature is about 5,500°C. A small part of the Sun's energy reaches Earth as light and thermal energy. ✓

Reading Check

1. Explain What is one effect of nuclear fusion?

______

______

______

Earth's Orbit

The motion of one object around another object is called **revolution.** Earth makes one complete revolution around the Sun every 365.24 days. *The path an object follows as it moves around another object is an* **orbit.** Earth orbits the Sun in an almost circular path. Earth's orbit is shown below.

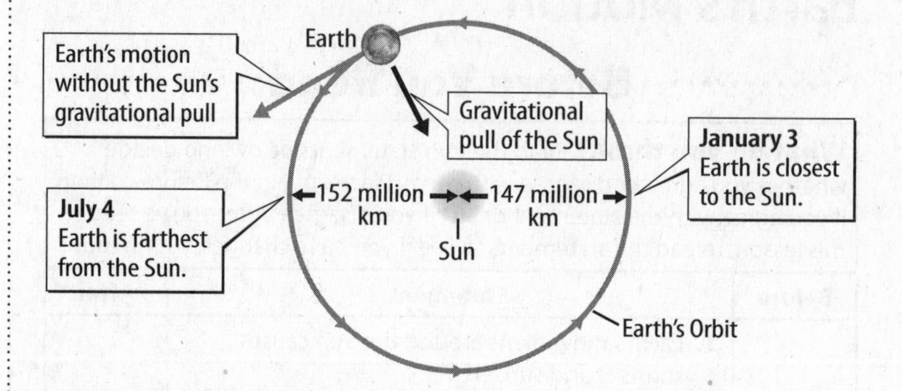

Visual Check

2. Describe What is the shape of Earth's orbit around the Sun? (Circle the correct answer.)

a. circle

b. near circle

c. a straight line

The Sun's Gravitational Pull

Earth orbits the Sun because the Sun's gravity pulls on Earth. The strength of gravity's pull between two objects depends on the masses of the objects and the distance between them. An object with more mass has a greater pull of gravity than an object with less mass. Likewise, gravity's pull is greater on objects that are closer together.

Earth's orbit around the Sun, shown above, is like the motion of an object twirled on a string. The string pulls on the object and moves it in a circle. If the string breaks, the object flies off in a straight line. The Sun's gravity is like the string. Gravity keeps Earth revolving around the Sun in a nearly circular orbit. If the pull of gravity between the Sun and Earth stopped suddenly, Earth would fly off into space in a straight line.

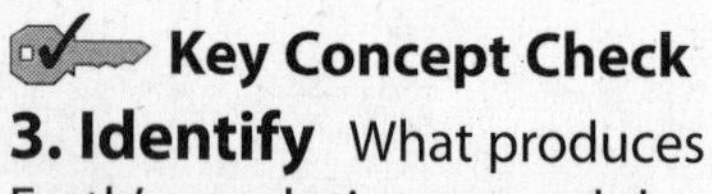

Key Concept Check

3. Identify What produces Earth's revolution around the Sun?

Earth's Rotation

As Earth revolves around the Sun, it spins. *A spinning motion is called* **rotation.** Earth rotates on an imaginary line that runs through its center. *The line on which an object rotates is the* **rotation axis.**

If you could look down onto Earth's North Pole, you would see that Earth rotates in a counterclockwise direction, from west to east. One complete rotation of Earth takes about 24 hours. One rotation completes Earth's cycle of day and night. It is daytime on the half of Earth that faces the Sun. It is nighttime on the half of Earth that faces away from the Sun.

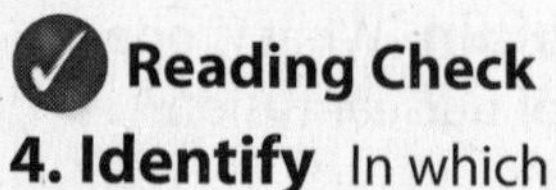

Reading Check

4. Identify In which direction does Earth rotate?

The Sun's Apparent Motion Each day, the Sun appears to move across the sky from east to west. It seems as if the Sun is moving around Earth. In fact, it is Earth's rotation that causes the Sun's apparent motion.

Earth rotates from west to east. This makes the Sun appear to move from east to west across the sky. The Moon and stars also seem to move from east to west across the sky due to Earth's west-to-east rotation. Earth's west-to-east rotation causes apparent east-to-west motion in the sky. ✓

The Tilt of Earth's Rotation Axis Earth's rotation axis is tilted, as shown in the figure below. The tilt of Earth's rotation axis does not change. During one-half of Earth's orbit, the north end of the rotation axis is toward the Sun. During the other half of Earth's orbit, the north end of the rotation axis is away from the Sun.

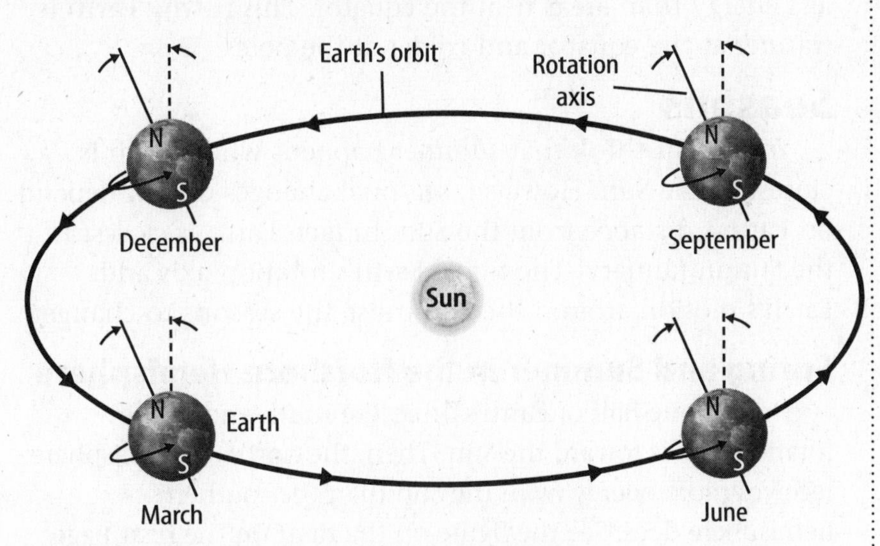

Temperature and Latitude

As Earth orbits the Sun, the Sun shines on the half of Earth that faces the Sun. Sunlight carries energy. The more sunlight that reaches a part of Earth's surface, the warmer that part becomes. Because Earth's surface is curved, different parts of Earth's surface get different amounts of the Sun's energy. ✓

Energy Received by a Tilted Surface

Suppose you shine a flashlight onto a flat card. The beam shines in a circle on the card. As you tilt the top of the card away from the beam of light, the light becomes more spread out on the card's surface. The energy that the light beam carries also spreads out more over the card's surface. An area on the surface within the light beam receives less energy when the surface is more tilted relative to the light beam.

✓ **Reading Check**

5. Describe What causes the Sun's apparent motion across the sky?

✓ **Visual Check**

6. Identify Between which months is the north end of Earth's rotation axis away from the Sun?

✓ **Reading Check**

7. Summarize Why do some parts of Earth's surface get more energy from the Sun than other parts?

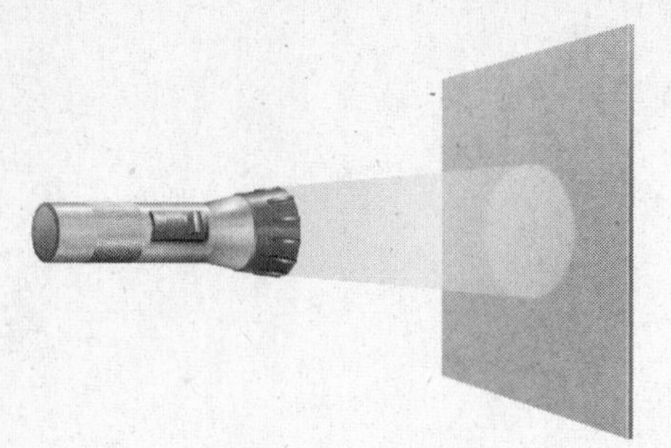

Surface is vertical.

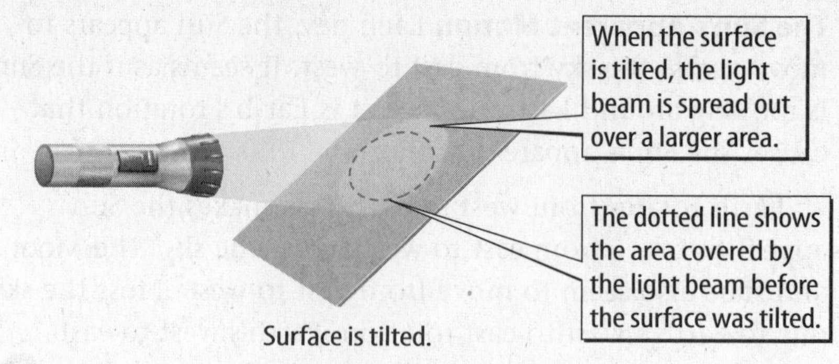

Surface is tilted.

Visual Check

8. Determine Is the light energy more spread out on the vertical or tilted surface?

____

____

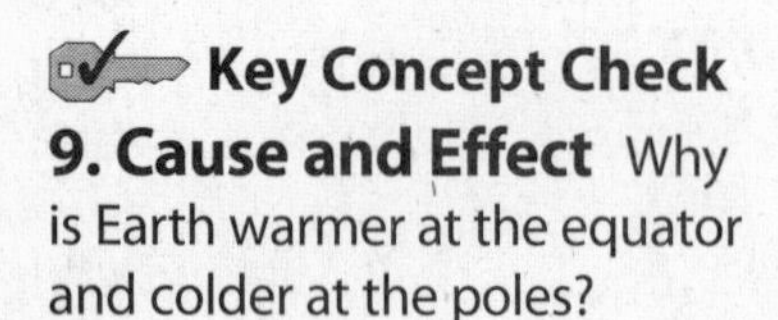

Key Concept Check

9. Cause and Effect Why is Earth warmer at the equator and colder at the poles?

____

____

____

Math Skills

When Earth is 147,000,000 km from the Sun, how far is Earth from the Sun in miles? To calculate the distance in miles, multiply the distance in km by the conversion factor.

$147{,}000{,}000 \text{ km} \times \frac{0.62 \text{ miles}}{1 \text{ km}}$

$= 91{,}100{,}000 \text{ miles}$

10. Convert Units When Earth is 152,000,000 km from the Sun, how far is Earth from the Sun in miles?

____

The Tilt of Earth's Curved Surface

Instead of being flat and vertical like the card shown above on the left, Earth's surface is curved and tilted, somewhat like the card on the right. Earth's surface becomes more tilted as you move away from the equator and toward the poles. As a result, regions of Earth near the poles receive less energy than areas near the equator. This is why Earth is warmer at the equator and colder at the poles.

Seasons

You might think that summer happens when Earth is closest to the Sun. However, seasonal changes do not depend on Earth's distance from the Sun. In fact, Earth is closest to the Sun in January! The tilt of Earth's rotation axis and Earth's motion around the Sun cause the seasons to change.

Spring and Summer in the Northern Hemisphere

During one-half of Earth's orbit, the north end of the rotation axis is toward the Sun. Then, the northern hemisphere receives more energy from the Sun than the southern hemisphere does. See the figure on the right on the next page.

Temperatures are higher in the northern hemisphere and lower in the southern hemisphere. Daylight hours last longer in the northern hemisphere. Nights last longer in the southern hemisphere. It is spring and summer in the northern hemisphere and fall and winter in the southern hemisphere.

Fall and Winter in the Northern Hemisphere

During the other half of Earth's orbit, the north end of the rotation axis is away from the Sun, as shown in the left figure on the next page. Then, the northern hemisphere receives less energy from the Sun than the southern hemisphere does.

Temperatures are cooler in the northern hemisphere and warmer in the southern hemisphere. It is fall and winter in the northern hemisphere. At the same time, spring and summer occur in the southern hemisphere.

North end of rotation axis points away from the Sun.

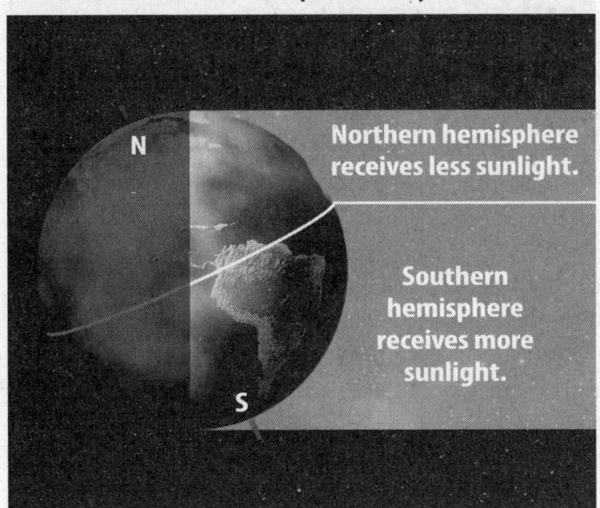

North end of rotation axis points toward the Sun.

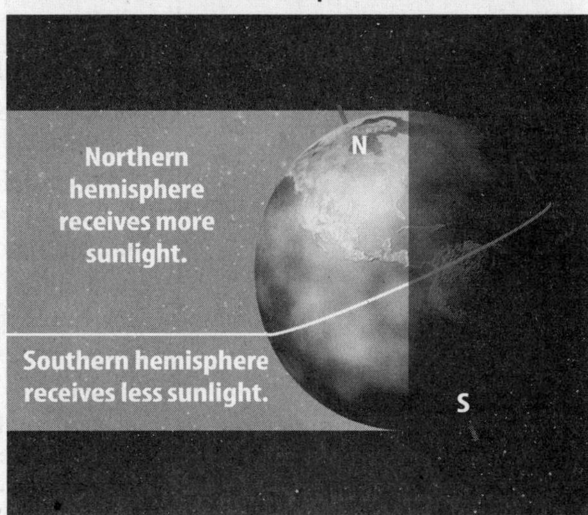

Solstices, Equinoxes, and the Seasonal Cycle

As Earth travels around the Sun, Earth's rotation axis always points in the same direction in space. But the amount that Earth's rotation axis is toward or away from the Sun changes. This causes the yearly cycle of the seasons.

There are four days each year when the direction of Earth's axis is special relative to the Sun. *A* **solstice** *is a day when Earth's rotation axis is the most toward or away from the Sun. An* **equinox** *is a day when Earth's rotation axis is leaning along Earth's orbit, neither toward or away from the Sun.*

March Equinox to June Solstice The north end of the rotation axis slowly points more and more toward the Sun. As a result, the northern hemisphere slowly receives more solar energy. Spring takes place in the northern hemisphere.

June Solstice to September Equinox The north end of the rotation axis still points toward the Sun but does so less and less. The northern hemisphere starts to receive less solar energy. This is summer in the northern hemisphere.

September Equinox to December Solstice The north end of the rotation axis points more and more away from the Sun. The northern hemisphere receives less and less solar energy. Fall takes place in the northern hemisphere.

December Solstice to March Equinox The north end of the rotation axis still points away from the Sun but does so less and less. As a result, the northern hemisphere starts to receive more solar energy. This is winter in the northern hemisphere. Earth's seasonal cycle is summarized in the table on the next page.

Visual Check

11. Label the left figure to indicate whether it is summer or winter in the northern hemisphere.

Make a bound book from two sheets of paper to organize information about each season and its solstices and equinoxes.

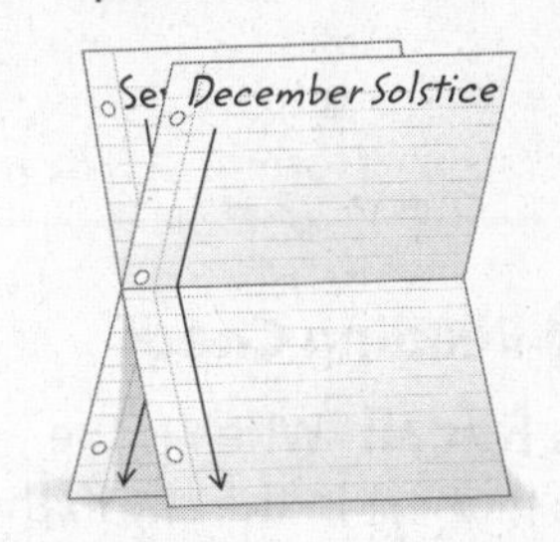

Key Concept Check

12. Conclude How does the tilt of Earth's rotation axis affect Earth?

Visual Check

13. Show Highlight in one color the seasons that begin on each solstice and each equinox in the northern hemisphere. Use a different color to highlight the seasons that begin on each solstice and equinox in the southern hemisphere.

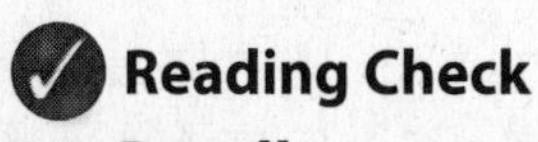

Reading Check

14. Recall When is the Sun highest in the sky in the northern hemisphere?

____

____

Earth's Seasonal Cycle

December Solstice	March Equinox
• The December solstice is on December 21 or 22. • The north end of Earth's axis is leaning farthest from the Sun. The south end is closest. • The northern hemisphere has its fewest number of daylight hours, and winter begins. • The southern hemisphere has its greatest number of daylight hours, and summer begins.	• The March equinox is on March 20 or 21. • Both ends of Earth's rotation axis are equal distances from the Sun. • There are about 12 hours of daylight and 12 hours of night everywhere on Earth. • Spring begins in the northern hemisphere. • Fall begins in the southern hemisphere.
June Solstice	**September Equinox**
• The June solstice is on June 20 or 21. • The north end of Earth's axis is leaning closest to the Sun. The south end is farthest away. • The northern hemisphere has the greatest number of daylight hours, and summer begins. • The southern hemisphere has the fewest number of daylight hours, and winter begins.	• The September equinox is on September 22 or 23. • Both ends of Earth's rotation axis are equal distances from the Sun. • There are about 12 hours of daylight and 12 hours of night everywhere on Earth. • Fall begins in the northern hemisphere. • Spring begins in the southern hemisphere.

Changes in the Sun's Apparent Path Across the Sky

As the seasons change, the Sun's apparent path across the sky also changes. In the northern hemisphere, the Sun's path through the sky is highest on the June solstice. Similarly, the Sun's path is lowest on the December solstice.

After You Read

Mini Glossary

equinox: a day when Earth's rotation axis is leaning along Earth's orbit, neither toward nor away from the Sun

orbit: the path an object follows as it moves around another object

revolution: the motion of one object around another object

rotation: a spinning motion

rotation axis: the line on which an object rotates

solstice: a day when Earth's rotation axis is the most toward or away from the Sun

1. Review the terms and their definitions in the Mini Glossary. Write a sentence that describes how solstices and equinoxes differ.

____

____

2. In the two boxes in the first column, draw Earth to show how it rotates and how it revolves. Then complete the other parts of the table.

	What is rotation?	**What does Earth's rotation cause?**
Earth's Rotation	Rotation is the spinning of an object around an axis.	________ ________ ________
	What is revolution?	**What does Earth's revolution cause?**
Earth's Revolution	________ ________ ________ ________	seasons

What do you think NOW?

Reread the statements at the beginning of the lesson. Fill in the After column with an A if you agree with the statement or a D if you disagree. Did you change your mind?

Log on to ConnectED.mcgraw-hill.com and access your textbook to find this lesson's resources.

CHAPTER 20
LESSON 2

The Sun-Earth-Moon System

Earth's Moon

Key Concepts
- How does the Moon move around Earth?
- Why does the Moon's appearance change?

Before You Read

What do you think? Read the two statements below and decide whether you agree or disagree with them. Place an A in the Before column if you agree with the statement or a D if you disagree. After you've read this lesson, reread the statements to see if you have changed your mind.

Before	Statement	After
	3. The Moon was once a planet that orbited the Sun between Earth and Mars.	
	4. Earth's shadow causes the changing appearance of the Moon.	

Study Coach

Make an Outline As you read, make an outline to summarize the lesson. Use the main headings in the lesson as the main headings in the outline. Complete the outline with the information under each heading. Use the completed outline to review the lesson.

Read to Learn

Seeing the Moon

Why does the Moon shine? Why does its shape seem to change? The Moon does not give off light. Unlike the Sun, the Moon is a solid object. You see the Moon because it reflects light from the Sun. Data about the Moon's mass, size, and distance from Earth are shown in the table below.

Moon Data

Mass	Diameter	Average distance from Earth	Time for one rotation	Time for one revolution
1.2% of Earth's mass	27% of Earth's diameter	384,000 km	27.3 days	27.3 days

1. Name the hypothesis that best explains how the Moon formed.

The Moon's Formation

The most widely accepted idea about how the Moon formed is the giant impact hypothesis. The giant impact hypothesis states that shortly after Earth formed, an object about the size of the planet Mars crashed into Earth. The impact caused a ring of vaporized rock to form around Earth. Eventually, the material in the ring cooled and clumped together and formed the Moon.

The Moon's Surface

Early in the Moon's history, various features formed on its surface. The main features on the Moon are craters, maria (MAR ee uh), and highlands.

Craters The Moon's craters formed when objects from space, such as large rocks, crashed into the Moon. Light-colored streaks called rays run outward from some craters.

Most of the Moon's craters formed about 3.5 billion years ago. This was long before dinosaurs lived on Earth. Earth also was hit by many large objects during this time. Wind, water, and plate tectonics erased these craters on Earth.

The Moon has no atmosphere, no water, and no plate tectonics. Without these forces, Moon's craters, formed billions of years ago, have not changed much.

Maria *The large, dark, flat areas on the Moon are called* **maria.** The maria (MAR ee uh) formed long after most impacts on the Moon's surface had ended. Maria formed when lava flowed through the Moon's crust and hardened. The hardened lava covered large portions of the Moon's surface, including many of its craters and other features. ✓

Highlands The light-colored areas on the Moon are the highlands. Highlands were not covered by the lava that formed the maria because they were too high for the lava to reach. Highlands are older than the maria and are covered with craters.

The Moon's Motion

As Earth revolves around the Sun, the Moon revolves around Earth. The pull of Earth's gravity causes the Moon to move in an orbit around Earth. Recall from Lesson 1 that if the pull of the Sun's gravity ended, Earth would fly in a straight line into space. The same would be true for the Moon if the pull of Earth's gravity ended. The Moon makes one revolution around Earth every 27.3 days.

Like Earth, the Moon also rotates as it revolves. One complete rotation of the Moon also takes 27.3 days. Notice that this is the same amount of time it takes the Moon to make one complete revolution around Earth.

The same side of the Moon always faces Earth because the Moon takes the same amount of time to orbit Earth and make one rotation. This side of the Moon that faces Earth is called the near side. The side of the Moon that cannot be seen from Earth is the far side of the Moon.

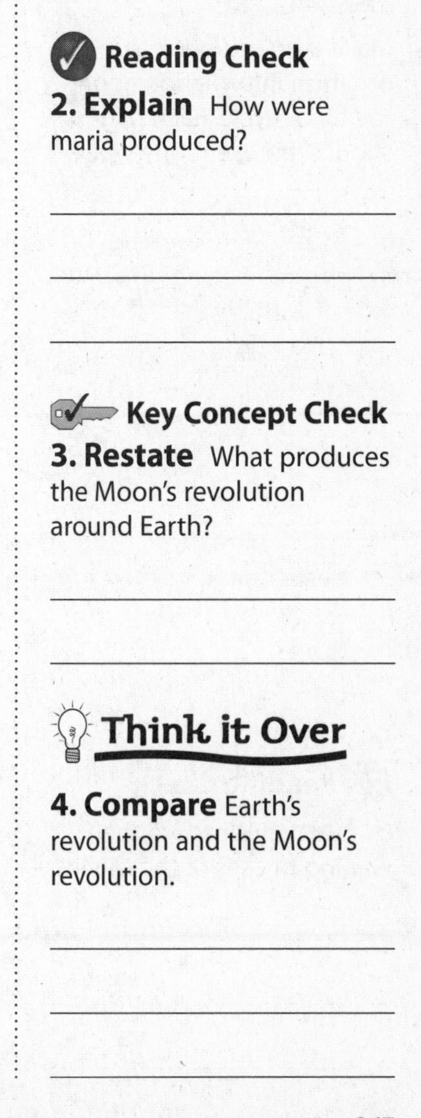

Reading Check

2. Explain How were maria produced?

Key Concept Check

3. Restate What produces the Moon's revolution around Earth?

Think it Over

4. Compare Earth's revolution and the Moon's revolution.

Phases of the Moon

The Sun is always shining on half the Moon, just as it is always shining on half of Earth. As the Moon moves around Earth, usually only one part of the Moon's near side is lit. *The lit part of the Moon or a planet that can be seen from Earth is called a* **phase.**

The motion of the Moon around Earth causes the phase of the Moon to change. The phases follow a regular pattern that is called the lunar cycle. One lunar cycle takes 29.5 days, or slightly more than four weeks, to complete.

Key Concept Check
5. Identify What produces the phases of the Moon?

Waxing Phases

The waxing phases occur during the first half of the lunar cycle. *During the* **waxing phases,** *more of the Moon's near side is lit each night.*

Week 1—First Quarter The lunar cycle begins. A sliver of light appears on the Moon's western edge. Each night, the lit part grows larger. By the end of the first week, the Moon reaches its first quarter phase. The entire western half of the Moon is now lit.

FOLDABLES®

Make a bound book to organize information about the lunar cycle. Each page should represent one week.

Week 2—Full Moon During the second week in the lunar cycle, more and more of the near side of the Moon becomes lit. By the end of the second week, the Moon's near side is completely lit. It is at its full moon phase.

Waning Phases

After the Moon waxes, it is said to wane. *During the* **waning phases,** *less of the Moon's near side is lit each night.* As seen from Earth, the lit part is now on the Moon's eastern side.

Week 3—Third Quarter During the third week in the lunar cycle, the lit part of the Moon becomes smaller. By the end of the third week, only the eastern half of the Moon is lit. This is the third quarter phase.

Week 4—New Moon During the fourth week in the lunar cycle, less and less of the near side of the Moon is lit. When the Moon's near side is completely dark, it has reached the new moon phase. The entire lunar cycle is summarized in the figure on the next page.

Reading Check
6. Name What are the waning phases of the Moon?

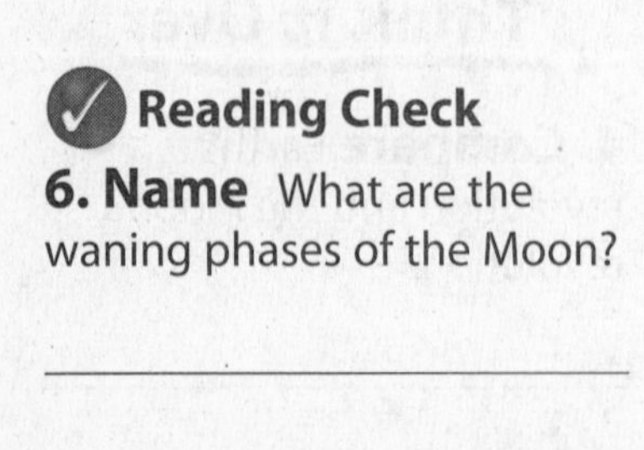

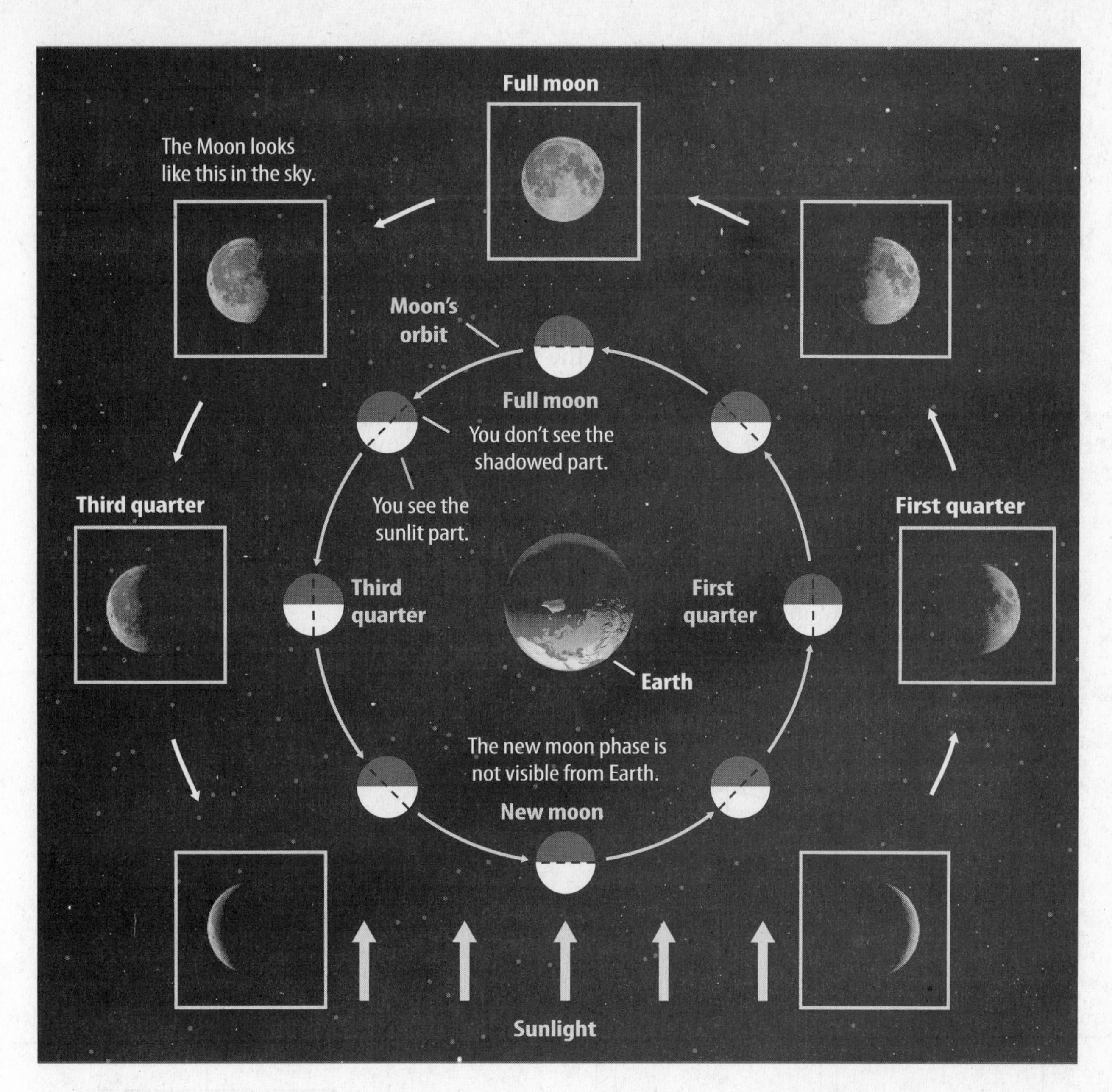

Visual Check

7. Circle the name of the moon phase that would be seen at the end of Week 2 of the lunar cycle.

After You Read

Mini Glossary

maria (MAR ee uh): the large, dark, flat areas on the Moon

phase: the lit part of the Moon or a planet that can be seen from Earth

waning phase: the part of the lunar cycle in which less of the Moon's near side is lit each night

waxing phase: the part of the lunar cycle in which more of the Moon's near side is lit each night

1. Review the terms and their definitions in the Mini Glossary. Write a sentence that describes the difference between waxing and waning.

____

____

2. Fill in the graphic organizers below to summarize the features of the Moon's surface and the phases of the Moon.

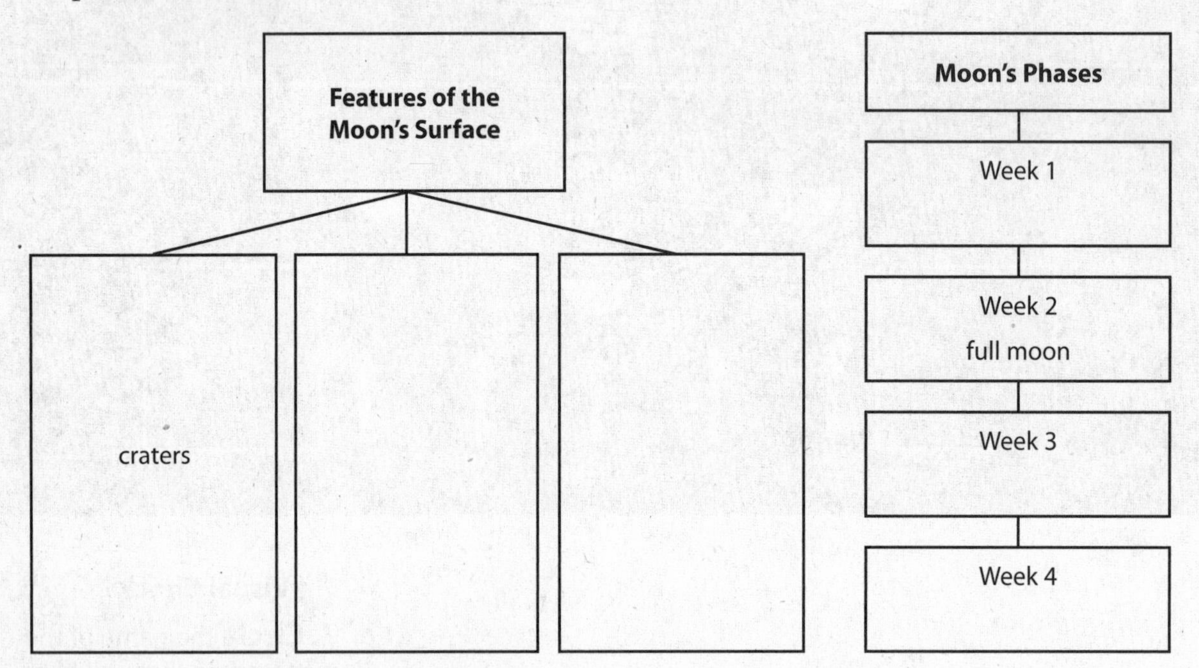

3. How did making an outline help you understand what you read?

____

____

What do you think NOW?

Reread the statements at the beginning of the lesson. Fill in the After column with an A if you agree with the statement or a D if you disagree. Did you change your mind?

Log on to ConnectED.mcgraw-hill.com and access your textbook to find this lesson's resources.

The Sun-Earth-Moon System

Eclipses and Tides

Before You Read

What do you think? Read the two statements below and decide whether you agree or disagree with them. Place an A in the Before column if you agree with the statement or a D if you disagree. After you've read this lesson, reread the statements to see if you have changed your mind.

Before	Statement	After
	5. A solar eclipse happens when Earth moves between the Moon and the Sun.	
	6. The gravitational pull of the Moon and the Sun on Earth's oceans causes tides.	

Key Concepts 
- What is a solar eclipse?
- What is a lunar eclipse?
- How do the Moon and the Sun affect Earth's oceans?

Read to Learn

Shadows—the Umbra and the Penumbra

A shadow forms when one object blocks the light that another object emits or reflects. For example, a tree blocks light from the Sun and casts a shadow. To stand in the shadow of a tree, you must place yourself with the tree in a line between you and the Sun.

Look carefully at a shadow on the ground on a bright, sunny day. You will notice that the edges of the shadow are not as dark as the rest of the shadow. Light from the Sun and other wide light sources casts shadows with two parts. *The* ***umbra*** *is the central, darker part of the shadow where light is totally blocked. The* ***penumbra*** *is the lighter part of a shadow where light is partially blocked.*

Mark the Text

Summarize As you read, underline words or phrases that summarize the information under each heading. Then, after you finish a section, reread the underlined parts to reinforce what you just read.

Solar Eclipses

The Sun shines on the Moon. The Moon casts a shadow that extends into space. Sometimes the Moon passes between Earth and the Sun. This can happen only during the new moon phase. When Earth, the Moon, and the Sun are lined up, the Moon casts a shadow on Earth's surface. *When the Moon's shadow appears on Earth's surface, a* ***solar eclipse*** *is occurring.*

Key Concept Check

1. Explain Why does a solar eclipse occur only during a new moon?

____

____

____

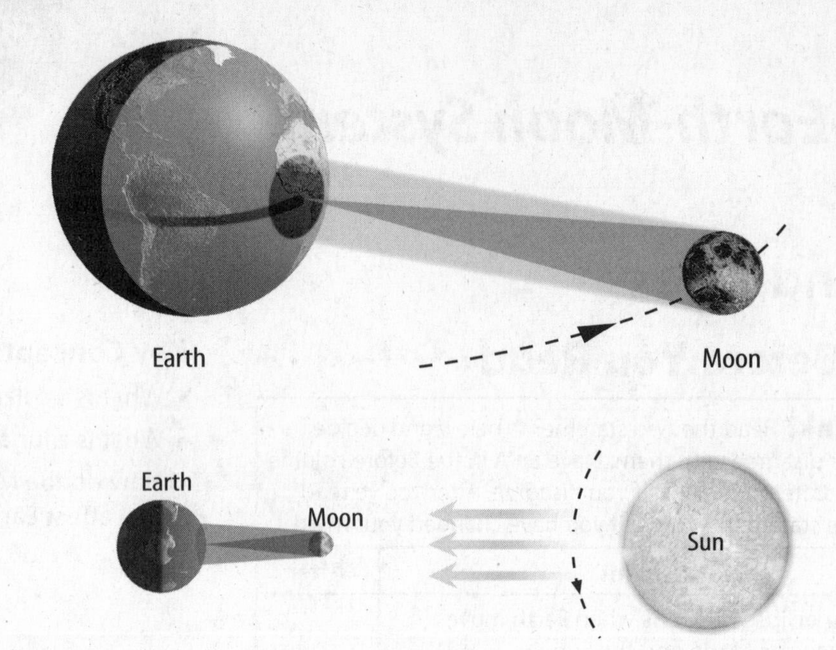

Visual Check

2. Identify Why would a person in North America not see the solar eclipse shown here?

Total Solar Eclipses

The positions of Earth, the Moon, and the Sun during a solar eclipse are shown above. Look at the shadow that the Moon casts on Earth. Notice that the umbra (the darker, inner part) is much narrower than the penumbra (the lighter, outer part). The type of eclipse you see depends on whether you are in the path of the umbra or the penumbra. If you are outside the umbra and penumbra, you cannot see the eclipse at all.

You can see a total solar eclipse only if you are within the Moon's umbra. During a total solar eclipse, the Moon appears to completely cover the Sun. The sky becomes dark, and you can see stars. A total solar eclipse lasts no longer than about 7 minutes.

Partial Solar Eclipses

If you are in the Moon's penumbra, you will see a partial solar eclipse. The Moon never completely covers the Sun during a partial solar eclipse.

Reading Check

3. Summarize Why don't solar eclipses happen every month?

Why don't solar eclipses occur every month?

Solar eclipses occur only during the new moon phase of the lunar cycle. During a new moon, Earth and the Sun are on opposite sides of the Moon. However, solar eclipses do not occur at every new moon phase. The Moon's orbit is slightly tilted compared to Earth's orbit. During most new moons, Earth is above or below the Moon's shadow. Only when the Moon is in a line between the Sun and Earth do solar eclipses take place.

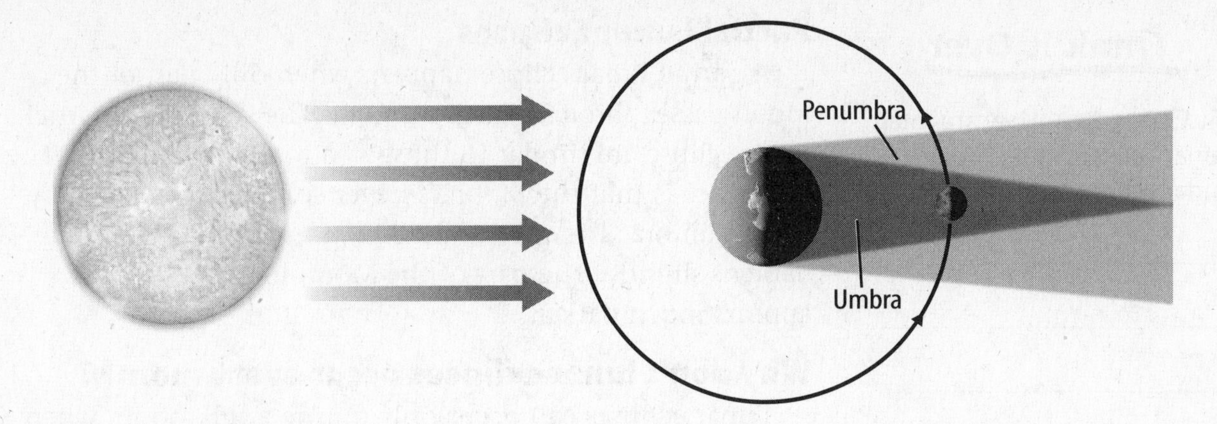

Lunar Eclipses

Just like the Moon, Earth casts a shadow into space. As the Moon revolves around Earth, it sometimes moves into Earth's shadow, as shown above. *A* **lunar eclipse** *occurs when the Moon moves into Earth's shadow.* During a lunar eclipse, Earth is in a line between the Sun and the Moon. A lunar eclipse can take place only during the Moon's full moon phase.

Like the Moon's shadow, Earth's shadow has an umbra and a penumbra. There are different types of lunar eclipses, depending on which part of Earth's shadow the Moon moves through. Unlike solar eclipses, all lunar eclipses can be seen from any place on Earth where it is nighttime.

Total Lunar Eclipses

When the entire Moon moves through Earth's umbra, a total lunar eclipse takes place. During a total lunar eclipse, the Moon's appearance changes slowly as it moves

- into Earth's penumbra, then
- into Earth's umbra, then
- back into Earth's penumbra, and then
- completely out of Earth's shadow.

You can still see the Moon when it is completely in Earth's umbra. Earth blocks most of the Sun's rays. Some of the rays, however, deflect off Earth's atmosphere and into Earth's umbra. This reflected sunlight has a reddish color and gives the Moon a reddish tint during a total lunar eclipse.

The deflection of some of the Sun's rays also explains why you can see the unlit part of the Moon on a clear night. Astronomers often call this Earthshine.

Visual Check

4. Analyze Why would more people be able to see a lunar eclipse than a solar eclipse?

____

____

____

Key Concept Check

5. Identify When can a lunar eclipse occur?

____

____

____

____

FOLDABLES®

Use a two-tab book to organize your notes on eclipses.

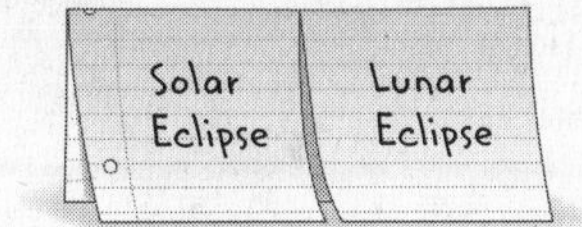

6. Compare How are total lunar eclipses and partial lunar eclipses similar?

Partial Lunar Eclipses

A partial lunar eclipse happens when only part of the Moon passes through Earth's umbra. The stages of a partial lunar eclipse are similar to those of a total lunar eclipse. The difference is that the Moon is never completely covered by Earth's umbra. The part of the Moon in Earth's penumbra darkens slightly. The part of the Moon in Earth's umbra appears much darker.

Why don't lunar eclipses occur every month?

Lunar eclipses can occur only during a full moon, when Earth is between the Sun and the Moon. Like solar eclipses, lunar eclipses do not occur every month. The Moon's orbit is slightly tilted in relation to Earth's orbit. During most full moons, the Moon is slightly above or slightly below Earth's penumbra.

Tides

The positions of the Moon and the Sun also affect Earth's oceans. Two times each day, the height of the water in Earth's oceans, or sea level, rises and falls. *The daily rise and fall of sea level is called a* **tide.** Tides are caused mostly by the effect of the Moon's gravity.

The Moon's Effect on Earth's Tides

Look at the figure below. In this view, you are looking down on Earth's North Pole. The figure shows that the strength of the Moon's gravity is a little stronger on the side of Earth closer to the Moon. The strength of the Moon's gravity is slightly weaker on the side of Earth opposite the Moon. The difference in the strength of the Moon's gravity causes tidal bulges in the oceans on opposite sides of Earth. High tides occur at the tidal bulges, and low tides occur between them. Low tides occur about six hours after a high tide.

7. Locate Highlight the tidal bulges that represent high tides in both figures below.

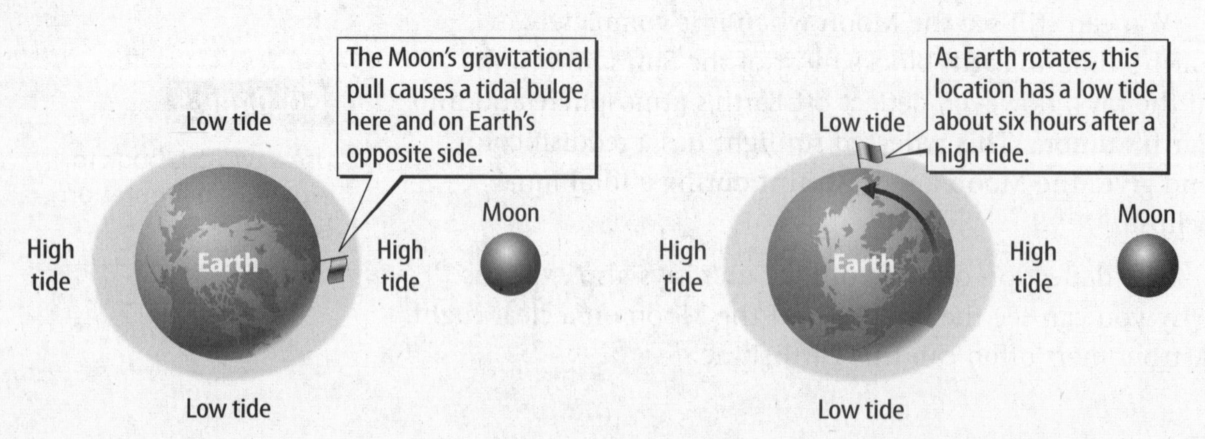

The Sun's Effect on Earth's Tides

Even though the Sun is much larger than the Moon, the Sun's effect on tides is about half that of the Moon. This is because the Sun is much farther from Earth than the Moon is.

Spring Tides Spring tides do not occur only in the season of spring. Spring tides occur during the full moon and new moon phases. The Sun's and the Moon's gravitational effects combine during spring tides. As a result, high tides are higher and low tides are lower.

Neap Tides A neap tide occurs one week after a spring tide. The Sun, Earth, and the Moon form a right angle. The Sun's effect on tides reduces the Moon's effect. During neap tides, high tides are lower and low tides are higher. The cycle of spring tides and neap tides is shown in the figure below.

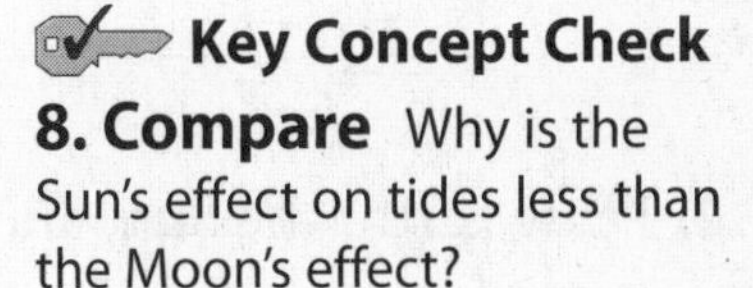

Key Concept Check

8. Compare Why is the Sun's effect on tides less than the Moon's effect?

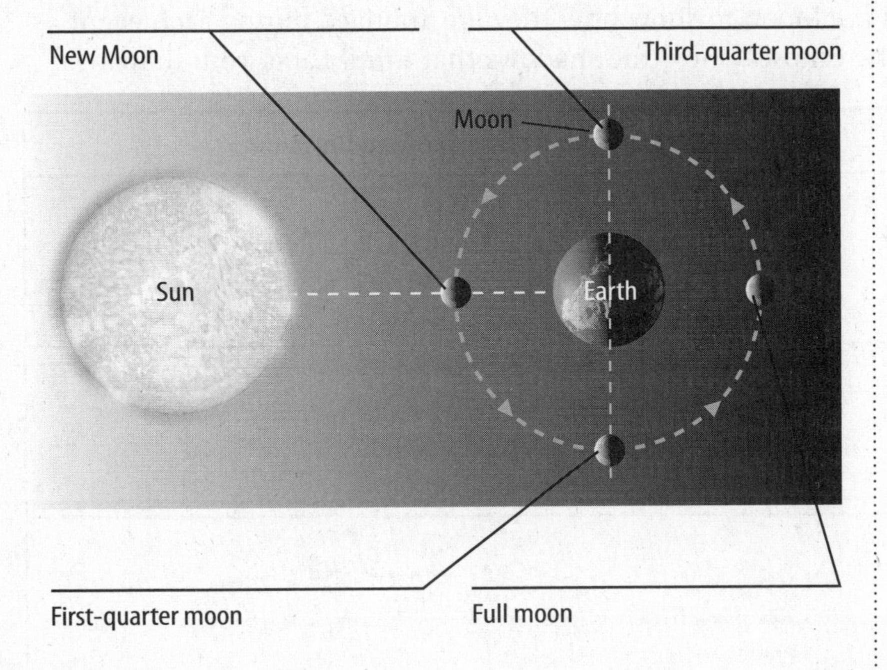

Visual Check

9. Label Based on the descriptions in the text, add the labels *Neap tide* and *Spring tide* to the diagram. Each label will be used twice.

After You Read

Mini Glossary

lunar eclipse: when the Moon moves into Earth's shadow

penumbra: the lighter part of a shadow where light is partially blocked

solar eclipse: when the Moon's shadow appears on Earth's surface

tide: the daily rise and fall of sea level

umbra: the central, darker part of a shadow where light is totally blocked

1. Review the terms and their definitions in the Mini Glossary. Write a sentence that explains the difference between an umbra and a penumbra.

____

____

2. Draw Earth, the Sun, and the Moon to show how they are arranged during each event named in the table. For the eclipses, show the shadows that form. Label your drawings.

Event	Positions of Earth, the Sun, and the Moon
Total solar eclipse	
Total lunar eclipse	
Spring tide	
Neap tide	

What do you think NOW?

Reread the statements at the beginning of the lesson. Fill in the After column with an A if you agree with the statement or a D if you disagree. Did you change your mind?

Log on to ConnectED.mcgraw-hill.com and access your textbook to find this lesson's resources.

Exploring Our Solar System

Our Solar System

Before You Read

What do you think? Read the two statements below and decide whether you agree or disagree with them. Place an A in the Before column if you agree with the statement or a D if you disagree. After you've read this lesson, reread the statements to see if you have changed your mind.

Before	Statement	After
	1. Our solar system has eight planets.	
	2. Earth's atmosphere is mostly oxygen.	

Key Concepts 

- How do objects in the solar system move?
- How did distance from the Sun affect the makeup of the objects in the solar system?
- What objects are in the solar system?

Read to Learn

Origin and Structure of Our Solar System

A solar system is a group of objects that revolve around a star. You might be familiar with our solar system—the eight planets and other objects that revolve around the Sun. Our Sun is not the only star in the universe that has a solar system. But our solar system is the only one scientists can study in detail.

Mark the Text

Ask Questions As you read, write questions you may have next to each paragraph. Read the lesson a second time and try to answer the questions. When you are done, ask your teacher any questions you still have.

Formation of the Solar System

Did you know there are clouds in space? Clouds in space are made mostly of hydrogen gas. Five billion years ago, our solar system formed from a spinning cloud of hydrogen gas and dust. When gravity caused the cloud to collapse, the cloud began to spin faster. The cloud also got hotter. When the center of the cloud became hot enough for nuclear reactions to occur, a star formed. That star was our Sun. The stars you see in the night sky formed in much the same way as the Sun formed.

As the cloud containing the Sun continued to spin, it flattened, with the Sun at its center. Small pieces of ice and rock orbiting the Sun clumped together and formed small, rocky or icy bodies called planetesimals (pla ne TE sih mulz). Gravity pulled some of the planetesimals together. The larger bodies that formed became planets, asteroids, and other objects. ✓

Reading Check

1. Summarize How did planetesimals become planets?

Solar System

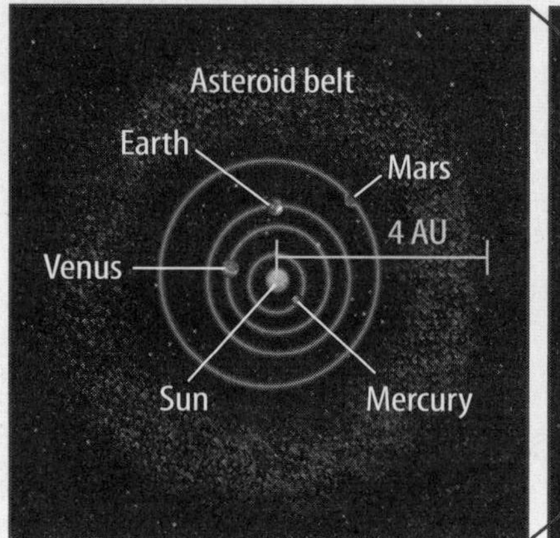

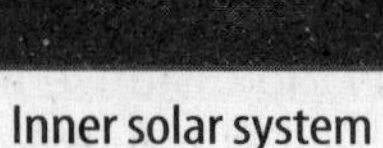
Inner solar system

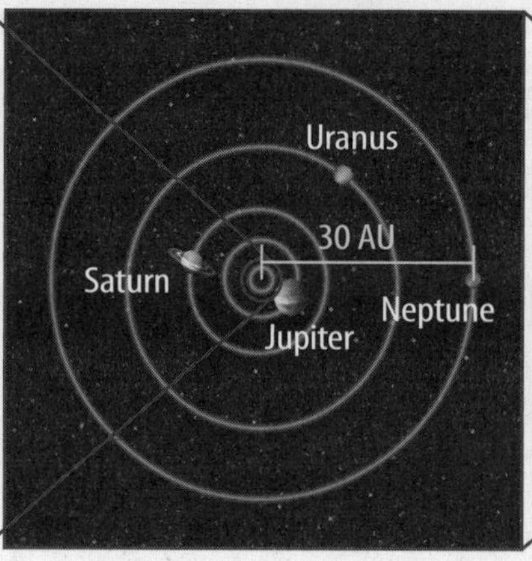

Planets of the outer solar system

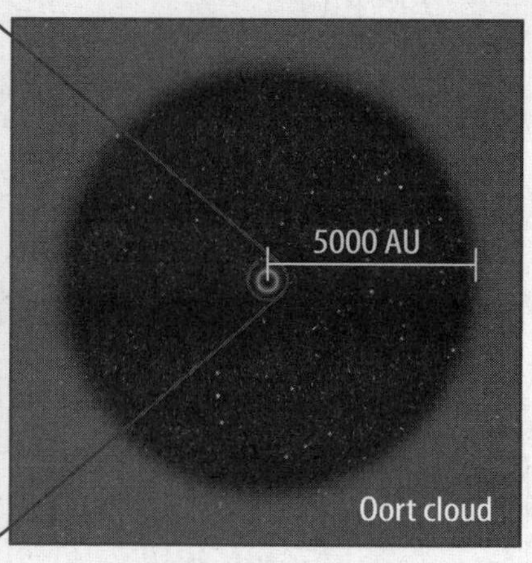

Outer solar system

Visual Check

2. Locate Which planet is farthest from the Sun?

____

Distances Within the Solar System

The solar system is so large that scientists rarely use kilometers to measure distances within it. Instead, they use astronomical units. *An* **astronomical unit** (AU) *is Earth's average distance from the Sun, nearly 150 million km.* Scientists often divide our solar system into regions, depending on distance from the Sun. Refer to the figure above as you read about regions in the solar system. The inner solar system extends to about 4 AU from the Sun. The inner solar system contains the planets closest to the Sun—Mercury, Venus, Earth, and Mars. It also contains the asteroid belt between Mars and Jupiter, 2–4 AU from the Sun.

The four planets farthest from the Sun—Jupiter, Saturn, Uranus, and Neptune—are part of the outer solar system. The outer solar system extends from Jupiter's orbit to the Oort (ORT) cloud. The Oort cloud is a large, spherical shell of icy planetesimals. Evidence suggests the Oort cloud orbits the Sun from about 5,000 AU to as far as 50,000 AU.

Reading Check

3. Recognize How far does the solar system extend?

____

____

____

Makeup of the Solar System

You live on a planet with a solid, rocky surface. But the solar system is made mostly of hydrogen gas. Ices, rocks, and metals make up less than 2 percent of our solar system's mass.

Hydrogen gas is the least-dense material in the solar system. The Sun, Jupiter, and Saturn contain most of the hydrogen in the solar system. They are the only bodies with enough mass to have large amounts of hydrogen.

Ices are denser than gases. They are made mostly of water, carbon dioxide, methane, or ammonia. Rocks are denser than ices and are made mostly of silicates. Metals, such as iron, are the solar system's densest materials.

REVIEW VOCABULARY

silicates
compounds composed mostly of silicon and oxygen, with smaller amounts of other atoms

Movement in the Solar System

The Sun contains 99 percent of the solar system's mass. Most other objects in the solar system revolve around the Sun. The Sun's enormous gravitational pull holds these objects in orbit.

Revolution is the movement of one object around another object. The closer an object is to the Sun, the faster it revolves. While most objects in the solar system revolve, they also rotate, or spin, on their axes. Both of these movements—rotation and revolution—are regular and predictable.

Direction of Motion

The motion of an object around the Sun is like the motion of an object whirled on a string. The string pulls on the object just as the Sun's gravity pulls on planets and objects that revolve around it.

Our solar system formed from a spinning cloud of gas and dust. Objects that formed from this material spun in the same direction. Planets and most other solar system objects still revolve around the Sun in the same direction. If you were far above Earth's North Pole and looked down on the solar system, you would see objects revolving in a counterclockwise direction. The Sun and six of the eight planets, including Earth, also rotate in a counterclockwise direction. Venus and Uranus rotate clockwise.

The View from Earth

Earth rotates on its axis once every 24 h. But on Earth, it seems as though you are standing still, and the Sun, the Moon, and stars move around you. While Earth rotates from west to east, objects in the sky appear to move from east to west. The same is true on a merry-go-round. As you move in one direction on a merry-go-round, objects around you seem to move in the opposite direction.

Objects in Our Solar System

Our solar system contains billions of objects. Scientists group these objects into categories based on their makeup, their size, their distance from the Sun, and whether they orbit the Sun or another object.

From Gas and Dust to Solids Recall that the solar system formed from a cloud of gas and dust that was extremely hot at its center, where the Sun formed. As regions beyond the Sun cooled, some of the gases solidified into ices, rocks, and metals.

FOLDABLES®

Use four sheets of paper to make an eight-layer book. Use it to identify and describe objects in the solar system.

Objects In The Solar System
Sun
Inner and Outer Planets
Dwarf Planets
Asteroids
Natural Satellites
Kuiper Belt Objects
Comets
Meteoroids, Meteors, Meteorites

Key Concept Check

4. Explain Why do most objects in our solar system move in the same direction?

____

____

____

____

Key Concept Check
5. Evaluate How did distance from the Sun affect the makeup of objects in the solar system?

Visual Check
6. Differentiate How do the inner and outer planets differ?

Ices, Rocks, and Metals Ices formed far from the Sun, where temperatures were extremely cold. Closer to the Sun, temperatures were too high for ices to form. Most gases there solidified into rocks and metals. The densest matter, the metals, sank to the centers of the largest objects throughout the solar system.

The Sun

The Sun is made mostly of hydrogen gas. It also contains helium and tiny amounts of other elements. The Sun is the only star in our solar system, and it is the largest object in the solar system. The Sun's diameter is 10 times that of Jupiter and more than 100 times that of Earth.

Planets

A **planet** *orbits the Sun, is large enough to be nearly spherical in shape, and has no other large object in its orbital path.* The figure below shows the eight planets in their order from the Sun.

The four inner planets—Mercury, Venus, Earth, and Mars—formed from rocks and metals. As shown in the figure, they are smaller than the outer planets. They have few or no moons and rotate slowly.

The four outer planets—Jupiter, Saturn, Uranus, and Neptune—formed mostly from gas and ice. They are large and have many moons. They rotate quickly and have rings. The table on the next page describes the planets.

The Planets

The Planets of the Solar System

Planet	Distance from the Sun and Diameter	Makeup and Atmosphere
Mercury	0.39 AU from the Sun 4,900 km diameter	Mercury has a large metal core under its small rocky mantle. Its surface is covered with craters and looks much like the surface of Earth's moon. Mercury has no permanent atmosphere.
Venus	0.72 AU from the Sun 12,100 km diameter	Venus is similar to Earth in size and makeup. Its rocky mantle surrounds a molten or partially molten metal core. Its thick carbon dioxide atmosphere traps thermal energy, making Venus's surface the hottest of all the planets.
Earth	1 AU from the Sun 12,800 km diameter	Most of Earth is covered by a thin layer of liquid water. Earth has a rocky mantle, a molten outer metal core, and a solid inner metal core. Its atmosphere is 80 percent nitrogen and 20 percent oxygen.
Mars	1.5 AU from the Sun 6,800 km diameter	Mars has a rocky mantle and a partially molten metal core. Its thin atmosphere is mostly carbon dioxide. Iron in its surface rock gives the planet a reddish color. The surface of Mars has ice but no liquid water.
Jupiter	5.2 AU from the Sun 143,000 km diameter	Jupiter has more mass than all the other planets combined. Under its atmosphere of hydrogen gas is a layer of liquid hydrogen. Rock and metal have sunk to its core. A thin, barely visible ring system surrounds it.
Saturn	9.6 AU from the Sun 121,000 km diameter	Saturn's makeup is similar to that of Jupiter, but its atmosphere is hazier. Saturn's rings, the most distinctive rings of all the outer planets, are made mostly of small particles of ice.
Uranus	19 AU from the Sun 51,100 km diameter	Uranus has a hydrogen gas outer layer; a fluid inner layer made of water, methane, and ammonia; and a rocky core. Uranus has a blue-green color because of the small amount of methane in its cloud layers. It has thin rings.
Neptune	30 AU from the Sun 49,500 km diameter	Neptune has more mass than Uranus but is slightly smaller. Neptune and Uranus have similar makeup. Neptune is a deeper blue because it has more methane in its atmosphere. Like the other outer planets, it has thin rings.

Interpreting Tables

7. Name the largest and smallest planets, by diameter.

____

____

Dwarf Planets

Dwarf planets *orbit the Sun and are nearly spherical in shape, but they share their orbital paths with other objects of similar size.* Eris is the largest known dwarf planet. All known dwarf planets, including Eris and Pluto, are smaller than Earth's moon, as shown in the figure on the right.

Dwarf Planets

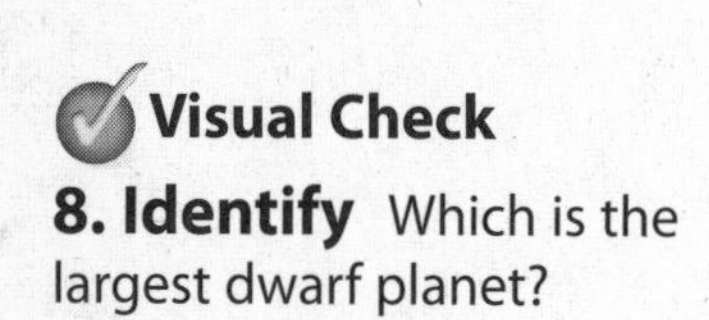

8. Identify Which is the largest dwarf planet?

____

Our solar system has at least five dwarf planets. However, scientists hypothesize that the solar system might contain hundreds of dwarf planets, most of them orbiting the Sun beyond Neptune. At least one dwarf planet, Ceres (SIHR eez), shown in the figure, orbits the Sun between the orbits of Mars and Jupiter, in the asteroid belt.

Asteroids

Asteroids are small, rocky or metallic objects that are left over from the solar system's formation. There are hundreds of thousands of asteroids in the asteroid belt, but they are so small that their total mass is less than the mass of Earth's moon. Though most asteroids exist in the asteroid belt, some exist elsewhere in the solar system. Most asteroids are irregularly shaped and have craters on their surfaces.

Reading Check

9. Locate Where do most asteroids exist?

____

____

Natural Satellites

A **satellite** *is an object that orbits a larger object other than a star.* Natural satellites are also known as moons. The solar system has over 170 moons. Most of them orbit planets or dwarf planets. However, some moons orbit smaller objects, such as asteroids. For example, the asteroid Ida is 50 km long. It is 20 times larger than Dactyl (DAK tul), its moon.

Just as planets have different makeups, satellites have different makeups depending on their location. The satellites of the inner planets are mostly rock. Jupiter's moons are a mixture of rock and ice. The moons around the three outermost planets are mostly ice.

Think it Over

10. Analyze Why are the moons around the outermost planets mostly ice rather than rock?

____

____

____

Kuiper Belt Objects

You read that the asteroid belt lies between the orbits of Mars and Jupiter. The Kuiper (KI pur) belt, shown in the figure below, is a similar but much larger belt of objects between 30 and 50 AU from the Sun. Like the asteroid belt, the Kuiper belt contains objects left over from the solar system's formation. While asteroids are mostly rock and metal, Kuiper belt objects are mostly ice. Pluto is the best-known object in the Kuiper belt.

Kuiper Belt and Comets

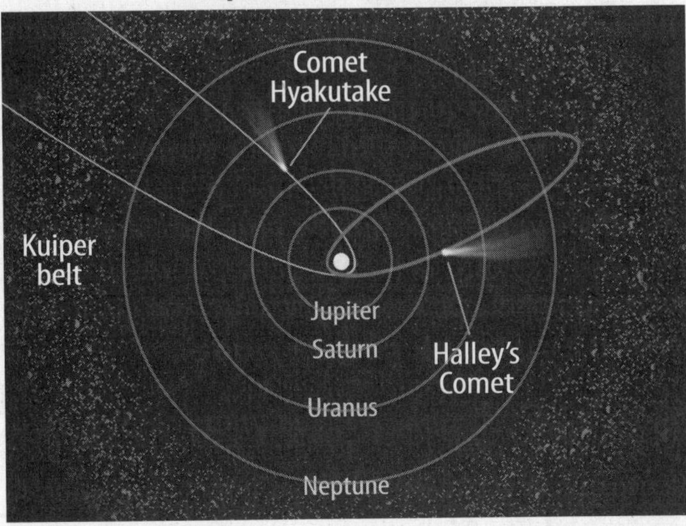

Reading Check

11. Contrast How do Kuiper belt objects differ from objects in the asteroid belt?

Visual Check

12. Observe Which direction does a comet's tail point?

Comets

Comets are small objects made mostly of ice. Most comets formed in the Kuiper belt or the Oort cloud. Comets revolve around the Sun in long, stretched-out orbits, as shown in the figure above. As a comet's orbit nears the Sun, some of the comet's ice becomes gas and forms the comet's tail. The tails of some comets extend millions of kilometers into space. A comet loses mass with each orbit. After a certain number of orbits, the comet breaks up.

Meteoroids, Meteors, and Meteorites

A ***meteoroid*** *is a solar system object that is smaller than an asteroid or a comet.* Some meteoroids are the result of collisions between asteroids. Some are debris from comets. *A* ***meteor*** *is the streak of light created when a meteoroid enters Earth's atmosphere.* Most meteoroids are no bigger than a grain of rice. They are visible only when they get hot and glow as they pass through Earth's atmosphere. A few large meteoroids become meteorites. *A* ***meteorite*** *is a meteoroid that strikes Earth.*

Key Concept Check

13. Name What objects are in our solar system?

After You Read

Mini Glossary

astronomical unit (AU): Earth's average distance from the Sun, nearly 150 million km

dwarf planet: an object that orbits the Sun and is nearly spherical in shape, but shares its orbital path with other objects of similar size

meteor: the streak of light created when a meteoroid enters Earth's atmosphere

meteorite: a meteoroid that strikes Earth

meteoroid: a solar system object that is smaller than an asteroid or a comet

planet: an object that orbits the Sun, is large enough to be nearly spherical in shape, and has no other large object in its orbital path

satellite: an object that orbits a larger object other than a star

1. Review the terms and their definitions in the Mini Glossary. Write a sentence that explains some differences between a planet and a dwarf planet.

____

____

2. Use the descriptions given in the graphic organizer to identify objects in our solar system.

Object	Description	Object	Description
	the only star in our solar system		result of collisions between asteroids or debris from comets
	small object made mostly of ice		orbits a larger object other than a star
	a meteoroid that strikes Earth		small, rocky or metallic object with an irregular shape and craters
	nearly spherical object that orbits the Sun but shares orbital path with other similarly sized objects		streak of light created when a meteoroid enters Earth's atmosphere

3. Does the Sun rise in the east or west on Venus? Explain your answer.

____

____

What do you think NOW?

Reread the statements at the beginning of the lesson. Fill in the After column with an A if you agree with the statement or a D if you disagree. Did you change your mind?

Log on to ConnectED.mcgraw-hill.com and access your textbook to find this lesson's resources.

Exploring Our Solar System

Life in the Solar System

Before You Read

What do you think? Read the two statements below and decide whether you agree or disagree with them. Place an A in the Before column if you agree with the statement or a D if you disagree. After you've read this lesson, reread the statements to see if you have changed your mind.

Before	Statement	After
	3. Earth's atmosphere protects life on Earth from dangerous solar radiation.	
	4. Scientists think conditions for life might exist on some moons in the solar system.	

Key Concepts 

- What conditions on Earth enable life to exist?
- What conditions on other bodies in the solar system might enable life to exist?
- Where might life possibly exist beyond Earth?

Read to Learn

Conditions for Life on Earth

Life exists in nearly every environment on Earth. Some environments have conditions so extreme that humans cannot live in them. These places might have extreme temperatures, high salt levels, total darkness, or little water. Even though humans cannot live in these places, other organisms can.

Even though some organisms live in extreme conditions, all of Earth's life-forms need the same basic things to survive: a source of energy, liquid water, and nourishment. Scientists have not yet discovered life anywhere else in the solar system. But by studying the conditions that support life on Earth, they are learning about conditions that might support life elsewhere. **Astrobiology** *is the study of the origin, development, distribution, and future of life in the universe.*

Energy from the Sun

The Sun is the source of almost all energy on Earth. Sunlight provides light and thermal energy. It also provides energy for plants, which are at the base of most food chains. However, a small percentage of organisms on Earth receive energy from chemicals or from Earth itself. For example, a variety of animals live in complete darkness near hot water jets in the ocean floor.

Study Coach

Make Flash Cards For each head in this lesson, write a question on one side of a flash card and the answer on the other side. Quiz yourself until you know all of the answers.

Key Concept Check

1. State What do organisms on Earth need to survive?

____

____

____

FOLDABLES®

Make a three-tab Venn book to compare and contrast the ability of Earth and the Moon to sustain life.

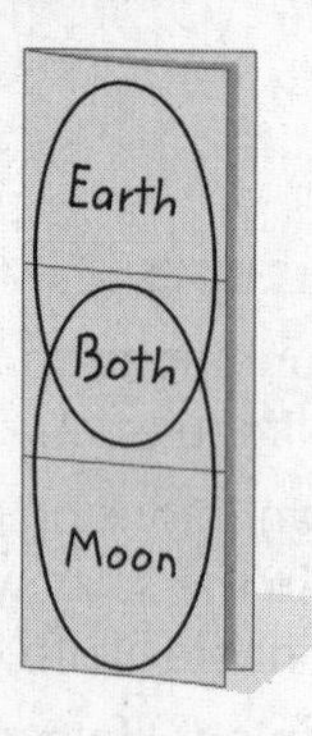

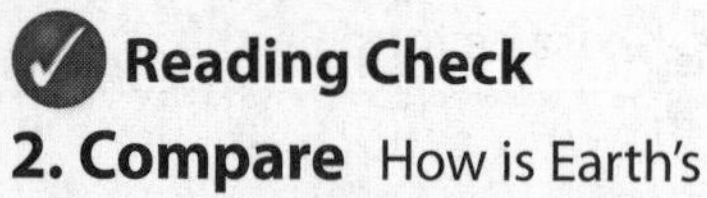

Reading Check

2. Compare How is Earth's atmosphere like a blanket?

____

____

____

____

Reading Check

3. Identify What kinds of radiation are absorbed by Earth's atmosphere?

____

____

____

Protection by the Atmosphere

Earth's moon receives about the same amount of sunlight as Earth. Yet conditions on the Moon are more extreme than they are on Earth. The Moon's surface temperature can rise to 100°C during the day and drop to −150°C at night.

Why do temperatures vary so widely on the Moon? Temperatures are extreme on the Moon because the Moon, unlike Earth, does not have an atmosphere.

Maintains Temperatures Earth's atmosphere is like a blanket around Earth. It absorbs sunlight during the day and keeps thermal energy from escaping into space during the night. It maintains Earth's average surface temperature at a comfortable 14°C. ✓

Absorbs Harmful Radiation Have you ever had a painful sunburn? The Sun's ultraviolet light causes sunburns. Even though you cannot see ultraviolet light, you can feel its effects. Too much radiation in the form of ultraviolet light can harm you. Fortunately, Earth's atmosphere absorbs most of the Sun's ultraviolet light.

In addition to ultraviolet light, Earth's atmosphere absorbs X-rays and other potentially harmful light waves from the Sun. The atmosphere also helps protect Earth from highly charged particles that erupt from the Sun in powerful storms. ✓

Burns Up Meteoroids Earth's atmosphere also protects Earth's surface from meteoroids. Millions of meteoroids enter Earth's atmosphere every day. But almost all of them burn up in the atmosphere before they reach Earth's surface.

Liquid Water

Liquid water is necessary for all life on Earth. Water dissolves minerals and transports molecules in cells. Without liquid water, cells could not function and life would not exist. Water exists as a liquid on Earth's surface because the atmosphere keeps pressures and temperatures within a certain range.

At Sea Level Depending on temperature and pressure on Earth, water is solid, liquid, or gas, as shown in the figure on the next page. At sea level on Earth (1 atm of pressure), water is liquid between 0°C and 100°C. Above 100°C, water boils and becomes water vapor. Below 0°C, water freezes into ice.

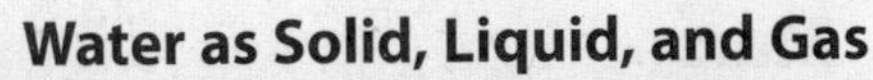

Water as Solid, Liquid, and Gas

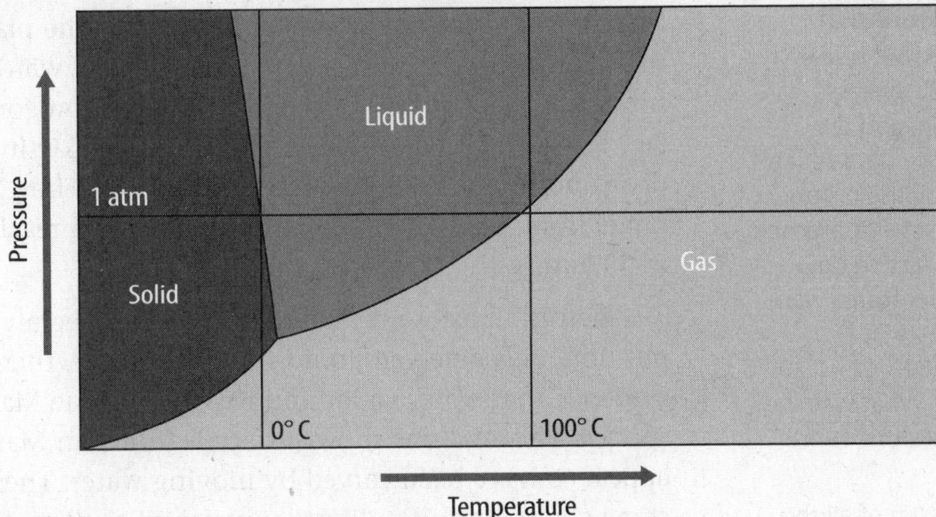

At Other Altitudes At different altitudes on Earth, such as on the top of a mountain, the boiling and freezing temperatures of water change slightly because the pressure of the atmosphere changes. Without Earth's atmosphere, pressures on Earth's surface would be too low for water to be liquid. Water would exist only as water vapor or ice.

Nourishment

Living things are nourished by nutrients they take from the air, water, and land around them. They use the nutrients for energy, growth, and other processes, such as reproduction and cellular repair. All molecules that provide nourishment for life on Earth contain carbon. They are organic molecules. **Organic** *refers to a class of chemical compounds in living organisms that are based on carbon.* Though inorganic life could exist elsewhere, astrobiologists are most interested in places where water is liquid and carbon is plentiful.

Looking for Life Elsewhere

In 1835, a New York newspaper published articles claiming that herds of bison and furry, winged bat-men had been observed on Earth's moon. The articles fooled many people. Today, people know the Moon is airless, and scientists have yet to find life there.

Because liquid water is essential for life on Earth, scientists look for places in our solar system where liquid water might exist or might have existed in the past. In 2009, scientists discovered water on the Moon. Although water might not exist on the surface of a planet or a moon, it might exist beneath the surface.

Visual Check

4. Observe What happens to water when temperatures are high?

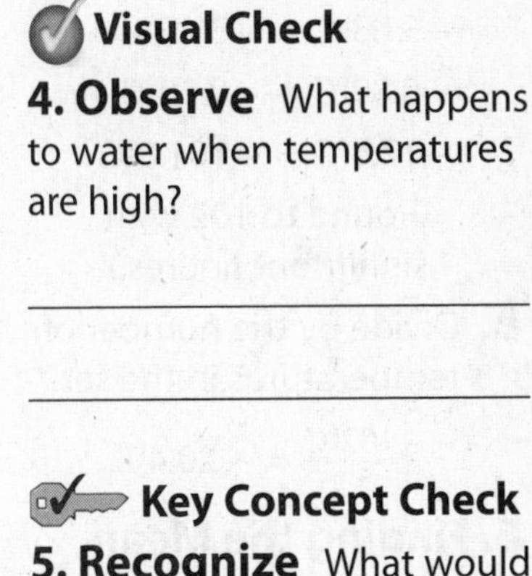

Key Concept Check

5. Recognize What would happen to water on Earth's surface if Earth had no atmosphere?

Math Skills

The mean of a set of data is the arithmetic average. To find the mean, add the numbers in the data set and then divide by the number of items in the set. For example, during one Martian day, surface temperatures were measured at −51.3°C, −31.9°C, −0.800°C, −0.200°C, and −17.6°C. What was the mean temperature during that day?

a. Find the sum of all the values.

−51.3°C + −31.9°C + −0.80°C + −0.20°C + −17.6°C = −101.8°C

(Round to 102 with significant figures.)

b. Divide by the number of temperatures in the set.

$$\frac{-102°C}{5} = -20.4°C$$

6. Finding the Mean The temperature at the Martian polar ice caps can drop to −143°C. The warmest spots on the planet can reach 20°C. What is the mean of these extreme temperatures?

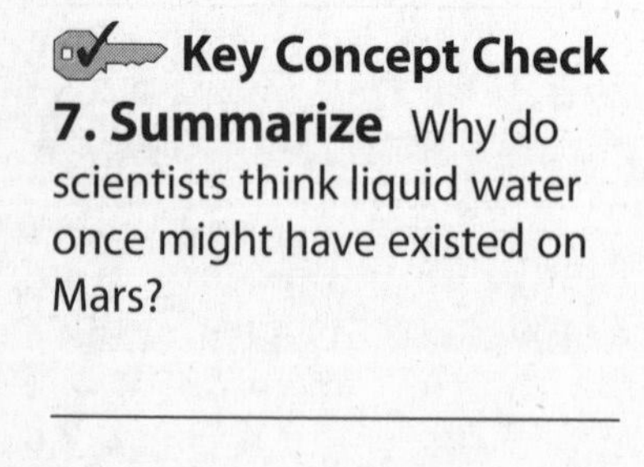

Key Concept Check

7. Summarize Why do scientists think liquid water once might have existed on Mars?

Mars

Scientific evidence suggests that Mars is the planet, other than Earth, that is most likely to have liquid water. On the surface of Mars, pressures are probably too low for water to be liquid. Water would likely evaporate quickly in the thin, dry atmosphere. Temperatures are also low. They generally range from −87°C to −5°C, though they can reach a high of 20°C during the Martian summer.

Scientists have sent many uncrewed spacecraft to Mars, but none has detected liquid water. However, there is much evidence that water vapor and ice exist on the Martian surface. Photographs show surface features on Mars that appear to have been carved by moving water. These photos show channels on the Martian surface. Scientists hypothesize that these Martian channels could be ancient streambeds. It is possible that water from an underground ocean seeped to the surface and flowed as rivers or floods before evaporating. Scientists still do not know how much water was in these channels or how long ago it flowed.

Other Planets

Mercury and Venus are too hot for water to be liquid on or near their surfaces. The four outer planets are too cold. The outer planets also are too gaseous. They have no solid surfaces on which liquid water could form. Though some liquid water might exist deep in the interiors of the outer planets, it is unlikely that the water could support life.

Natural Satellites

Scientists continue to look for further evidence of water on Earth's moon and on the moons of other planets. Even though temperatures in the outer solar system are extremely cold, scientists have observed that as a satellite orbits a massive planet, the planet's gravity can cause the satellite's interior to heat. This might provide enough thermal energy to allow liquid water to exist near the satellite's icy surfaces.

Surface Features Several moons in the outer solar system have surface features that indicate the presence of liquid water not far below. For example, scientists suggest that the ridges on Europa (yuh ROH puh), one of Jupiter's moons, could be cracks in the ice where liquid water has seeped to the surface and frozen solid. Callisto and Ganymede, two other moons of Jupiter, and Titan, a moon of Saturn, show similar surface features.

Geysers Several other moons in the outer solar system, including Enceladus (en SEL uh dus), a moon of Saturn, and Triton, a moon of Neptune, show evidence of geysers (GI zurz). *A* ***geyser*** *is a warm spring that sometimes ejects a jet of liquid water or water vapor into the air.* The massive geysers on Enceladus are hundreds of kilometers high. Two other moons of Saturn, Tethys (TEE thus) and Dione (di OH nee), also have geyserlike plumes. These geysers are evidence that liquid water might exist beneath the icy surfaces of these moons.

Key Concept Check

8. Identify Where might life exist in the solar system beyond Earth?

After You Read

Mini Glossary

astrobiology: the study of the origin, development, distribution, and future of life in the universe

geyser (GI zur): a warm spring that sometimes ejects a jet of liquid water or water vapor into the air

organic: a class of chemical compounds in living organisms that are based on carbon

1. Review the terms and their definitions in the Mini Glossary. Write a sentence explaining what an organic compound is.

2. Use the graphic organizer to summarize how Earth's atmosphere helps sustain life on Earth.

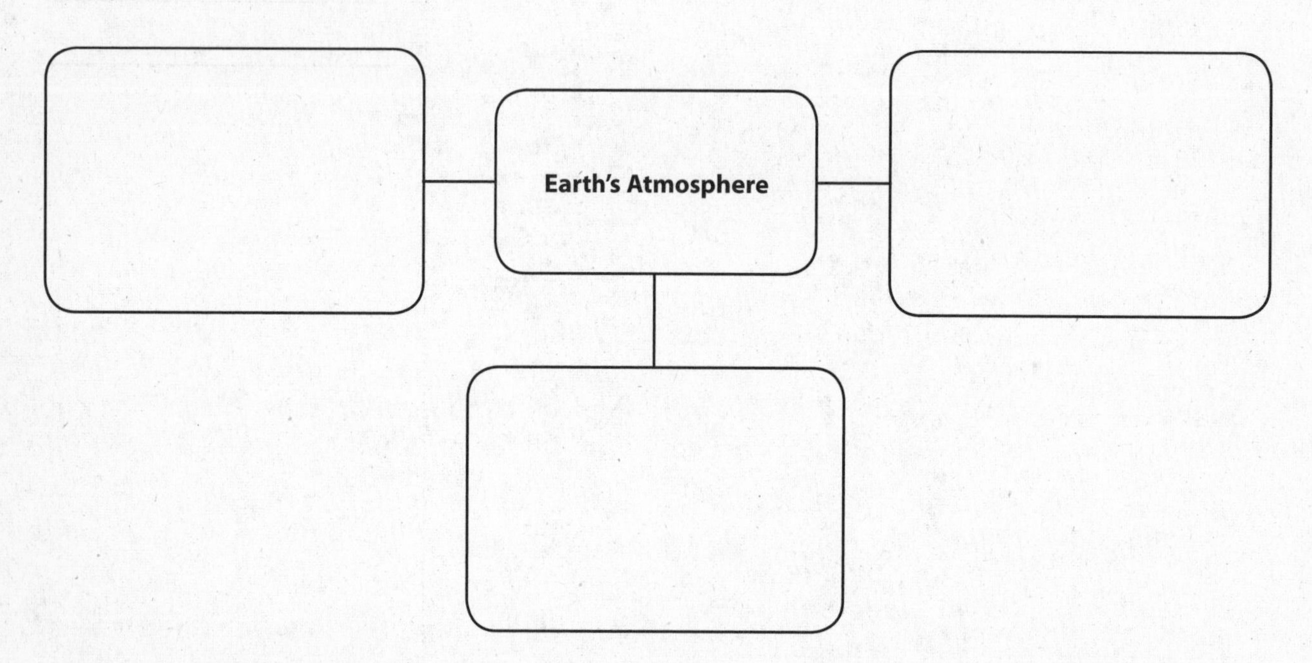

3. Select and define a word from one of the flash cards you created as you read the lesson.

What do you think NOW?

Reread the statements at the beginning of the lesson. Fill in the After column with an A if you agree with the statement or a D if you disagree. Did you change your mind?

Log on to ConnectED.mcgraw-hill.com and access your textbook to find this lesson's resources.

Exploring Our Solar System

Human Space Travel

Before You Read

What do you think? Read the two statements below and decide whether you agree or disagree with them. Place an A in the Before column if you agree with the statement or a D if you disagree. After you've read this lesson, reread the statements to see if you have changed your mind.

Before	Statement	After
	5. Astronauts float in space because there is no gravity above Earth's atmosphere.	
	6. The United States is the only country with a human space-flight program.	

Key Concepts 

- What technology has allowed humans to explore and travel into space?
- What factors must humans consider when traveling into space?

Read to Learn

Technology and Early Space Travel

You have lived your entire life in the space age. In 1957, the former Soviet Union launched *Sputnik I*. Most people consider this event to be the beginning of the space age. *Sputnik I* was the first artificial satellite sent into orbit around Earth. *An* **artificial satellite** *is any human-made object placed in orbit around a body in space.*

Today, hundreds of artificial satellites operate in orbit around Earth. Some artificial satellites are communication satellites. Some observe Earth. A few observe stars and other objects in distant space. ✓

Mark the Text

Sticky Notes As you read, use sticky notes to mark information that you do not understand. Read the text carefully a second time. If you still need help, write a list of questions to ask your teacher.

Escaping Gravity

How do artificial satellites and other spacecraft reach space? You know that when you jump up, you land back on the ground because of Earth's gravity. But if you could jump fast enough and high enough, you would launch into space!

Only a rocket can travel fast enough and far enough to offset Earth's gravity. *A* **rocket** *is a vehicle propelled by the exhaust made from burning fuel.* As its exhaust is forced out, the rocket accelerates forward, acting against Earth's gravity. Most rockets that travel long distances carry two or more tanks of fuel to be able to travel far enough to counter Earth's gravity.

✓ **Reading Check**

1. Describe How are artificial satellites used today?

______

______

______

______

FOLDABLES®

Make a vertical four-tab book to organize your notes on the challenges for humans in space.

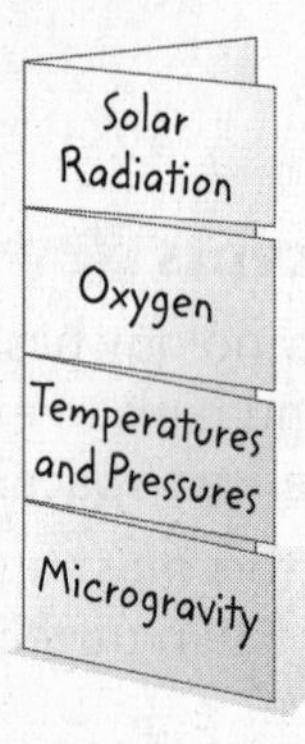

ACADEMIC VOCABULARY

transmit
(verb) to send something from one person, place, or thing to another

Robotic Space Probes

The Moon is the farthest object from Earth that humans have visited. However, scientists have sent robotic missions to every planet and also to some moons, asteroids, dwarf planets, and comets. *A* **space probe** *is an uncrewed vehicle that travels to and obtains information about objects in space.* There are three main types of space probes:

- Flybys travel to one or more distant space objects and fly by without orbiting or landing.
- Orbiters travel to a distant space object and are placed into orbit around the object.
- Landers travel to a distant space object and land on the surface.

Probes do not return to Earth. They carry cameras and scientific instruments that transmit data back to Earth.

There are many reasons to send probes instead of people into space. Probes cost less than crewed vehicles, and there is less risk to humans. Objects in space are very far away. A visit to Mars and back would take more than a year. A round trip to Saturn could take 15 years. Robotic missions do have risks. Only half of the missions sent to Mars have been successful. Space probes that do arrive at their destinations undergo harsh conditions and often do not survive long.

The National Aeronautics and Space Administration (NASA) is the U.S. government agency responsible for most space missions and space-flight technology. Other nations also have space programs. Astronauts from more than 30 countries have traveled to space. Several countries have sent robotic missions to the Moon and beyond.

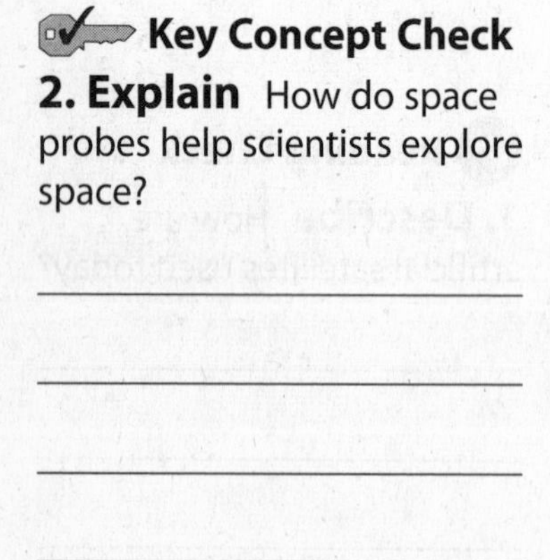

Key Concept Check

2. Explain How do space probes help scientists explore space?

Challenges for Humans in Space

When astronauts travel into space, they must take their environments and life-support systems with them. Otherwise, they could not withstand the temperatures, the pressures, and the other extreme conditions that exist in space.

Solar Radiation

One threat to astronauts is harmful radiation from the Sun. You read that Earth's atmosphere protects life on Earth from most of the Sun's dangerous radiation. However, as astronauts travel in space, they move far beyond Earth's atmosphere. They must rely on their spacecraft and spacesuits to shield them from dangerous solar radiation and solar particles.

Oxygen

Humans must have oxygen. Outside Earth's atmosphere, there is not enough oxygen for humans to survive. Air circulation systems inside spacecraft supply oxygen and keep carbon dioxide, which people breathe out, from accumulating. The air humans breathe on Earth is a mixture of nitrogen and oxygen. For short trips into space, spacecraft carry tanks of oxygen and nitrogen, which are mixed into the proper proportions onboard. For long trips, oxygen is supplied by passing an electric current through water. This separates water's hydrogen and oxygen atoms.

Temperature and Pressure Extremes

Most places in the solar system are either extremely cold or extremely hot. Pressures in space also are extreme. In most places, pressure is much lower than the pressure humans experience on Earth. Environmental control systems in spacecraft protect astronauts from temperature and pressure extremes. Outside their spacecraft, astronauts wear Extravehicular Mobility Unit (EMU) suits. EMU suits provide oxygen, protect astronauts from radiation and meteoroids, and make it possible for astronauts to talk to each other. ✓

Microgravity

You might think astronauts are weightless in space. But astronauts in orbit around Earth are subjected to almost the same gravity as they are on Earth's surface. Then why do astronauts float inside their spacecraft? As their spacecraft orbits Earth, the astronauts inside are continually falling toward Earth. But because their spacecraft is moving, they do not fall. They float. If their spacecraft suddenly stopped moving, they would plunge downward.

The space environment that astronauts experience is often called microgravity. In microgravity, objects seem to be weightless. This can be an advantage. No matter how much something weighs on Earth, astronauts can move it easily in space. Microgravity also makes some tasks, such as turning a screwdriver, more difficult. If an astronaut is not careful, instead of the screw turning, he or she might turn instead. ✓

On Earth, working against gravity helps keep your muscles, bones, and heart strong and healthy. But in microgravity, astronauts' bones and muscles don't need to work as hard, and they begin to lose mass and strength. Astronauts in space must exercise each day to keep their bodies healthy.

✓ Reading Check

3. Recognize What are the purposes of an EMU suit?

✓ Reading Check

4. Describe microgravity.

Key Concept Check

5. Summarize What factors must humans consider when traveling into space?

Living and Working in Space

Even when they are protected from the extremes of space, astronauts still face many challenges when living and working in space. Life in space is dramatically different from life on Earth.

International Space Station

The *International Space Station (ISS)* is a large, artificial satellite that orbits Earth. It is the ninth and largest space station to be built in space. Eight pairs of solar panels provide power for the *ISS*. Up to seven people work and live in pressurized modules on the *ISS* for up to six months at a time. The *ISS* was constructed by astronauts from over 15 nations, and crews have occupied the *ISS* continuously since the first crew arrived in 2000. ✓

✓ Reading Check

6. Identify What is the *International Space Station?*

The *ISS* crew conducts scientific and medical experiments. These include experiments to learn how microgravity affects people's health and how it affects plants. People living in space for long periods might need to grow plants for food and oxygen. In the future, in addition to being an orbiting research laboratory, the *ISS* might serve as a testing and repair station for missions to the Moon and beyond.

Living in space is not easy. For example, astronauts must place a clip on a book to hold it open to the right page. They eat packaged food, using magnetized trays and tableware. Toilets flush with air instead of water. And astronauts must be strapped down while they sleep. Otherwise, they would drift and bump into things.

Think it Over

7. Consider Name one other everyday activity that might be very different in space than on Earth.

Transportation Systems

Space transportation systems are the rockets, the shuttles, and the other spacecraft that deliver cargo and humans to space. Early rockets and spacecraft, such as those used to transport astronauts to the Moon, were used only once. Since then, NASA has developed other transportation systems, some of which can be used more than one time.

Space Shuttle NASA's first reusable transportation system was the space shuttle. It left Earth attached to a rocket, but it landed like an airplane. The space shuttle was first launched in 1981. It was designed to transport astronauts to the *International Space Station* and to service uncrewed satellites, such as the *Hubble Space Telescope*. ✓

✓ Reading Check

8. Differentiate How is the space shuttle different from early rockets and spacecraft?

Orion NASA is designing a new space transportation vehicle called *Orion* (uh RI uhn). *Orion* is part of NASA's Project Constellation, a human space-flight program. The goal of this program is to send astronauts to the Moon and eventually to Mars. A new rocket system called *Ares* will launch *Orion*. Another, larger rocket—*Ares V*—will launch heavier, nonhuman cargo. 

Outposts on the Moon and Mars

Orion might make its first flight to the Moon as early as 2020. It will carry from four to six astronauts. Once on the Moon, astronauts will build an outpost where they can stay for up to six months. The astronauts will learn how to survive in a harsh environment.

If astronauts visit Mars, they probably will continue the search for life. Human exploration on Mars might help scientists learn if life ever existed on Mars, or if life exists today beneath the planet's surface.

Reading Check

9. Name What space transportation vehicle will carry astronauts to the Moon in the future?

____

After You Read

Mini Glossary

artificial satellite: any human-made object placed in orbit around a body in space

rocket: a vehicle propelled by the exhaust made from burning fuel

space probe: an uncrewed vehicle that travels to and obtains information about objects in space

1. Review the terms and their definitions in the Mini Glossary. Write a sentence in your own words to explain what an artificial satellite is.

____

____

2. Use the graphic organizer to identify the three types of space probes and describe how each type differs from the others.

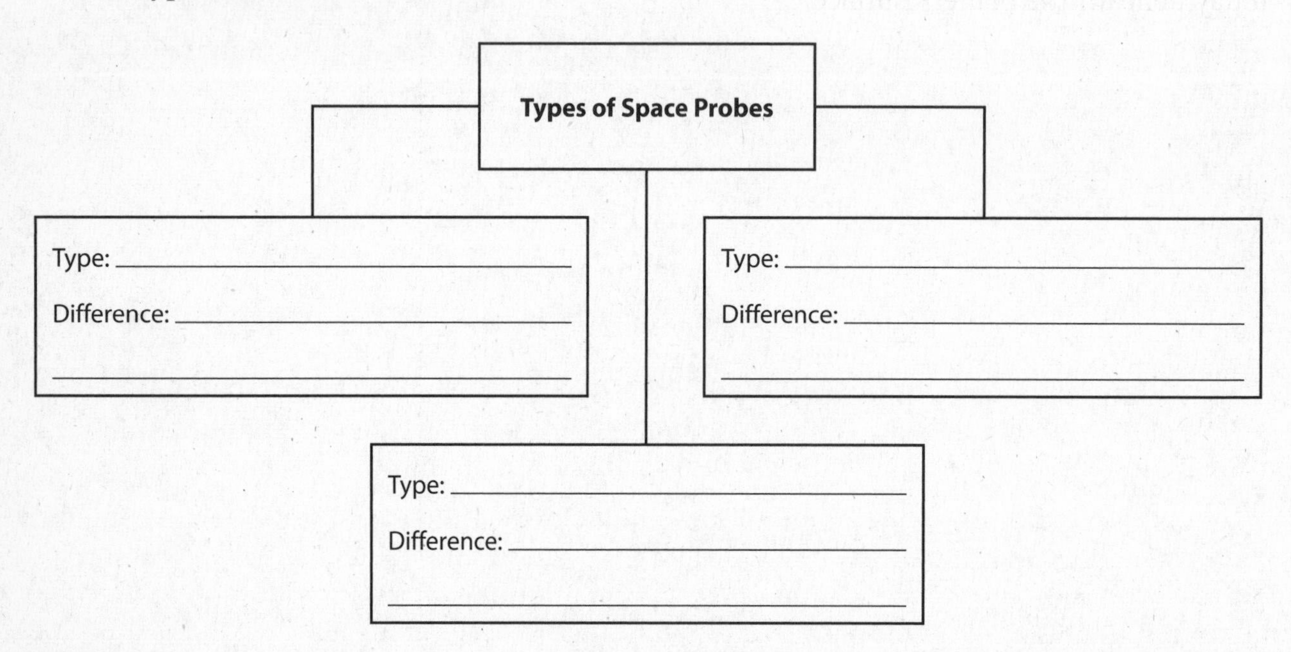

3. What do you consider to be the most difficult challenge humans face in space? Why?

____

____

What do you think NOW?

Reread the statements at the beginning of the lesson. Fill in the After column with an A if you agree with the statement or a D if you disagree. Did you change your mind?

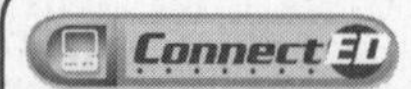

Log on to ConnectED.mcgraw-hill.com and access your textbook to find this lesson's resources.

PERIODIC TABLE OF THE ELEMENTS

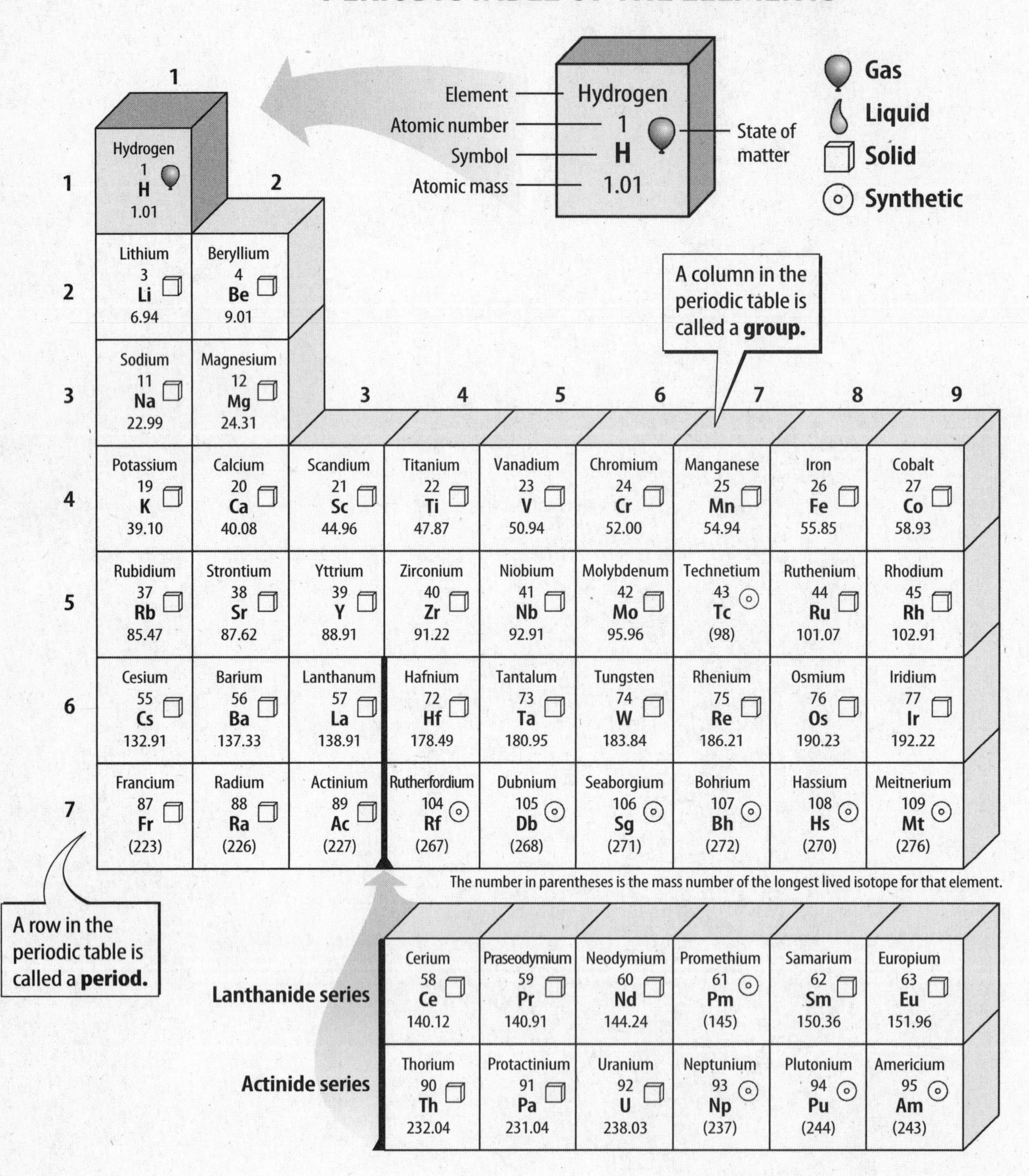